Conjugated Polymeric Materials: Opportunities in Electronics, Optoelectronics, and Molecular Electronics

NATO ASI Series

Advanced Science Institutes Series

A Series presenting the results of activities sponsored by the NATO Science Committee, which aims at the dissemination of advanced scientific and technological knowledge, with a view to strengthening links between scientific communities.

The Series is published by an international board of publishers in conjunction with the NATO Scientific Affairs Division

A	**Life Sciences**	Plenum Publishing Corporation
B	**Physics**	London and New York
C	**Mathematical and Physical Sciences**	Kluwer Academic Publishers
		Dordrecht, Boston and London
D	**Behavioural and Social Sciences**	
E	**Applied Sciences**	
F	**Computer and Systems Sciences**	Springer-Verlag
G	**Ecological Sciences**	Berlin, Heidelberg, New York, London,
H	**Cell Biology**	Paris and Tokyo

Series E: Applied Sciences - Vol. 182

Proceedings of the NATO Advanced Research Workshop on
Conjugated Polymeric Materials: Opportunities in Electronics,
Optoelectronics, and Molecular Electronics
Mons, Belgium
September 3–8, 1989

Library of Congress Cataloging in Publication Data

ISBN 0–7923–0751–8

Published by Kluwer Academic Publishers,
P.O. Box 17, 3300 AA Dordrecht, The Netherlands.

Kluwer Academic Publishers incorporates the publishing programmes of
D. Reidel, Martinus Nijhoff, Dr W. Junk and MTP Press.

Sold and distributed in the U.S.A. and Canada
by Kluwer Academic Publishers,
101 Philip Drive, Norwell, MA 02061, U.S.A.

In all other countries, sold and distributed
by Kluwer Academic Publishers Group,
P.O. Box 322, 3300 AH Dordrecht, The Netherlands.

Printed on acid-free paper

Conjugated Polymeric Materials: Opportunities in Electronics, Optoelectronics, and Molecular Electronics

edited by

J. L. Brédas
University of Mons, Mons, Belgium

and

R. R. Chance
Exxon Research and Engineering Co.,
Annandale, New Jersey, U.S.A.

Kluwer Academic Publishers

Dordrecht / Boston / London

Published in cooperation with NATO Scientific Affairs Division

TABLE OF CONTENTS

PREFACE

This book constitutes the Proceedings of the NATO Advanced Research Workshop on Conjugated Polymers held at the University of Mons, Belgium, during the first week of September 1989. The Workshop was attended by about fifty scientists representing most of the leading research groups within NATO countries, that have contributed to the development of conjugated polymeric materials. The program was focused on applications related to electrical conductivity and nonlinear optics. The attendance was well balanced with a blend of researchers from academic, industrial, and government labs, and including synthetic chemists, physical chemists, physicists, materials scientists, and theoreticians.

The Workshop provided an especially timely opportunity to discuss the important progress that has taken place in the field of Conjugated Polymers in the late eighties as well as the enormous potential that lies in front of us.

Among the recent significant developments in the field, we can cite for instance:

(i) The discovery of novel synthetic routes affording conjugated polymers
- that are much better characterized, especially through control of the molecular weight;
- that can be *processed* from solution or the melt; the early promise that conducting polymers would constitute materials combining the electrical conductivities of metals with the mechanical properties of plastics is now being realized;
- that can reach remarkably high conductivities.

(ii) The exploitation of the high nonlinear optical responses that are inherent to conjugated materials, either for second-order or third-order effects. The detailed investigation of the dynamics of the excited state relaxation has proven proves to be an area of important basic and applied interest.

(iii) The novel chemistry and physics of the polyanilines and other families of polymers where a simple protonation process can lead to an insulator-to-conductor transition. The influence of ring motions in the transport properties is a topic of growing interest. Furthermore, polyaniline fibers appear to display most promising mechanical properties.

(iv) The achievement of electronic devices such as MISFET's in which conjugated polymers in the semiconducting state play an active role in the device physics.

Five working groups were organized, that during the week actively discussed the prospects of conjugated polymers in the areas of synthesis and processability, technological potentials, exploitation of the electronic properties, nonlinear optics, molecular electronics, and the key needs in terms of theoretical efforts. The reports of the working groups, which are collected at the end of the Proceedings, highlight the exciting opportunities for conjugated polymers into the 1990's.

That the Workshop ran smoothly and efficiently was due to the collaboration and hard work of many people. Thanks are due to the University of Mons, especially the President of the University, Professor Y. Van Haverbeke, and the Administrator, Mr. J. Quenon, who fully supported the endeavor; the staff of the Restaurant Universitaire; and the staff of the Cité Universitaire.

Nothing could have been accomplished without the tremendous help from the members of the Laboratory for Chemistry of Novel Materials at the University of Mons: especially Ms. F. Meyers who took care of all the organizational details from the very beginning and served as Scientific Secretary of the Workshop, Mr. J.M. Toussaint, Dr. G. Lambin, Dr. R. Lazzaroni, and Dr. J. Orszagh as well as Mrs. M. Cornez and Mrs. J. Sauvenière.

The generous financial support from the NATO Division of Scientific Affairs was what made the Workshop possible. The genuine interest in the Workshop of the NATO team, especially Professor J. Ducuing, Dr. C. Sinclair, and Professor L. Da Cunha is gratefully acknowledged. Support was also received from Exxon Research & Engineering Company.

Finally, we wish to stress the fact that *all* the participants that presented a talk or a poster, did write a chapter for these Proceedings, thus making this book very comprehensive. It would also be impossible not to mention the fruitful late-evening sessions around special Belgian beers (simply the best) and the high-tech touch brought by Professor MacDiarmid. He used his own portable machine to fax his working group report to his secretary at Penn in order to have it typed overnight. Never forget the human factor, though! Professor MacDiarmid had to phone her back and dictate the report because she was not able to read his handwriting!

J.L. Brédas, R.R. Chance
December 1989

THEORY OF CONJUGATED POLYMERS AND MOLECULAR CRYSTALS

Robert Silbey
Department of Chemistry
and
Center for Materials Science and Engineering
Massachusetts Institute of Technology
Cambridge, Mass 02139

I. <u>Introduction</u>:   The electronic, spectral and conducting properties of molecular crystals, such as anthracene, naphthalene, and benzene have been studied by modern methods and theories for almost 40 years[1]. In spite of this attention, there are still interesting effects being discovered.   The study of the same properties of conjugated polymers, such as polyacetylene and polypyrrole, is only about 10 years old.   These systems are surprising and wonderful in many regards, not the least of which is the strong interdisciplinary nature of the scholarship devoted to them.

In this paper, I will review the field from the perspective of a physical chemist who has thought about molecular crystals and conjugated polymers from a theoretical viewpoint always (I hope) with an eye on the experimental details.

II. <u>Molecular Crystals</u>:   Molecules in a molecular crystal are only slightly perturbed from their gas phase structure, so that theory always begins with a description of the crystal as a frozen gas of such molecules. The prototypical (for our purposes) system is naphthalene with its planar structure and large number(48) of vibrational modes .   Because of its anisotropic shape, the intermolecular forces in the solid are also anisotropic; in addition, they are weak.   Because of this, defect structures have low energies, and are important in the description of the properties. The crystal packing is the "herring bone" pattern with two molecules in a unit cell, a structure often seen in planar aromatic systems.   The structure is soft: there are low frequency torsional modes as well as

1

J. L. Brédas and R. R. Chance (eds.), Conjugated Polymeric Materials:
Opportunities in Electronics, Optoelectronics, and Molecular Electronics, 1–9.
© 1990 *Kluwer Academic Publishers. Printed in the Netherlands.*

the low frequency acoustic and optical phonons. Because of the anisotropy of the intermolecular forces and their short range, the properties often exhibit strong anisotropies or dimensionalities. All of these lead to the prediction that, in most of these systems, there is no single dominant interaction. On the contrary, many interactions compete, and the final result often depends on small details.

A good example of this is the 20 year struggle to understand the conductivity of electrons and holes in naphthalene crystals[2]. It has taken 20 years of hard work to produce crystals pure and defect-free enough to obtain reliable experimental results over a wide temperature range. The theory of the conduction in naphthalene (and other aroomatic crystals) is still not fully worked out, largely, I believe, because a number of phonon and librational modes strongly interact with the charge carriers, making a simple model inadequate.

The hamiltonian for a molecular crystal can be written

$$(1) \quad H = \Sigma \, h_n \; + \; \Sigma \, V_{nm}$$

where h_n is the hamiltonian for the single molecule and V_{nm} is the interaction between molecules n and m. Even in the simplest cases, we muct treat this in an approximate manner, by introducing a basis set which is based on the eigenfunctions of the first term in (1). If we are considering the excited electronic states of the system, we often concentrate on a single band of states arising from one electronic state of the molecule. Then, we find

$$(2) \quad H = \Sigma \varepsilon_n \, a_n^+ a_n \; + \Sigma \, t_{nm} \, a_n^+ a_n + H_{vib}$$

where ε_n is the energy of the electronic state in the absence of the intermolecular forces while t_{nm} represents the excitation transfer matrix element between sites n and m. Finally H_{vib} represents the phonon and vibrational hamiltonian. In the case of hole or electron conduction, a similar effective hamiltonian would be used, with perhaps the index n augmented with a spin index, σ. If there is electron-phonon or electron vibration interaction, we expand the matrix elements ε_n and t_{nm} about the equilibrium positions of the lattice and keep linear (and sometimes quadratic) terms. The linear electron phonon terms can be written

$$(3) \quad H_{ep} = \Sigma\, g_{n\lambda}\, a_n^+ a_n\, (b_\lambda + b_{-\lambda}^+) + \Sigma\, f_{nm\lambda}\, a_n^+ a_m\, (b_\lambda + b_{-\lambda}^+),$$

where b_λ (b_λ^+) destroys(creates) a phonon of index λ, and g and f are coupling constants, describing site diagonal and site off-diagonal interactions. Finally, there is a term in the hamiltonian, important for electron and hole bands, but not for exciton bands, reperesenting the interaction between electrons or excitations:

$$(4) \quad H_{corr} = \Sigma\, U\, N_n(N_n\text{-}1) \quad + \Sigma\, V(n\text{-}m)\, (N_n\text{-}1)(N_m\text{-}1)$$

where U represents the on-site interaction between electrons, $V(n\text{-}m)$ represents the intersite interaction between electrons, and N_n represents the number of electrons on site n. In this description of the system, we have neglected static disorder, which can be made small, but never removed from these systems.

In the nearest neighbor approximation, relevant for some of the low lying electronic states (especially the triplet states) of these crystals, the transfer matrix elements, t_{nm}, are replaced by the values to the near neighbors, and all others are set to zero. Since the t_{nm} are not equal in all directions, the reduced dimensionality of the exciton bands becomes evident at this point, and the bandwidth, B, is equal to $2\Sigma t_i$, where the sum is over the nearest neighbor directions, and t_i is the transfer matrix element in the ith direction. For one-dimensional bands, the bandwidth is 4t, etc. The bandwidth term in the hamiltonian tends to delocalize the wavefunction, while the electron phonon terms, which represent scattering of the electron by the phonons (or vibrations) tend to localize the wavefunction and decrease the effective bandwidth. In addition, in low dimensionality the latter may distort the structure of the lattice (Peierls transition)[3]. The correlation terms add another complication. U tends to keep electrons apart, spoiling the simple band or molecular orbital picture, and when large, can decrease the effective bandwidth.

This hamiltonian has been discussed by a number of authors for the case of exciton bands in molecular crystals [4,5,6], and has

4

been applied to study a variety of properties in these systems. As mentioned above, the application to electron and hole conduction in these systems has been less successful (but see ref [7,8]).

III. <u>Polyenes, polyacetylene, and all that</u>: The polyenes, short chains of alternating C-C single and double bonds have been studied by chemists for about 50 years[9,10]. Of particular interest has been the evolution of the optical band gap, or the energy difference between the ground and first <u>dipole allowed</u> excited state. The theoretical description of these molecules has gone from the free electron model, to the Huckel model, through the Pariser Parr Pople (PPP) semi-empirical model, and finally, at least for short chains, to ab-initio quantum chemical calculations. Along the way, it has become dogma, at least to the physical chemists studying these molecules, that the simple description afforded by, for example, Huckel theory, is inadequate. This is so largely because of the inability of the Huckel model to describe the other excited states of these molecules, in particular the triplet or radical states (where the spin densities are given poorly) and the famous second $^1A_{1g}$ excited state (discussed by Kohler in these proceedings). The latter can be described in a one electron molecular orbital picture only as a double excitation with a large amount of configuration interaction. To treat these correctly, it has become necessary to consider the correlation term carefully.

The Huckel model corresponds to keeping equ(2) as the Hamiltonian for a polyene, where only the p_z (π) electrons are considered, one per site, and the sigma electrons form part of the core which is unaltered upon excitation of the π electrons. The addition of the electron phonon coupling, equ (3), yields the Su-Schrieffer-Heeger (SSH) [11] Hamiltonian for polyacetylene. Addition of equ (4) to equ (2) yields the Hubbard (V=0) or extended Hubbard models($V\neq 0$). The sum of equ (2), (3) and (4) for a given geometry (i.e. so the phonon variables are fixed) is the PPP model, recently treated exactly (albeit, numerically) by Soos[12] for chains up to 14 carbon atoms. At the present time, there is no evidence to believe that it is wrong to extrapolate the results of the small molecule calculations (or experiment) to the long chains (presumably in polyacetylene samples). This implies that to understand the properties of polyacetylene, one must treat the correlation terms explicitly and carefully. Thus polyacetylene becomes a system with three competing interactions: the electron bandwidth B, the electron phonon coupling energy S, and the correlation energy, U[13,14]. This is a hard problem: for example the optical band gap with both correlation effects and Peierls effects (bond alternation) is not known in general, but is known numerically for short chains, and perturbatively for long chains. In addition, the bond alternation as a function of correlation energy is a non monotonic function of U, indicating the complexity of the problem.

IV. <u>Conjugation Lengths and Optical Properties:</u> In studying non-conjugated molecules , it has been possible to treat most properties as bond-additive. That is, the polarizability, for example, of a small molecule can be found by adding (tensorially) the polarizabilities of all the bonds in the molecule. This is not true for conjugated molecules. The strong coupling of the p electrons in an all trans planar polyene leads to a qualitatively different polarizability from the sum of single and double bond polarizabilities found from non-conjugated molecules. In fact, for short polyene chains the polarizability increases non-linearly with the size of the chain, n. This can not go on indefinitely as n increases: on some length scale, the p electrons are no longer strongly correlated, and the polarizability increases linearly with the size of the system. This length scale can be associated with the conjugation length. This raises interesting questions: How do we define conjugation length?, What are the physical factors limiting conjugation length?, Is there a different conjugation length for different properties? In this section, we give partial answers to these questions; however, a general formulation of these has still not been worked out.

First, we present a very qualtitative argument, due to Ducuing [15], for the length dependence of the polarizabilities, linear and non-linear, for these one dimensional systems. The perturbation formula foor the zero frequency polarizability along the chain (x) axis is

$$(5) \quad \alpha_{xx}(0) = (2e^2/\hbar) \ \Sigma \ |\langle 0| \ \hat{x} \ |f\rangle|^2 \ / \ \omega_{fo}$$

where $\langle 0| \ \hat{x} \ |f\rangle$ is the matrix element of $\hat{x}$ ($= \Sigma \ x_i$, the sum being over all electrons) between the ground state and the fth excited electronic eigenstate of teh molecule in the absence of the field, and

ω_{fo} is the frequency (energy) difference between those states. The problem is to find the dependence of α_{xx} on n, which is both the number of electrons and the length of the chain. Both the matrix element and the frequency are functions of n. A useful approximation is to replace ω_{fo} by an "average", $\bar{\omega}(n)$, and then do the sum, to find

$$(6) \quad \alpha_{xx}(0) = (2e^2/\hbar\,\bar{\omega}(n)) \{<0| \hat{x}^2 |0> - <0| \hat{x} |0>^2\}$$

For small n, the ground state will be extended over the entire chain, so we expect $<0| \hat{x}^2 |0>$ to be $O(n^2)$, while for large n, we expect it to be $O(n)$. In addition, we expect that $\bar{\omega}(n) = \omega_0 + A/n$, since the band gap scales in this manner. This suggests that $\alpha_{xx}(0) \sim n^2 - n^3$ at small n and $\sim n$ at large n. If we manipulate the oscillator strength sum rule

$$(7) \quad \Sigma \; |<0| \hat{x} |f>|^2 \, \omega_{fo} = (n\hbar/2m)$$

in the same way, we have

$$(8) \quad \bar{\omega}(n) <0| \hat{x}^2 |0> = (n\hbar/2m)$$

Solving (8) for $<0| \hat{x}^2 |0>$ in terms of $\bar{\omega}(n)$ and substituting into (6) yields

$$(9) \quad \alpha_{xx}(0) \approx (e^2/m) \{ n/[\bar{\omega}(n)]^2 \}$$

which gives $\alpha_{xx}(0) \sim n^3$ at small n and $\sim n$ at large n (if $\omega_0 \neq 0$). The length at which it saturates can be called the conjugation length.

A similar argument for γ_{xxxx}, the second hyperpolarizability yields

$$(10) \quad \gamma_{xxxx} \sim \{<0| \hat{x}^4 |0> - 2<0| \hat{x}^2 |0>^2 \} /[\bar{\omega}(n)]^3$$

which is $O(n^4)$ to $O(n^7)$ at small n to $O(n)$ at large n. The length scale at which saturation occurs is the conjugation length for γ. These arguments are very qualitative; however, as a guide to what to expect from the calculation, they are surprisingly sensible.

In the last few years, a number of groups [16,17,18,19] have calculated α_{xx} and γ_{xxxx} for all trans polyenes, some using the PPP model (in various approximations[16,17] or exact numerical methods[18]) and some using ab-initio techniques [19]. The results are interesting, and for the most part in agreement. It was found that $\gamma \sim n^\mu$ with $\mu \sim$ 4-4.5 in agreement with our qualitative guide above, and in quite good agreement with the experimental values of χ ($\sim \gamma/n$). These calculations show that saturation sets in for α at about 15 double bonds. However, the calculation of γ shows little evidence for saturation up to n = 20 (with different authors interpreting these signs slightly differently). So the theoretical situation at present is that perfect all trans chains of polyenes have conjugation lengths larger than about 15 double bonds, but perhapos not much larger.

Another important theoretical issue is the effect of various motions of the chain on these properties. Although this has yet to be addressed directly, recently a calculation appeared[20] of the average angle between p_z orbitals down an all trans chain as a function of n, T and torsional potential. The basic physical picture is that the torsional oscillations of the chain will make, on average, the p_z orbitals be less and less conjugated at large separations. Using the available experimental data on torsions, it was found that the angle or a function of this angle, fell off with distance as $\exp(-L/L_c)$. This length scale, L_c, was associate with the conjugation length. For polyenes, the prediction is that this length is on the order of 15 double bonds at room temperature, and smaller as T is raised. Note that the coincidence of this conjugation length with the saturation length in the polarizability is fortuitous, as there is no hard connection made between them. For polymers such as polythiophene and polypyrrole, L_c is much smaller, on the order of a few rings at room temperature. Details can be found in ref [20]. If the connection between this length and the saturation length of the polarizabilty and hyperpolarizability can be made firmly, this suggests a simple way to describe the correlations in these conjugated systems. However, it is clear that the role of torsions

8

and other motions is important and perhaps limiting. Clearly they should be the focus of more study.

V. <u>The Theoretical Outlook</u>: In my view, the theoretical problems for the study of non-linear optical properties in one dimensional conjugated polymers in the immediate future are: i) the effects of substitution by electron donating or withdrawing groups, ii) the effects of geometric and electronic defects, including solitons, polarons, and bipolarons, iii) vibronic effects, including a more careful study of torsional motion, and iv) the effect of the solid state (local field effects and chain-chain interactions). Because of the competing interactions and one dimensionality of these systems, all of these must be taken into account before a true understanding of the non-linear properties can be found.

Acknowledgements: This work has been supported in part by a grant from the NSF.

References:

[1] A.S. Davydov, <u>Theory of Molecular Excitons</u>, McGraw Hill N.Y., 1962.
[2] W. Warta and N. Karl, Phys. Rev <u>B32</u>,1172 (1985): M. Pope and C.E. Swenberg, Ann. Rev. Phys. Chem. <u>35</u>, 613 (1984).
[3] R. Peierls, <u>Quantum Theory of Solids</u> (Oxford Univ. Press, 1955)
[4] J.Jortner and S. A. Rice in <u>Physics and Chemistry of the Organic Solid State</u>, edited by D. Fox and M. Labes (John Wiley, 1967)
[5] Y. Toyozawa, in <u>Relaxation of Elementary Excitations</u>, ed. R. Kubo and E. Hanamura (Springer, NY, 1980)
[6] R. Silbey, Ann. Rev. Phys. Chem.<u>27</u>, 203 (1976)
[7] H. Sumi J. Chem. Phys. <u>70</u>, 3775 (1979); <u>71</u>, 3403 (1979)
[8] V. Kenkre, J. Andersen, D. H. Dunlap, and C. B. Duke, Phys. Rev. Lett. <u>62</u>, 1165 (1989)
[9] L. Salem, <u>The Molecular Orbital Theory of Conjugated Systems</u> (Benjamin, NY 1966).
[10] B. Hudson, K. Schulten, and B.E. Kohler, <u>Excited States</u>, vol 6, ed. E.C. Lim (Academic, NY 1982)
[11] W. Su, R. Schrieffer, A. Heeger, Phys. Rev. Lett. <u>42</u>, 1698 (1979)
[12] Z. Soos and G. Hayden, in Electroresponsive Molecular and Polymeric Systems, ed T.A. Skotheim (M. Dekker, NY, 1988)

[13] A.Ovchinnikov, I. Ukrainski and G. Kventsel, Soviet Physics Uspekhi 15, 575 (1973)
[14] J. Hirsch, Phys. Rev. B31, 6022(1985)
[15] see for example J. Ducuing in Proceedings of the International School of Physics "Enrico Fermi" Course 64: Non-Linear Spectroscopy, edited by N. Bloembergen (North Holand, N.Y.) 1977
[16] A. Garito, K. Wong, and O. Zamani-Khamiri, in Non-Linear Optical and Electroactive Polymers, ed by P. Prasad and D. Ulrich (Plenum, NY, 1988)
[17] C. DeMelo and R. Silbey, Chem. Phys. Lett. 140, 537(1987); J. Chem. Phys. 88, 2567 (1988)
[18] Z. Soos and S. Ramesesha, J. Chem. Phys. 90, 1067 (1989); Chem. Phys. Lett. 153, 171 (1988).
[19] G. Hurst, M. Dupuis, and E. Clementi, J. Chem. Phys. 89, 385 (1988); B. Kirtman and H. Hasan Chem. Phys. Lett (in press)
[20] G. Rossi, R. R. Chance, R. Silbey, J. Chem. Phys. 90, 7594 (1989)

SURVEY OF ELECTRICALLY CONDUCTING ORGANIC MATERIALS

Dr. H. NAARMANN
BASF Plastics Research Laboratory
6700 Ludwigshafen
F.R.G.

ABSTRACT.   This article concentrates on the most important
routes for synthesizing polymers containing conjugated
-C=C- bonds. It begins with oxidative coupling, which at
the start of the 60s developed through the different
stages of polyaromatics and polyheteroaromatics to
culminate in electrically conducting polymers. Subsequent
chapters embrace Wittig reactions, cyclo-condensations and
the formation of charge-transfer complexes, e.g. of
polyaromatics. Complex formation has been found to
dramatically increase electrical conductivity. Ziegler-
Natta catalysts and other similar systems containing
metals permit a very wide variety of polyenes to be
synthesized. The application of, for example, metathesis
reactions has led to soluble prepolymers and finally to
characterized polymers. The article finishes by
considering the synthesis of polymers that have been
oriented by mechanical stretching, the direct preparation
of oriented polymers by matrix polymerization and the
preparation of transparent materials.

1.   Oxidative coupling

1.1   CHEMICAL

In 1962, Kovacic and Kyriakis described the preparation of
polyphenylenes with a low number of constituents from
benzene in the presence of $AlCl_3/CuCl_2$ (1). This method
was expanded in terms of reaction conditions and monomers
employed with the result that dehydrogenation
polymerization came to be recognized as the general
principle for synthesizing polyaromatic, polyheterocycles
and polymetal complexes (2, 3), leading to electrically
conducting systems.

J. L. Brédas and R. R. Chance (eds.), Conjugated Polymeric Materials:
Opportunities in Electronics, Optoelectronics, and Molecular Electronics, 11–51.
© 1990 *Kluwer Academic Publishers. Printed in the Netherlands.*

It is interesting to note that varying the reaction
conditions opens up a broad spectrum of products of
different properties (see Fig. 1).

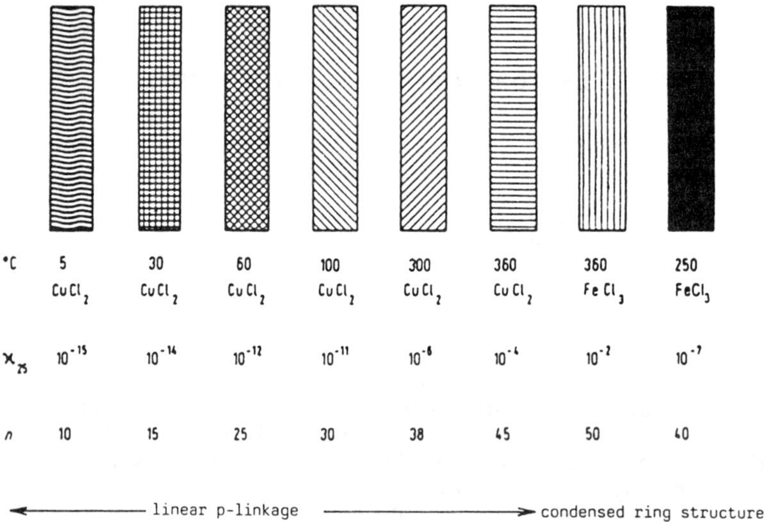

$^\circ C$	5	30	60	100	300	360	360	250
	$CuCl_2$	$CuCl_2$	$CuCl_2$	$CuCl_2$	$CuCl_2$	$CuCl_2$	$FeCl_3$	$FeCl_3$
$\varkappa_{25}$	10^{-15}	10^{-14}	10^{-12}	10^{-11}	10^{-6}	10^{-4}	10^{-2}	10^{-7}
n	10	15	25	30	38	45	50	40

Fig. 1.   Conductivity of polyphenylene as a function of
number of constituents

It can be seen that an increase in the number of
constituents, n*, is accompanied by a deepening in colour
and an increase in electrical conductivity** and cross-
linkages.

The stepwise synthesis of polyphenylenes is shown in
the following reaction scheme (Fig. 2).

This cation radical starts and gives rise for further
branching and crosslinking to layer lattices (graphite).

An indirect method for determining the molar mass has
been reported (4). It consists in alkylation to convert
the polyphenylene to a soluble product.

---

*   The number of constituents, n, was determined from IR
     spectra by calculating the ratio of monofunctional
     end groups to bifunctional centre groups and
     comparing them with the known lower constituents:
     terphenyl, quaterphenyl etc.

**   Unless otherwise stated, the term conductivity refers
     to the "dark conductivity" as measured in a cell by
     Beck (3), p. 559, at 300 bar and room temperature

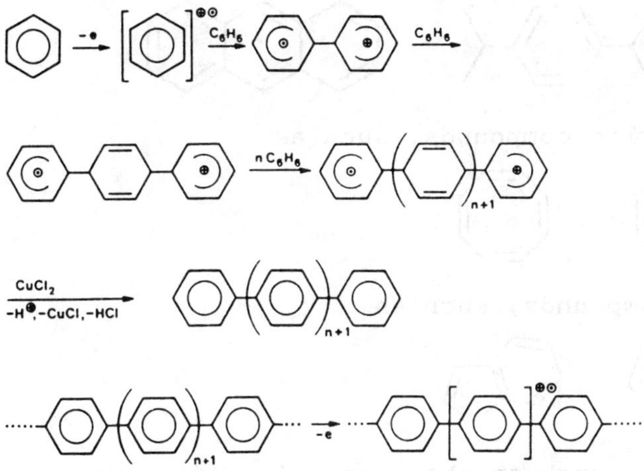

Fig. 2   Reaction scheme for oxidative coupling

The products of oxidative coupling are insoluble and can more or less only be converted to test specimens by sintering under pressure (> 200 °C, > 200 bar).

IR studies have confirmed that, although linking occurs primarily in the para position, ortho and meta branching and crosslinking also take place (4, 5).

In principle, all Friedel-Crafts catalysts and the usual dehydrogenation agents can be used for oxidative coupling (2).

The degree of suitability is shown below in decreasing order from left to right.

(1) $AlBr_3 > AlCl_3 > FeCl_3 > MoCl_6 > NbCl_5 > TaCl_4 > TiCl_4 > SnCl_4 > BF_3 > ZnCl_2$

as tested on benzene/$CuCl_2$

(2) $Pd^{II} > Co^{III} > Mn^{III} > Cu^{II} > Fe^{III} > chloranil > V^{III} > I_2$

as tested on benzene/$AlCl_3$

To test the scope of this method, the following types of compounds were employed.

(a)   Aromatic compounds, substitution products, condensed systems, such as

14

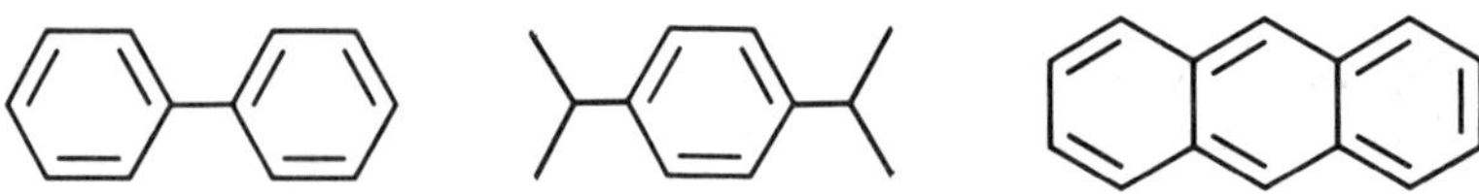

(b)   Unsaturated cyclic compounds, such as

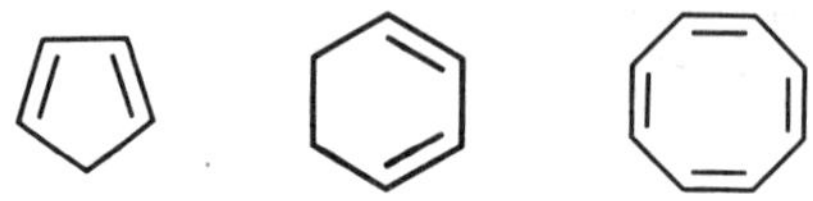

(c)   Heterocyclic compounds, such as

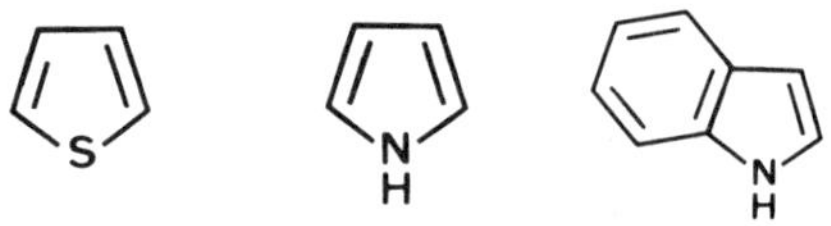

(d)   Metal complexes, such as phthalocyanine, ferrocene, and

(e) combinations of (a) and (b).

In all cases, the products were dark coloured, mainly insoluble and had electrical conductivities of up to 0.5 S/cm (2, 6 a, 6 b, 6 c, 6 d, 6 f)*.
    Since all of these reactions yield crosslinked, uncharacterized structures (see Fig. 3) attempts have been made to obtain soluble and thus characterizable products by e.g.

(f)   starting from characterized prepolymers (Fig. 3)

Fig. 3:

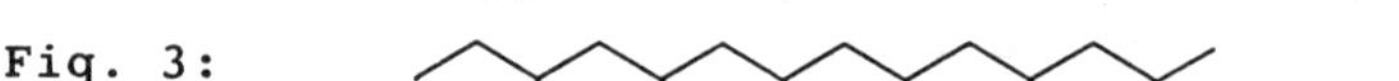

Poly(vinyl methyl ketone)

Poly(cyclohexenone) (6g)

---

*   Polyphenylenes have enjoyed a renaissance since the 70s (6h)

and performing Michael addition reactions:

(g)

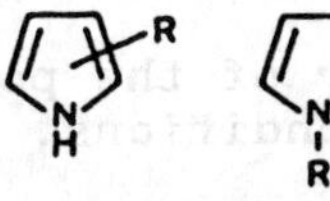

n = 4 - 25

1.2 ELECTROCHEMICAL

Under the right conditions of current, electrolyte and electrodes, anodic electrochemical oxidation is an effective means of performing selective dehydrogenation and polycondensation reactions (7).

Classic examples include the electrochemical preparation of polyheterocycles, polypyrrole, polythiophene and related compounds, and polyaniline. Electrochemical polymerization was first described in 1957 (8).

In 1979 Diaz (9) published the results of studies on the "electrochemical synthesis of polypyrrole". Developments in this field are chronologically treated in the article "From Powder to Plastics" (9).

Conductive polypyrrole* films are obtained directly in the anodic polymerization of pyrrole in organic solvent. They are black and, if the reaction conditions are suitable, can be detached from the anode in the form of self-supporting films. Some of the conductive salt used in the synthesis is contained in the film as a counterion.

Film produced under the conditions quoted above is brittle and has a conductivity of about 30 S/cm.

The polymerization reaction is a very complicated one. The main reaction steps are shown in Fig. 5.

---

*   It should be noted that PPY represents a number of electrochemically polymerizable pentacyclic heterocycles such as

Thiophene   Furan   or substituted pyrroles
                                            and oligomers

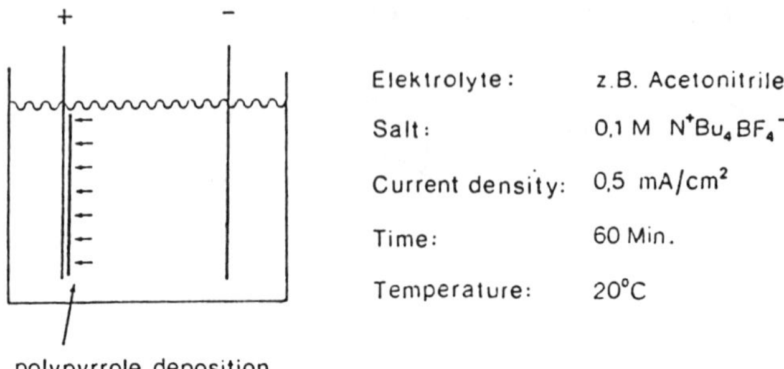

Fig. 4   Schematic diagram for the anodic deposition of polypyrrole. In this electrolytic process, the conductive salt used is incorporated in the polymer and conductive polypyrrole is obtained directly

Fig. 5   Electropolymerization of pyrrole (10)

The initial oxidation step in which a cation radical is formed, is followed by coupling, deprotonation and single electron oxidation to regenerate aromaticity. An important feature is the terminating reaction in which -O- is introduced to form a labile cyclic -CO-NH group.

<u>Scope available for varying the synthesis</u>

The quality of the polymers is greatly influenced by the reaction conditions.

If the monomers are of the same purity, the properties of the polymers are largely reproducibly influenced by the synthesis conditions.

The decisive factors are the electrolyte/conductive salt and the current density but particularly the conductive salt because it is incorporated in the polymer. Thus flexible and smooth films having conductivities of up to 200 S/cm can be produced by judicious selection of the conductive salt, e.g. of $\phi\text{-}SO_3\text{-}N^+Bu_3H$ instead of the BF_4 salt. Some typical conductive salts are shown in Fig. 6.

These counterions allow specific effects to be introduced into the polypyrrole. It can thus be understood that it is precisely the anion - or its size, geometry, charge, etc. - that governs the properties of the polymers. In general, one anion is incorporated into 3 pyrrole units.

Some general considerations apply to the choice of anion. If it is organic, flexible smooth films that can be readily detached from the anode are generally obtained. If it is hydrophilic, e.g. ClO_4^- or BF_4^-, the polypyrrole formed is also hydrophilic. This is in contrast to polypyrroles in which a polymeric counterion is used, i.e. poly(styrenesulphonic acid).

Changing the synthesis conditions allows different types of product with different surface morphology, e.g. an open porous structure, to be prepared.

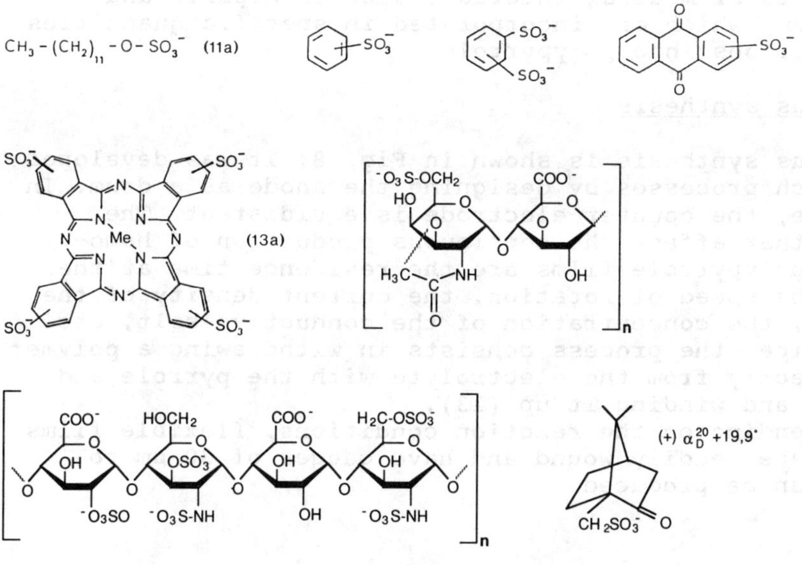

Fig. 6   Conductive salts that can be incorporated as counterions in polypyrrole (11)

Pyrrolesulphonic and thiophenesulphonic acids (12) are
interesting variants of conductive salts (Fig. 7).

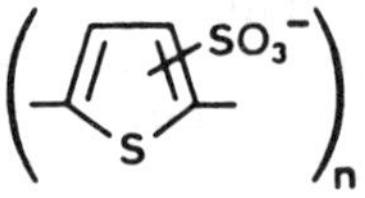

Thiophenesulphonic
acid and
oligothiophene-
sulphonic acid

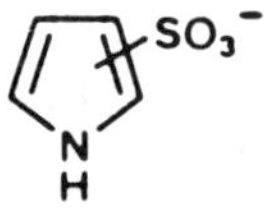

Pyrrolesulphonic
acid

Fig. 7   Functionalized monomers as counterions

In this instance, the counterion is coupled direct to the
monomer. The result is that, under ideal polymerization
conditions, there is one counterion (x⁻) for every monomer
unit. The counterion is not "externally" incorporated but
is attached, i.e. chemically bound, to the monomer.
    This exclusion of counterions can be performed
specifically, and offers, for instance, the possibility of
releasing optically active counterions or active
ingredients of medical interest, such as heparin and
monobactam, which are incorporated in specific quantities
as counterions into polypyrrole.

<u>Continuous synthesis</u>

Continuous synthesis is shown in Fig. 8. It was developed
from batch processes by designing the anode as a drum. In
each case, the counter-electrode is equidistant. The
factors that affect the continuous production of homogeneous polypyrrole films are the residence time at the
anode, the speed of rotation, the current density of the
monomers, the concentration of the conductive salt, etc.
In practice, the process consists in withdrawing a polymer
film directly from the electrolyte with the pyrrole and
the salt and winding it up (13).
    Depending on the reaction conditions, flexible films
that can be readily wound and have gauges of 30 μm to
150 μm can be produced.

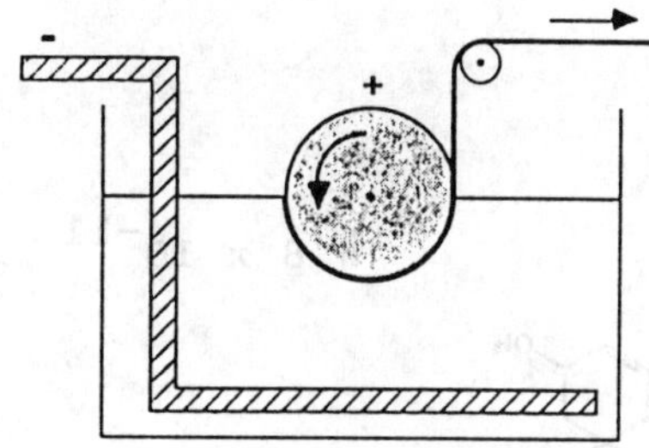

Fig. 8   Continuous production of polypyrrole.

2.   Stepwise synthesis of –C=C– bonds

The reaction between carbonyl compounds and activated methylene groups is an elegant method of systematically building up –C=C– bonds.

As can be seen from Fig. 10, increasing the number of –C=C– units shifts the absorption to a longer wavelength. It is interesting to note that the electrical conductivity increases at the same time (Fig. 9).

Fig. 9   Correlation between structure, –C=C– units and electrical conductivity

Structure	$(C=C)_n$ units	S/cm*
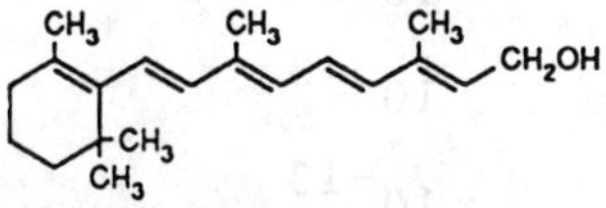 **Vitamin A**	5	2.8×10^{-14}
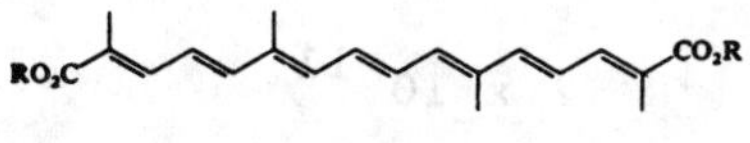 **Dimethylcrocetin**	7	5×10^{-13}

---

* The values quoted were determined as dark conductivity by BASF on undoped, compressed powders under 300 atmospheres at 25 °C

All-trans-β-carotene   11   6.8 x 10^{-11}

Eschscholtzxanthin, 3,3'-Dihydroxyretro-β-carotene
                                              12   9.0 x 10^{-11}

                                              13   6.5 x 10^{-10}
Torulene, 3'4'-Dechydro- -carotene

Structure	$(C=C)_n$ units	S/cm*
$R-C_6H_5-(CH=CH)_nCOOR$		
R = tert. butyl	1	10^{-13}
	2	10^{-13}
	3	10^{-13}
	4	10^{-13}
	5	10^{-11}
	6	2 x 10^{-11}
	8	5 x 10^{-10}
	10	8 x 10^{-10}

The values shown in Fig. 9 prove the correlation that
exists between the number of -C=C- units and the
electrical conductivity.

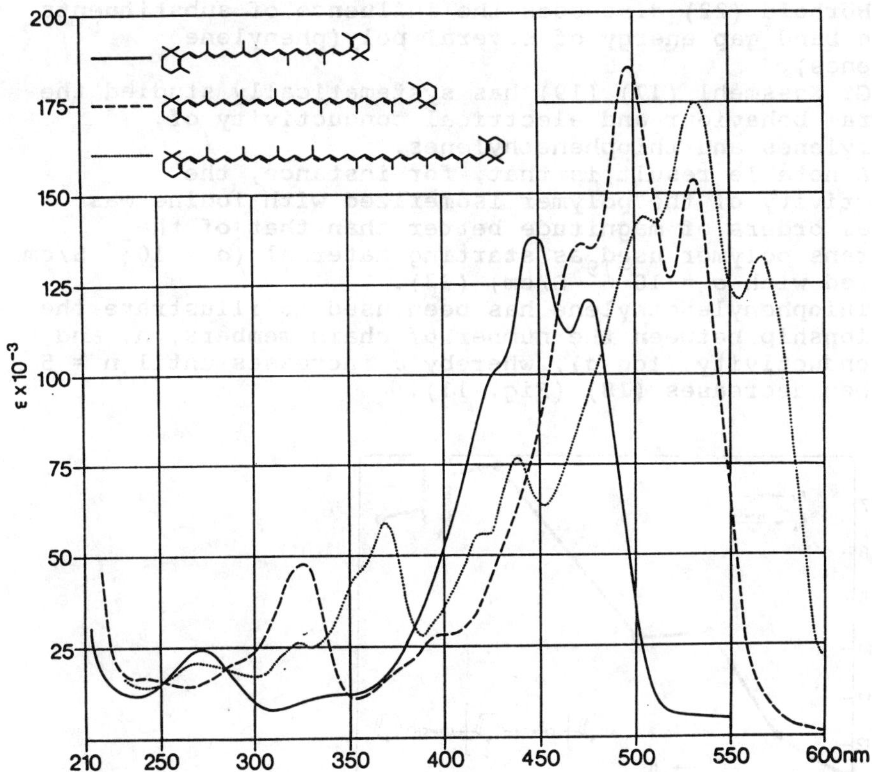

Fig. 10   Ultraviolet and visible light absorption spectra
of ß-carotene (—————), decapreno-ß-carotene (– – – –)
and dodecapreno-ß-carotene (.........) (in petroleum
ether)

Methyl-branched polymers have a lower conductivity
than unbranched systems. The substituent influence exerted
by the CH_3 group is minor. The influence of CN, $CONH_2$, and
$COOCH_3$ groups on -C=C- is of especial interest in regard
to storage stability and subsequent modifications (15).
Another variant makes use of the Wittig reaction to
introduce aromatic and heteroaromatic side chains onto
polyenes via carbonyl derivatives (16), e.g.:

H.H. Hörhold (22) discusses the influence of substituents on the band gap energy of several poly(phenylene vinylenes).

G. Kossmehl (17) (19) has systematically studied the spectral behaviour and electrical conductivity of, polyarylenes and thiophenethylenes.

A notable result is that, for instance, the conductivity of the polymer isomerized with iodine was several orders of magnitude better than that of the cis/trans polymer used as starting material ($\sigma = 10^{-4}$ S/cm compared with $\sigma = 10^{-19}$ S/cm) (33).

Thiophenylenethylene has been used to illustrate the relationship between the number of chain members, n, and the conductivity (log σ), whereby σ increases until n = 5 and then decreases (18) (Fig. 11).

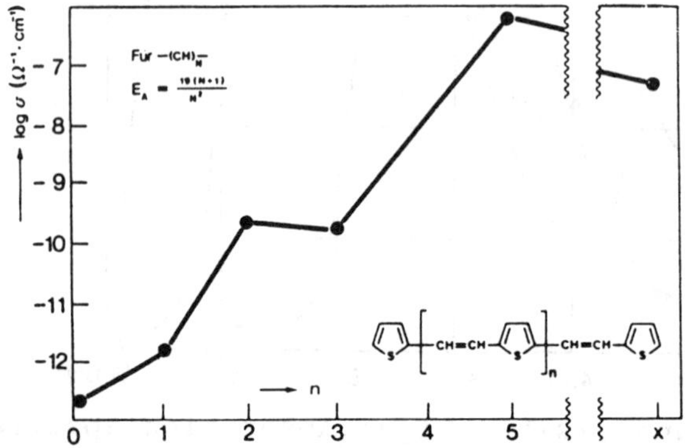

Fig. 11   Conductivity as a function of the length of the conjugated system

As first shown by McDonald and Campbell (20), the Wittig reaction can be used for synthesizing the polyxylylidene system. Drefahl et al. (21) have developed the synthesis reaction further.

H.H. Hörhold's article entitled "Development of an electroactive polymer material from unmeltable powder to transparent film" describes the progress that has been made in this sector.

3.   Ring-forming condensation and addition reactions

Condensations

Ring closure of vinyl methyl ketone (VMK) and poly-VMK to form structures of the type has already been discussed in Chapter 1. (f) and (g).

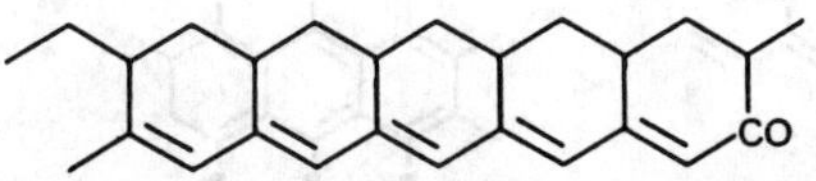

An analogous reaction is the cyclization of polyacrylonitrile (PAN) via polydihydropyridin and subsequent high-temperature aromatizing to the polyacene system (23).

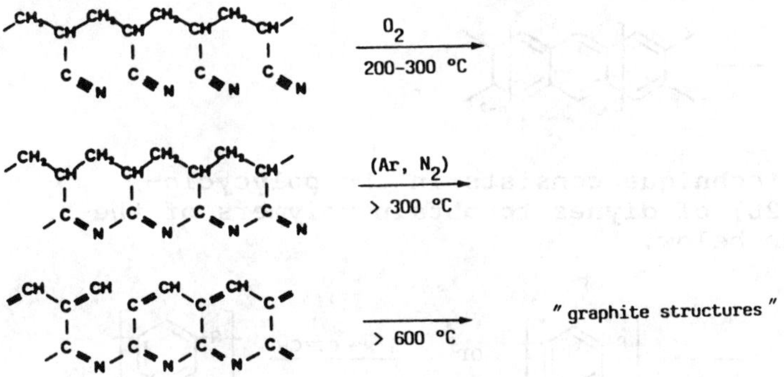

Fig. 12   Transformation of PAN into graphite-like lattices

Correspondingly, poly(1,2-vinyl butadiene) can be cyclized and dehydrated (24).

1,6-Heptadiyne can be converted in the presence of Ziegler-Natta catalysts (Al-trialkyl/TiX$_4$) to soluble products having the cyclohexene structure (25).

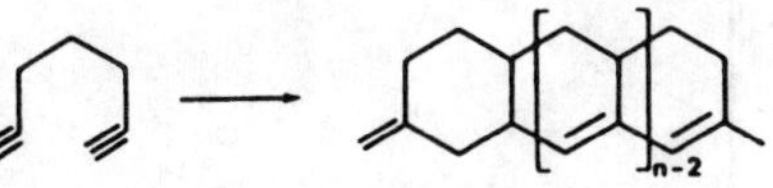

Under the same conditions, 1,7-octadiyne and 1,8-nonadiyne can be polymerized but afford partially crosslinked products containing 50 % and 22 % soluble fractions.

Studies by Gibson and Epstein (26) have shown that 1,6-heptadiyne can be converted to greenish gold, self-supporting films. X-ray defraction studies show that the films are completely amorphous and much more sensitive to oxygen than is polyacetylene.

1,4-Butadiyne can be polymerized from the gas phase onto polymer surfaces, e.g. polyethylene and Teflon (26 a). Structures with triple-bond side groups or polyacenes are suggested for the polymer.

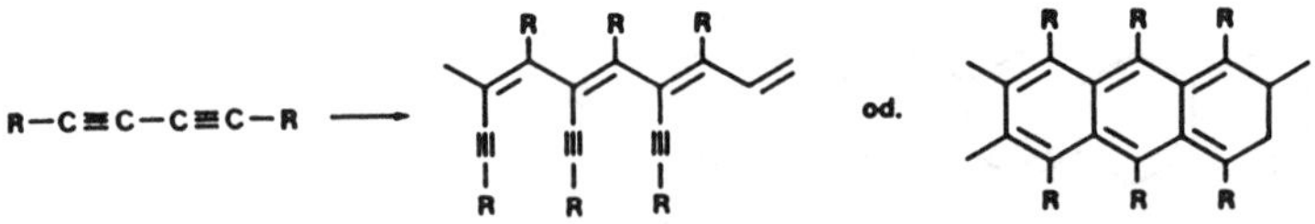

Conjugated polymer systems of the structure shown below
can, for instance, be produced by polycondensation of
1-chlorobutene (27).

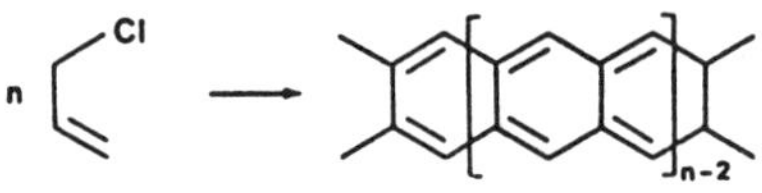

An interesting technique consists in the polycyclo-
trimerization (28) of diynes to obtain polymers of the
structures given below.

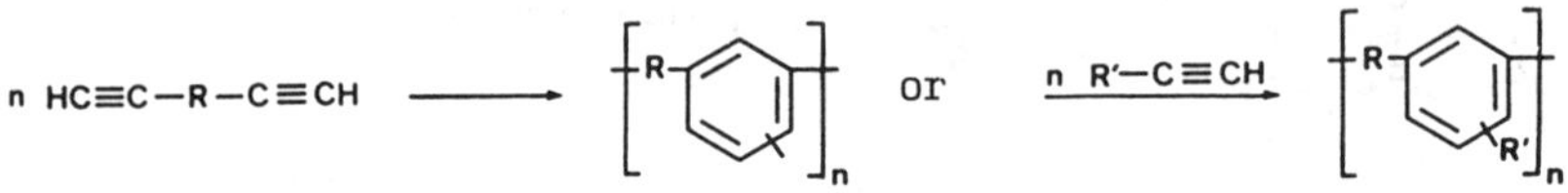

Structures of the following type

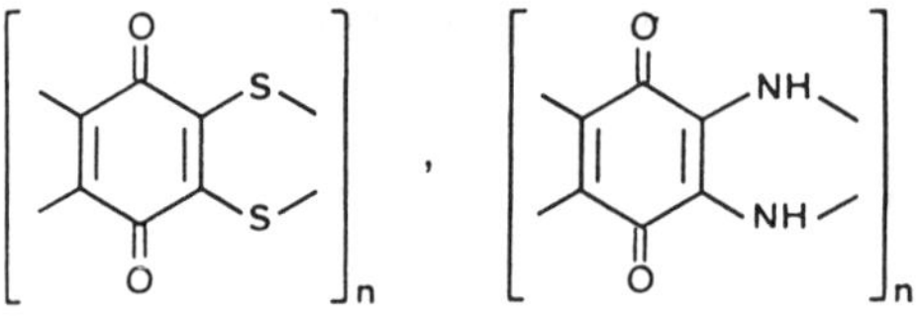

are obtained by condensation of chloranil with e.g. Na_2S
or $NaNH_2$ (29).
    The products are insoluble, difficult to purify and
largely uncharacterized.
    The idea behind the synthesis of the types of
structures shown above is to combine electron donor and
acceptor units in one molecule.
    Polymers of the following structures have been
synthesized in order to obtain characterized polymers
(30).

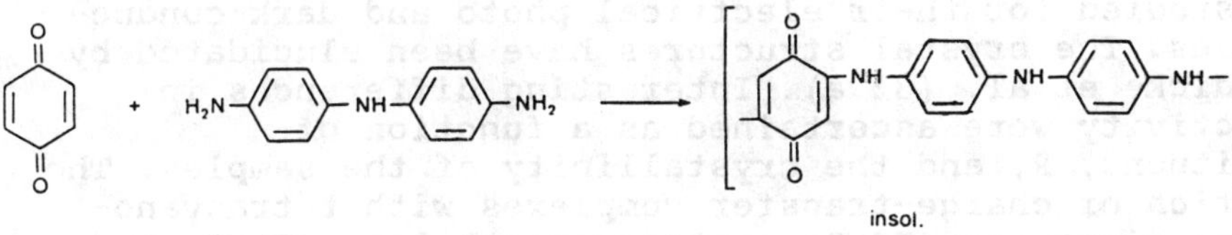

Another method consists in exploiting the Diels-Alder reaction, which proceeds bifunctionally and gives rise to characterized polymers in accordance with the following reaction scheme (31).

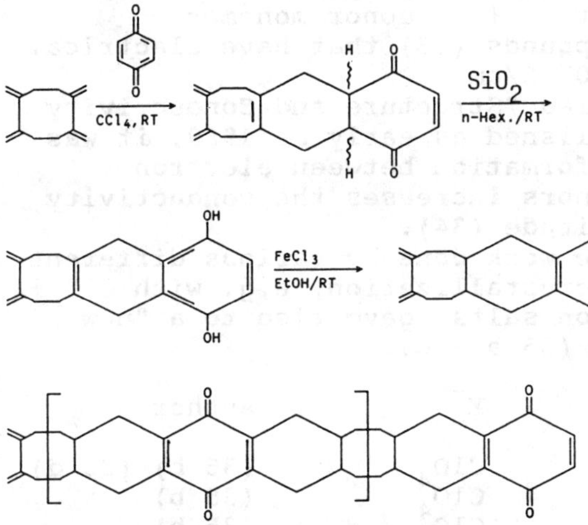

Fig. 13   Reaction scheme for repetitive Diels-Alder reaction

4.   Charge-transfer complexes

In the search for easy-to-manufacture, highly stable compounds with a characterized number of double bonds, perylene derivatives of the imide-type structure I and imidazole-type structure II (32)

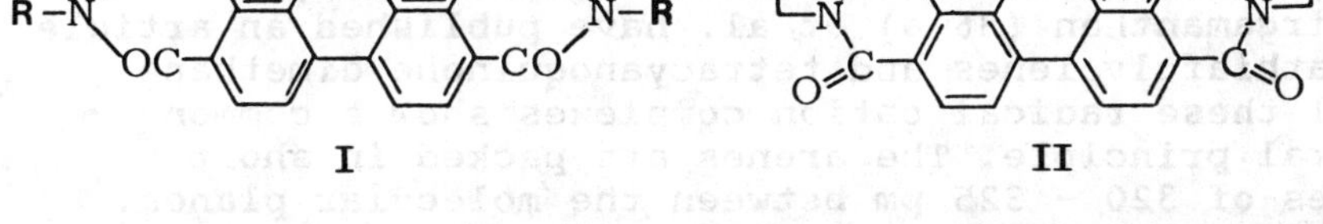

I

II

were studied for their electrical photo and dark conductivities. The crystal structures have been elucidated by E. Hädicke et al. (32 a). Interesting differences in conductivity were ascertained as a function of substituent, R, and the crystallinity of the samples. The formation of charge-transfer complexes with tetracyanoquinone dimethane (TCNQ), tetracyanoethylene (TCNE) and iodine (I) increased the conductivity by a factor of 1 000, thus achieving the conductivity of graphite of 10^{-1} S/cm in some cases.

Translating the system to polymeric charge-transfer complexes of the type

polymer with donor	+	acceptor monomer
polymer with donor	+	polymer with acceptor
polymer with acceptor	+	donor monomer

led to a new class of compounds (33) that have electrical conductivities of up to 10^2 S/cm.

In the article entitled "Structure and Conductivity of Organic Polymers", published as early as 1969, it was pointed out that complex formation between electron acceptors and electron donors increases the conductivity by several orders of magnitude (34).

At the same time, the work done by various different working groups on electrocrystallization, e.g. with formation of radical cation salts, gave rise to a "new family of organic metals" (35 a - c).

Aryl	X^-	Author
Pyrene	ClO_4	(35 b) (35 d)
Perylene	ClO_4	(35 b)
Azulene	ClO_4	(35 b)
Benzene and derivatives	ClO_4	(35 c)
Naphthalene	PF_6	
Biphenyl	AsF_6	
Anthracene	SbF_6	
Fluorene	BF_4	
Fluoroanthene	PF_6	(35 a)
Triphenylene	AsF_6	(35 d)

The conductivities have values of ca. 10^{-1} S/cm. In December 1979, K. Bechard et al. (36) discovered supraconductivity at 4.6 K in the $(TMTSF)_2PF_6$ system (ditetramethyl tetraselenafulvalene hexafluorophosphate). P. Kathirgamanthan (36 a) et al. have published an article on tetrathiafulvalenes and tetracyanoquinone dimethane.

All these radical cation complexes show a common structural principle. The arenes are packed in short distances of 320 - 325 pm between the molecular planes.

The anions are located in the channels between the stacks (35).

The enormous developments and progress that have taken place in this field were the subject of the ICSM 88 in Santa Fe (37).

5. Metal-catalyzed polymerization reactions

5.1 ZIEGLER-NATTA SYSTEMS

Polymerizing acetylene in the presence of Ziegler-Natta catalysts is the most effective method of preparing poly-acetylenes of different morphology (powder, gel or film) and with variable properties such as density, alignability and dopability (38).

5.2 OTHER CATALYST SYSTEMS CONTAINING METALS

Attention is particularly drawn to the work performed by the Japanese groups of Hatano, Ikeda and Shirakawa (see Table of initiator systems, Fig. 15, Nos. 8, 10, 12, 16).

Polyacetylene (PAC) exists in various geometrical isomeric forms

cis-transoid (cis-PAC)

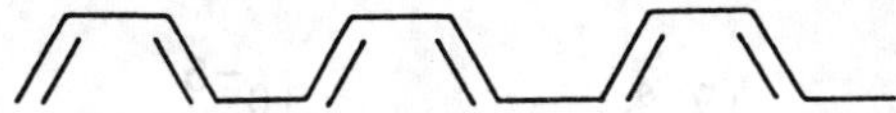

trans-cisoid

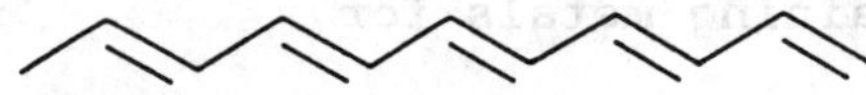

trans-transoid (trans-PAC)

cis-PAC is thermodynamically relatively unstable and reverts by isomerization to the thermodynamically stable trans-PAC via the metastable trans-cisoid PAC.

cis-cisoid PAC has not yet been prepared in pure form.
Model reactions, however, have shown that cyclic and
helical structures are possible (39).

The stability of the polyacetylene depends on defect
sites (sp^3 fractions) in the polymer with a close
correlation existing between conductivity and
crystallinity (40; see Fig. 14 and Chapter VIII).

Fig. 14   Correlations between structure and conductivity
of polyacetylene

Catalyst	sp^3 Amount	Crystallinity	Conductivity S/cm
$Al(C_2H_5)_3/Ti(OC_4H_9)_4$ – Shirakawa –	1 %	80 %	10^3
$NaBH_4/Co(NO_3)_2 6H_2O$ – Luttinger –	11 %	70 %	10^{-1}
$Ni(acac)_2$ – Cuprene –	18 %	50 %	10^{-3}
$Al(C_2H_5)_3/TiCl_4$ – Hatano –	40 %	19 %	10^{-5}

Fig. 15   Initiator systems containing metals for
synthesizing polyenes (38 b)

No.	Catalyst system	Authors	Literature
1.	$FeCl_3/C_6H_5MgBr$	A. Job. et al., Compt. Rend. <u>189</u> (1929) 1089	

No.	Catalyst system	Authors	Literature
2.	$Ni(CO)_4/P(C_6H_5)_3$	W. Reppe et al., Ann. Chem. $\underline{560}$ 1, (1948) 104	
3.	$Al(alkyl)_3/Ti^{IV}alkoxides$	G. Natta et al., Atthi. Accad. naz. Lincei, Rend. Classe Sci. fis. mat. nat. (1958) 25, 3	
	$Al(alkyl)_3/VCl_3$ and $TiCl_3$	Ital. Pat. 530753	
4.	$Al(alkyl)_3/TiCl_4$	B. Franzus et al., J. Am. Chem. Soc. $\underline{81}$ (1959) 1514	
5.	Fe and Co carbonyls	W. Hubel et al., Chem. Ber. $\underline{93}$ (1960) 103	
6.	$NaBH_4/Ni$ salt (complex)	L.B. Luttinger, Chem. and Ind., 3rd September (1960) 1135	
	$NaBH_4/Ni$ or Co complex	M. L. H. Green et al., Chem. and Ind. $\underline{3rd}$ September (1960) 1136	
7.	$Ni(CO)_2((C_6H_5P)_2)_2$	E. C. Colthup et al., J. Org. Chem. $\underline{26}$ (1961) 5155	
8.	$Al(C_2H_5)_3$ Ti-butylate	M. Hatano J. Polym. Sci. $\underline{51}$ (1961) 26	
	$Al(C_2H_5)_3/TiO(acac)_2$ $VO(acac)_2$ $Cr(acac)_2$ $Fe(acac)_3$ $Al(C_2H_5)_3/TiCl_4$ $VOCl_3$ VCl_4 VCl_3		
9.	$TiCl_4/Al(C_4H_9)_3/LiC_4H_9$	W. H. Watson et al., J. Polym. Sci. $\underline{55}$ (1961) 137	
10.	$Al(C_2H_5)_3/Fe(acac)_3$ $/V(acac)_3$ $/VO(acac)_3$	H. Noguchi, Polym. Letters Vol. $\underline{1}$ (1963) 553	

No.	Catalyst system	Authors	Literature

No.	Catalyst system	Authors Literature
11.	$NiBr_2/2\ P(C_6H_5)_3$ $/2\ PO(C_6H_5)_3$ $CoBr_2/2\ P(C_6H_5)_3$	W. E. Daniels, J. Org. Chem. __29__ (1964) 2936
12.	$TiCl_4/Al(CH_3)_3$ $TiCl_3/Al(C_2H_5)_3$ $Ti(OC_4H_9)_4$ $Al(C_2H_5)_3$ $TiCl_2(acac)_2$	S. Ikeda, A. Tamak; Inter. Symp. on Makromol. Chem. Tokyo-Kyoto 1966 I 124
13.	Metal carbonyls	DBP. 1298 708, BASF 26.11.64/12.03.70
14.	$Fe(acac)_3/Al(alkyl)_3$	F. Ciardelli et al., Makromol. Chem. __103__ (1967) 1
15.	Fe glyoxinate/$Al(C_2H_5)_3$ "Gel"	F. D. Kleist et al., J. Polym. Sci. Part, A-17, (1969) 3419 H. Noguchi et al., J. Polym. Sci B. __1__, 553 (1963)
16.	a) $Al(alkyl)_3$/Ti-alkoxides high concentration "Films" b) $Al(alkyl)_3Ti(acac)_3$ $Al(alkyl)_3/Cr(acac)_3$ "Powder" c) $Al(alkyl)_3/Mn(acac)_3$ $Al(alkyl)_3/Co(acac)_3$ "Powder"	H. Shirakawa et al., JA 32581 (1973) J. Polym. Sci., Polymer Chem. Ed. __12__ (1974) 11, Polym. J. Vol. __2__/2 (1971) 231
17.	Arene metal carbonyls	P.S. Woon et al., J. Polym. Sci: Polymer, Chemistry Ed. __12__ (1974) 1749
18.	Ti-cyclopentadienyl complex	S. L. Hsu et al., J. Chem. Phys. __69__ (1) (1974)

No.	Catalyst system	Authors	Literature
19.	a) Cyclopentadienyl dititanium complex	Showa Denko K. K., Tokio	
	b) Al(Et)$_3$/Ti(OBu)$_4$ "Gel"	JA 3 6284 (1979), 28.03.80/07.05.81, G. Wnek et al., J. Polym. Sci-Polym. Lett <u>17</u> (1979) 779	
20.	TiCl$_3$/Mg (C$_2$H$_5$)$_2$/Na	M. Zikmund et al., Chem. Zvestic (5) (1980) 618	
21.	PdCl$_2$/DMF	W. Deits et al., Am. Chem. Soc. Org. Coat. Plast. Chem. <u>43</u> (1980) 867 ACS Polymer Preprints <u>22</u> 1, (1981) 197	
22.	Me carbonyls/UV	T. Masuda et al., Polym. Bulletin <u>2</u> (1980) 823, DAS 1495215 BASF, 26.11.64/13.02.69 T. Masuda et al., Polym. Journal <u>13</u> 3 (1981) 301	
24.	Me carbonyl complex	J. Levisalles et al., J.C.S. Chem. Comm. 1981, 1055	
25.	Al$_2$O$_3$	N. Kurokawa et al., J. Polym. Sci., Polym. Letters, Ed. Vol. <u>19</u> (1981) 355	

The literature coverage of this table is not comprehensive. The criterion adopted in selection was the chronological development in catalysts and polymerization.

Many questions concerning structure and special conductivity remain unanswered. This may be due to the intractability of high-molecular-weight polyacetylene, which is insoluble in all solvents, has no characterized melting point, decomposes gradually at elevated temperatures, and reacts readily with oxygen to yield oxygenated material. Owing to the polymer's insolubility, the polyacetylene samples obtained are generally in the form of powders, gels and films of various thicknesses, orientation and conductivity and are difficult to characterize.

32

The polyacetylenes are crosslinked and insoluble.
Some of them contain cyclic structures such as cyclooctatetraene, benzene and ethylbenzene. Substituted
acetylenes, however, mainly yield soluble, linear polyacetylenes and benzene derivatives (41).
An interesting method is the polymerization of
butenyne (Fig. 15, No. 13 and 22)

$$n \quad \xrightarrow[\Delta T, h\nu]{Me - Carbonyle} \quad \left(\diagup\diagdown \right)_n$$

in which a polyene is formed by isomerization.

Chemical modification of $(-CH=CH-)_x$ (41 a)

Interesting chemical modifications are cyclo-additions on
the $(CH)_x$ backbone, e.g. with chlorosulfonyl isocyanate.
The ring of the adduct thus formed can be opened by
alkalis.

Cyclo-addition of chlorosulfonyl isocyanate and
ring-opening to substituted hydrophilic polyacetylene:

$$-(CH{=}CH)_x- \; + \; \underset{ClSO_2}{\overset{}{N{=}C{\diagdown}_O}} \xrightarrow{100°C} \; -(CH{-}CH)_x- \xrightarrow[100°C]{OH^-} \; -(CH{-}CH)_x-$$

The dominant reaction involving 3-chloroperbenzoic acid is
the formation of oxirane structures, which can react
further. Metal carbonyls, e.g. $Fe_3(CO)_{12}$, react only with
cisoid units. Otherwise, the metal atoms combine with two
moles of the en-component, or isomerization occurs and
leads to cis-configurations. Both types of reaction are
confirmed by IR spectroscopy.
CO-insertion can also be observed in molybdenum
carbonyls. Cyclo-addition with maleic anhydride (MA) and
3,4-dichloromaleic anhydride (DCMA) leads to the following
adducts. The adduct formed by DCMA is worth mentioning
because it gives rise to a fusible polyacetylene
(165 – 180 °C).

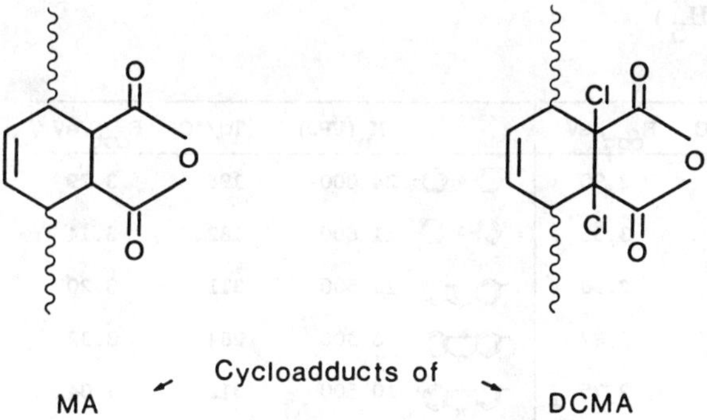

Cycloadducts of

MA   DCMA

6. Soluble prepolymers

The lack of processing scope and of perfect
characterization were the reasons behind the search for
systems that proceed from characterized precursors and
characterized, soluble and processable prepolymers to the
polyenes.

Pioneer work in this field was performed by Stille
(42) and Marvel (43).

1,4-Dipolar addition of a dilactone to diethynyl
benzene affords quantitative yields of
phenylene-oligomers; dicyclopentadienone reacts similarly.

Marvel's synthesis (43) proceeds in accordance with
the following scheme:

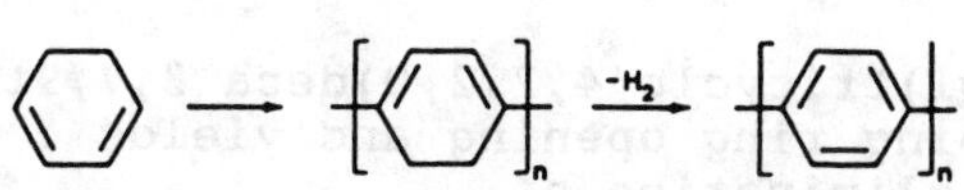

H.H. Hörhold (22) describes the polycondensation of
soluble polymers of the type $(-Ar-CR=CR-)$ where $R = C_6H_5$
using a wide variety of aromatic components.

Fig. 16   Substituted poly(phenyl vinylene) (22)

M_n was determined by vapour pressure osmometry and T_g by
DSC. E_{opt} is the energy value at which the absorbance has
fallen to 1/10 of that at the wavelength maximum.

$$[-Ar-CR=CR-]_n \qquad (R = C_6H_5)$$

Ar	M_n(VPO)	T_g/°C	E_{opt}/eV	Ar	M_n(VPO)	T_g/°C	E_{opt}/eV
(1,4-phenylene)	23 000	250	2,90	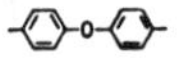 (4,4'-oxydiphenylene)	24 000	196	3,29
(1,3-phenylene)	14 000	192	3,35	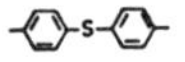 (4,4'-thiodiphenylene)	11 600	182	3,14
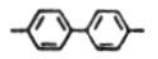 (4,4'-biphenylene)	25 000	303	2,98	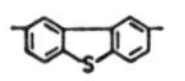 (dibenzothiophene-diyl)	23 500	311	3,20
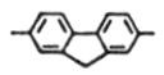 (fluorene-diyl)	15 400	247	2,87	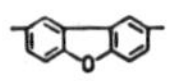 (dibenzofuran-diyl)	6 300	284	3,32
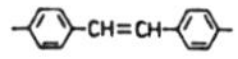 (4,4'-stilbenylene, –CH=CH–)	9 600	246	2,75	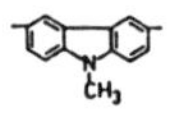 (N-methylcarbazole-diyl)	10 500	312	3,04

Other methods for synthesizing polyenes via characterized
soluble precursors include ring-opening reactions.

The Feast method (44) (44 a) for producing "Durham
PAC" proceeds according to the following scheme.

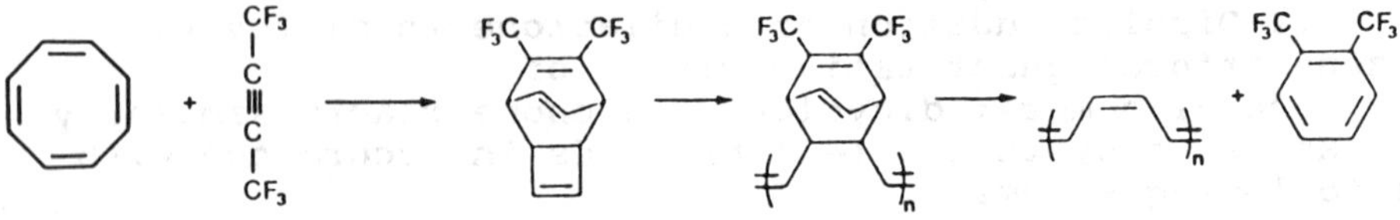

7,8-Bis(trifluoromethyl)tricyclo(4,2,2,0)deca 3,7,9triene
polymerizes by undergoing ring opening and yields
polyacetylene through elimination of
1,2-bis(trifluoromethyl)benzene.

In the Grubbs method (45), polybenzvalene is
isomerized in the presence of $HgCl_2$ to PAC.

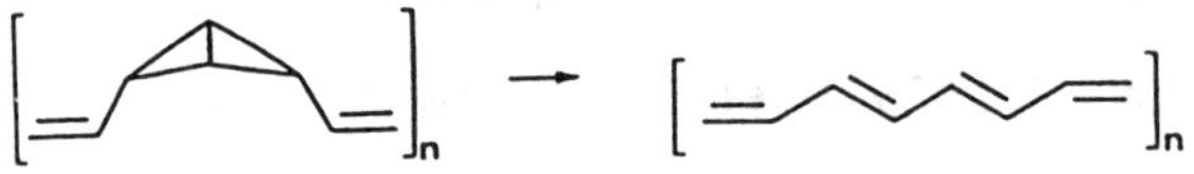

Both of these methods start off with certain monomers that
are converted to soluble pre-polymers that then yield
insoluble perconjugated polymers after thermal treatment.

Another interesting route to a conjugated polymer
with the following structure

involving an isomerization step has been proposed by Feast
(44 a).

The synthesis of polyenes has quite a long history.
Bohlmann (46) reported in 1956 that diacetylenes and
higher homologues could be thermally or photo-catalytically
converted to polymers and that the reaction only proceeds
in the crystalline state. In other words, the chains of
the starting compounds must be oriented in parallel
beforehand. Since then, this work has been intensively
developed under the key phrase "polydiacetylenes".
Elucidation of the structure by Hädicke et al. (47)
created the conditions for the interesting field of matrix
polymerization.

7. Elimination reactions between polymers

Apart from polycondensations, which proceed by elimination
of H_2 during oxidative coupling (see Section 1.), of $\emptyset_3 PO$
during Wittig condensations (Section 2.) and of water
during carbonyl condensations, such as

$$\tag{48}$$

There exist elimination reactions between the polymers
that will be given special treatment here (49).
The general reaction scheme is shown in Fig. 17.

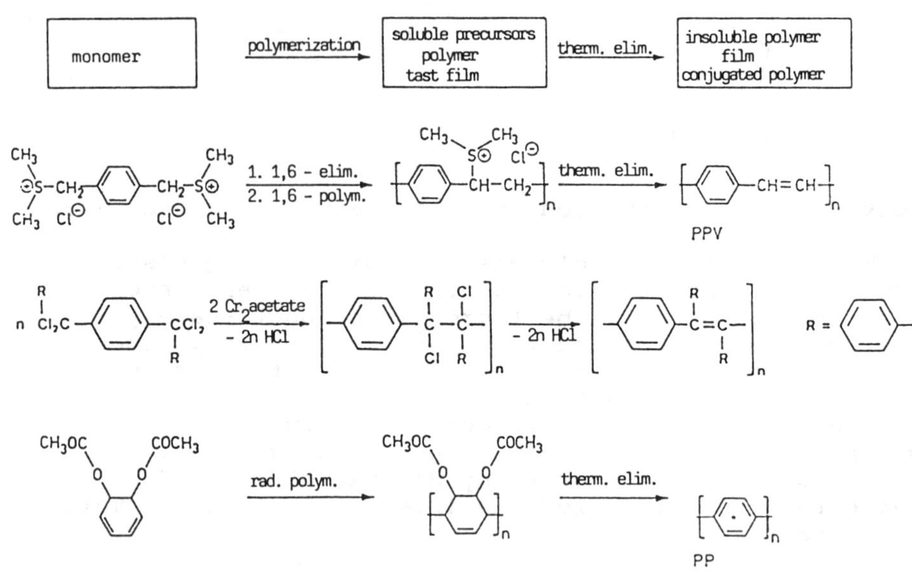

Fig. 17   Reaction scheme for elimination reactions (22)

Another interesting synthesis for doubly phenyl-
substituted polyxylylidenes has been described by Smets
et al. (50). It consists in acid-catalyzed, stepwise
condensation of aromatic bisdiazoalkanes.
    Dehydrohalo polymerizations (51), HCl cleavage from
poly(vinyl chloride) (52) and dehydration of poly(vinyl
alcohol) (53) yield poorly characterized products. Ohnishi
et al. have used sulfonium salts as precursors to
synthesize highly conducting poly(p-phenylene vinylene)
and the corresponding 2,5-thienylene-vinylene polymers
(54) with conductivities of approx. 10^3 S/cm after
stretching and 10^5 S/cm after pyrolysis.

8.   Synthesis of oriented polymers

8.1   MECHANICAL ALIGNMENT AFTER POLYMERIZATION

The simplest method of achieving orientation in polymers
consists in mechanical stretching in a preferred
direction, provided that the polymer admits of such
treatment (55a).

As mentioned in the previous sections, most electrically conducting polymers are insoluble and crosslinked. Orientation of such products in a preferred direction is almost impossible or occurs to a very minor extent.

This is due to the crosslinking sites in the polymer, the number of which can be ascertained by C^{13} NMR determination of the sp^3 fraction. The synthesis of polyacetylene (55) that is almost totally free from defects was the first time that high stretching rates could be achieved.

For instance, polyacetylene that has been produced in silicone oil with a specially treated catalyst shows surprisingly high conductivities after being doped with iodine (CCl_4 saturated with iodine at 25 °C):

Fig. 18   Comparative conductivities of several metals and different types of polyacetylenes (*doped with I_2)

Metal	Conductivity (volume)	Density	Conductivity (weight)
Hg	10 365 S/cm	13.546 g/cm³	767 S cm² g⁻¹
Pt	101 522	21.45	4 733
Fe	102 986	7.87	13 085
Ag	671 140	10.491	63 972
Cu	645 000	8.94	72 147
S-(CH)$_x$ (56)	520	before/after doping 0.40 1.23	422*
(S-CH)$_x$ oriented in LC matrix (57)	1 600 – 1 800*	0.50 1.10	1 750*
N-(CH)$_x$ oriented (55)	18 000 – 28 000*	0.85 1.12	~ 20 000*
ARA method N-(CH)$_x$ oriented (55)	120 000*	0.90 1.15	104 347*

38

The Shirakawa $(CH)_x$ (Fig. 19) is crosslinked and contains
approx. 2 % sp^3. The polyacetylene shown in Fig. 20 has
been prepared by the new BASF technique. It is linear (no
sp^3), is thus highly orientable (up to 660 %) and has a
conductivity of more than 100 000 S/cm.

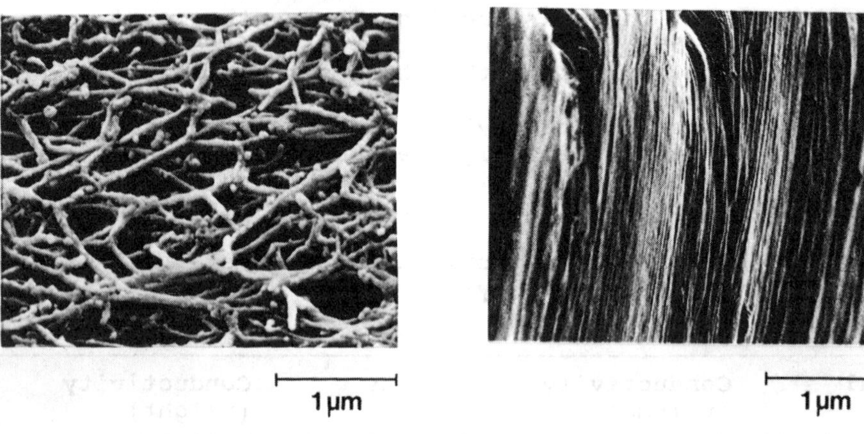

Fig. 19   Fig. 20

In general, the stability of the polyacetylene depends on
the method of preparation. Some of the differences are
illustrated in Fig. 21 (55).

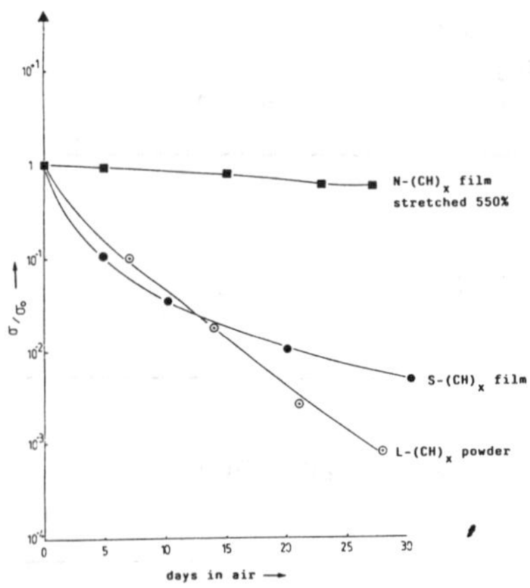

Fig. 21

Order of stability: Luttinger L-(CH)$_x$ < Shirakawa S-(CH)$_x$ < new BASF N-(CH)$_x$.

The greater stability of the N-(CH=CH)$_x$ samples is probably due to the absence of defect sites at high crystallinity.

A parallel phenomenon is that of high anisotropy in the stretched samples. Fig. 22 shows the dependency of conductivity on the degree of stretching (||) and the direction of stretching (⊥).

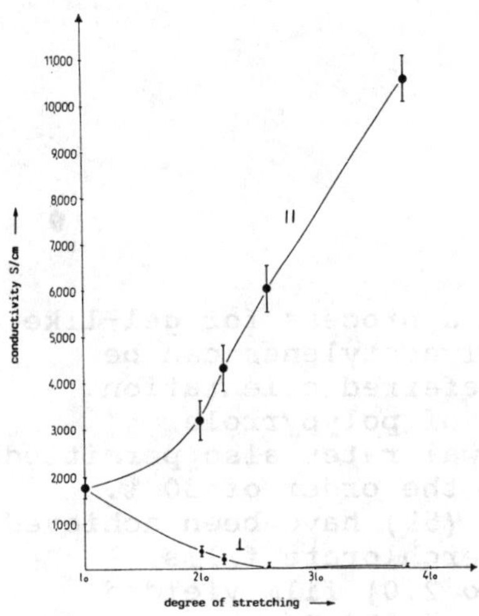

Fig. 22

The anisotropy of the stretched polyacetylenes permits the construction of a polarizer. When polyacetylene (PA) strips are laid across each other, polarized light is extinguished in the region of overlap characterized by ABCD of the anisotropic polymer.

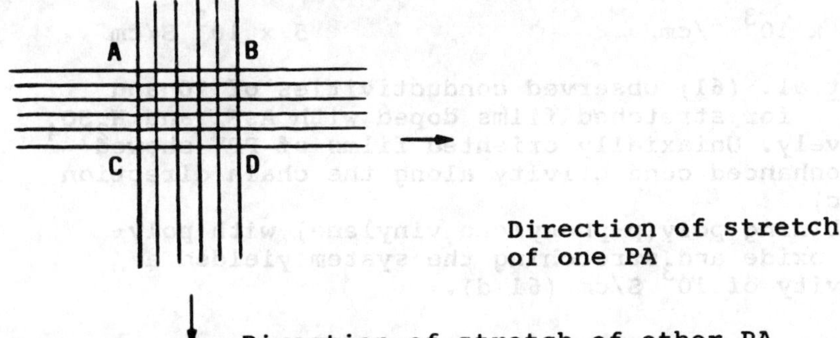

If the polyacetylenes have been prepared by the usual
Ziegler-Natta process (56) (38 b), the possible stretching
rates are small, comparable to that obtained when a
parallelogram is stretched diagonally.

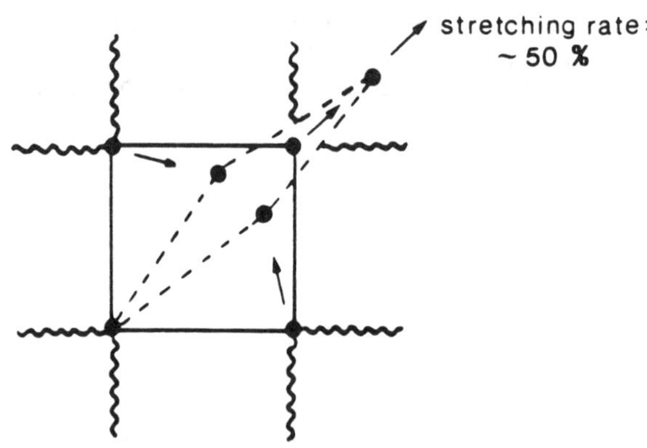

The Showa Denko K.K. has developed a process for gel-like
polyacetylenes in which swollen polyacetylenes can be
compressed to form films of the preferred orientation.

In the continuous manufacture of polypyrrole
(Section 1.2) (13), strong withdrawal rates also permitted
orientation of the polymer film of the order of 30 %.

Much greater stretching rates (59) have been achieved
by Yamaura et al. on polypyrrole-perchlorate films.
Uniaxially stretched (stretch ratio 2.0) film yielded
conductivities of 770 S/cm (σ ‖) and 150 S/cm
($\sigma \perp$).

Biaxially stretched (stretch ratio 1.75/1.75) film
yielded conductivities of 800 S/cm (σ ‖) and 290 S/cm
($\sigma \perp$).

$$-\!\!\left[\!\!\bigcirc\!\!\right]_{S}\!\!-CH\!\!=\!\!CH- \qquad \text{and} \qquad -\!\!\left[\!\!\bigcirc\!\!\right]\!\!-CH\!\!=\!\!CH-$$

2.5 x 10^3 S/cm $\qquad\qquad\qquad\qquad$ 5 x 10^3 S/cm

Karasz et al. (61) observed conductivities of 10 and
100 S/cm^{-1} for stretched films doped with AsF$_5$ and H$_2$SO$_4$
respectively. Uniaxially oriented films of PPV showed
greatly enhanced conductivity along the chain direction
(61 a - c).

Combining poly(p-phenylene vinylene) with polyethylene oxide and stretching the system yielded a
conductivity of 10^3 S/cm (61 d).

The use of the Langmuir-Blodgett technique for orienting polymers

A noteworthy reaction is the manufacture of substituted polypyrroles of high anisotropy (60) by means of the Langmuir-Blodgett technique. The conductivity parallel and perpendicular to the direction of stretch is 10^{-1} S/cm and 10^{-11} S/cm respectively. Further work has been performed by Nakahara et al. (60 a) and Watanabe et al. (60 b) to achieve high orientation of, for example, polythiophenes and ferrocenes by means of the Longmuir-Blodgett spreading technique. The technique can also be used to process charge-transfer complexes, as has been shown by Matsumoto (60 c).

8.2 ORIENTATION IN PREFERRED DIRECTION OF STRETCH DURING POLYMERIZATION IN LIQUID CRYSTAL MATRICES

The aim is to create preferred directions of stretch in order to obtain anisotropy. In this method of preparation, polymerization is performed in an oriented matrix (e.g. influence of strong magnetic field) consisting of liquid crystals (57).

Highly oriented polyacetylene films are synthesized by three different methods in which nematic liquid crystals are used as an ordered matrix solvent: Method 1, polymerization of acetylene is carried out in a quiescent nematic solution in which a $Ti(OBu)_4$ AlEt$_3$ Ziegler-Natta catalyst is dissolved homogeneously; Method 2, macroscopic orientation is attained by gravity flow of the nematic liquid crystal-catalyst system; and Method 3, the nematic liquid crystal-catalyst solution is oriented under a magnetic field. Results indicate that films prepared in this way have highly oriented fibrils.

The equipment for this technique is shown in the Fig. 23 (62).

Another variant is the use of liquid crystals during the electrochemical preparation of polyheterocycles (69).

The products are, e.g. polypyrroles, whose conductivity anisotropy is as much as 5 (σ) : 1 (σ).

Polymerization of extremely thin polyacetylene films on crystal surfaces by epitaxial growth, e.g., on frozen benzene, terphenyl etc., has been demonstrated by A.G. MacDiarmid (64) and also serves to induce orientation phenomena in the deposited polymer layer.

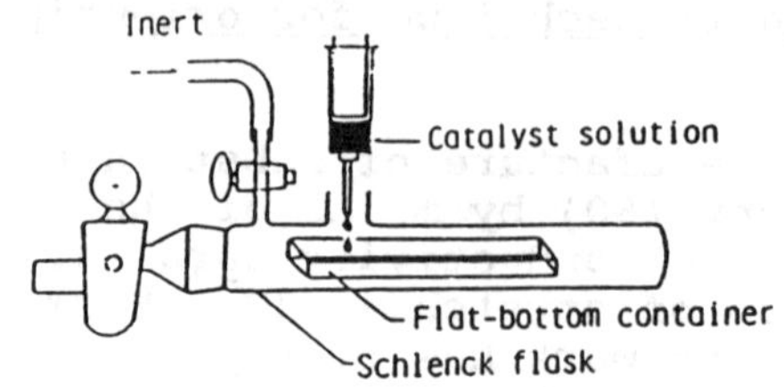

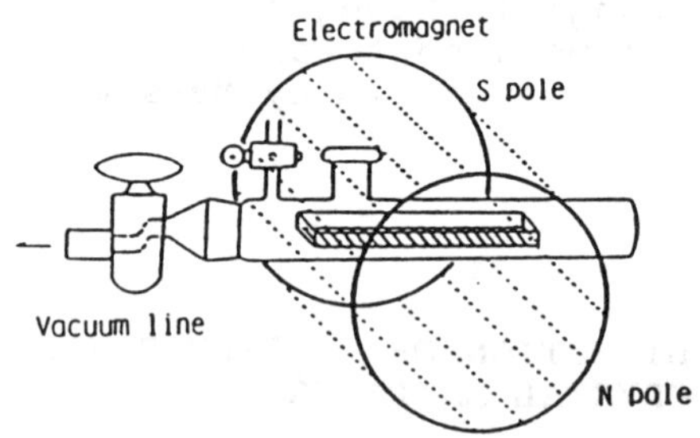

Fig. 23   Apparatus and procedure used in the magnetic field method

9.   Transparent electrically conducting polymers

The preparation of transparent and moreover electrically conducting polymers is a challenge to the chemist.

Fig. 24

One method consists in preparing extremely thin films, e.g. of oriented polyacetylene.

Prepared with silicone oil as catalyst (55), such polyacetylenes can be stretched to thicknesses of < 0.5 μm. The films are transparent (65) (see Fig. 24) and have conductivities in excess of 5 000 S/cm. Chemical coatings (Fig. 25), e.g. of polypyrrole also give rise to transparent materials (65).

Fig. 25

H.H. Hörhold has also described transparent, electrically conducting poly(phenylene vinylene) films and reported their photoconductivity spectrum (22).

Transparent films and coatings based on alkyl-substituted polythiophenes (66, 66 a, 66 b) have been prepared and have conductivities of up to 10^{-1} S/cm.

Another method consists in converting polyacrylo-nitrile to the corresponding amide oxime, which is then applied as a film and subsequently, for instance, complexed with colourless, polyvalent metal ions (67).

Acknowledgements

We are indebted to H. Heckmann for REM, H. Haberkorn for X-ray, P. Simak and M. Passlack for i.r., J. Schlag for conductivity- and R. Voelkel for C^{13}-measurements. Thanks to G. Köhler and K. Penzien for $(CH)_x$ preparation.
   This work is part of the project No. 03 C134-0 sponsored by BMFT. The author thanks the Bundesministerium für Forschung und Technologie for support of this work.

10.   References

(1)   P. Kovacic and A. Kyriakis, Tetrahedron Letters (London) <u>1962,</u> 467

(2)   H. Naarmann, GdCH Halbleitertagung, Munich 12.10.64, Naturwissenschaften, <u>56</u> Heft 6, (1969) 308

(3)   F. Beck, Ber. der Bunsengesellschaft Vol. <u>68,</u> No. 6, (1964) 558

(4)   M.B. Jones et al., Polym. Chem. Ed., Vol. <u>19</u> (1981) 89

   P. Kovacic et al., J. Appl. Polym. Sci. <u>12</u> (1968) 1735

(5)   I.G. Speight et al., J. Macromol. Sci., C <u>5</u> (1971) 295
   M.B. Jones et al., Polym. Prep. Am. Chem. Soc. Div.
   Polym. Chem. <u>21</u> (2) (1980) 259

(6 a)   DB Pat 1 195 497, 01.02.63/17.02.66

(6 b)   DB Pat 1 178 529, 11.04.63/24.09.65

(6 c)   DB Pat 1 298 710, 16.10.63/05.03.70

(6 d)   DB Pat 1 256 421, 30.05.64/14.12.67

(6 e)   DB Pat 1 092 137, 11.03.65/22.11.67

(6 f)   DB Pat 1 000 679, 11.04.63/11.08.65

(6 g)   DB Pat 1 214 402, 19.10.63/14.04.66
   BASF AG, H. Naarmann, E.G. Kastning

(6 h)   Handbook of Conducting Polymers, edited by
         T.A. Skotheim, Marcel Dekker, N.Y. 1986
         "Phenylene-Based Conducting Polymers"
         R.L. Elsenbaumer and L.W. Shacklette p. 214 - 258

(7)   Organic Electrochemistry
         M.M Baizer and H. Lund, Marcel Dekker, N.Y. 1893
         (p. 552, Chap. 5, 13 - 17, 21, 25)

(8)   H. Lund, Acta Chem. Scand., 11, 1323 (1957)
         A. Stanienda, Z. Naturforsch. 225, 1107 (1967)

         H. Lund, Elektrodenreaktioner i Oraganisk
         Polarografi og Valtammetri, Aarhus
         Stiftsbogtrykkerie, Aarhus 1961, A. Dall'Olio
         et al., C.R. Akad. Ser. C 267, p. 433 (1968)

(9)   A.F. Diaz et al., J. Chem. Soc., Chem. Commun.
         854, 635 (1979)
         A.F. Diaz and J. Bargon "Electrochemical Synthesis
         of Conducting Polymers" in Handbook of Conducting
         Polymers Vol. 1, p. 82 - 114 (1986) edited by
         T.A. Skotheim, Marcel Dekker Inc., N.Y.

         G.B. Street, "From Powder to Plastics" in Handbook
         of Conducting Polymers, Vol. 1, p. 266 - 291
         (1986)

(10)   E.M. Genies, G. Bidan and A.F. Diaz,
         J. Electrochem. Soc. 129, 1685 (1982)

(11)   H. Naarmann, Makromol. Chem., Macromol. Symp. 8,
         1 - 15 (1987)

(11 a)   EP 129 070, 25.05.83/27.12.84, BASF AG, G. Wegner
         and W. Wernet

(12)   BASF, DE-OS 3 425 511, Jul. 11, 1984/Jan. 16,
         1968, H. Naarmann and G. Köhler
         A. Heeger and F. Wudl: Electrochemistry of Polymer
         Layers, Internat. Workshop, Duisburg, F.R.G.,
         Sept. 15, 1986.
         A.O. Patil, Y. Ikenoue, F. Wudl, and A. Heeger,
         J. Am. Chem. Soc., 109, 1858 (1987)

(13)   US Pat 4 468 291, 28 August 1984, DE 3 226 278,
         14 July 1982, BASF AG, H. Naarmann, G. Köhler and
         J. Schlag

(13 a)   H.Li. and Th. F. Guarr, J. Chem. Soc. Chem. Comm.
         1989, p. 832

(14)   Spectroscopic Methods of Carotenoids, W. Vetter
         et al., p. 200 in "Carotenoids", edited by
         O. Isler (1971)
         Birkhäuser Verlag, Basle
         Substituent influence on the band gap energy of
         poly(phenylene vinylenes) (22)

(15)   EP 0 100 509, 25.07.83/21.05.86
         DOS 3 114 342 A 1, 09.04.81/04.11.82
         BASF AG Ludwigshafen, F.R.G.

(16)   DOS 3 244 899, 04.12.82/07.06.84
         BASF AG Ludwigshafen, F.R.G.

(17)   G. Kossmehl et al., Die Makromolek. Chemie 131
         (1970) 197 Table 3
         G. Kossmehl et al., Die Makromolek. Chem. 131
         (1970) 15
         H.H. Hörhold et al., Makromol. Chem. 182/11 (1981)
         3185

(18)   M. Härtel et al., Angew. Makromol. Chem. 29/30
         (1979) 307
         G. Kossmehl et al., Makromol. Chem. 131 (1970) 37

(19)   G.A. Kossmehl in Handbook of Conducting Polymers
         Vol. 1 p. 351 - 398 (1986), edited by
         T.A. Skotheim, Marcel Dekker, N.Y.

(20)   R.N. McDonald et al., J. Amer. Chem. Soc. 82
         (1060) 4669

(21)   G. Drefahl et al., Chem. Ber. 94 (1961) 907

(22)   H.H. Hörhold et al., Z. Chem. 27. Jg. (1987)
         Heft 4, p. 126 - 137

(23)   Preparative Methods of Polymer Chemistry, (1968)
         p. 237
         Verlag Intersc. Publ. N.Y.
         A.M. Butkus et al., Synthetic Metals, 3 (1981) 151

(24)   A. Rembau, et al., J. Polym. Sci. 61 1 (1962) 155

(25)   J.K. Stille et al., J. Amer. Chem. Soc. 83 (1961)
         1967
         Molar mass (in benzene) M_n 13500

(26)   H.W. Gibson et al., J. Chem. Soc. Chem. Comm.
          1980, 426

(26 a)   A.W. Snow, Nature Vol. 292, 2 July (1981) p. 40

(27)   S. Aftergut et al., WADC-TR. 59 - 469 Part III
          Columbus, Ohio, May (1962), Wright-Patterson Air
          Force Base

(28)   V. Korshak et al., J. Appl. Polym. Symp. $\underline{26}$,
          (1975) 237
          Polyacenes obtained when conjugated diynes are
          used.

(29)   DB Pat 1 179 715, 10.10.63/15.10.64
          DB Pat 1 197 228, 10.10.63/22.07.65
          DB Pat 1 179 716, 10.10.63/15.06.65
          BASF AG Ludwigshafen, F.R.G.

(30)   BRITE Report RI IB 0109-DB (1989)
          BASF AG Ludwigshafen, F.R.G.

(31)   P 3 835 029.7, 14.10.88
          BASF AG Ludwigshafen, F.R.G.

(32)   H. Naarmann, Ber. Bunsenges. Phys. Chem. 83, 431
          (1979)
          H. Geser, DAS 2 451 781, 08.01.76

(32 a)   G. Klebe, F. Graser, E. Hädicke and J. Berndt,
          Acta Cryst. B $\underline{45}$, 69 - 77 (1989)

(33)   DOS 1 953 898, 25.10.69/06.05.71
          BASF AG Ludwigshafen, F.R.G.

(34)   H. Naarmann, Angew. Chem. Int. Ed. Vol. 8 (1969)
          No. 11, p. 915

(35)   (a)   Ch. Kröhnke, V. Enkelmann, G. Wegner,
                Angew. Chem. Int. Ed. Vol. 19 (1980)
                No. 11, p. 912
          (b)   T.C. Chiang, A.H. Reddoch, D.F. Williams,
                J. Chem. Phys. $\underline{54}$, (1971), 2051

          (c)   H.P. Fitz, H. Gebauer, Z. Naturforsch. B33,
                (1978) 498 - 506; ibid 702 - 707

          (d)   M.E. Peover, B.S. White, J. Electroanal.
                Chem. $\underline{13}$, 93 (1967)

(36)   Organic super conductors, New Scientist
          10 July 1980, p. 104 - 107
          D. Jerome, M. Ribault and K. Bechard

(36 a)   P. Kathirgamanath, St. A. Mucklejohn,
          D.R. Rosseinsky
          J.C.S. Chem. Comm. 1979, p. 86; ibid 1980, p. 839

(37)   Charge Transfer Salts, I-II
          Synthetic Metals Vol. 27/1 - 2 A, (1988)
                             Vol. 27/3 - 4 B, (1988)

(38)   (a)   Chr. Kröhnke and G. Wegner, Polymerisation
                von Alkinen in Houben-Weyl 20/2,
                1212 - 1364 (1986), Georg Thieme Verlag,
                Stuttgart

          (b)   Initiatorsysteme für die Polymerisation von
                Acetylen und subst. Acetylenen
                H. Naarmann, Die Angew. Makromol. Chemie,
                109/110 (1982) 317 - 319

(39)   Polyacetylene: Helices I. Bozovic, Modern Physics
          Letters B Vol. 1/3, 81 - 88 (1987)
          H. Naarmann, Synth. Metals 17, 225 (1987),
          Q. Minxie et al., Synth. Metals 30, 1 (1984)

(40)   H. Haberkorn et al., Synthetic Metals 5, 51 - 71
          (1982)
          H. Haberkorn et al., Eur. Polym. J. Vol. 24, 5
          p. 497 (1988)
          I.M. Pochan, The Oxidative Stability of Dopable
          Polyenes in Handbook of Conducting Polymers
          Vol. 2, p. 1383, edited by T.A. Skotheim,
          Marcel Dekker, N.Y.

(41)   E.F. Lutz, J. Am. Chem. Soc. 83 (1961) 2551
          S. Ikeda et al., J. Polym. Sci. Polym. Lett. Ed.
          4 (1966) 605
          E. Lombardi et al., Redn. accad. Nazl. Lincei (8),
          25, (1958) 514
          G. Wegner, Angew. Chem. 93, 356 (1981)

(41 a)   H. Naarmann, "New Materials" - Conference,
          Oct. 12, 1988, Osaka, Japan

(42)   I.K. Stille, Makromol. Chem. 154 (1972) 49
          I.K. Stille et al., Macromolecules 5, (1972) 541

(43)   C.S. Marvel, J.A.C.S. 81 (1959) 448

(44)   H.H. Edwards, W.J. Feast, Polym. <u>21</u>, 595 (1980)

(44 a)   W.J. Feast, Synthesis of Conducting Polymers
         <u>Vol. 1</u>, 35 (1986) in Handbook of Conducting
         Polymers. Edit. T.A. Skotheim, Marcel Dekker
         Inc., N.Y.

(45)   T.M. Swager, D.A. Dougherty, R.H. Grubbs,
         J. Am. Soc. <u>110</u>, 2973 (1988), ibid <u>111</u>, 4413
         (1989)

(46)   F. Bohlmann, Angew. Chemie <u>69</u>, 82 (1957)
         F. Bohlmann, E. Inhoffen, Chem. Ber. <u>89</u>, 1276
         (1956)
         DAS 1 091 657 28.08.57/27.10.60, Farbwerke
         Hoechst, Frankfurt
         Inventors: F. Bohlmann, H. Dexheimer, O. Fuchs,
         H. Krämer

(47)   E. Hädicke, E.C. Merz and C.H. Krauch and
         G. Wegner and I. Kaiser, Angew. Chem. 83/7 (1971)
         253

(48)   I. Schopov, C. Iossifov, Polymer <u>23</u>, 613 (1982)

(49)   H.H. Hörhold, Z. Chemie <u>12</u>, 41 (1972)

(50)   G. Smets, L. de Koninck, J. Polym. Sci. A1/7, 3313
         (1969)

(51)   Y.M. Paushkin et al., in Organic Polymeric Semi-
         conductors, Wiley N.Y. (1974)
         T. Imoto et al., J. Polym. Sci. Part A <u>2</u> (1964)
         4573 describe the formation of polymers of
         structure $(-CH=CH-)_n$ CHO by making acetaldehyde
         react with triethylamine under pressure.

(52)   J.P. Roth et al., J. Polymer. C <u>4</u> (1964) 1347

(53)   S.B. Mainthia et al., J. Chem. Physics <u>37</u> (1962)
         2509

(54)   Highly conducting poly(2,5-thienylenevinylene) by
         a precursor route; I. Murase, T. Ohnishi,
         T. Noguchi and M. Hirooka, Polym. Comm. <u>1987</u>,
         Vol. 28, p. 229
         poly(p-phenylenevinylene) ibid <u>1985</u>, Vol. 26,
         p. 362, Synth. Metals <u>14</u>, 207 (1986)

(55)   BMFT-Forschungsbericht "Entwicklung von
         elektrisch leitfähigen Alternativ-Polymeren"
         03 C 134-0, 01.08.1982 - 31.07.1985, BASF AG
         Ludwigshafen; Chap. II, p. 1 - 75
         H. Naarmann, Synth. Metals $\underline{17}$, 223 (1987)
         H. Naarmann, N. Theophilou, Synth. Metals $\underline{22}$, 1
         (1987)
         Th. Schimmel et al., Solid State Comm. Vol.
         $\underline{65}$/11, 1311 (1988)

(55 a)   G. Lugli, U. Pedretti, G. Perego, J. Polym. Sci.
         Polym. Lett. Ed. $\underline{23}$, 129 (1985)

(56)   H. Shirakawa et al., J. Polym. Sci., Polym. Chem.
         Ed. $\underline{12}$, 11 (1974)

(57)   H. Shirakawa et al., Synthetic Metals, $\underline{14}$, 173
         and 199 (1986)
         M. Aldissi, J. Polym. Sci. Polym. Lett. Ed. $\underline{23}$,
         167 (1985)

(58)   G.B. 2 072 197, 21.03.80/30.09.81
         Showa Denko K.K. Tokyo Japan

(59)   M. Yamaura, T. Hagiwara, M. Hirasaka, T. Demura
         and K. Iwata, Synth. Metals $\underline{28}$ (1989) C 157 - 164

(60)   T. Shimidzu et al., Langmuir 1987/1, 1169

(60 a)   H. Nakahara et al., Thin solid films, Vol. $\underline{60}$
         (1988) p. 87 and p. 153

(60 b)   J. Watanabe et al., Synth. Metals $\underline{28}$, (1989) C 473

(60 c)   M. Matsumoto et al., ibid p. 61

(61)   D. Gagnon, J. Capistran, F. Karasz, R. Lenz,
         Polym. Prepr. $\underline{25}$ (1984) 284 - 285

         (a)   D. Gangnon, F. Karasz, E. Thomas, R. Lenz,
               Synth. Metals $\underline{20}$, (1987) 85 - 95

         (b)   F. Karasz, J. Capistran, D. Gagnon,
               Mol. Cryst. Liq. Cryst., 118 (1987) 327

         (c)   D. Gagnon, J. Capistran, F. Karasz,
               R. Lenz, Polym. Bull. 12 (1984) 293 - 298

         (d)   Personal communication

(62)   H. Shirakawa et al., J. Macromol. Sci. Chem. A 25
        (5 - 7) p. 643 (1988)
        H. Shirakawa et al., Synth. Metals, 28 (1989) D 1
        A. Montaner et al., Synth. Metals, 28 (1989) D 19

(63)   DOS 3 533 252 A1, 18.05.85/19.03.87
        BASF AG Ludwigshafen, F.R.G.

(64)   A.G. MacDiarmid, T. Woerner, A.G. Heeger,
        A. Feldblum, J.
        Polym. Sci., Polym. Lett. Ed. Vol. 3, 305 (1982),
        ibid Vol. 22, 119 (1984)

(65)   BASF Plastics, Research and Development KVX 8611e,
        10.1986 p. 37/38

(66)   R.L. Elsenbaumer et al., Synth. Metals 15,
        (1986) 169
        M. Sato et al., ibid 18, (1987) 229
        S.D.D.V. Rughooputh et al., ibid 21, (1987) 41

(66 a) G. Gustafson et al., Synth. Metals 28, (1989)
        C 427
        J.E. Österholm et al., ibid p. 435
        further information ibid to p. 486

(66 b) M. Feldhues et al., Synth. Metals 28, (1989) C 487

(67)   DOS 1 669 705 and Cd. 813 983, 06.07.67/30.10.69
        BASF AG Ludwigshafen, F.R.G.

NEW DEVELOPMENTS IN THE SYNTHESIS AND DOPING OF POLYACETYLENE AND POLYANILINE

A. G. MACDIARMID
Department of Chemistry
University of Pennsylvania
Philadelphia, PA 19104-6323, USA

and

A. J. EPSTEIN
Department of Physics
The Ohio State University
Columbus, Ohio 43210-1106, USA

ABSTRACT. A study of *cis*-rich N-(CH)$_x$ and ARA-(CH)$_x$ forms of B.A.S.F. polyacetylene involving elemental analysis of the pristine polymer and conductivities of the I$_2$/CCl$_4$ doped polymer are reported. Even though significant quantities of catalyst residue are present, the conductivity of the doped polymer is comparable to that of doped Shirakawa (CH)$_x$. The synthesis of polyaniline in its fully oxidized, fully reduced and selected average intermediate oxidation states is described. The processing of polyaniline films and fibers by thermal stretching to give conductivities up to ~100 S/cm is reported. Both doped and undoped polyaniline fibers have tensile strengths approaching those of commercial polymers.

1. Introduction

Polyacetylene and polyaniline represent two very different classes of conducting polymers - the former being doped to the metallic state by redox processes involving either partial oxidation or partial reduction of the pi-system; the latter in the emeraldine oxidation state, by a non-redox process involving protonation of the polymer in which the total number of electrons associated with the polymer remain unchanged. Some recent studies on these two classes of polymers are reported below.

2. Polyacetylene

As is well known, the conductivity of polyacetylene, (CH)$_x$, is determined by a variety of parameters including, but not limited to the number of defects in the polymer chain, the degree of alignment of the polymer chains, the type of dopant and method of doping, etc. The present investigation was carried out for the purpose of studying the conductivity of I$_2$-doped B.A.S.F. polyacetylene, both the "New", i.e. N-(CH)$_x$ and the "ARA", i.e."Additional Reducing Agent", viz., "ARA-(CH)$_x$" forms in their non-aligned state

J. L. Brédas and R. R. Chance (eds.), Conjugated Polymeric Materials:
Opportunities in Electronics, Optoelectronics, and Molecular Electronics, 53–63.
© 1990 *Kluwer Academic Publishers. Printed in the Netherlands.*

thereby eliminating one of the variables which has a pronounced effect on conductivity, in order to delineate more clearly the effect of synthetic conditions and purity of the polymer on the magnitude of conductivity. Both forms of B.A.S.F. $(CH)_x$ were synthesized and studied by Dr. N. Theophilou in collaboration with Mr. D. Swanson, using similar methods and apparatus [1] to those employed by the former person in his original collaborative investigations of the polymer at B.A.S.F. Company.

Representative data on which our overall conclusions are based are presented in Table 1. The undoped, non-aligned *cis*-rich $(CH)_x$ of both types contained significant amounts of residual catalyst as illustrated by many elemental analyses. Elemental analyses of samples of S-$(CH)_x$ synthesized in the present study and also in a previous study [2] by the Shirakawa method, are also given in Table 1. No elemental analysis having as high a total carbon and hydrogen value were obtained from any sample of N-$(CH)_x$ or ARA-$(CH)_x$ synthesized in the present study.

TABLE 1. Representative Analytical Data [a] and Conductivities of I_2/CCl_4 - Doped Non-Stretched $(CH)_x$ films.

Type of $(CH)_x$	Analysis %				Conductivity (I_2-doped, 4probe, S/cm)		
	C%	H%	Total (C+H)%	Ash% [d]	High	Low	Avg[h]
cis-S-$(CH)_x$ [b,c]	89.89	7.89	97.78	1.04	930	638	737
cis-S-$(CH)_x$ [e]	92.16	7.81	99.97	–	–	–	500 [f]
cis-N-$(CH)_x$	87.24	7.88	96.12	1.53	646	361	525
cis-N-$(CH)_x$ [g]	82.49	7.83	90.32	5.40	1069	546	768
cis-N-$(CH)_x$	81.73	7.55	89.28	4.74	1221	733	978
cis-N-$(CH)_x$	84.25	7.42	91.76	4.13	637	512	575
cis-N-$(CH)_x$	–	–	–	8.10	2400	800	–
cis-ARA-$(CH)_x$	73.55	7.34	80.89	9.52	517	470	489
cis-ARA-$(CH)_x$	76.39	7.42	83.81	6.00	587	499	548
cis-ARA-$(CH)_x$	–	–	–	4.60	8,100	500	900

[a] Data are for different preparations of the polymer.

[b] *cis*-rich (>80% *cis*) films.

[c] S = Shirakawa $(CH)_x$.

[d] Ash percentage calculated form total weight of polymer sample used in a given C, H analysis.

[e] From reference [2]. Doping was performed with I_2 vapor.

[f] Only one value reported. Not cited as an average value.

[g] Washed for 30 minutes in 6% HCl in methanol. In one instance no ash was obtained, but the sum of the C and H analyses was only ~98%, indicating the presence of oxygen, chlorine, or other impurities.

[h] Average of many separate determinations.

The infrared spectra of the pristine, as-formed, *cis*-rich polymers showed significant absorptions in the 3000 to 2840 cm^{-1} region representative of C-H stretching vibrations of sp^3 hybridized carbon, as well as other absorptions in the 800-1100 cm^{-1} region, not associated with $(CH)_x$. Four hundred and twenty-five mass spectra were taken during a period of 10 minutes as a sample of N-$(CH)_x$ having the above bands in its FTIR spectrum was heated from 50°C to 285°C. The following principle fragments, with possible assignments, were observed: $(C_2H_3O)^+$, $(C_4H_9)^+$, $(CH_3O)^+$, $(C_2H_5)^+$ and $(C_4H_5O)^+$. The FTIR spectrum of a sample of the polymer, when heated in a similar manner, showed a decreased intensity of these impurity peaks suggesting that the above fragments came from the thermal decomposition of the impurities. It is concluded that both forms of $(CH)_x$ contained butoxide and/or ethyl groups attached to catalyst residues.

Doping was performed as previously described [1] for the B.A.S.F. forms of $(CH)_x$ by immersing the polymer film in a saturated solution of iodine in carbon tetrachloride, CCl_4, for one hour at room temperature followed by two or three ~5 second washes with fresh CCl_4 to remove excess dopant solution followed by pumping in a dynamic vacuum for one hour to remove CCl_4 and any excess iodine on the surface of the film. Care must be taken to use only short washing times otherwise iodine dopant will be removed resulting in reduction of conductivity. Extended pumping times in a dynamic vacuum, which sublimes iodine from the doped polymer, must also be avoided.[3]

The 4-probe room temperature *in-situ* resistance of a sample $(CH)_x$ film using pressed platinum contacts in a Teflon device, with stainless steel screws to adjust and maintain the pressure of the contacts on the film, was measured while the film was immersed in the I_2/CCl_4 solution. The resistance fell immediately and became constant after ~1 hour indicating that *either* the film was doped uniformly throughout its thickness *or* that doping was not homogenous and that equilibrium was being approached very slowly, after ~1 hour. The resistance of the film both in the dry, doped state and in the doping solution decreased as the pressure was increased and then remained constant even when the pressure was increased by further tightening of the screws in the device.

The conductivity of a Shirakawa-type $(CH)_x$ non-aligned film synthesized in the present study and doped with I_2/CCl_4 as described above fell in the range 638 to 930 S/cm. The conductivity of similarly doped N-$(CH)_x$ fell in the range 361 to 2,400 S/cm while that of ARA-$(CH)_x$ varied between 470 and 8,100 S/cm. The previously published maximum conductivity values for I_2/CCl_4-doped non-stretched N-$(CH)_x$ and ARA-$(CH)_x$ are 2,000 and 5,000 S/cm respectively.[1] It should be noted that the conductivity of the B.A.S.F. doped $(CH)_x$ may depend critically [4] on the type and manufacturer of the silicone oil used in its synthesis. This may be caused by the presence of traces of desirable or undesirable impurities in the oil. It is remarkable that even though the B.A.S.F. films of N-$(CH)_x$ synthesized in the present study contained significantly more impurities than the Shirakawa-type films that their conductivity was not diminished. In this respect it is interesting to note that it has been suggested [5] that the presence of Ti catalyst residues in $(CH)_x$ may contribute to the electronic structure of the $(CH)_x$ by promoting the formation of positive solitons.

A few studies were carried out on stretched B.A.S.F. films. The conductivities of the films, I_2/CCl_4-doped after stretching, were: N-$(CH)_x$, $l/l_o = 6$, σ max. = 43,200 S/cm, σ min. = 3,950 S/cm, σ avg. = 12,900 S/cm. Values previously reported are: N-$(CH)_x$, $l/l_o = 6.5$; $\sigma(max) = 28,000$ S/cm and ARA-$(CH)_x$, $l/l_o = 5.0$; $\sigma = 147,286$ S/cm.[1]

Undoped *trans*-N-$(CH)_x$ and ARA-$(CH)_x$ have similar band gap, photo-induced and magnetic transport properties [6] to Shirakawa-type $(CH)_x$ but somewhat different A.C. conductivity dependencies [7]. The iodine-doped *trans*-N-$(CH)_x$ has a Pauli susceptibility similar to that of iodine-doped Shirakawa *trans*-$(CH)_x$ but a somewhat smaller thermopower which has a more linear temperature dependence [8]. It is apparent that even though twelve years have passed since the discovery of its doping, $(CH)_x$ is still not completely understood and still offers many challenges to chemists and physicists alike!

3. Polyaniline

The term "polyaniline" as commonly employed today [9,10] refers to a class of polymers consisting of up to 1,000 or more (ring-N-) repeat units which can be considered as being derived from a polymer, the base form of which has the generalized composition

$$\left[\left[-\!\!\left\langle\bigcirc\right\rangle\!\!-\!\overset{H}{N}\!-\!\left\langle\bigcirc\right\rangle\!\!-\!\overset{H}{N}\!\right]_{y}\left(-\!\left\langle\bigcirc\right\rangle\!\!=\!N\!-\!\left\langle\bigcirc\right\rangle\!\!=\!N\!\right)_{1-y}\right]_{x}$$

and consists of alternating reduced,

$$-\!\left\langle\bigcirc\right\rangle\!\!-\!\overset{H}{N}\!-\!\left\langle\bigcirc\right\rangle\!\!-\!\overset{H}{N}\!-$$

and oxidized,

$$-\!\left\langle\bigcirc\right\rangle\!\!-\!N\!=\!\left\langle\bigcirc\right\rangle\!\!=\!N\!-$$

repeat units. In principle "y" can be varied continuously from one to give the completely reduced polymer,

$$\left[-\!\left\langle\bigcirc\right\rangle\!\!-\!\overset{H}{N}\!-\!\left\langle\bigcirc\right\rangle\!\!-\!\overset{H}{N}\!-\!\left\langle\bigcirc\right\rangle\!\!-\!\overset{H}{N}\!-\!\left\langle\bigcirc\right\rangle\!\!-\!\overset{H}{N}\!-\right]_{x}$$

to zero to give the completely oxidized polymer,

$$\left[-\!\left\langle\bigcirc\right\rangle\!\!-\!N\!=\!\left\langle\bigcirc\right\rangle\!\!=\!N\!-\!\left\langle\bigcirc\right\rangle\!\!-\!N\!=\!\left\langle\bigcirc\right\rangle\!\!=\!N\!-\right]_{x}$$

The imine atoms in any of the species can be protonated in whole or in part to give the corresponding salts, the degree of protonation of the polymeric base depending on its oxidation state and on the pH of the aqueous acid.

The terms "leucoemeraldine", "emeraldine" and "pernigraniline", used in the following discussion will refer to the different <u>oxidation</u> <u>states</u> of the polymer where y = 1, 0.5 and 0 respectively, either in the base form, e.g. emeraldine base or in the protonated salt form, e.g. emeraldine hydrochloride. [9,10] It seems highly likely that the true average emeraldine oxidation state where y is <u>exactly</u> equal to 0.5 may never have been synthesized from aniline. The term "emeraldine" in the following discussion will therefore refer to an average oxidation state where y is <u>approximately</u> equal to 0.5 in the generalized formula of the polyaniline bases given above.

3.1. SYNTHESIS OF POLYANILINE IN THE EMERALDINE OXIDATION STATE

The partly protonated emeraldine hydrochloride salt can be synthesized easily as a partly crystalline black-green precipitate (dark green by transmitted light) by the oxidative polymerization of aniline, $(C_6H_5)NH_2$, in aqueous acid media by a variety of oxidizing agents, the most commonly used being ammonium peroxydisulfate, $(NH_4)_2S_2O_8$, in aqueous HCl.[9,11] As noted in a later section, we have recently shown that the pernigraniline oxidation state is first formed and that this is subsequently converted to the emeraldine oxidation state.

Emeraldine hydrochloride may also be deposited by "*in-situ* adsorption polymerization" in a few minutes as strongly adhering films on a variety of substrates such as natural and synthetic fibers and textiles, [12] plastic, glass, silver chloride pellets etc. [13] by immersing the substrate in a freshly mixed acidic aqueous solution of aniline and oxidizing agent such as ammonium peroxydisulfate. As noted in a later section, the pernigraniline oxidation state is first formed and is then converted to the emeraldine oxidation state. It is believed that a reactive intermediate, possibly an oligomeric radical cation of aniline is first adsorbed which subsequently polymerizes. [12] The emeraldine hydrochloride films are deprotonated instantly by ammonium hydroxide to give corresponding films of the emeraldine base. Such films, because they react so rapidly with a variety of reagents and because their electronic and IR spectra can be easily obtained, are useful in rapidly screening reactions of the polymer.

3.2. LEUCOEMERALDINE AND PERNIGRANILINE OXIDATION STATES

3.2.1. *Synthesis*. Leucoemeraldine, the completely reduced form of polyani-

line, [chemical structure], was first synthesized in 1910.[14] It can

be conveniently prepared as an analytically pure, off-white powder by the reduction of emeraldine base, [14] the most commonly used reducing agents being phenyl hydrazine or hydrazine.[15] Similarly to emeraldine base, free-standing films can be cast from NMP solutions.[15]

Pure pernigraniline, the completely oxidized form of polyaniline,

[chemical structure], has recently been synthesized for the first

time.[16] It was reported in 1910 [14] to be formed in an impure state by the oxidation of emeraldine base but that it was unstable and rapidly decomposed, especially when wet. Synthesis of the analytically pure dark purple, partially crystalline powder has been accomplished by the controlled oxidation of emeraldine base with m-Cl(C$_6$H$_4$)C(O)OOH/N(C$_2$H$_5$)$_3$ in NMP containing a trace of CrCl$_3$. Free-standing, lustrous, copper-colored films can be cast from this solution and subsequently leached in a methanol/acetone mixture to remove excess oxidizing agent.

We have found recently that polyaniline in the pernigraniline oxidation state is the first formed product in the common method of synthesizing the emeraldine oxidation state involving the oxidative polymerization of aniline by acidic (HCl) aqueous (NH$_4$)$_2$S$_2$O$_8$.[13] This is the case both for the *in-situ* deposition of polyaniline as thin films on substrates as described in a previous section and for the synthesis of the polymer as a bulk powder.

When the synthesis is carried out at ~0°C using excess aniline [9,11] the initial oxidation potential of the reaction system increases from ~0.66V (vs. SCE) immediately after mixing the reactants to ~0.75V after ~3 minutes where it stays essentailly constant for approximately 6 minutes whereupon it decreases rapidly to ~0.45V, a potential characterisitc of the emeraldine oxidation state [17] in this medium during the next ~10 minutes. The temperature increases noticeably during this latter period. The above times and potentials are, of course, dependant on the temperature at which the reaction is studied. Treatment of the *in-situ* deposited films when the potential of the system is ~0.75V with aqueous alkali gives electronic and I/R spectra characterisitc of pernigraniline base. If the reaction mixture at this potential is poured with aqueous NaOH solution, pernigraniline base powder is obtained as confirmed by spectroscopic studies.

58

Our working hypothesis at the present time to rationalize the above observations is as follows: The potential at which aniline commences to undergo electrochemical oxidation in aqueous acid is ~0.7V (vs. (SCE).[18] The potential of the ~0.05M solution of $(NH_4)_2S_2O_8$ in 1M aqueous HCl before adding aniline is ~1.05V and that of pernigraniline and emeraldine bases in 1M HCl are 0.83V [16] and 0.43V [17] respectively. It therefore appears not unreasonable that the $[(NH_4)_2S_2O_8 + HCl]$ system employed might be expected to produce, at least initially, polyaniline in its highest oxidation state i.e., the pernigraniline oxidation state. Furthermore, it seems highly likely that pernigraniline, if formed in the presence of excess aniline, would oxidatively polymerize the aniline to the emeraldine oxidation state while it, itself was reduced and also converted to the emeraldine oxidation state. If this should be the case, then the final polymer product having the emeraldine oxidation state would be formed, under the experimental conditions employed, by two different mechanisms:

(1) $C_6H_5NH_2 + (NH_4)_2S_2O_8 + HCl \rightarrow$ pernigraniline (faster reaction)

(2) pernigraniline + $C_6H_5NH_2 \rightarrow$ emeraldine (slower reaction)

Thus part of the final polymer formed in the emeraldine oxidation state would come from the pernigraniline state which had undergone reduction in step 2 and part would come from the aniline which had been polymerized in step 2. This is qualitatively consistent with the bimodal molecular weight distribution curve [19,20] observed for emeraldine base synthesized in the ususal way. The presence of a bimodal molecular weight distribution curve in a polymer suggests that the polymer may have been formed by two different reaction pathways.

The above conclusions are consistent with the observation that when a film of pernigraniline polymer formed by *in-situ* polymerization on conducting glass was placed in a fresh solution of aniline in 1M aqueous HCl that its potential fell within ~2 minutes to ~0.43V, characteristic of the emeraldine oxidation state, consistent with step 2 above. The electronic spectrum of the resulting film was identical to that of emeraldine. HCl.

In summary, experimental evidence at the present time strongly suggests that aniline undergoes oxidative polymerization in the commonly used $[(NH_4)_2S_2O_8 + HCl]$ medium to give at first polyaniline in its highest (pernigraniline) oxidation state and that as the concentration of the oxidizing agent, $(NH_4)_2S_2O_8$ is decreased by the reaction occuring in step 1, the polymer in the pernigraniline oxidation state then reacts more extensively with the excess aniline according to step 2.

The fact that the two extreme members of polyaniline bases - the completely reduced and completely oxidized - have been synthesized should now make it easier to interpret future systematic studies of intermediate members of the series.

3.3.3. *Reaction of Leucoemeraldine with Pernigraniline.* It has recently been observed[21] that the completely reduced base form of polyaniline, (leucoemeraldine), spontaneously reduces, and is itself oxidized, when its solution in NMP is mixed with an NMP solution of the completely oxidized base form of polyaniline, (pernigraniline). The reaction which occurs when equimolar solutions are mixed is given below:

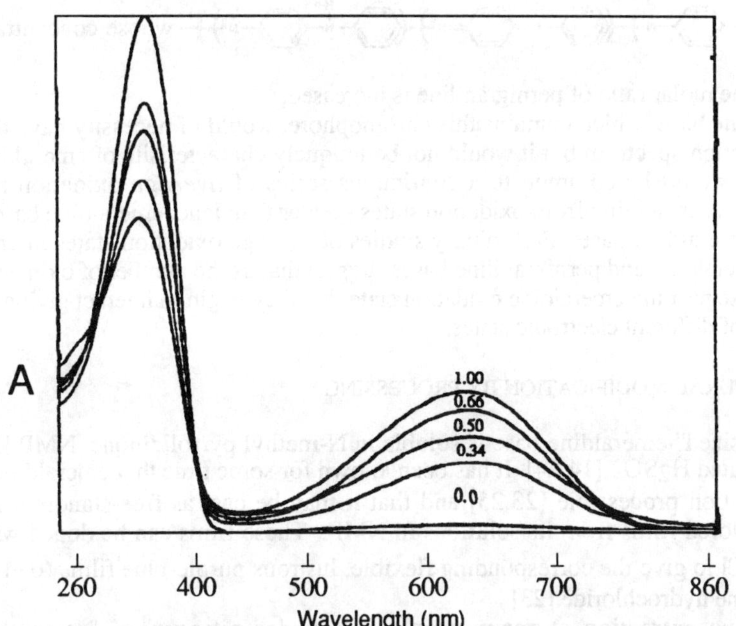

Under the experimental conditions employed, equilibrium appears to be attained at room temperature in ~14 hours. The electronic spectra of a solution of pure leucoemeraldine and of mixtures of leucoemeraldine and pernigraniline at various molar ratios in NMP were examined after ~14 hours. They are shown in Figure 1.

Figure 1 - Electronic spectra (A = absorbance) of leucoemeraldine and pernigraniline in NMP in molar ratios (pernigraniline/leucoemeraldine) of: 0.34; 0.50; 0.66 and 1.00 after ~14 hours. (0.0 = pure leucoemeraldine).

It can be noted that a peak at ~633nm (1.96eV) grows continually with increasing concentration of pernigraniline. This peak is virtually absent in pure leucoemeraldine base. The peak at 343nm (3.61eV) characteristic of leucoemeraldine base decreases with increasing concentration while at the same time an absorption at 327nm (3.79eV) begins to appear. The spectrum containing the equilibrated mixture of a 1:1 initial ratio of leucoemeraldine and pernigraniline bases is identical to that of pure emeraldine base in NMP. A set of two isosbestic points is observed at the two points where the spectra of leucoemeraldine base and emeraldine base cross each other. The invariant position of the 633 nm absorptions in all spectra and the isosbestic point set can be interpreted at the present time in one or the other of the two different ways outlined below: (1) no evidence is obtained for the formation of discreet electronic states in NMP solution, such as the postulated [14] protoemeraldine ($y = 0.75$) oxidation state, intermediate between the leucoemeraldine and emeraldine oxidation states. This leads to the conclusion that leucoemeraldine base is converted in one step, involving no intermediate species, to emeraldine base, i.e. the system consists of a physical mixture of leucoemeraldine base and emeraldine base molecules or of long segments of leucoemeraldine and emeraldine base in the same molecule. This is qualitatively consistent with cyclic voltammetry/spectroscopic studies of oligomers and of *solid* polyaniline which suggest that polyaniline may exist locally in only three discreet oxidation states although the average oxidation state of the bulk polymer may be intermediate between them.[22] *or* (2) the 633nm peak is characteristic of a chromophore consisting of one oxidized repeat unit surrounded on each side by a reduced repeat unit, e.g.

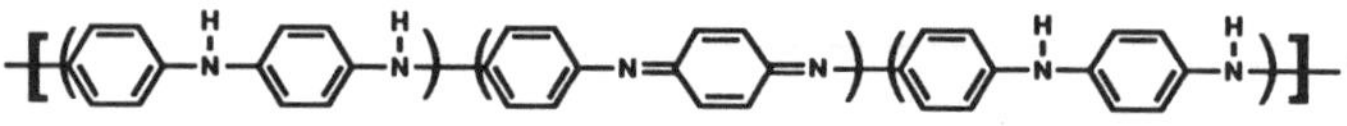

$$\left[\left(\!\!\!\bigcirc\!\!\!-\overset{H}{\underset{N}{}}-\bigcirc\!\!\!-\overset{H}{\underset{N}{}}\right)\!\!\left(\bigcirc\!\!-N=\bigcirc=N\right)\!\!\left(\bigcirc\!\!-\overset{H}{\underset{N}{}}-\bigcirc\!\!-\overset{H}{\underset{N}{}}\right)\right]$$

— whose concentration in-

creases as the molar ratio of pernigraniline is increased.

Emeraldine base, which contains this chromophore, would of necessity have this peak in its absorption spectrum but it would not be uniquely characterisitc of emeraldine base but instead would be common to a continuous series of *average* oxidation states of polyaniline bases ranging from oxidation states greater than leucoemeraldine base to, and including emeraldine base. Preliminary studies of average oxidation states intermediate between emeraldine and pernigraniline bases suggest that as the number of oxidized repeat units increase past the emeraldine oxidation state that they begin to interact giving rise to a continuum of different electronic states.

3.3. STRUCTURAL MODIFICATION BY PROCESSING.

"As-synthesized" emeraldine base is soluble in N-methyl pyrrolidinone, NMP [23], and in concentrated H_2SO_4.[14,24] It has been known for some time that emeraldine base is readily solution-processible [23,25] and that it may be cast as free-standing, flexible, coppery-colored films from its solutions in NMP. These films can be doped with ~1M aqueous HCl to give the corresponding flexible, lustrous purple-blue films (σ~1-4S/cm) of emeraldine hydrochloride.[23]

Exhaustive extraction at room temperature under nitrogen of "as-synthesized" essentially amorphous emeraldine base, first with tetrahydrofuran, THF, then with NMP results in the removal of ~20 weight% of oligomeric material and impurities.[19,20,26] The resulting "processed" emeraldine base powder has an excellent elemental analysis, and is up to ~50% crystalline (orthorhombic).[27,28] It is insoluble in NMP and in concentrated sulfuric acid under the same experimental conditions in which the "as-synthesized" emeraldine base powder is "soluble" in these solvents. Increased crystallinity, not unexpectedly, tends to result in decreased solubility.

It is interesting to note that the above partly crystalline emeraldine base powder can be repeatedly interconverted to its amorphous form, and then back to its crystalline form. For example, protonation with HCl to give the corresponding emeraldine hydrochloride (σ~18 S/cm.;4-probe compressed pellet) followed by deprotonation with NH_4OH results in the amorphous form. Treatment of this amorphous form with THF, in which it is essentially insoluble results in some solvation of the polymer giving it sufficient mobility to rearrange at least in part back to the more thermodynamically stable, partly crystalline form.

A solution of the amorphous "processed" emeraldine base in NMP exhibits a bimodal molecular weight distribution (by g.p.c; polystyrene standard), the maximum molecular weight fraction corresponding to approximately 325,000 [19,29] consistent with earlier studies.[19,26] Lower molecular weights in concentrated H_2SO_4 solution of emeraldine base synthesized in a slightly different manner have also been reported.[24] Partly crystalline films of the insoluble protonated polymer are formed when thin layers of these concentrated H_2SO_4 solutions are exposed to air.[24]

3.3.1. *Oriented Films.* Oriented partly crystalline emeraldine base films are obtained by simultaneous heat treatment and stretching of films formed from "as-synthesized" emeraldine base containing a plasticizer such as NMP.[30,31] Samples are observed to elongate by up to 400% when held above the glass transition temperature $[\geq\sim110^oC]$.[30,31] The resulting films have an anisotropic x-ray diffraction and optical response, with a misorientation of only a few degrees.[28] Oriented polyaniline may also be obtained by evaporating (in the present of a heat lamp) a solution of emeraldine base in NMP on polyethylene, polyacetylene or other substrates while the substrate is being mechanically stretched.[32]

3.3.2. *Oriented Fibers.* Fibers of emeraldine base can be formed by drawing a ~20 weight % "solution" of any form of amorphous emeraldine base in NMP in a water/NMP solution.[20,29] They can be thermally stretch-oriented up to 400% in a similar manner to emeraldine base films.[20] X-ray diffration studies show directional enhancement of the Debye-Scherrer rings. They differ from the oriented, partly crystalline fibers prepared from solutions of emeraldine base in concentrated sulfuric acid [24] by coagulation in water in that they consist of undoped polymer. Doping with 1M aqueous HCl results in a significant increase in the conductivity parallel to the direction of stretching (to 40 - 170 S/cm;Av.,85 S/cm) [29] similar to that observed with fibers formed from concentrated H_2SO_4 solution.[24] If desired, the emeraldine base "solution" in NMP may also be drawn in aqueous HCl which results in direct formation of the doped fiber.

The thermally stretch oriented base fibers have tensile strengths [366(best), 318 (av.)MPa; initial modulus, 8.6 (best), 8.1 (av.)GPa; 1 inch gauge length] which falls within the lower limit of the tensile strengths of commercial fibers (tensile strengths: Nylon 6, 200-905; Nylon 66, 231-985; polyesters, 236-1,165 MPa). Doping of such fibers with 1M HCl reduces the tensile strength somewhat [176 (best), 150 (av.) MPa; initial modules, 5.0 (best), 4.6 (av.) GPa]. This is possibly caused by the somewhat reduced degree of observed crystallinity.[29] These early results suggest that further work will result in tensile strengths of the doped polymer at least equal to that of commercial fibers.

4. Conclusions

Considerable effort has been given to studying the effect of alignment, but almost none to studying the effect of chemical purity on the conductivity of $(CH)_x$. It would seem highly desireable that more attention be directed towards this important variable which has such a significant effect on the conductivity of ordinary metals. Although extensive advances have been made during the last four years in obtaining an understanding of the fundamental electronic, magnetic, spectroscopic, transport and theory relating to the large and diverse polyaniline class of conducting polymers it is apparent that the field has not yet been clearly delineated. The fact that aniline is a relatively inexpensive compound, that its polymerization can be readily accomplished by simple chemical procedures and that some of its conducting derivatives apparently exhibit good thermal and environmental stability strongly suggest it may have many good technological uses. The richness of the chemistry, electrochemistry, and physics of the polyanilines indicates they will continue to serve as a focus for challenging interdisciplinary research in the future.

5. Ackowledgements

The authors wish to thank Mssrs. P.B. Friese and S.K. Manohar in particular for their invaluable assistance in preparing the manuscript. Most of the work reported by the authors was supported by the Defense Advanced Research Projects Agency through a contract monitored by the Office of Naval Research.

6. References

[1] H. Naarmann and N. Theophilou, <u>Synth. Met.</u> 22:1 (1987); H. Haberkorn, W. Heckmann, G. Köhler, H. Naarmann, J. Schlag, P. Simak, N. Theophilou and R. Voelkel, <u>Eur. Polym. J.</u> 24:497 (1988); Th. Schimmel, W. Reiss, J. Gmeiner, M. Schwoerer, H. Naarmann and N. Theophilou, <u>Solid State Comm.</u> 65:1311 (1988); N. Basecu, Z.-X. Liu, D. Moses A.J. Heeger, H. Naarmann and N. Theophilou, <u>Nature</u> 327:403 (1987).

[2] A.G. MacDiarmid and A.J. Heeger, "Molecular Metals", NATO Conference Series, ed. W.E. Hatfield, Plenum Press, New York; (1979), p. 161.

[3] A.G. MacDiarmid and A.J. Heeger, <u>Synth. Met.</u> 1:101 (1979/1980).

[4] Dr. H. Naarmann stated during the discussion of this paper that the highly conducting B.A.S.F. $(CH)_x$ was preferentially obtained using <u>Tegiloxan AV1000</u>® (Wacker Chemie, Theodore Goldschmidt) silicone oil.

[5] J.C.W. Chien, J.M. Warakomski, F.E. Karasz, W.L. Chia and C.P. Lillya, <u>Phys. Rev. B</u> 28:6937 (1983).

[6] N. Theophilou, D.B. Swanson, A.G. MacDiarmid, A. Chakraborty, H.H.S. Javadi, R.P. McCall, S.P. Treat, F. Zuo and A.J. Epstein, <u>Synth. Met.</u> 28:D35 (1989).

[7] G. Du, A.J. Epstein, N. Theophilou, D.B. Swanson and A.G. MacDiarmid, to be published (1989).

[8] H.H.S. Javadi, A. Chakraborty, C. Li, N. Theophilou, D.B. Swanson, A.G. MacDiarmid and A.J. Epstein, to be published (1989).

[9] J-C. Chiang and A.G. MacDiarmid, <u>Synth. Met.</u> 13:193 (1986).

[10] A.G. MacDiarmid, J-C. Chiang, A.F. Richter and A.J. Epstein, <u>Synth. Met.</u> 18:285 (1987).

[11] A.G. MacDiarmid, J-C. Chiang, A.F. Richter, N.L.D. Somasiri and A.J. Epstein in: "Conducting Polymers", ed. L. Alcacér (Reidel Publications, Dordrecht, 1987).

[12] R.V. Gregory. W.C. Kimbrell and H.H. Kuhn, <u>Synth. Met.</u> 28:C-823 (1989).

[13] S.K. Manohar, A.G. MacDiarmid and A.J. Epstein, <u>Bull. Am. Phys. Soc.</u> 34:582 (1989); S.K. Manohar, A.G. MacDiarmid and A.J. Epstein, 1989, unpublished observations.

[14] A.G. Green and A.E. Woodhead, <u>J. Chem. Soc. Trans.</u> 97:2388 (1910); A.G. Green and A.E. Woodhead, <u>J. Chem. Soc. Trans.</u> 101:1117 (1912).

[15] J. Masters, S.K. Manohar, A. Ray, A.G. MacDiarmid and A.J. Epstein, unpublished results; A. Ray Ph.D. dissertation, University of Pennsylvania, (1989).

[16] Y. Sun, A.G. MacDiarmid and A.J. Epstein, unpublished observations, (1989).

[17] W-S. Huang, A.G. MacDiarmid and A.J. Epstein, <u>J. Chem.Soc., Chem. Commun.</u>, 1784, 1987.

[18] W-S. Huang, B.D. Humphrey and A.G. MacDiarmid, <u>J. Chem. Soc. Faraday Trans. 1</u> 82:2385 (1986).

[19] A.G. MacDiarmid, G.E. Asturias, D.L. Kershner, S.K. Manohar, A. Ray, E.M. Scherr, Y. Sun, X. Tang and A.J. Epstein, <u>Polymer Preprints</u> 30-1:147 (1989).

[20] X. Tang, E. Scherr, A.G. MacDiarmid and A.J. Epstein, <u>Bull. Am. Phys. Soc.</u> 34:583 (1989).

[21] J.G. Masters, Y. Sun, A.G. MacDiarmid and A.J. Epstein, unpublished observations (1989).

[22] L.W. Shacklette, J.F. Wolfe, S. Gould and R.H. Baughman, <u>J. Chem. Phys.</u> 88:3955 (1988).

[23] M. Angelopoulos, G.E. Asturias, S.P. Ermer, A. Ray, E.M. Scherr, A.G. MacDiarmid, M. Akhtar, Z. Kiss and A.J. Epstein, <u>Mol. Cryst. Liq. Cryst.</u> 160:151 (1988).

[24] A. Andreatta, Y. Cao, J-C. Chiang, A.J. Heeger and P. Smith, <u>Synth. Met.</u> 26:383 (1988).

[25] M. Angelopoulos, A. Ray, A.G. MacDiarmid and A.J. Epstein, <u>Synth. Met.</u> 21:21 (1987).

[26] X. Tang, A.G. MacDiarmid, A.J. Epstein and Y. Wei, unpublished observations, (1989).

[27] M.E. Jozefowicz, R. Laversanne, H.H.S. Javadi, A.J. Epstein, J.P. Pouget, X. Tang and A.G. MacDiarmid, <u>Phys. Rev. B, Rapid Commun.</u> 39:12958 (1989).

[28] A.J. Epstein and A.G. MacDiarmid, <u>in:</u> "Electronic Properties of Conjugated Polymer", eds. H. Kuzmany, M. Mehring and S. Roth, Springer-Verlag, Berlin (1989) in press.

[29] X. Tang, A.G. MacDiarmid and A.J. Epstein, unpublished observations, (1989).

[30] K.R. Cromack, M.E. Jozefowicz, J.M. Ginder, R.P. McCall, A.J. Epstein, E.M. Scherr and A.G. MacDiarmid, <u>Bull. Am. Phys. Soc.</u> 34:583 (1989).

[31] E.M. Scherr, A.G. MacDiarmid and A.J. Epstein, unpublished observations, (1989).

[32] N. Theophilou, A.G. MacDiarmid, D. Djurado, J.E. Fischer and A.J. Epstein, <u>in:</u> "Electronic Properties of Conjugated Polymers," eds. H. Kuzmany, M. Mehring and S. Roth, Springer-Verlag, Berlin, (1989) in press.

SYNTHESIS, CHARACTERIZATION, AND APPLICATIONS OF SUBSTITUTED POLYACETYLENES DERIVED FROM RING-OPENING METATHESIS POLYMERIZATION OF CYCLOOCTATETRAENES

E. J. GINSBURG,[1a] C. B. GORMAN,[1a] R. H. GRUBBS,[*1a]
F. L. KLAVETTER,[1a] N. S. LEWIS,[1a] S. R. MARDER,[1b]
J. W. PERRY,[1b] M. J. SAILOR[1a]

Division of Chemistry and Chemical Engineering
California Institute of Technology
Pasadena, CA 91125

Jet Propulsion Laboratory
California Institute of Technology
Pasadena, CA 91109

ABSTRACT.   Polyacetylene and partially substituted derivatives of polyacetylene are synthesized via the ring-opening metathesis polymerization (ROMP) of cyclooctatetraene (COT) and its derivatives. The physical properties and characterization of these polymers are reported. In particular, certain poly-COT derivatives afford soluble, highly conjugated polyacetylenes. Nonlinear optical properties are highlighted by the fact that, in several cases, scattering losses are greatly minimized. These polymers have also been used in device applications such as Schottky barrier diodes and solar cells.

1. Introduction

Polyacetylene, the simplest fully-conjugated organic polymer, has captured the attention of chemists, physicists and materials scientists alike since it displays optical and electrical properties not typically seen in hydrocarbons. Ideally, polyacetylene can be described as a one-dimensional semi-conductor capable of supporting charge carriers which can migrate through the chain and has the potential to display highly anisotropic properties.[2]   In actuality, polyacetylene is an insoluble, unprocessable material with a morphology which is largely fixed during its synthesis. Moreover, polyacetylene is often produced with defects due to reduction and/or crosslinking reactions. Morphology and the nature of any defects in the polymer are very important to the properties of polyacetylene, particularly its electrical conductivity,[3] thus new methods for its synthesis are of interest.[4]   Moreover, any modifications to the structure of polyacetylene which provide solubility and thus processability whilst maintaining extended conjugation are very significant.

65

J. L. Brédas and R. R. Chance (eds.), Conjugated Polymeric Materials:
Opportunities in Electronics, Optoelectronics, and Molecular Electronics, 65–81.
© 1990 *Kluwer Academic Publishers. Printed in the Netherlands.*

Ring-opening metathesis polymerization (ROMP) of cyclooctatetraene (COT) produces poly-cyclooctatetraene, a new form of polyacetylene (Figure 1). By extension of this method to the ROMP of substituted COTs, we have developed a route into partially substituted polyacetylenes that are soluble in several cases.

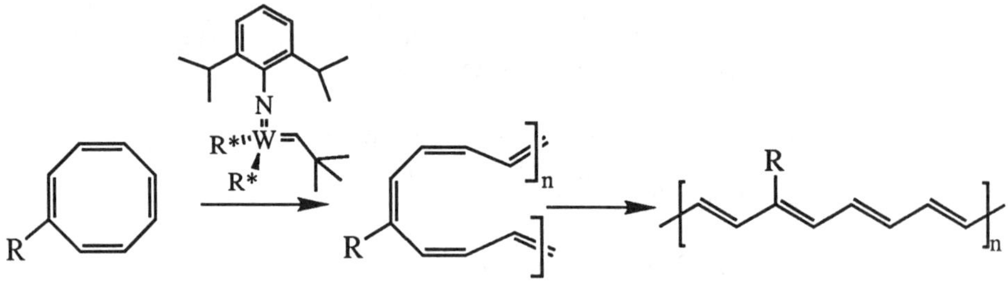

$R^* = (CF_3)_2(CH_3)CO$; R = (See Text)

Figure 1. Polymerization of Cylooctatetraenes

2. Synthesis

Polymerization is readily accomplished on gram scales in a nitrogen drybox. In a typical polymerization, the tungsten catalyst[5] (2 mg, 2.5 µmol) is dissolved in a solution containing 20 µL of tetrahydrofuran and the monomer (yellow liquid, 100 mg, 0.6 mmol). The yellow solution polymerizes over the course of 1-2 minutes, during which time it may be cast onto a variety of substrates. Typically, it is transferred by pipette onto a glass slide where it spreads out to form a film which is 20-200 µm thick depending on the viscosity of the reaction mixture at the time of the transfer. Polymerization in dilute solution is avoided since a decrease in the monomer concentration at the catalyst center encourages "back-biting" reactions to produce benzene and/or substituted benzenes (Figure 2). This side reaction does not terminate the polymerization, but it does reduce the molecular weight , and is not a significant factor in concentrated solution polymerizations.

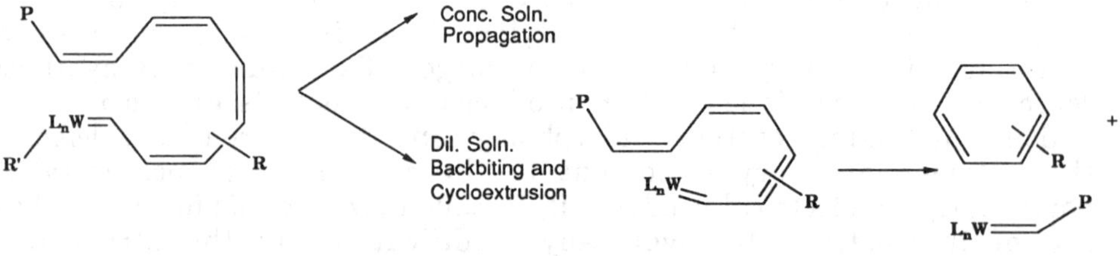

Figure 2. Cycloextrusion in Dilute Solution Polymerization of COTs

3. Properties of Poly-Cyclooctatetraene

Poly-cyclooctatetraene has many of the same physical properties as polyacetylene produced by the polymerization of acetylene by the Shirakawa route (Shirakawa polyacetylene or S-PA, Table 1).[6] Unlike Shirakawa polyacetylene which has a fibrillar morphology,[7] poly-COT possesses a smooth surface morphology as seen by scanning electron microscopy. Cross sections of poly-COT films reveal the interior of the film to be much more fibrillar, which explains the similar densities between S-PA and poly-COT. The wide angle powder X-ray pattern of poly-COT is similar to that observed for S-PA. Elemental analysis is consistent with the molecular formula $(CH)_x$.

Table 1. Properties of poly-cyclooctatetraene

Property	Poly-COT	Shirakawa PA
Appearance	shiny, silvery smooth surface	shiny, silvery; fibrillar surface[8]
Surface Area (m^2/g)	31 ± 3	66 [9]
Density, bulk (g/cc)	0.40 ± .04	0.4-0.5 [10]
Density, flotation (g/cc)	1.12 ± .01	1.13 [9]
X-ray spacing (d, in Å)	3.90 ± .05	3.80-3.85 (*cis*)[11]
Conductivity, undoped (ohm^{-1} cm^{-1})	< 10^{-8}	10^{-5} (*trans*)[12] 10^{-9} (*cis*)
Conductivity, iodine-doped (ohm^{-1} cm^{-1})	50-350	160 (*trans*)[11] 550 (*cis*)
Solid state CP-MAS ^{13}C NMR (ppm)	126.4, 132.2	126-129 (*cis*)[13]
....After Thermal Isom.	135.9	136-139 (*trans*)
Major Infrared peaks (cm^{-1})	930, 980 765	1015 (*trans*)[10] 740 (*cis*)
....After Thermal Isom.	1015	1015 (*trans*)
DSC exotherm (temp., ^{o}C)	150	150 [14]
Elemental Analysis: found (expected)		
Carbon	91.95 (92.26)	C+H[15]
Hydrogen	7.66 (7.74)	99%+

Nascent poly-COT has a high cis content. Differential scanning calorimetry reveals an irreversible exotherm at 150 °C corresponding to cis/trans isomerization in polyacetylene. Integration of the exotherm indicates the enthalpy change for poly-COT is -50 ± 1 cal/g. Tober and Ferraris have found that the heat evolved during isomerization (h) is linear with respect to cis content:[16]

$$\% \text{ cis } = (1.3 \text{ g/cal})h + 1$$

The observed value for poly-COT corresponds to a cis content of 66% for the pristine film. Three of the four cis bonds in the monomer are expected to retain their geometric configuration during polymerization to give a polymer with at least 75% cis configuration. However, although cis/trans isomerization is slow at room temperature,[17] the polymerization is exothermic and may induce some isomerization.

4. Copolymers of COT

Two types of linear copolymers containing COT can be prepared through metathesis polymerization (Figure 3). A random copolymer results when COT is co-polymerized with 1,5-cyclooctadiene, a monomer of similar reactivity. The gradation in color and Raman shift (Table 2) demonstrate that the conjugation length of polyene sequences in these films increases with increasing COT content. Maximum conductivities of I_2-doped random copolymers range from 10^{-10} to 10^2 $\Omega^{-1}\text{cm}^{-1}$ when the COT content in the COD/COT mixture is varied from 60 to 100%. Thus, incorporation of COD into the polymer interrupts conjugation, allowing one to vary the distribution of conjugation lengths. This control has been used to study the dependence of the third order nonlinear optical properties on conjugation length (*vide infra*). When norbornene, a much more reactive monomer, is copolymerized with COT, a tapered block copolymer is produced. These block copolymers exhibit extended conjugation even for a low mole fraction of COT. Films produced from 20% COT/80% norbornene mixtures have a shiny golden surface and Raman shift (half-height range of C=C stretch) of 1480-1540 cm^{-1}.

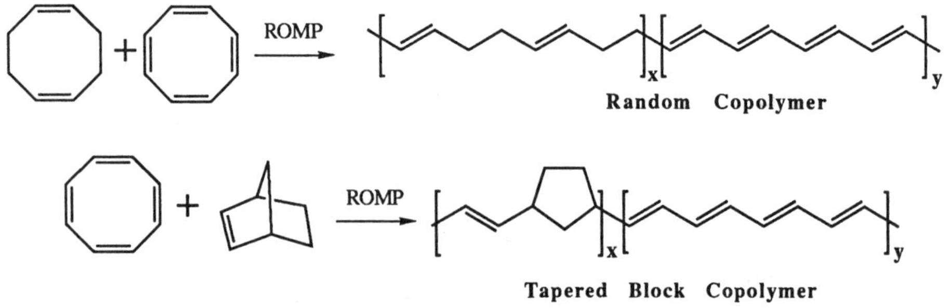

Figure 3. Copolymers of COT

Table 2. COD/COT Copolymer Characteristics

%COT in Mixture	Film Appearance	Raman C=C stretch, (half-height range), cm^{-1}
0	colorless	1659-1673
20	orange	1517-1538
40	red	1509-1544
60	dark red	1494-1544
80	red-black	1483-1542
90	dark silver	1479-1545
100	silver	1463-1531

5. Properties of Partially Substituted Polyacetylenes

Polymerization of substituted COT derivatives results in partially substituted polymers which, in several cases, are soluble and still highly conjugated.[18,19] Substitution of polyacetylene via the polymerization of substituted acetylenes results in materials with low effective conjugation lengths as evidenced by their high-energy visible absorption spectra and comparatively low iodine-doped conductivities.[20] This low conjugation length is presumably due to twisting around the single bonds in polyacetylene resulting from steric repulsions of the side groups.[21,22]

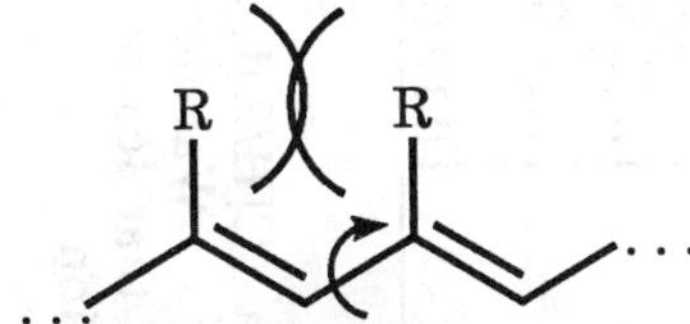

Figure 4. Chain Twisting in Substituted Polyacetylene

The poly-RCOT films can be iodine doped to a conductive state (Table 3). Moreover, in contrast to poly-cyclooctatetraene which shows large optical scattering losses due to its crystallinity, the alkylCOT polymers are amorphous and show low scattering losses (*vide infra*).[23,24]

Whereas poly-methylCOT, like polyacetylene, is brittle and completely insoluble, films of poly-RCOT with straight chain alkyl substituents of at least four carbons are flexible and somewhat soluble in solvents such as tetrahydrofuran (THF) and methylene chloride. Red/brown THF solutions can be prepared from nascent poly-*n*-butyl-, poly-*n*-octyl- or poly-*n*-octadecylCOT. These solutions turn blue upon sitting overnight. Solutions stored under argon at -50°C in the dark do not turn blue. We propose that the color change is due to the cis double bonds in the nascent polymer isomerizing to give a material with trans double bonds, although aggregation phenomena can not be ruled out at this time.[25]

Table 3. Conductivity and Spectroscopic Data

Polymer	σ (S/cm)	y^a	Abs. Max. T = 15 min[b]	Abs. Max. T $\approx$ 6-12 hrs.[b]
Poly-MethylCOT	15-44	----	522 [c]	----
Poly-n-ButylCOT	0.25-0.7	0.10-0.13	462	614
Poly-n-OctylCOT	15-50	0.11-0.19	480	632
Poly-n-OctadecylCOT	0.60-3.65	0.13-0.16	538	630
Poly-PhenylCOT	0.3-0.6	0.19-0.27	522	620
Poly-t-ButylCOT	$<10^{-8}$	≈ 0.03	302	432
Poly-TMSCOT	0.2[d]	0.12	380	512
Poly-NeopentylCOT	0.2-1.5	0.10-0.18	412, 628	634
after recasting from solution:	15-21	0.12-0.17	---	---

[a] Based on the molecular formula $(C[H/R]I_y)_x$
[b] Spectra taken in tetrahydrofuran. Wavelength in nm. T is the time after catalyst addition
[c] Very small amount of material that leached out of the film
[d] After recasting from THF solution

Similar changes have been observed in the optical spectra of films of polyacetylene.[26] Also, red-shifts in the absorption spectra of short polyenes upon cis-trans isomerisation have been observed previously.[27] Polymers of n-alkylCOT are only completely soluble when first synthesized. After 2-3 hours at room temperature, films are found to be much less soluble. For example, 1-2 mg/mL solutions of cis polymer gel or precipitate, if not diluted at least tenfold before isomerization. The predominantely cis isomer of these polymers is soluble whereas the trans isomer is much less so. This conclusion is consistent with previous observations of oligomeric polyenes.[27]

Nascent poly-neopentylCOT is completely soluble, but also precipitates upon isomerization to the trans form if not diluted. Blue suspensions that result from isomerization of concentrated solutions, however, can be recast into gold films that display iodine-doped conductivities two orders of magnitude higher than the original material. The aromatic side-chain derivative poly-phenylCOT is partially soluble in the cis form, but almost completely insoluble in the trans form.

Solubility in both the cis and trans forms is afforded by placing tertiary substituents off of the polyacetylene backbone. Poly-t-butylCOT is a freely soluble polymer but is yellow-orange in color indicating a low effective conjugation length. However, poly-trimethylsilylCOT is freely soluble and highly conjugated. The photochemically induced cis/trans isomerization of this polymer has been monitored by UV/Vis spectroscopy (Figure 5) and ^{1}H NMR.[28] These data are also consistent with the hypothesis that poly-TMSCOT is undergoing a cis/trans isomerization upon photolysis in solution. Highly reflective soluble polymer films can be cast from a photolyzed purple solution.[29]

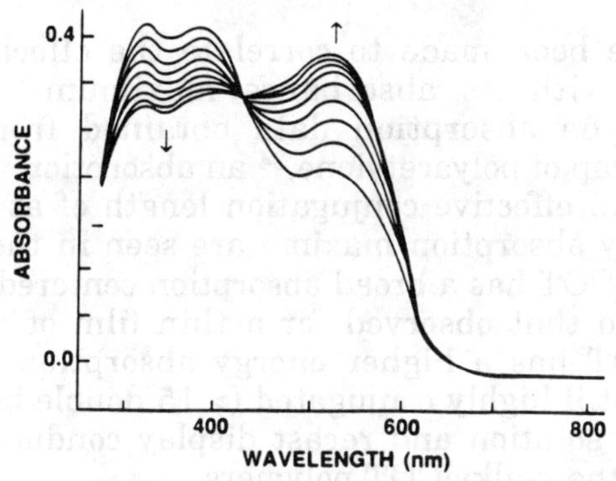

Figure 5. Cis/Trans Isomerization of poly-TMSCOT

Gel permeation chromatography on solutions of the polymers indicates that the solutions contain high molecular weight material[19,30] although molecular weight distributions are very broad in several cases. The retention times of the polymers generally increase after isomerization, suggesting an increase in hydrodynamic radius.The polymers were examined by ^{1}H NMR and Raman spectroscopy. The NMR spectra of all the polymers show broad multiplets between 6 and 7 ppm and broad signals in the regions expected for the side groups, indicative of a polymeric, olefinic chain. Raman spectroscopy provides further evidence of a conjugated chain. Two peaks, analogous to the A_g C-C stretch (v_1) and A_g C=C stretch (v_2) in *trans*-polyacetylene[31] are observed in all the spectra (Table 4), except for that of poly-phenylCOT, which displays a complicated spectrum dominated by phenyl stretches. No peaks corresponding to the cis form of polyacetylene are seen, possibly because no precautions were taken to prevent isomerisation by the incident 488 nm laser light. In unsubstituted polyacetylene the infrared C-H out of plane deformations at 740 cm^{-1} and 1015 cm^{-1} are diagnostic for cis and trans material, respectively. The infrared spectra of the substituted polyacetylenes are not as useful, since the vibrations of the side chains obscure those of the main chain.

Table 4. Raman Shifts of Polyacetylene Derivatives (cm^{-1})

Polymer	v1	v2
Polyacetylene[32]	1126, 1104	1507
Poly-MethylCOT	1126-1132	1516
Poly-*n*-ButylCOT	1132	1514
Poly-*n*-OctylCOT	1114-1128	1485
Poly-*t*-ButylCOT	1147	1539-1547
Poly-NeopentylCOT	1131	1509
Poly-TMSCOT	1132	1532

Several attempts have been made to correlate the effective conjugation length of a polyene with its absorbance maximum. Based on the extrapolation of polyene absorption data obtained from a variety of workers[33] to the band gap of polyacetylene,[34] an absorption maximum of 600 nm (Table 3) implies an effective conjugation length of at least 25 double bonds.[35] Lower energy absorption maxima are seen in the solid state. A thin film of poly-*n*-octylCOT has a broad absorption centered around 650 nm which is comparable to that observed for a thin film of polyacetylene.[34] Although poly-TMSCOT has a higher energy absorption maximum than poly-*n*-octylCOT, it is still highly conjugated (> 15 double bonds), and films that are isomerized in solution and recast display conductivities that are comparable to those of the p-alkylCOT polymers.

6. Nonlinear Optical Properties of COD/COT Copolymers and Poly-RCOTs

Recent experimental[36] and theoretical[37] studies indicate that extended conjugation leads to large cubic susceptibilities. The optical spectra of COD/COT copolymer solutions (*vide supra*), an example of which is shown in Figure 6, indicate that the distribution of conjugation lengths is dominated by segments of 5 double bonds with a significant number of segments of 9 double bonds, but there are quite few segments with 13 or more double bonds.[38] Nonlinear optical properties of the polymer mixtures were studied as a function of the composition. The third-order susceptibilities of the copolymer solutions were determined using wedged cell third harmonic generation (THG) techniques.[39,40] The 1907 nm Raman shifted (H_2 gas) output from a Q-switched Nd:YAG laser was used as the fundamental for THG measurements. Wedge THG interference fringes were observed by translating the cell normal to the laser beam. Table 5 summarizes the results of the THG studies. The third-order susceptibility of the polymer solution, $|\chi^{(3)}{}_p|$, and the hyperpolarizability per monomer unit, γ_p, values listed in Table 5 are based on the total mole fraction of monomer incorporated into polymer.

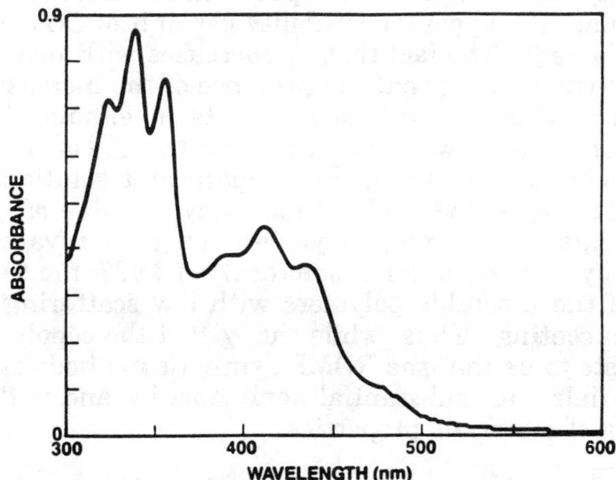

Figure 6. UV-visible spectrum of copolymer derived from 50% cyclooctatetraene, 50% 1,5-cyclooctadiene monomer mixture.

Table 5. Summary of composition and third-order optical nonlinearities of cyclooctatetraene and 1,5-cyclooctadiene copolymers.

Initial mol. fract. COT	Mol. fract.COT in polymer	mMol COT in polymer	$\chi^{(3)}_p$ 10^{-14} esu[a,b]	γ_p 10^{-36} esu[a]	γ'_p 10^{-36} esu[a]
0.10	0.08	0.064	33	20	210
0.20	0.15	0.108	60	36	220
0.40	0.27	0.159	130	81	280
0.50	0.32	0.166	160	100	300

a) Nominal uncertainty ± 25 %

b) We have independently measured the $\chi^{(3)}$ of cyclooctatetraene, 1,5-cyclooctadiene and $(CH_3)_3CC(O)H$ and have determined that correcting for their residual contributions only alters the solution $\chi^{(3)}$ values by 1-2%.

The $\chi^{(3)}$ and γ_p values of the copolymers increase substantially with increasing fraction of COT. This increase reflects both the increasing concentration of conjugated units and the increasing conjugation length with higher fraction of COT. It is expected that the nonlinearity of the units of COD in the polymer is negligible compared to the units of COT; assuming so, one can calculate the hyperpolarizability per unit of COT in the polymer, γ'_p, as listed in Table 5. The fact that γ'_p increases with increasing fraction of COT in the polymer shows that the presence of the increased conjugation lengths (segments of 9 double bonds) results in enhanced nonlinearity. From the solution results, we have estimated the $\chi^{(3)}$ of a copolymer film with 32% COT to be ~2 x 10^{-12} esu. By comparison, a solution measurement on β- carotene (11 conjugated double bonds) gave a value of $\chi^{(3)}$ = 9 x 10^{-11} esu. Measurements on neat polyacetylene have given a value of 1.3 x 10^{-9} esu (enhanced by three-photon resonance) at 1907 nm.[6] Transparent uniform films of these soluble polymers with low scattering losses can be prepared by spin coating. Thus while the $\chi^{(3)}$ of the copolymer is modest, this work suggests to us that the ROMP synthetic methodology can be used to produce materials with substantial nonlinearities and is flexible enough to allow tailoring of materials properties.

Accordingly, we studied the nonlinear optical properties of some partially substituted polyacetylenes prepared by ROMP. The linear and nonlinear optical properties of films of poly-n-butylCOT were examined. These films were typically prepared by polymerizing the neat monomer and casting the polymerizing mixture either between glass slides, resulting in films of about 20 μm thickness, or between the fused silica windows of a 100 μm pathlength demountable optical cuvette. Films cast between substrates were easily handled in air and were very stable for long periods of time (months). In addition, such assemblies were convenient for examination of the optical properties.

THG measurements on poly-n-butylCOT films, referenced to a bare fused silica plate, were made using 1064 nm pulses. These measurements showed that the $|\chi^{(3)}|$ of films of poly-n-butylCOT, ~1×10^{-10} esu were comparable to that for unoriented polyacetylene at the same wavelength.[36c] However, comparison of the linear transmission spectra of these materials in the near infra-red shows that the partially substituted polyacetylene has greatly improved optical quality.

Absorption spectra of polyCOT films show high optical density (1-3 for 20 µm thick films) even below the true absorption edge[41] in the near IR. The apparent absorption decreases with increasing wavelength but extends out beyond 2000 nm. This apparent absorption is actually due to scattering as shown by laser light scattering observations. As an example, we estimate the loss coefficient of such films to be > 500 cm^{-1} at 1500 nm. The origin of this scattering is certainly due to internal optical inhomogeneities in the polymer associated with the semi-crystalline, fibrillar morphology. In contrast, films of poly-n-butylCOT show very clean transmission in the near IR. Films 100 µm thick show a sharp absorption edge at ~900 nm and very little absorption beyond 1000 nm. For poly-n-butylCOT films, we estimate the loss coefficient to be < 0.2 cm^{-1} at 1500 nm. The greatly reduced scattering loss indicates that partial substitution of polyacetylene with n-butyl groups has resulted in a more homogeneous morphology, approaching that of an amorphous polymer.

We have also examined films of poly-TMSCOT. As discussed above, this polymer is completely soluble and can be converted to a fully trans conformation in solution. Films of the trans form of the polymer are then easily produced from solution by casting or spin-coating. THG measurements at 1064 nm on films of poly-TMSCOT give $|\chi^{(3)}| = 2 \pm 1 \times 10^{-11}$ esu. This value is somewhat lower than that of poly-n-butylCOT or polyacetylene, consistent with the reduced effective conjugation length inferred from the energy of the absorption maximum, as discussed earlier. The films of poly-TMSCOT prepared from solution are of good optical quality and show low scatttering losses at least as low as the poly-n-butylCOT films.

The results presented here bear out the idea that the ROMP method can be used to produce processable, conjugated polymers with high optical nonlinearities and low scattering losses. Given the ability to fabricate uniform high quality films with optical nonlinarity comparable to that of polyacetylene, these polymers may be of interest for nonlinear waveguiding experiments.

7. Device applications of poly-COT and poly-RCOT

The discovery of this new class of polymeric materials has led to a variety of interesting applications. For instance, ROMP has been used to grow layered structures of different polymers. Addition of a few drops of cyclooctadiene (COD) to the surface of a freshly polymerized film of poly-COT results in growth of a layer of polybutadiene on top of the polyacetylene layer. Apparently the catalyst residues on the surface of the film are still active for polymerization. The process can be continued to add another layer of polyacetylene, resulting in a polyacetylene/polybutadiene/poly-acetylene sandwich. Unlike polyacetylene, poly-butadiene is nonconjugated and cannot be doped to a highly conducting state. Thus, upon iodine doping, this structure becomes an "organic capacitor".

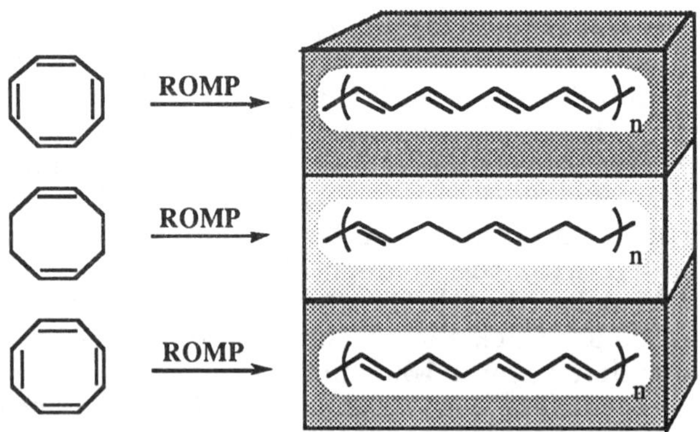

Figure 7. An Organic Capacitor

Another approach to forming layered structures takes advantage of the solubility of the substituted poly-COT derivatives. Smooth films of poly-TMSCOT can be made by placing a few drops of a THF solution of the polymer on a glass slide. The resulting films can be made insoluble in THF by doping them with iodine. This allows the subsequent casting of a second polymer layer without dissolving the original one. Multilayered structures of different poly-COT derivatives have been fabricated in this fashion.

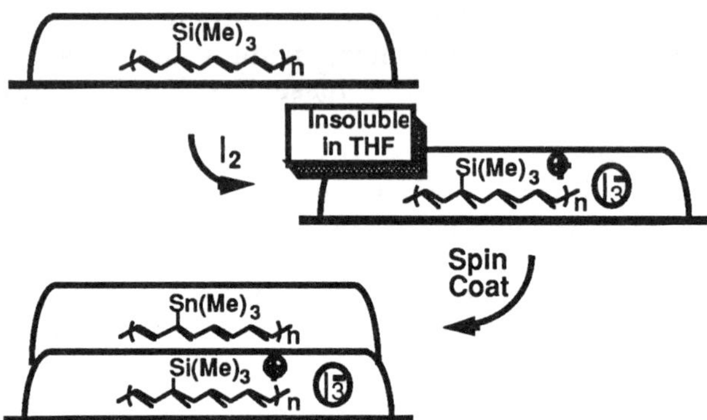

Figure 8. Multilayerd Polyacetylene Structure

The ability to cast thin films of conducting polymers has led to the investigation of their use in solar cell applications.[42] One of the simplest solar cells to manufacture is a Schottky barrier cell, which consists of a thin transparent metal overlayer on a semiconductor substrate such as silicon. An n-Si cell based on the Schottky design has been constructed, but with a film of doped poly-TMSCOT replacing the metal overlayer. Surprisingly, these cells generate higher open-circuit voltages than metal-based Schottky solar cells, indicating that they are more efficient at separating charge than the conventional devices. The factor responsible for the improvement observed in the polymer-based system is unclear at this time, although it must be related to the nature of the interface that forms between the semiconductor and the conducting overlayer. Metal-silicon interfaces are limited by Fermi level pinning, and the implication with the polymer based system is that this mechanism is no longer operative, or at least significantly suppressed.

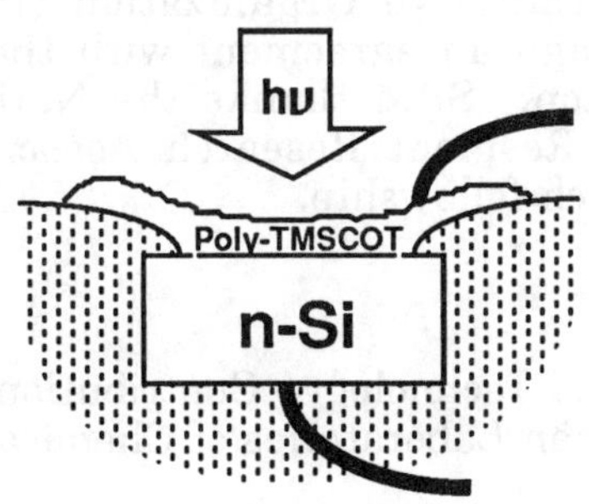

Figure 9. A Poly-TMSCOT/n-Si Solar Cell

The silicon/polyacetylene interface was also studied as a function of polymer dopant. A unique feature of polyacetylene is that it can be doped either oxidatively or reductively. In this sense the material has a "tunable" work function. That is, the ease with which an electron can be removed from polyacetylene is dependent on the dopant. Conventional metals are not this versatile. Ideally, the barrier to charge transport across a semiconductor-metal interface depends on the work function of the metal, and the tunability of the polyacetylene work function has been used to study this phenomenon. One of the long-standing problems associated with silicon/metal contacts has been Fermi level pinning at the interface. This pinning results in a fixed barrier to charge transport regardless of the metal work function. The silicon/polyacetylene interface is apparently not subject to this constraint, because a significant change in barrier height is observed on changing the polymer dopant from iodine to potassium.[43] As with the silicon/poly-TMSCOT solar cell described above, it is clear that the conducting polymer does not behave as a conventional metal in this application.

8. Conclusions

Ring-opening metathesis polymerization produces polyacetylene from cyclooctatetraene. Copolymerization with monomers of similar or increased reactivity allows for the synthesis of random or block copolymers, respectively. Extension of this method to substituted cyclooctatetraene derivatives allows for the synthesis of partially substituted polyacetylenes, many of which afford solubility to the polymer without significantly decreasing the conjugation length due to chain twisting. Successful investigations of the nonlinear optical properties and device applications of these polymers have been made.

Acknowledgements.
CBG, EJG, and RHG acknowledge financial support from the Office of Naval Research. NSL and MJS acknowledge financial support from NSF grant CHE-8814694. This work was performed in part by the Jet Propulsion Laboratory, California Institute of Technology and was sponsored in part by the Strategic Defense Initiative Organization (Innovative Science and Technology Office) through an agreement with the National Aeronautics and Space Administration. SRM thanks the National Research Council and NASA for a NRC Resident Research Associateship at JPL. EJG thanks IBM for a research fellowship.

------

[1] (a) California Institute of Technology. Contribution #8348 from the Arnold and Mabel Beckman Laboratories of Chemical Synthesis (b) Jet Propulsion Laboratory

[2] See discussions in: (a) *Handbook of Conducting Polymers*; Skotheim, T. A., Ed.; Marcel Dekker: New York, 1986; 2 vols. (b) *Electroresponsive Molecular and Polymeric Systems;* Skotheim, T. A., Ed.; Marcel Dekker: New York, 1988; Vol. 1.

[3] (a) Basescu, N.; Liu, Z.-X.; Moses, D.; Heeger, A. J.; Naarman, H.; Theophilou, N. *Nature (London),* **1987**, *327*, 403. (b) Naarman, H.; Theophilou, N. *Synth. Met.,* **1987**, *22*, 1.

[4] Much work has been done on precursor routes to polyacetylene: (a) Bott, D. C.; Brown, C. S.; Chai, C. K.; Walker, N. S.; Feast, W. J.; Foot, P. J. S.; Calvert, P. D.; Billingham, N. C.; Friend, R. H. *Synth. Met.,* **1986**, *14*, 245. (b) Swager, T. M.; Grubbs, R. H. *J. Am. Chem. Soc.,* **1989**, *111*, 4413.

[5] Schaverien, C.; Dewan, J.; Schrock, R. R. *J. Am. Chem. Soc.,* **1986**, *108*, 2771.

[6] Klavetter, F. L.; Grubbs, R. H. *J. Am. Chem. Soc.,* **1988**, *110*, 7807.

[7] Chien, J. C. W. In *Polyacetylene*; Academic: Orlando, FL, 1984; p. 121

[8] Ito, T.; Shirakawa, H.; Ikeda, S. *J. Polym. Sci., Polym. Chem. Ed.,* **1974**, *12*, 11.

[9] Ref 7, p. 150.

[10] Gibson, H. W.; Pochan, J. M. *Encyclopedia of Polymer Science and Engineering*, Wiley-Interscience: New York, 1985; Vol. 1, pp 87-130.

[11] Shirakawa, H.; Ikeda, S. *Polym J. (Tokyo)*, **1971**, *2*, 231.

[12] MacDiarmid, A.; Heeger, A. *Synth. Met.*, **1980**, *1*, 101.

[13] Ref 7, p. 249.

[14] Ito, T.; Shirakawa, H.; Ikeda, S. *J. Polym. Sci., Polym. Chem. Ed.*, **1975**, *13*, 1943.

[15] Ref 7, p. 37.

[16] Tober, R. L.; Ferraris, J. P. *Polymer Commun.*, **1987**, 28(12), 342.

[17] Chien, J. C.; Karasz, F. E.; Wnek, G. E. *Nature (London)*, **1980**, 285.

[18] Ginsburg, E. J.; Gorman, C. B.; Marder, S. R.; Grubbs, R. H. *J. Am. Chem. Soc.*, **1989**, *111*, 7621.

[19] Gorman, C. B.; Ginsburg, E. J.; Marder, S. R.; Grubbs, R. H. *Angew. Chem.*, In Press.

[20] a) Zeigler J. M., U.S. Patent Appl. US 760 433 AO, 21 November, 1986; *Chem. Abstr*, **1986**, *20*, 157042. (b) Zeigler J. M. *Polym. Prepr.*, **1984**, *25*, 223. (c) Okano, Y; Masuda, T.; Higashimura, T. *J. Polym. Sci.: Polym. Chem. Ed.*, **1984**, *22*, 1603. (d) Masuda, T.; Higashimura, T. *Adv. in Polymer Science*, **1987**, *81*, 121.

[21] Leclerc, M.; Prudhomme, R. E. *J. Polym. Sci: Polym. Phys. Ed.* **1985**, *23*, 2021.

[22] Chien has prepared copolymers of acetylene and methyl-acetylene, Chien, J. C W.; Wnek, G. E.; Karasz, F. E.; Hirsch, J. A. *Macromolecules*, **1981**, *14*, 479. However, extension of this method to other copolymerizations requires mixing a gas (acetylene) and a liquid (R-acetylene), and this two phase system is not expected to be well-behaved.

[23] Perry, J. W.; Marder, S. R.; Gorman, C. B.; Ginsburg, E. J.; Grubbs, R. H. unpublished results.

[24] Only amorphous halos are observed in the wide angle X-ray profile of these polymers.

[25] Batchelder, D. N. *Contemp. Phys*, **1988**, *29*, 3.

[26] Ref 7, p. 225.

[27] Schrock, R. R.; Krouse, S. A.; Knoll, K.; Feldman, J.; Murdzek, J. S.; Yang, D. C. *J. Mol. Cat.*, **1988**, *46*, 243. and references contained therein.

[28]NMR data for the *cis*-poly(TMSCOT) were obtained from a sample containing both isomers of the polymer. *cis*-Poly(TMSCOT): ^{1}H NMR (THF-d^8, 400 MHz) δ 0.15 (s), 5.8-7.1 (br m). *trans*-Poly(TMSCOT): ^{1}H NMR (THF-d^8) δ 0.23 (s, 9H), 6.5 (br s, 5H), 7.0 (br m, 2H). ^{13}C {^{1}H} NMR (methylene chloride-d^2) δ -0.25 (s), 128-145 (m). Elemental analysis, calculated for $(C_{11}H_{16}Si)_n$: C, 74.93; H, 9.14. Found: C, 74.2; H, 8.9.

[29]The polymer is treated as an air-sensitive material. After exposure of a 20- to 30- μm-thick film to air for two hours, approximately 20% of the material is no longer soluble in tetrahydrofuran.

[30] Typical number average (M_n) and weight average (M_w) molecular weights of these polymers are as follows: Poly-*n*-ButylCOT, M_n = 33,200, M_w = 153,400; Poly-OctylCOT, M_n = 35,900, M_w = 369,000; Poly-*t*-ButylCOT, M_n = 58,300, M_w = 71,900; Poly-NeopentylCOT, M_n = 23,400, M_w = 27,100; Poly-PhenylCOT, M_n = 30,600, M_w = 121,000; Poly-TMSCOT, M_n = 59,000, M_w = 77,000. Except for p-trimethylsilylCOT and poly-*t*-butylCOT, these data are reported for the predominantly cis polymers.

[31] S. Lichtmann, PhD. Thesis, Cornell University, 1980.

[32] Lefrant, S.; Lichtmann, L. S.; Temkin, H.; Fitchen, D. B.; Miller, D. C.; Whitwell II, G. E.; Burlitch, J. M. *Solid State Comm.*, **1979**, *29*, 191. Values are reported for an exciting wavelength of 514.5 nm which is the laser line closest to the one employed by us. ν_2 varies from 1540-1470 cm^{-1} for exciting wavelengths of 457.9 nm, 514.5 nm, and 605 nm, possibly due to resonance enhancement of different conjugation lengths at each wavelength.

[33] (a) Bohlmann, M. *Chem. Ber.*, **1952**, *85*, 387. (b) Bohlmann, M. *Chem. Ber.*, **1953**, *86*, 63. (c) Bohlmann and Kieslich, *Chem. Ber.*, **1954**, *87*, 1363. (d) Nayler, P.; Whiting, M. C. *J. Chem. Soc. Chem. Comm.*, **1955**, 3037. (e) Sondheimer, F.; Ben-Efriam, D.; Wolovsky, R. *J. Am. Chem. Soc*, **1961**, *83*, 1675. (g) Karrer and Eugster, *Helv. Chim. Acta*, **1951**, *34*, 1805. (h) Winston, A.; Wichacheewa, P. *Macromolecules* , **1973**, *6*, 200.

[34] Patil, A. O.; Heeger, A. J.; Wudl, F. *Chem. Rev.*, **1988**, *88*, 183.

[35] Similar results have been obtained by R. Chance, Exxon Corp. Unpublished data.

[36] (a) Sauteret, C.; Hermann, J. P.; Frey, R.; Predere, F.; Ducuing, J.; Baughman, R. H.; Chance, R. R. *Phys. Rev. Lett.*, **1976**, *36*, 956. (b) Carter, G. M.; Chen, Y. J.; Tripathy, S. K. *Appl. Phys. Lett.*, **1983**, *43*, 891. (c) Kajzar, F.; Etemad, S.; Messier, J.; Baker, G. L. *Synth. Met.*, **1987**, *17*, 563.

[37] (a) Agrawal, G. P.; Cojan, C.; Flytzanis, C. *Phys. Rev. B: Solid State*, **1978**, *15*, 776. (b) Beratan, D. N.; Onuchic, J. N.; Perry, J. W. *J. Phys. Chem.*, **1987**, *91*, 2696. (c) Garito, A. F.; Heflin, J. R.; Wong, K. Y.;

Zamani-Khamiri, O. in *Nonlinear Optical Properties of Polymers, Materials Research Society Symposium Proceedings, 109*, Heeger, A. J.; Orenstein, J.; Lurich, D. R. Eds.; Materials Research Society: Pittsburgh, 1988. p. 91.

[38] Marder, S. R.; Perry, J. W.; Klavetter, F. L.; Grubbs, R. H. *Chem. Mater.*, **1989**, *1*, 171.

[39] Meredith, G. R.; Buchalter, B.; Hanzlik, C. *J. Chem. Phys.*, **1983**, *78*, 1543.

[40] Kajzar, F.; Messier, J. *J. Opt. Soc. Am. B.*, **1983**, *4*, 1040.

[41] Weinberger, B. R.; Roxlo, C. B.; Etemad, S.; Baker, G. L.; Orenstein, J. *Phys. Rev. Lett.*, **1984**, *53*, 86.

[42] For other examples of conducting polymer-based solar cells, see: Kanicky, J., in ref 1 and references contained therein; Garnier, F.; Horowitz, G., *Synthetic Metals*, **1987**, *18*, 693; Frank, A. J.; Glenis, S.; Nelson, A. J., *J. Phys Chem.* **1989**, *93*, 3818.

[43] Sailor, M. J.; Klavetter, F. L.; Grubbs, R. H.; Lewis, N. S. Manuscript in preparation.

SYNTHESIS AND CHARACTERIZATION OF A WATER SOLUBLE PPV
DERIVATIVE

Songqing Shi and Fred Wudl
Institute for Polymers and Organic Solids
Department of Physics and Department of Chemistry
University of California, Santa Barbara, CA 93106

ABSTRACT. A water soluble poly (p-phenylenevinylene) (PPV) precursor polymer was
prepared in eight steps from p-methoxyphenol. The precursor polymer was of very high
molecular weight (ca. 10^6 Daltons) and was converted to the conjugated polymer both in
solution and in the solid state. The latter method produces a crosslinked material which is
insoluble in all common solvents and water. The former method affords a water soluble
PPV. Some physical and chemical properties such as in situ spectroscopy and
conductivity measurements are presented.

1. Introduction:

Conjugated polymeric materials, with poly(acetylene) as their simplest member have had a
major impact in the fundamental research on organic solids as a result of their accessible,
reversible redox properties. Their enormous potential for battery electrode materials,
electrochromic devices and semiconductor devices was considered almost concomitantly
with Shirakawa's discovery of film formation conditions in the Ziegler-Natta
polymerization of acetylene. With the exception of polyaniline, those potentials have yet
to be realized.
   The first stumbling block to realize the technological potentials of conjugated polymeric
materials was their intractability; i.e., non-processibility of stiff rod macromolecules.
This problem was ameliorated considerably in the recent past with the introduction of
flexible chain substituents in the poly(thiophenes) and in situ polymerization during gel-
processing of poly(acetylene).
   Another problem for the potential application of electrochromism was speed of the
electrochromic effect which has as its rate limiting step the migration of large ions through
the bulk of the polymer in concert with electron removal or addition to the backbone. With
the discovery of poly(alkylthiophenes) by Yoshino[1] and Elsenbaumer[2], a door was
opened to the possibility of producing poly(thiophenes) with widely varying substituents.
The substituents, for example could have a functional group attached at a remote site. We
considered various functional groups which would present the backbone with counterions,
should the backbone be forced to shed electrons and hence be positively charged. This led
to the design of the sulfonic acid substituted poly(thiophenes), which turned out to be
water soluble.

J. L. Brédas and R. R. Chance (eds.), Conjugated Polymeric Materials:
Opportunities in Electronics, Optoelectronics, and Molecular Electronics, 83–89.
© 1990 *Kluwer Academic Publishers. Printed in the Netherlands.*

The first water-soluble conducting polymers based on poly(thiophenes) were prepared in 1987[3]. In these polymers, the counterions are covalently bound to the polymer backbone, leading to the self-doping concept.

In this paper we present the most recent results on the first water soluble PPV derivative[4]. The introduction of an alkoxysulfonate group to the monomer not only made the resulting PPV easily water soluble and self-doped, but also lowered its band gap and enhanced its electrical conductivity, as in the case of alkoxy substitution of PPV.

2. Results and Discussion:

Scheme I shows the synthetic route for the monomer **7**. Yields of the various intermediates were: 66% for **1**, 65% for **2**, 93% for **3**, 88% for **4**, 90% for **5**, 92% for **6** and 76% (quantitative, crude) for **7**. The synthetic procedures for compounds **2**, **3** and **4** were adopted from that of the thiophene analogue as reported before[3]. Upon treatment with dimethylsulfide or tetrahydrothiophene, **6** gave the hygroscopic bis(sulfonium) chloride **7** which could be purified by recrystallization from methanol/acetone.

Scheme I

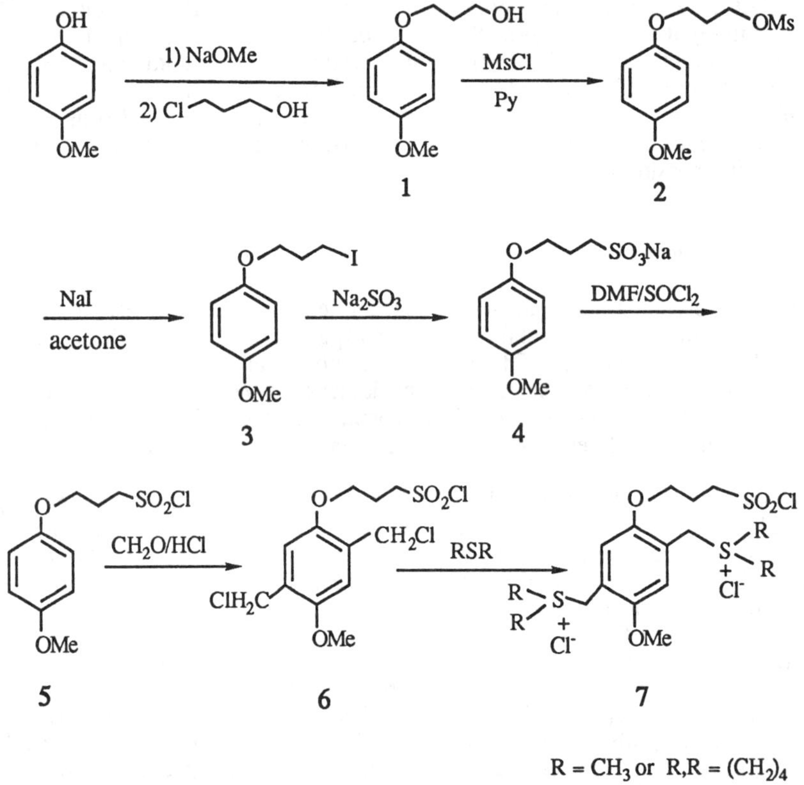

In the previous paper[4] we reported two possible ways to carry out the polymerization of monomer **7**. Scheme II shows the polymerization procedure we developed more recently for monomer **7**. The monomer **7** was polymerized either in methanol or in water with sodium methoxide or sodium hydroxide, respectively. A viscous gel or "chewing-gum" precursor polymer could be obtained depending on the concentration of the base used in the polymerization processes. As in all other poly(p-phenylenevinylene) derivatives, because the monomer **7** goes through a reactive p-xylylene intermediate, which undergoes an anionic[5] or radical addition polymerization[6], the purity of the monomer is crucial to produce narrow-polydispersity, high molecular weight polymer. To get pure monomer **7**, we purified the precursor **6**, followed by conversion to monomer **7** by reaction with a large excess of dimethylsulfide or tetrahydrothiophene. The resulting monomer **7** was further purified by recrystallization from methanol/acetone.

Scheme II

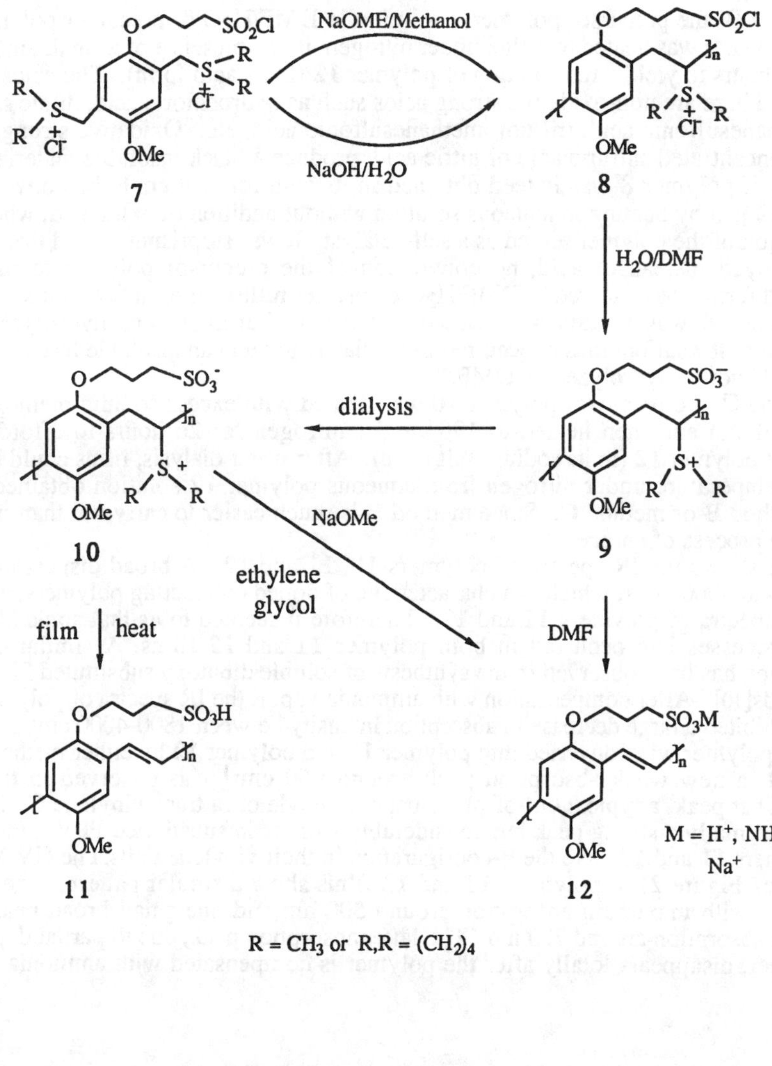

We found that polymer **8** obtained from either water or methanol could be hydrolyzed under vigorous condition by refluxing with DMF/water to give polymer **9**.

The resulting precursor polymer was usually colorless, but if more than one equivalent of base was used in the previous polymerization process, the resulting polymer **9** solution was strongly fluorescent pale yellow-greenish, due to partial elimination of the sulfonium group. Also under these conditions, polymer **9** was produced in its sodium salt form instead of acid form. After the precursor polymer **9** solution was dialysized against deionized water, a zwitterionic precursor polymer **10** solution was obtained.

There are three ways to convert the precursor polymer into fully conjugated PPV (see Scheme II). In method **A**, the precursor polymer film is in vacuum at 200° for 4 hours to give the fully conjugated PPV derivative, polymer **11**. This method is the general one employed to prepare PPV[4] and its derivatives[7,8]. Method **B** and method **C** are based on a method developed in our laboratory for the preparation of soluble dihexyloxy substituted PPV[9]. A slight difference is that, here, acid or base is also used beside the heat treatment.

In method **B**, the precursor polymer **9** solution in DMF/H_2O or precursor polymer **10** solution in water was heated to reflux under nitrogen in the presence of a small amount of acid for 4 hours to yield a red solution of polymer **12** (in its acid form). The acids which can be used here are non-oxidative strong acids such as hydrochloric acid, dilute sulfuric acid, methanesulfonic acid, trifluoromethanesulfonic acid, etc. Oxidative strong acids, such as concentrated sulfuric acid or nitric acid, produce a black insoluble material. We also noted if polymer **8** was indeed obtained in its acid form, it could be converted to polymer **12** just by heating in aqueous solution without addition of extra acid, where the acidic proton of the polymer served as a self-catalyst. It was surprising to find that if base was employed instead of acid, no conversion of the precursor polymer to its fully conjugated form was observed in DMF/H_2O even after refluxing for a few hours; also, if the precursor **8** was treated with concentrated acid before it was hydrolyzed, the elimination of the sulfonium salt occurred immediately to yield an insoluble black polymer which could not be hydrolyzed by DMF.

In method **C**, the precursor polymer **10** was treated with excess sodium methoxide in ethylene glycol and then heated to 190° under nitrogen for 26 hours to afford a red solution of polymer **12** (in its sodium salt form). After water dialysis, films could be cast at room temperature under nitrogen from aqueous polymer **12** solution obtained from either method **B** or method **C**. Since method **B** is much easier to carry out than method **C**, it is the process of choice.

Figure 1 shows the IR spectra of polymers **10**, **11** and **12**. A broad dispersion peak from 1800 to 4000 cm^{-1}, which is a characteristic of doped conducting polymers, appears in the IR spectra of polymers **11** and **12**. Therefore it seemed to us that some kinds of doping processes had occurred in both polymer **11** and **12** films. A similar doping phenomenon has been observed in the synthesis of soluble dibutoxy substituted PPV with strong acids[10]. After compensation with ammonia vapor, the IR spectra of polymers **11** and **12** exhibited a large decrease in absorption intensity between 1800-4000 cm^{-1}. As the precursor polymer was converted into polymer **11** and polymer **12** by either method **A** or method **B**, a new weak absorption peak around 960 cm^{-1} was observed in their IR spectra. That peak, a typical out-of-plane bending mode of a trans vinylene C-H group which is normally a strong peak but considerably weaker in substituted PPV's, indicates that polymers **11** and **12** have the E-configuration in their vinylene units. The UV-Visible spectra (see Figure 2) of polymers **11** and **12** films show a similar pattern; one major broad peak with maximum absorption around 500 nm and one small broad peak with maximum absorption around 720 nm. The latter absorption peak, due to partial doping of the polymer, disappears totally after the polymer is compensated with ammonia vapor.

The π-π^* transition onset in both polymer films is around 600 nm (2.07 eV), which is identical to that of alkoxy substituted derivatives of PPV[6,7] but red-shifted with respect to that of the parent PPV by about 0.43 eV. The UV-Visible spectrum of an aqueous solution of polymer 12 shows a maximum absorption around 508 nm with a sharp onset at 595 nm and a small tail extending up to 900 nm.

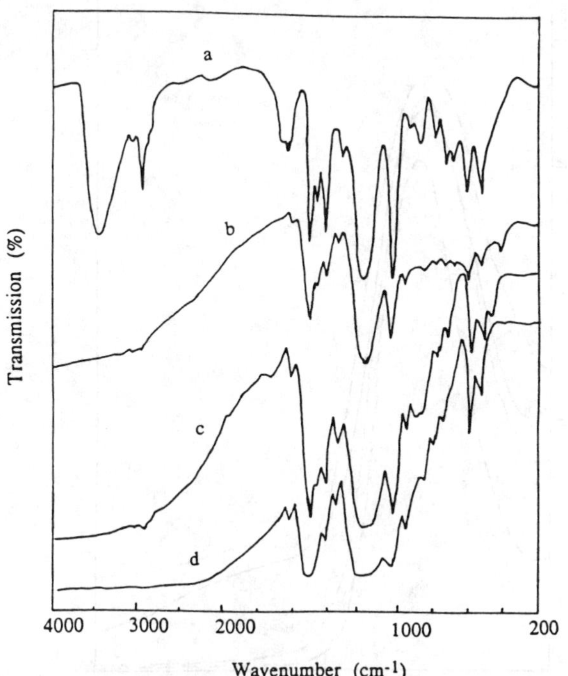

Figure 1: IR spectra of (a) precursor polymer film, (b) polymer 11 film after compensation with NH_3, (c) polymer 12 film, (d) polymer 11 film.

The conductivity of the polymer 11 film is around 2 x 10^{-6} Scm^{-1} (300K, air), yet the conductivity of the polymer 12 film, which is relative humidity dependent, ranges from 10^{-4}-10^{-2} Scm^{-1}. As is observed with other sulfonated polyelectrolytes[11], both polymers are very hygroscopic. Elemental analysis indicates polymer 11 contains 0.43 mol water and 0.32 mol sodium per repeat unit (see Scheme III), while polymer 12 has 1.45 mol water and 0.78 mol ammonium per repeat unit, where the ammonium ion resulted from the deliberately added ammonia to stabilize the polymer. It has been estimated from GPC, using pulluan as a standard, that the weight average molecular weight (M_w) of polymer 12 is around 1.12 x 10^6.

88

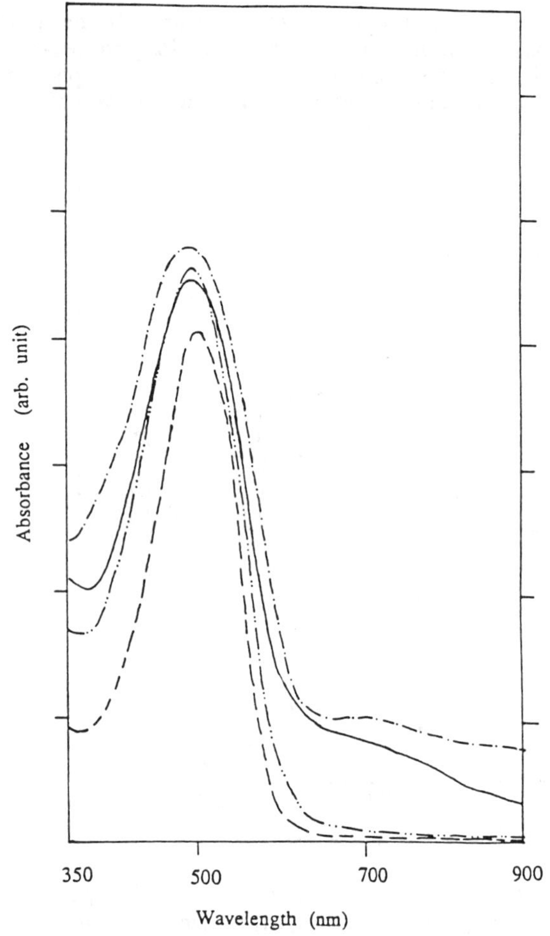

Figure 2: UV spectra of polymer 12 film (— · —), polymer 11 film (——), polymer 11 film after compensation with NH3 (— ·· —), polymer 12 aqueous solution (——).

Scheme III

11

12

Although the polymer **12** films cast from aqueous solution can be easily redissolved in water, polymer **11** films are found to be insoluble in any solvent. The reason could be that the high temperature treatment in the solid state during the formation of polymer **11** may cause some crosslinking, which makes polymer **11** insoluble. The fact that, mechanically, polymer **11** films are much stronger than polymer **12** films may be taken as evidence that some crosslinking occurred in polymer **11** films.

References:

1. Yoshino, K.; Hayashi, S. and Sugimoto, R. (1986) Jpn. J. Appl. Phys. 23, L899.
2. Jen, K. Y. and Elsenbaumer, R. L. (1986) J. Chem. Soc. Chem. Commun. 1346.
3. (a) Patil, A.O., Ikenoue, Y., Wudl, F. and Heeger, A.J. (1987) J. Am. Chem. Soc., 109, 1858.   (b) Patil, A.O., Ikenoue, Y., Basescu, N., Colaneri, N., Chen, J., Wudl, F. and Heeger, A.J. (1987) Syn. Mtls., 20, 151.   (c) Havinga, E.E., v. Horssen, L.W., ten Hoeve, W., Wynberg, H. and Meijer, E.W. (1987) Polimer Bull., 18, 277.   (e) Reynolds, J.R., Ruiz, J.P., Wang, F., Jolly, C.A., Nayak, K. and Marynick, D. (1989) Synth. Met., 28, C621 and references therein.
4. This work was reported in part at the 196th National Meeting of the American Chemical Society, Los Angeles, California, September 1988: Shi, S. and Wudl, F. (1988) Proceedings of the ACS Division of Polymeric Materials: Science and Engineering, 59, 1169.
5. Lahti, P.M., Modarelli, D.A., Denton III, F.R., Lenz, R.W. and Karasz, F.E. (1988) J. Am. Chem. Soc., 110, 7259.
6. (a). Wessling, R.A. and Zimmerman, R.G. U. S. Patent, 3401152 (1968); 3404132 (1968); 3532643 (1970); 3705677 (1972).   (b). Wessling, R.A. (1985) J. Polym. Chem.; Polym. Symp., 72, 55.
7. Murase, I., Ohnishi, T., Noguchi, T. and Hirooka, M. (1984) Polym. Commun., 25, 327.
8. Murase, I., Ohnishi, T., Noguchi, T. and Hirooka, M. (1985) Polym. Commun., 26, 362.
9. Askari, S.H., Rughooputh, S.D. and Wudl, F. (1988) Proceedings of the ACS Division of Polymeric Materials: Science and Engineering, 59, 1068.
10. Han, C.C., Jen, K.Y. and Elsenbaumer, R.L. (1988) Proceedings of the International Conference on synthetic Metals, Santa Fe, NM, (1989) Syn. Metals, 30,123.
11. For a Detailed Review See: (1984) Water-soluble Synthetic Polymers, CRC Press, Boca Raton, Florida.

MOLECULAR ORGANIZATION AND ELECTRICAL PROPERTIES OF MIXED LANGMUIR-BLODGETT MULTILAYER THIN FILMS OF POLYPYRROLE

J. CHEUNG, R. B. ROSNER, M. F. RUBNER
Department of Materials Science and Engineering, Massachusetts Institute of Technology, Cambridge, MA 02139 USA

X. Q. YANG, J. CHEN, T. A. SKOTHEIM
Brookhaven National Laboratory, Upton, NY 11973 USA

ABSTRACT. The molecular organization of Langmuir-Blodgett multilayer thin films containing electrically conductive polypyrrole chains dispersed throughout insulating domains of surface active pyrrole molecules was examined by several thin film spectroscopic techniques. Reflection-absorption FTIR and NEXAFS spectroscopy have revealed that the 3-ODOP (3-octadecanoyl pyrrole) surface active pyrrole molecules are highly oriented within the film with their fully extended hydrocarbon chains tilted away from the surface normal with an unusually large tilt angle of about 55°. Evidence for preferred orientation of the polypyrrole chains was also found. The multilayer films were found to exhibit very large dielectric constants (>100) at low frequencies and enormous conductivity anisotropies. These unusual electrical properties can be directly attributed to the molecular organization of the film which consists of polypyrrole chains sandwiched between well ordered layers of 3-ODOP molecules.

1. Introduction

Recent improvements in the environmental stability and processibility of conducting polymers have stimulated renewed excitement in this very interesting class of materials. New synthetic methodologies based on precursor polymers and derivatized polymers have resulted in materials that can be readily manipulated into a variety of useful forms without severely compromising their electrical properties [1]. In addition, the availability of a number of new processible polymers means that it is now possible to control, via suitable processing techniques, the molecular organization and ordering of these materials. This, in turn, is expected to dramatically improve their overall electrical and mechanical properties. For example, the manipulation of processible conjugated polymers into films and fibers in which the polymer chains are fully extended along a preferred axis has already been shown to produce materials with enhanced mechanical strengths and dramatically increased electrical conductivities [1]. In order to truly exploit the novel electrical and optical properties of these materials, however, it is also necessary to develop processing techniques that can be utilized to form them into highly ordered thin films with controllable molecular organizations. Indeed, the realization of many molecular electronic and thin film sensor schemes is strongly linked to the availability of such films.

91

J. L. Brédas and R. R. Chance (eds.), Conjugated Polymeric Materials:
Opportunities in Electronics, Optoelectronics, and Molecular Electronics, 91–99.
© 1990 *Kluwer Academic Publishers. Printed in the Netherlands.*

The Langmuir-Blodgett (LB) technique provides one of the few ways to manipulate organic materials into uniform thickness, thin film forms with controllable molecular architectures. Essentially, one simply disperses suitable molecules onto the air-water interface of a Langmuir trough and, by reducing the area available to the molecules, forces them to form an ordered condensed monolayer at the water surface. The monolayers are then sequentially transferred onto a substrate to produce a multilayer thin film; the thickness and molecular organization of which are precisely controlled by the number and type of monolayers deposited. Thus, using this approach, it is possible to form complex molecular architectures comprised of ordered layers of functionally different molecules exhibiting properties simply not found with the parent molecules in their normal molecular organizations.

To date, we have developed a number of different strategies that can be utilized to manipulate conducting polymers into multilayer LB films. All of these variations can be conveniently divided into two main categories: the manipulation of surface active conjugated polymers and the manipulation of non-surface active conjugated polymers. In the former case, the monomers or polymers used to make the LB films are structurally modified to create amphiphilic molecules that form true monolayers at the air-water interface. For the surface active monomers, polymerization is typically carried out at the air-water interface by adding an oxidizing agent into the water subphase resulting in an electrically conductive monolayer that is subsequently transferred into multilayers. It should be noted that it is also possible to first build multilayers from the surface active monomers and then electrochemically polymerize them as has been shown by Shimidzu and coworkers [2]. Alternatively, the surface active monomers can be prepolymerized and then directly manipulated via the LB technique. In the second main approach, soluble, non-surface active polymers are mixed with suitable portions of traditional surface active molecules such as stearic acid to from stable mixed monolayers that can be easily formed into LB films. This latter technique does not require any structural modification of the polymer and is generally applicable to any of the soluble conjugated polymers. A variation of this theme involves the adsorption of a non-surface active conducting polymer onto the monolayer of a surface active molecule and the transfer of this entire mixed bilayer assembly into a multilayer thin film.

Examples of the above processing schemes include the LB manipulation of mixed monolayers of stearic acid and the poly(3-alkyl thiophenes). In this case, it has been found [3] that mixed monolayers containing as much as 80 mole% of the conjugated polymer can be fabricated into multilayer thin films with well defined layer structures. The final molecular organization of these multilayer films consists of well ordered, highly aligned stacks of stearic acid molecules throughout which relatively disordered domains of the conducting polymer are dispersed. These films are, in fact, very reminiscent of biological membranes in which large protein molecules are dispersed throughout highly ordered bilayers of surface active phospholipid molecules. The conjugated polymer domains, however, do not achieve the level of ordering provided by nature in the protein molecules. The polymerization of pyrrole in the presence of a surface active pyrrole molecule at the air-water interface of an LB trough is an example of another approach that has successfully resulted in multilayer thin films of conducting polymers [4]. This system also produces a mixed monolayer, in this case, comprised of electrically conductive polypyrrole chains and insulating surface active pyrrole molecules.

In order to develop the full potential of these new thin film structures, new analytical approaches are necessary which allow detailed studies of the relationship between the molecular architecture on the one hand and the supermolecular ordering on the other. A number of techniques such as reflection/absorption FTIR spectroscopy [5] and Near Edge X-ray Absorption Fine Structure spectroscopy [6] (NEXAFS) have already been shown to be effective in determining the level of molecular order and orientation in insulating LB films. We are currently applying these techniques [7] along with other synchrotron radiation techniques to the study of these new electrically conductive LB films.

In this paper, we examine the structure and properties of LB films formed from the polypyrrole based system. It will be shown that by comparing the results of a number of different structural probes, it is possible to establish the structure/property relationships in these very interesting thin films.

2. Experimental

We have previously demonstrated [4] that electrically conductive polypyrrole films can be formed at the air-water interface of an LB trough by simply dispersing a solution containing a surface active pyrrole monomer and a large excess of pyrrole onto a subphase containing ferric chloride ($FeCl_3$). The ferric chloride acts to both polymerize the mixture at the air-water interface and simultaneously oxidize the resultant polymer thereby rendering it electrically conductive. Pyrrole monomer is needed to facilitate polymerization at the air-water interface as neither the surface active pyrrole monomer nor pure pyrrole will polymerize independently under the conditions used to prepare the films. A large molar excess of pyrrole monomer is used due to the high degree of water solubility exhibited by this material. Two types of surface active pyrrole monomers have been examined; 3-octadecyl pyrrole (3ODP) and 3-octadecanoyl pyrrole (3ODOP).

The key to creating electrically conductive polypyrroles that can be readily manipulated into multilayer structures using the LB technique lies with the proper choice of surface active pyrrole. For both surface active monomers, we find that electrically conductive films are only created at the air-water interface when a reaction is carried out between pyrrole and the surface active pyrrole with a mole ratio close to 5000/1. FTIR studies [4] clearly demonstrate that the 3ODP molecules are copolymerizing with the added pyrrole whereas the 3ODOP molecules only serve to promote the formation of polypyrrole homopolymer. Due to the higher reactivity of the former system, the reaction creates relatively thick (about 200 Å) surface films of polypyrrole rich copolymer that are extremely difficult to process into LB multilayer thin films. The latter system, on the other hand, produces uniform monolayer films at the air-water interface that can be easily transferred into multilayers using a conventional vertical lifting technique. Thus, the pyrrole/3ODOP system is preferred for fabricating multilayer thin films.

The synthesis of the surface active pyrrole monomers used in this study as well as the polymerization procedure have been described in detail elsewhere [4]. Polymerization was carried out on the water surface of a Lauda film balance at 20°C by spreading a solution of pyrrole and substituted pyrrole onto a subphase containing 1 wt % ferric chloride. Monolayers were transferred onto solid substrates as Y-type LB films at 25 mN/m and 20°C. A dipping speed of 5 mm/min was used to transfer the films. Drying times of at least two hours were used between the first and second dips. This time was reduced to one hour for all subsequent dips. Details of the electrical measurements will be published elsewhere.

3. Results and Discussion

The level of molecular order and orientation in LB films can be readily probed by reflection/absorption FTIR spectroscopy. In this technique, LB films are deposited onto

94

two different substrates; an infrared transparent ZnSe plate and an infrared reflecting platinum coated glass slide. Transmission spectra are then recorded on the sample deposited on ZnSe and grazing incidence reflection spectra (with an incident angle of $8°$ to the substrate surface) are recorded on the sample deposited on the platinum coated slide. In the transmission mode, the electric field vector is exclusively polarized in the plane of the substrate and hence only those molecular vibrations that have a component of their transition dipole moment in the plane of the film will be activated. In the reflection mode, however, the electric field vector is polarized essentially perpendicular to the substrate therefore only probing molecular vibrations that have a component of their dipole moment normal to the substrate plane. Thus, by comparing the infrared spectra obtained in these two experiments, it is possible to ascertain the type and level of molecular orientation present in the film.

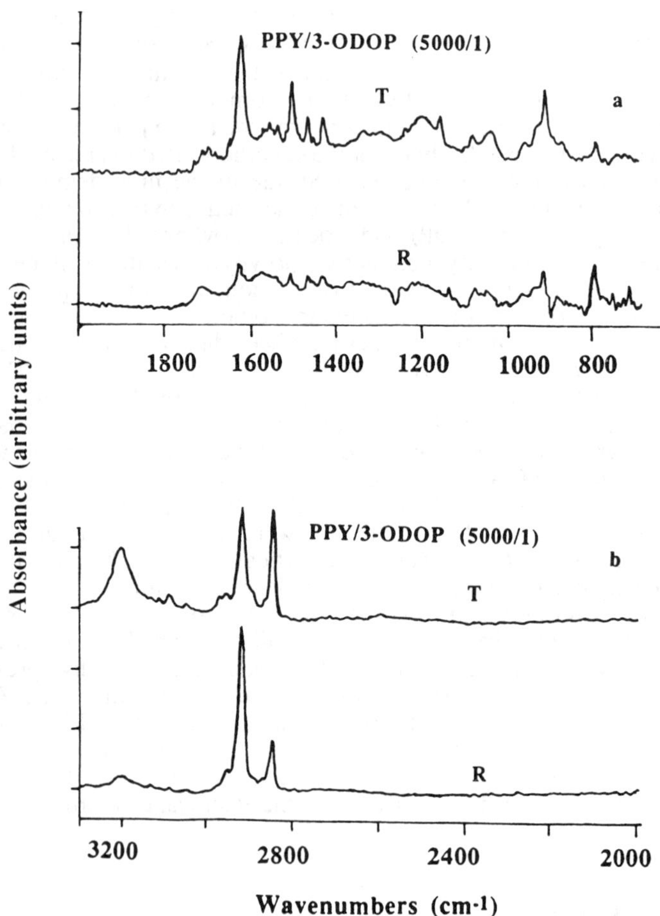

Figure 1.   Transmission (T) and Reflectance (R) spectra of polypyrrole/3-ODOP multilayer thin films.

Figure 1 shows the reflection (R) and transmission (T) spectra of the polypyrrole/3ODOP system in two separate regions of the infrared spectrum. In Figure 1b, it can be seen that the absorption due to the hydrogen bonded N-H stretch (around 3200 cm^{-1}) of the 3ODOP pyrrole rings is significantly stronger in transmission than in reflection implying that the pyrrole rings are preferentially lying flat in the plane of the multilayer film i.e., the rings are oriented nearly parallel to the substrate. In addition, a clear polarization dependence is observed in the C-H stretching region (between 2800-3000 cm^{-1}) indicating some preferred orientation of the hydrocarbons tails of the 3ODOP molecules. Specifically, the asymmetric CH_2 stretch (2921 cm^{-1}) is seen to be about equal in intensity to the symmetric CH_2 stretch (2844 cm^{-1}) in transmission but of much larger intensity in the reflection spectrum. Since the asymmetric and symmetric CH_2 vibrations have dipole moments that are orthogonal with respect to each other and are both contained within the plane that is perpendicular to the fully extended hydrocarbon chain axis, it would appear that the hydrocarbon tail groups of the surface active 3ODOP molecules are tilted away from from the substrate normal with a unusually large tilt angle. Although a complete analysis of these data has not been conducted, these results suggest a tilt angle of a least 45° from the substrate normal. This conclusion is further supported by low angle x-ray scattering results obtained from a multilayer of pure 3-ODOP which indicate a bilayer repeat distance of about 28Å. To achieve this bilayer stacking, the molecules would have to pack with a tilt angle of about 55° from the surface normal. Thus, the 3ODOP molecules are well ordered within the electrically conductive mixed LB film, essentially retaining the level and type of molecular orientation present in a pure 3-ODOP multilayer film.

The infrared spectra depicted in Figure 1a also show a strong polarization dependence. The absorption band due to the carbonyl stretching vibration of the 3ODOP molecules at about 1630 cm^{-1} is significantly stronger in transmission than in reflection. In addition, the major pyrrole ring vibrations of the 3ODOP molecules between 1600 and 1400 cm^{-1} are also much stronger in transmission than in reflection. These observations are consistent with the pyrrole rings of the 3ODOP molecules being oriented preferentially in the plane of the substrate (assuming that the carbonyl and pyrrole ring are coplanar). If the carbonyl is not coplanar with the ring, it is possible that the rings may be oriented edge on to the substrate; a determination of the exact orientation of the head groups awaits complete assignments of all of the in-plane and out-of-plane pyrrole ring vibrations. It can also be seen that the polypyrrole chains of this mixed LB film tend to be aligned with their heterocyclic backbones lying parallel to the plane of the substrate. This is indicated by the fact that the absorption bands characteristic of oxidized polypyrrole (between 1400 and 1000 cm^{-1}) are much more intense in transmission than in reflection. Thus, both the 3ODOP molecules and the polypyrrole chains show evidence of preferred orientation in the multilayer structure.

X-ray absorption studies also provide valuable information about molecular orientation in LB films. These studies focus on the near edge structure of the core level absorption spectrum. In NEXAFS spectroscopy, core level electrons are excited by the incident photons to unoccupied orbitals near the ionization threshold. This technique takes advantage of the high degree of polarization of the x-ray beam emanating from a synchrotron storage ring. The excitations examined in this work are core level excitations to the unoccupied σ^* and π^* orbitals. By scanning the incident photon energy and monitoring the Auger electrons resulting from the relaxation processes of the excitation, the empty states, such as C-C σ*, C=C σ* and π*and (C-H)*, can be identified and the direction of the bonds relative to the substrate plane can be determined.

Figure 3 shows the carbon *k*-edge NEXAFS spectra obtained from a multilayer of the polypyrrole/3ODOP system. Unlike the spectra obtained from a multilayer thin film of 3-octadecyl pyrrole monomer (3ODP), where a strong polarization dependence was clearly observed [7], the two spectra of polypyrrole/3-ODOP obtained with different polarizations are essentially identical. Generally, such a result would imply that the hydrocarbon tails of the 3ODOP molecules are disordered within the multilayer thin film. However, an alternative explanation is that the fully extended hydrocarbon tails are oriented perpendicular to the substrate direction with an unusually large tilt angle between 45 to 55 degrees from the surface normal. In light of the FTIR results, which show a clear indication of preferential orientation of the surface active pyrrole molecules, it is reasonable to conclude that the latter explanation is the correct one. Thus, the combined techniques of reflection-absorption FTIR spectroscopy, NEXAFS and low angle x-ray diffraction provide valuable insights into the structure of the conducting LB films.

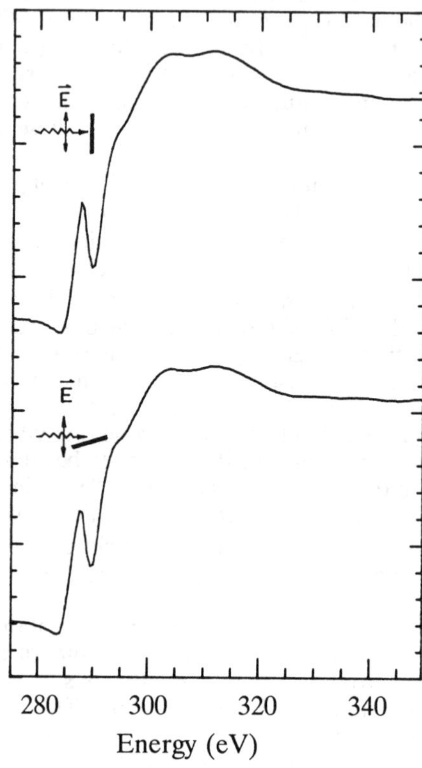

Figure 2.   Carbon *K*-edge NEXAFS spectra at normal and glancing angle incidence for a 20 layer film of polypyrrole/3-ODOP

The above structural studies suggest that the polypyrrole/3ODOP multilayer thin films contain ordered regions of the 3ODOP molecules and partially oriented regions of electrically conductive polypyrrole molecules. The exact supermolecular organization of these two separate domains, unfortunately, can not be easily deduced from these measurements. A clue to how the two domains are organized with respect to each other within the multilayer structure, however, is provided by electrical measurements. For example, measurement of the in-plane electrical conductivity of typical multilayer films reveals a value of about 10^{-2} S/cm. The transverse conductivity, on the other hand, was found to be less than 10^{-10} S/cm. This remarkable conductivity anisotropy over 10^8 clearly indicates that the structure of the film is also anisotropic.

Based on structural studies and electrical measurements, the current picture that emerges is that the multilayer thin films are comprised of polypyrrole chains sandwiched between well ordered insulating domains of 3ODOP molecules. In such an an organization, conduction could occur readily in the plane of the film since the conducting polypyrrole chains form a continuous or near continuous layer. Conduction across the film thickness, however, would be restricted by the presence of the insulating domains of 3ODOP.

In order to evaluate the dielectric properties of the multilayer thin films fabricated from the polypyrrole/3ODOP system, monolayers were transferred onto a platinum coated glass slide to create a thin film with two steps of 44 and 60 layers. The thin film therefore consisted of two sections, each with a different film thickness; one about 1980 Å thick and the other about 2700 Å thick (each deposited monolayer is 45 Å thick). After drying the film in vacuum for two days to remove residual water, an array of aluminum electrodes having active areas of 0.02 and 0.04 cm^2 was deposited onto the film via standard vacuum evaporation techniques. Each capacitor thus represents a sandwich structure of the type Pt/ LB film /Al.

The dielectric constant plotted as a function of frequency is presented in Figure 2. As can be seen, the dielectric constant displays a strong frequency dependence. At low frequencies, the dielectric constant is very large (around 150), it then drops to an apparent plateau in the range between 10^4 and 10^5 Hz with a value of about 50 and finally decays to values less than 5 in the MHz range. The frequency dispersion observed in this system reflects the influence of frequency on the movement of carriers and possibly ionic species to the various interfaces present in the sample. At low frequencies, the movement of these species across their respective domains is fast enough to allow them to easily follow the oscillating electric field. With progressively higher frequencies, however, the motion of some of the less mobile species is frozen out thereby eliminating their contribution to the dielectric constant.

It is tempting to attribute the large dielectric constant of this heterogeneous LB film at low frequencies to the accumulation of space charges at the interfaces of the conducting and insulating domains. As indicated earlier, the molecular organization in this case is best described as domains of conducting polypyrrole chains sandwiched between layers of the insulating hydrocarbon tail groups of the surface active 3ODOP molecules. Thus, carriers move through the polypyrrole domains, most likely via the hopping of positively charged bipolarons, until they encounter an interface with an insulating region of the film at which point they become trapped.

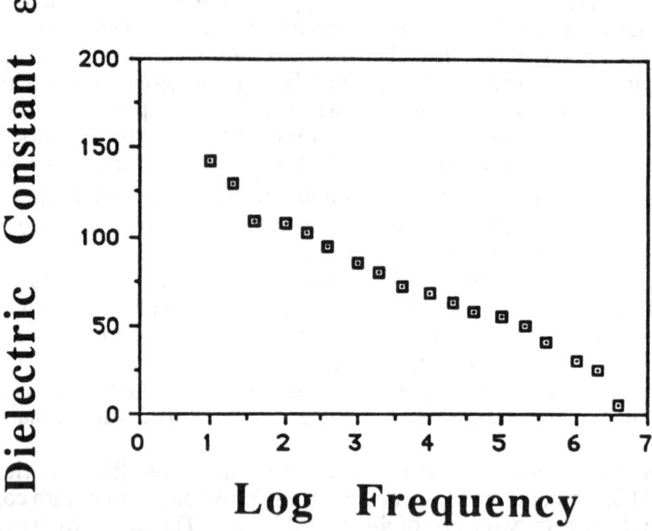

Figure 3.   The dielectric constant of the polypyrrole/3ODOP multilayer thin film as function of frequency

4. Conclusions

Reflection-absorption spectroscopy and NEXFS studies have shown that LB films created from mixed monolayers of electrically conductive polypyrrole and 3-ODOP surface active molecules form multilayer structures comprised of polypyrrole domains dispersed throughout well ordered domains of 3-ODOP molecules. The resultant heterogeneous structures give rise to unusually large dielectric constants due to the trapping of mobile charge carriers at the various interfaces present in the film.

5. Acknowledgements

Partial support of the MIT research was provided by the National Science Foundation, the MIT Center for Materials Science and Engineering and the U. S. Army Electronics Technology and Devices Laboratory (via the Army Research Office Scientific Services Program administered by Batelle). Work at the Brookhaven National Laboratory was partially supported by the U. S. Department of Energy, Division of Materials Science, under contract No. DE-AC02-76CH00016.

6. References

1.   See for example, *Proc. Synth. Metals, International Conference on Science and Technology of Synthetic Metals,* (1989) Vol. 28 and 29.

2.   Shimidzu, T., Iyoda, T., Ando, M., Ohtani, A., Kaneko, T. and Honda, K. (1988) *Thin Solid Films* *160*, 67.

3.   Watanabe, I., Hong, K. and Rubner, M. F. (1989) *J. Chem. Soc., Chem. Commun.* 123. Watanabe, I., Hong, K., Rubner, M. F. and Loh, I. H. (1989) *Synth. Met. 28*, C473. Watanabe, I., Hong, K.and Rubner, M.F. *Thin Solid Films*, in press. Watanabe, I., Hong, K. and Rubner, M.F., submitted to *Langmuir*.

4.   Hong, K. and Rubner, M.F. (1988) *Thin Solid Films 160* , 187. Hong, K. and Rubner, M.F. *Thin Solid Films*, in press. Hong, K., Rosner, R. B. and Rubner, M.F. submitted to *Chemistry of Materials*.

5.   Rabolt, J. F., Burns, F. C., Schlotter, N. E. and Swalen, J. D. (1983) *J. Chem. Phys.* 78, 96.

6.   Stohr, J., Outka, D. A. , Baberschke, K., Arvanitis, D. and Horsley, J. A. (1987) *Phys. Rev. B* 36, 2976.

7.   Skotheim, T. A., Yang, X. Q., Chen, J.; Hale, P. D., Inigaki, T., Samuelsen, L., Tripathy, S., Hong, K., Rubner, M. F., den Boer, M. L. and Okamoto, Y. (1989) *Synth. Metals 28*, C229. Yang, X. Q., Chen, J., Hale, P. D., Inigaki, T., Skotheim, T. A., Okamoto, Y., Samuelsen, L., Tripathy, S., Hong, K., Rubner, M. F., and den Boer, M. L. (1989) *Synth. Metals 28*, C251. Yang, X. Q., Chen, J., Hale, P. D., Inigaki, T., Skotheim, T. A., Okamoto, Y., Samuelsen, L., Tripathy, S., Hong, K., Watanabe, I., Rubner, M. F. and den Boer, M. L. *Langmuir*, in press.

THE ELECTRONIC AND CHEMICAL STRUCTURE OF POLY(3-HEXYL-THIOPHENE) STUDIED BY PHOTOELECTRON SPECTROSCOPY

W. R. SALANECK, R. LAZZARONI, N. SATO*, M. LÖGDLUND, B. SJÖGREN
Department of Physics, IFM,
Linköping University,
S-581 83 Linköping,
Sweden, and

M. P. KEANE, S. SVENSSON, A. NAVES de BRITO, N. CORREIA†, and
S. LUNELL**
*Department of Physics and **Department of Quantum Chemistry,*
Uppsala University,
S-752 21 Uppsala,
Sweden

Permanent addresses:
* *Department of Chemistry*   † *Department of Physics*
  Faculty of Arts and Sciences   *University of Brasilia*
  University of Tokyo, Japan   *70910 Brasilia - DF - Brasil*

ABSTRACT. The processability of conjugated polymers has lead the the possibility of performing better studies of their chemical and electronic structure. In this contribution, the thermochromism in a processable polyalkylthiophene, namely poly(3-hexylthiophene), is reviewed, with an emphasis on the proposal that, at high temperatures, soft conformational defects, which lead to electronic localization effects, are responsible for determining the reversible, temperature-dependent color changes which have been observed in this conjugated system in the insulating (non-doped) state. In addition, some new spectroscopic evidence for electronic localization, based upon studies of model molecules in the gas phase, is presented.

1. Introduction

Recent progress in the area of electrically conducting organic polymers includes the development of processable conducting polymers, i. e., conjugated polymer materials which can be processed by conventional, non-conjugated-polymer processing techniques, such as high temperature injection molding, drawing, or even solution casting. The

101

J. L. Brédas and R. R. Chance (eds.), Conjugated Polymeric Materials:
Opportunities in Electronics, Optoelectronics, and Molecular Electronics, 101–113.
© *1990 Kluwer Academic Publishers. Printed in the Netherlands.*

progress, however, is not unidirectional. Not only has the science which has been done on these materials lead to processability, but the new processable polymer materials, in turn, have allowed better science to be done. An illustration of this principle involves the substituted polythiophenes.

Electrochemically prepared polythiophene is non-soluble and non-fusable, i.e., non-processable {1}. Research on the substituted polythiophenes lead to the discovery that, with flexible alkyl ($-C_nH_{2n-1}$) groups attached at the 3- (or β-) positions, these polymer materials are both solution {2,3} and melt {4,5} processable when n > 1. The electronic structure of electro-chemically prepared polythiophene {6} and poly(3-methylthiophene) {7} has been studied by photoelectron spectroscopy. Due to the processes of removing samples from an electro-chemical cell and the subsiquent washing procedure required to remove residual chemicals from the electro-chemical process, however, electro-chemically made material, may contain surface contamination, which may contribute unwanted features to the photoelectron spectra. For the alkyl-substituted polythiophenes, however, it has been found that high quality solution-cast films can be made, the surfaces of which become very clean simply by heating in ultra high vacuum (UHV), enabling higher quality photoelectron spectra to be obtained {8,9}.

Subsequently, new physics has been uncovered during the course of studies of these high quality, ultrathin, solution-cast films of the poly(3-alkylthiophene)s. Undoped poly(3-alkylthiophene) was found to exhibit thermochromism in the solid state (when the alkyl-group contains at least 2 or more carbon atoms); the first "conducting polymer" to do so, *albeit* in the non-conducting (undoped) state {10}. Also, poly(3-hexylthiophene), doped to saturation *in situ* (in inert atmosphere) using $NOPF_6$, was found to exhibit a finite density-of-states at the Fermi energy {11}, indicating the existence of a polaron lattice, which is unstable in strictly one dimension {12}. This latter observation implies the presence of higher (than one) dimensional interactions, which were unexpected in the polythiophenes {11}, and is discussed elsewhere in these proceedings {13}.

In this contribution, the molecular origin of the thermochromism in one of the substituted polythiophenes, namely poly(3-hexylthiophene), or P3HT for convenience, is reviewed. The discussion is based upon the results of studies by X-ray and ultra violet photoelectron spectroscopy (XPS and UPS), as well as optical absorption spectroscopy. Because of the nature of the spectroscopies involved, some details of the data analysis must be described. Chemical preparation and characterization, sample preparation, as well as other technical experimental details can be found in the referenced literature {5,8,9,10}.

2. Spectroscopies

2.1 MOLECULAR PHOTOABSORPTION

The fundamental process of photon absorption of a photon of energy $h\upsilon$ by a molecule M is represented by

$$h\upsilon + M_O \rightarrow \Sigma_i M_O^*(i), \tag{1}$$

where M_O represents the neutral molecule in the ground state, and $\Sigma_i M_O^*(i)$ represents the manifold of all possible excited states (i) of the neutral molecule (excitations among the valence electrons) which may be generated in a single photon absorption event. Photoionization of the molecule is represented by

$$h\upsilon + M_0 \rightarrow e^-(E_k) + \Sigma_j M_+^*(j) + \Sigma_{j'} M_+^{**}(j') + ... , \qquad (2)$$

where $e^-(E_k)$ is the photoelectron of kinetic energy E_k, and $\Sigma_j M_+^*(j)$ represents the manifold of possible singly-excited (single-hole) states (j) of the molecular *ion*, and the $\Sigma_{j'} M_+^{**}(j')$ represents all of the 2-hole-1-electron excited states of the *ion*. These latter excited states correspond to excitations among the valence electrons in the presence of a core-hole. Other terms, such as higher order multiple excitations, as well as shake-off transitions to the continum, although important in a full analysis of a photoelectron spectrum {16}, can be ignored for the present purposes. For sufficiently high photon energies, as is the case in these discussions, E_k is large enough that the final state is free electron-like, and the wave functions of the electron and of the molecular ion are not coupled. In the taking of a spectrum, the *final states* which are investigated in optical absorption spectroscopy are the $\Sigma_i M_0^*(i)$, while in photoelectron spectroscopies, UPS and XPS, the *final states* $\Sigma_j M_+^*(j)$ and $\Sigma_{j'} M_+^{**}(j')$ are investigated. The details, as well as the additional terms in Eq. (2), are discussed in the references {14-16}.

2.2 ONE-ELECTRON APPROXIMATION

It is common and most convenient to *represent* the molecular final states in terms of one-electron molecular orbitals, despite the fact that, even in most simple cases, many-electron effects dominate the spectra {14}. The one-electron energy level diagram for the frontier levels in a hypothetical molecule are shown in Fig. 1.

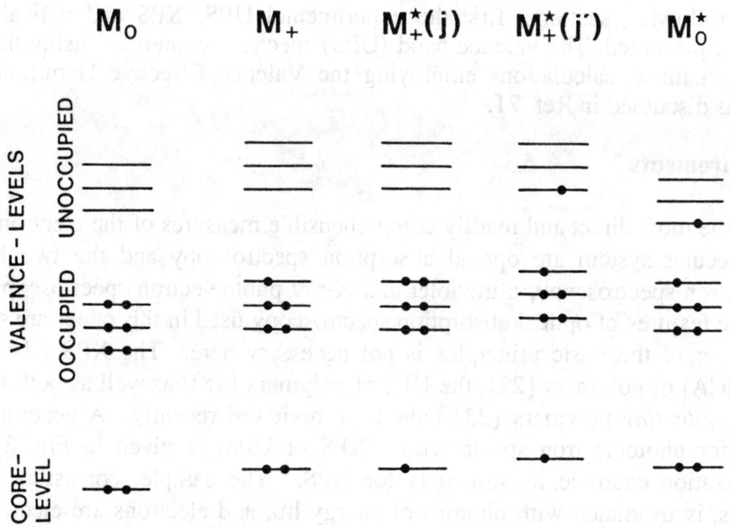

Figure 1. Illustration of the single-electron energy levels of the variuos states of the molecule, as discussed in the text.

In terms of one-electron energy levels, M_O represents the neutral molecule, M_+ the lowest energy *cation* state, M_O^* the lowest (photon) energy optically excited state, while the $M_O^*(i)$, the $M_+^*(j)$, and the $M_+^{**}(j')$ are described above. The $M_+^*(j)$, representing the one-electron energy levels in the molecular ion, give rise to peaks in the UPS (and XPS) valence band spectra. These peaks are usually interpreted in terms of the eigenstates of the neutral molecule, the $M_O^*(i)$, through Koopmans' theorem {17}. For large molecules and polymers, this is often a reasonable approximation, with certain reservations {18}. The $M_+^{**}(j')$ result in so-called *shake-up* satellite peaks in the photoelectron spectrum (usually) on the high binding energy (lower kinetic energy) side of the main photoelectron peak. In the systems to be discussed here, the $M_+^{**}(j')$ represent π-π^* transitions (in the presence of a core-hole) in the thiophene rings. These shake-up, or simply s.u., peaks are a measure of the *electronic localization* of the π-system under investigation. The photoionization events which result in electrons ejected from core-electron states (Fig. 2) occur under what are called non-adiabatic conditions. Although the phenomenon is a purely quantum mechanical one, a classical picture can be used for discussion purposes. In the "sudden" process of removing an electron from a localized core-state, the total charge near the core of the atom, as seen by the valence electrons, is suddenly changed. The remaining valence electrons are no longer in eigenstates of the (now ionic) system. An electronic relaxation process occurs {19}, which leads to electronic transitions from the "old" eigenstates (of the atom before ionization) and the "new" states in the molecular ion. Some of these electronic transitions result in the population of π^*-levels *in the molecular ion*. Riga *et al* {20} have shown experimentally, in a study of the s.u. satellites in the C1s XPS spectra of the fused-ring aromatic molecules, that as the π-electrons become more localized, the s.u. satellite structure in the C1s spectra become more intense, and the center-of-gravity moves further away from the main C1s peak.

In the following sections, first the experimental UPS, XPS and optical absorption spectra are presented. The valence band (UPS) spectra are analysed using the results of quantum chemical calculations employing the Valence Effective Hamiltonian (VEH) method, as discussed in Ref. 21.

3. Measurements

Three of the most direct and readily comprehensible measures of the electronic structure of a molecular system are optical absorption spectroscopy and the two branches of photoelectron spectroscopy, ultraviolet and X-ray photoelectron spectroscopy, UPS and XPS. The features of optical absorption spectroscopy used in this paper are so basic that a discussion of the basic principles is not necessary here. The XPS (also sometimes called ESCA) of polymers {22}, the UPS of polymers {18}, as well as both the XPS and UPS of *conducting* polymers {23} have been reviewed recently. A general conceptual diagram for photoelectron spectroscopy (XPS or UPS) is given in Fig. 2, where the photoionization example illustrated is for XPS. The sample, consisting of polymer molecules, is irradiated with photons of energy $h\upsilon$, and electrons are ejected from (the surface of) the sample with a kinetic energy, E_k, characteristic of the apparent binding energy, E_B, of the eigenstate from which the electron was photoexcited.

If the one-electron energy levels, V_i in the valence band and C_i for the core levels, are equated directly with the peaks in the photoelectron spectrum, as in Fig. 2, then, neglecting an instrumental reference correction, $E_B = h\upsilon - E_k$. This is a statement of Koopmans' theorem {17}, where the *intra-* and *inter-*molecular relaxation effects, illustrated in the figure, are neglected. This turns out to be a good approximation in large molecular systems (such as polymers), where the removal of an electronic charge in the photoionization event is only a minor perturbation of the system, and as a consequence the energy-ordering of the molecular eigenstates (almost always) remains unchanged {18}.

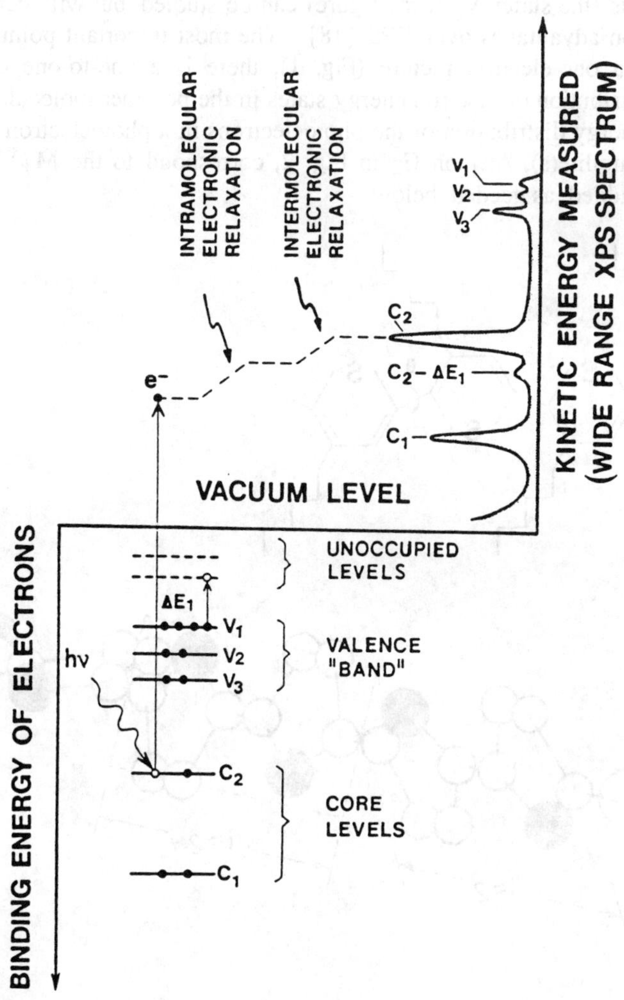

Figure 2: A schematic representation of a one-electron model of photoelectron spectroscopy.

Then the UPS spectrum is a direct measure of the so-called valence band density-of-states, but with a (usually) rigid shift of the entire spectrum towards lower binding energy, which is a direct result of *inter*molecular relaxation {19}. The *intra*molecular relaxation effects, on the other hand, correspond to the s.u. events discussed above. We return to the s.u. structure in XPS spectra below.

With soft X-ray photons (XPS), for example 1253.6 eV photons from $Mg(K_\alpha)$ radiation source or 1486.7 eV photons from an (often monochromatized) $Al(K_\alpha)$ source, both electrons in the atomic core-levels (C_i in Fig. 1) and valence electrons (V_i) can be studied. With ultra violet radiation (UPS), for example 21.2 eV or 40.8 eV photons for the HeI and HeII lines respectively, typical of a helium gas discharge source, only the valence electrons (the states V_i in the figure) can be studied, but with certain resolution and cross section advantages over XPS {18}. The most important point is that, within the context of a one-electron picture (Fig. 1), there is a one-to-one correspondence between the distribution of electron energy states in the polymer molecule and the peaks in the kinetic energy distribution of the photoelectrons in a photoelectron spectrum (Fig. 2). The s.u. satellite(s), ΔE_1 on C_2 in Fig. 2, correspond to the $M_+^{**}(j')$ in Fig. 1. Details will be added, as needed, below.

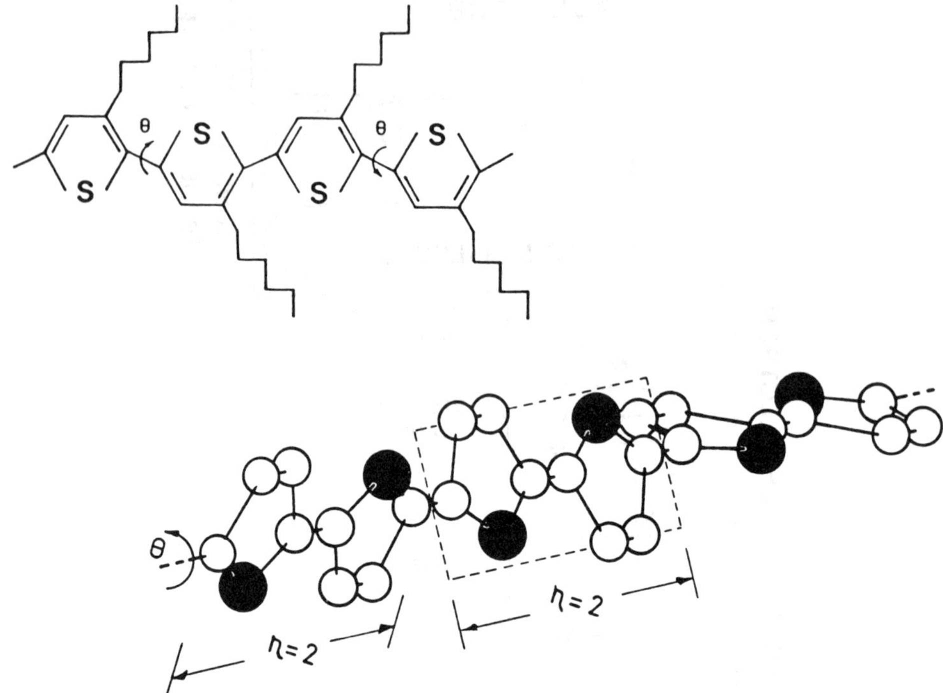

Figure 3. top: The chemical structure of poly(3-hexylthiophene); Bottom: The "dimer model" (η = 2; $\theta = \pi/4$) for structure at high-T {8,9}.

The experimental details pertaining to the various spectra reproduced here are to be found in the original papers {8,9,10,25}, and will not be discussed here. Note, however, that the XPS spectra for the polymers in the solid state were obtained in Linköping using non-monochromatic Mg(K_α) radiation {8,9}, while the XPS spectra of molecules in the gas phase were obtained with a special high-resolution instrument in Uppsala, employing monochromatic Al(K_α) radiation {25}.

4. Electronic Structure of Poly(3-hexylthiophene)

4.1 VALENCE ELECTRONIC STRUCTURE

The structural formula (geometry) of P3HT is shown at the top of Fig. 3. The aliphatic hexyl groups are attached at the "3" position of the thiophene ring. Since the hexyl groups contain no π-electrons, the presence of these alkyl groups does not directly affect the π-system of the thiophene rings. The presence of the alkyl groups, however, leads to the effects to be described below, which are based upon changes in the π-electron system as a function of temperature.

At low temperatures, the molecular geometry is planar {8,9,21}, which is the case for unsubstituted polythiophene, or PT, at all temperatures {24}.

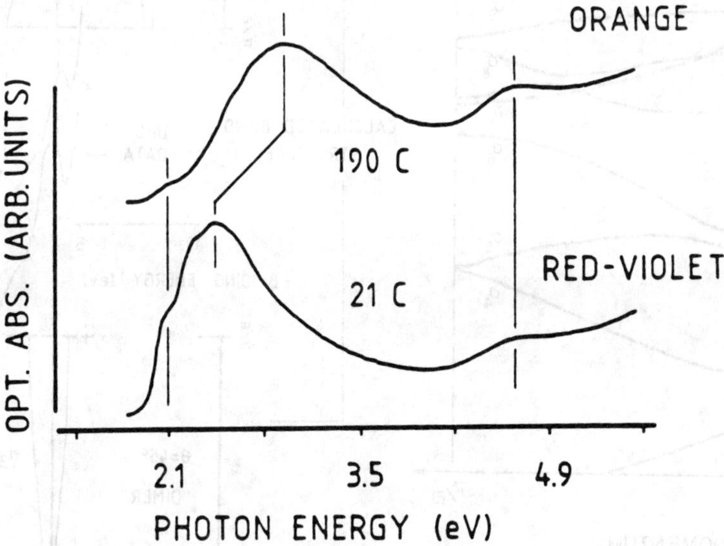

Figure 4. Optical absorption spectra of P3HT at low- and high-T {10}.

The band structure of planar PT, shown in Fig. 5-left {6}, is identical to that of P3HT for low temperatures {21}. To the right in Fig. 5is shown the lowest binding energy portion of the π-band region of the same P3HT band structure, turned upon the side in order to enable a better comparison with the UPS spectra, along the low binding energy region of the UPS (density-of-states) spectra for P3HT taken at two different temperatures {8}.

108

The UPS data for T = -60 C clearly shows structure (peaks) corresponding to the π_3-band edge and the π_2-π_3 flat-band region, in agreement with the calculated band structure for the isolated, planar polymer chain. At high-T (190 C in this case), the UPS spectrum indicates a collapse of the π_3-band into essentially one stronger peak (intensity is conserved), which shows up as an increase in the density-of-states as well as a slight shift in energy of the peak in the vicinity of the π_2-π_3 flat-band region. This collapse of the π_3-band edge, along with the simultaneous shift observed in the optical absorption spectrum (equivalent to a change in the electron energy gap, E_g) are accounted for **simultaneously** by a *dimer model* of the geometrical structure of P3HT (Fig. 3) at the elevated temperatures, as calculated using the VEH method {21}.

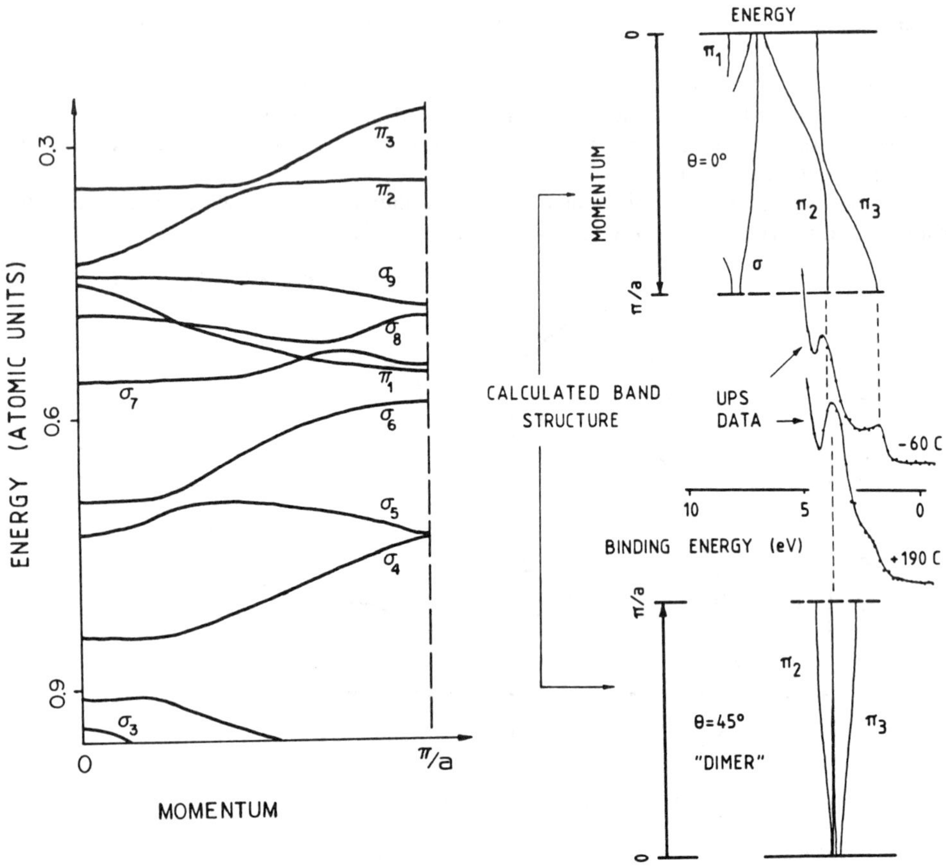

Figure 5. Left: The one-dimensional band structure of polythiophene, as calculated using the VEH method {6}. Right: The Low binding energy UPS density-of-states of P3HT is shown at low- and high-T {8,9}, and compared with the VEH band structure of the low energy π-bands for both planar (top) and dimer-model (bottom) geometries {21}.

The *dimer model*, i.e. two coplanar thiophene rings twisted together through a torsion angle of $\theta = \pi/4$ relative to the nearest neighbors (Fig. 3, Bottom), is only one of several models of "soft conformation defects" which are theoretically equivalent in accounting for the thermochromic behavior of P3HT; monomers (with $\theta \approx \pi/5$) and trimers (with $\theta \approx \pi/3$) are equally valid within the context of the VEH calculations {21}. The details concerning these calculations have been discussed previously {8,9,21}, and are not repeated here. These discrete soft conformational (geometrical) defect models simply are (idealised) special cases, which represent the basis for actual polymer geometry at high-T. In a mathematical sense, the true polymer geometry at high-T may be described by a "linear combination" of the idealised soft conformational defects.

The significance of the rotations associated with each defect, however, is that each rotation represents a reduced degree of conjugation between neighboring thiophene rings, i.e., a decrease in overlap of the atomic p_z orbitals of neighboring carbon atoms (on either side of the torsion angles θ indicated in Fig. 3). For $\theta = 0$, there is maximum atomic overlap. For $\theta = \pi/2$, however, the overlap of the neighboring p_z atomic orbitals is zero, and the π-conjugation is completely broken at that point. Values of $0 < \theta < \pi/2$ result in intermediate values of the p_z overlap and a corresponding weakening of the π-conjugation. As θ approaches zero, the π-electrons become completely localized on the η coplanar rings located inbetween the twists. Intermediate values of θ, as illustrated for the dimer model ($\eta = 2$; $\theta = \pi/2$) in Fig. 3, correspond to intermediate degrees of electronic localization of the π-electrons. Thus, these soft conformation defects are geometrical models, which, within the context of VEH level calculations, reproduce the observed effects on the π-electronic structure, as seen in the T-dependent optical absorption and UPS results.

4.2 XPS CORE LEVEL SPECTRA OF P3HT

Some XPS C1s spectra for P3HT at both low-T (- 80 C) and high-T (190 C) {8} are shown in the lower half of Fig. 6, along with the s.u. portion of spectra of several small oligomer molecules, as recorded in the gas phase {25}. Also included in the figure is a high-T spectrum (250 C) of poly(3-butylthiophene), or P3BT, which is included to illustrate that the spectra are not unique to only hexyl-substituted polythiophene {9}. Discussions about the thermochromism in other alkyl-substituted polythiophenes can be found in the cited literature {26}.

Starting at the bottom of the figure, in order to explain the weak "s.u.-like" structure at about 3 eV below the main C1s line, it must be borne in mind even in a limited aromatic system, such as benzene, the inelastic energy-loss peaks in the photoelectron spectrum which correspond to the optical π-π* absorption energy (defined by the M_O^* in Fig. 1) fall at the same energy as the strongest s.u. transition {27}. In the case of a fully conjugated polymer chain, the core ionization causes only a very small perturbation to the delocalised π*-band. Therefore, in the case of the solid polyalkylthiophenes at low temperatures, or in general polymers which are fully conjugated at any temperature, it is to be expected that a structure will be found in the XPS core-level spectra with an energy, relative to the main line, equal to the optical band gap, i.e. to the optical absorption band edge energy. This structure is due to both s.u. transitions and inelastic scattering processes. In the case of the solid polymer, however, these two cases can not be easily distinguished.

In general, the core-hole generated in the core-level photoionization event is always completely electronically screened by the remainder of the electrons in the system {14}. If the electronic system is infinite, then the complete screening is accomplished without a major perturbation of the remaining electronic system, and the strengths of s.u. transitions become weak. If, on the other hand, the electronic system is localized, in order that the complete screening of the core hole can occur, interband transitions (s.u. transitions corresponding to the set of $M_+^{**}(j')$ in Fig. 1, which correspond to the ΔE_i in Fig.2) become necessary {14}.

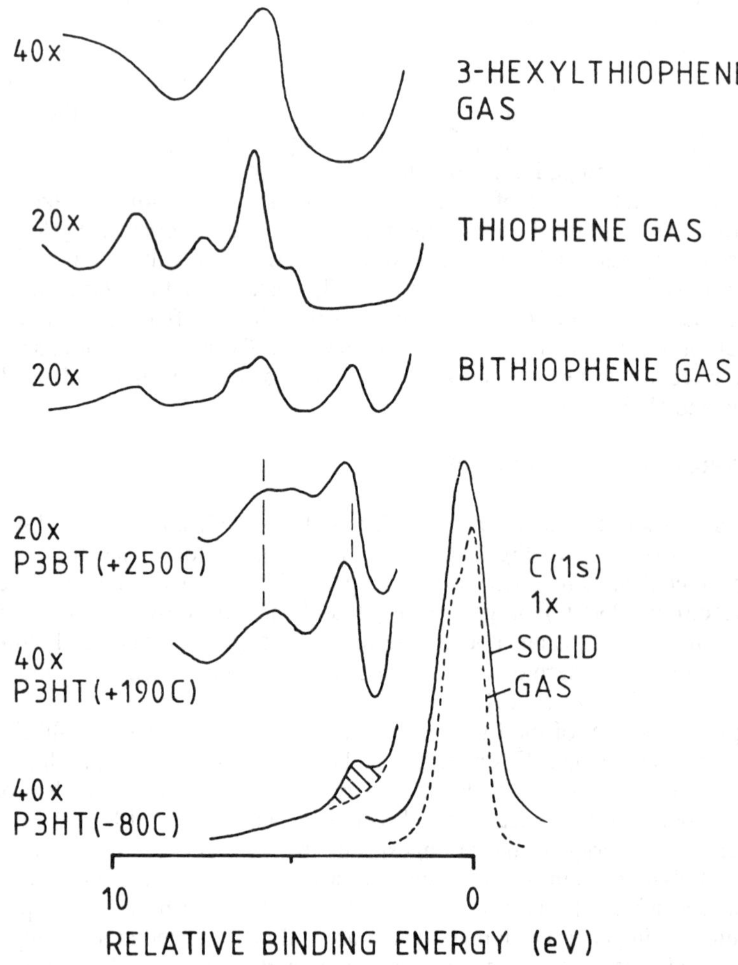

Figure 6: The s.u. portions of a series of XPS C1s spectra for two poly(alkylthiophene)s and several small oligomers of polythiophene are shown {25}.

These interband transitions, in the ionic state, yield the observable s.u. satellite structure seen in Fig. 6.

In the upper half of Fig. 6 is shown the s.u. portions of the C1s spectra for three small oligomers of P3HT, namely, from the top downwards, 3-hexylthiophene, thiophene, and bithiophene, obtained on the molecules in the gas phase {25}. Note the following, strictly experimental, observations: (i) the s.u. spectra for the two monomers have no peak at a binding energy of 31/2 eV (relative to the main C1s peak) as seen in the polymer spectra; (ii) the spectrum for the hexyl-substuted monomer is very similar to that of the unsubstituted monomer (since the hexyl groups contain no π-electrons, only an indirect effect, not discussed here, occurs); and (iii) the structure in the s.u. spectrum for the dimer in the gas phase agrees very well with that of the polymer spectra in the energy region of about 31/2 to 8 eV relative binding energy. Note also that the solid state spectra would exhibit an intense and broad "tail" for energies greater than about 8 eV (not shown), due to inelastic scattering of some of the primary photogenerated electrons as they leave the surface region of the solid sample. Because of the enormous difference in molecular density between a solid sample and a sample consisting of molecules in the gas phase, this inelastic "tail" is essentially absent in the gas phase spectra, especially when data are obtained, as in the present case, at sufficiently low gas pressures {25}.

4.3 XPS SPECTRA OF OLIGOMERS OF P3HT

The structure in the s.u. region of the C1s spectra of the small molecules in the gas phase has been analysed in detail using appropriate INDO/CI quantum chemical calculations {25}. Here, we state only that the peak in the dimer spectrum (near 31/2 eV), which is noticably absent in the two monomer spectra, is due to electronic transitions between adjacent thiophene rings (charge transfer transitions within the $M_+^{**}(j')$ of Fig. 1) within the coplanar dimer molecule. The strength of the s.u. features (relative to that of the main C1s line), as well as the distribution over relative binding energy, in the gas phase dimer spectrum are consistent with the fact that the π-electronic system in the dimer is localized (by definition) to the dimer molecule. The similarities between the s.u. features in the C1s spectrum of the dimer in the gas phase and those in the C1s spectrum for the polymers at high-T in the solid state, are evidence that the π-electronic systems of the alkyl-substituted polythiophenes at high-T are *electronically localized* to dimer-like entities. This *electronic* localization is consistent with, but does not prove the existance of, the *geometrical* soft conformational defects that come out of the VEH modelling of the temperature-dependence of the UPS and the optical absorption spectra of the polymers.

In other words, in the soft-conformational defect geometrical model, the polymer system evolves, with increasing temperature, from longer conjugated chain segments to a system represented equally well by conjugated monomers, dimers or trimers at high-T {8,9,21}. On the other hand, the XPS s.u. spectra indicate the growth of predominently dimer-like electronic localization as a function of increasing temperature, without evolution through intermediate conjugation segment lengths.

The connection between these two apparent manifestations of the temperature-dependent *geometric* and *electronic* structure of the alkyl-substituted polythiophenes is the subject of present and future studies of the structure of these and other conjugated polymers.

5. Summary

The alkyl-substutited polythiophenes are processable conjugated polymers, whose processability allows the preparation of high quality thin films, which make possible the studies of important phenomena within the area of conducting polymer research. This contribution contains an overview of the modelling of the thermochromic effect in one particular polyalkylthiophene, namely poly(3-hexylthiophene). The electronic structure has been studied at various temperatures using optical absorption spectroscopy and photoelectron absorption spectroscopy. Strong evidence exists indicating that electronic localization effects occur at elevated temperatures which lead to the color change observed. Furthermore, geometric modelling of the thermochromic effect leads to the suggestion that certain soft conformational defects account for the observed spectroscopic changes induced by temperature. New photoelectron spectra of small oligomers of the polythiophenes, studied in the gas phase, lend strong support to the concept of *electronic* localization at elevated temperatures as the origion of the thermochromism. However, the s.u. spectra do not necessarily confirm (nor dispute) the *geometrical* models, which are derived from the simultaneous temperature-dependent optical absorption spectra, the UPS spectra, and the results of VEH calculations

6. Acknowledgements

This research on polymers, conjugated polymers and organic molecules in general, has been supported by grants from the Swedish Board for Technical Development (STU), and the Swedish Natural Sciences Research Council (NFR). One of us (NS) is grateful to the Ministry of Education, Science and Culture, of Japan, which made possible an extended visit to Linköping, through the Japan-Sweden International Scientific Research Program (63044147). ANB and NC thank the CNPq and CAPES, respectively, in Brazil for financial support.

7. References

1. K. Kaneto, K. Yoshino and Y. Inuishi, *Jap. J. Appl. Phys.* **21**, L657 (1982).
2. K. Y. Jen, R. Oboodi and R. L. Elsenbaumer, *Polym. Materials: Sci. Eng.* **53**, 79 (1985).
3. M. Sato, S. Tanaka and K. Kaeriyama, *J. Chem. Soc., Chem. Commun.* **295**, 873 (1986).
4. K. Yoshino, S. Nakajima, M. Fujii, and R. Sugimoto, *Polym. Commun.* **28**, 309 (1987).
5. J. -E. Österholm, J. Laakso, P. Nyholm, H. Isotalo, H. Stubb, O. Inganäs, and W. R. Salaneck, *Synth. Met.* **28**, C435 (1989).
6. C. R. Wu, J. O. Nilsson, O. Inganäs, W. R. Salaneck, J. -E. Österholm, and J. -L. Brédas, *Synth. Met.* **21**, 197 (1987).
7. Y. Jugnet, G. Tourillon and Tran Min Duc, *Phys. Rev. Lett.* **56**, 1862 (1986).
8. W. R. Salaneck, O. Inganäs, B. Thémans, J. O. Nilsson, B. Sjögren, J. -E. Österholm, J. -L. Brédas, and S. Svensson, *J. Chem. Phys.* **89**, 4613 (1988).
9. W. R. Salaneck, O. Inganäs, J. O. Nilsson, J. -E. Österholm, B. Thémans, and J. -L. Brédas, *Synth. Met.* **28**, C451 (1989).

10. O. Inganäs, W. R. Salaneck, J. -E. Österholm, and J. Laakso, SYNTH. MET. **22**, 395 (1988).
11. M. Lögdlund, R. Lazzaroni, W. R. Salaneck, S. Stafström, J. O. Nilsson, X. Shuang, J. -E. Österholm, and J. -L. Brédas, *Proc. 3rd Intern. Winter School Elec. Prop. Polym.*, H. Kuzmany, M. Mehring and S. Roth, Ed's (Springer-Verlag, Berlin, 1989), in press.
12. S. Stafström and J. -L. Brédas, *Phys. Rev.* **B38**, 4180 (1987).
13. R. Lazzaroni, M. Lögdlund, S. Stafström, W. R. Salaneck, D.D.C. Bradley, R. H. Friend, N. Sato, E. Orti, and J. -L. Brédas, these proceedings.
14. G. Wendin, *Photoelectron Spectra*, STRUCTURE AND BONDING 45 (Springer-Verlag, Berlin, 1981).
15. H. -J. Freund and R. W. Bigelow, *Physica Scripta* **T17**, 50 (1987).
16. H. -J. Freund, E. W. Plummer, W. R. Salaneck, and R. W. Bigelow, *J. Chem. Phys.* **75**, 4275 (1981).
17. T. Koopmans, *Physica* **1**, 104 (1934).
18. W. R. Salaneck, *CRC Rev. Sol. State. Mat. Sci.* **12**, 267 (1986).
19. C. B. Duke, W. R. Salaneck, T. J. Fabish, J. J. Ritsko, H. R. Thomas, and A. Paton, *Phys. Rev.* **B18**, 5717 (1978).
20. J. Riga, J. J. Pireaux, R. Caudano, and J. J. Verbist, *Physica Scripta* **16**, 346 (1977).
21. B. Thémans, W. R. Salaneck and J. -L. Brédas, *Synth. Met.* **28**, C359 (1989).
22. D. T. Clark, in *Photon, Electron and Ion Probes of Polymer Structure and Properties*, D. W. Dwight, T. J. Fabish and H. R. Thomas, Ed's (Am. Chem. Soc., Washington, DC, 1981) Chap. 17.
23. W. R. Salaneck, in *Handbook of Conducting Polymers*, T. Skotheim, Ed. (Marcel Dekker, New York, 1986) Vol. 2, Chap. 37.
24. Z. Mo,K. B. Lee, Y. B. Moon, M. Kobayashi, A. J. Heeger, and F. Wudl, *Macromolecules* **18**, 1972 (1985).
25. M. P. Keane, S. Svensson, A. N. de Brito, N. Correia, S. Lunell, B. Sjögren, O. Inganäs, and W. R. Salaneck, Submitted.
26. G. Gustavsson, O. Inganäs, and J. -O. Nilsson, *Synth. Met.* **28**, C427 (1989).
27. D. Nordfors, A. Nilsson, N. Mårtensson, S. Svensson, U. Gelius, and S. Lunell, *J. Chem. Phys.* **88**, 2630 (1988):

POLY-1,2-AZEPINES BY THE PHOTOPOLYMERIZATION OF PHENYL AZIDES: A NEW ROUTE TO PROCESSIBLE CONDUCTING POLYMERS

E.W. MEIJER*, S. NIJHUIS, F.C.B.M. VAN VROONHOVEN, AND E.E. HAVINGA
Philips Research Laboratories
P.O.Box 80.000, 5600 JA Eindhoven
The Netherlands.

ABSTRACT. Processibility of conjugated and conducting polymers is one of the major topics in the chemistry and applications of these polymers. In this paper we describe the gasphase photopolymerization of phenyl azides, that yield thin films of poly-1,2-azepines on a variety of substrates. The photochemistry occurs at the interface of substrate and gas-phase and the polymer is grown at the surface of the substrate. High resolution patterns of poly-1,2-azepines are formed when a photomask is used in this gasphase photopolymerization. Upon oxidative doping with iodine or arsenic pentafluoride these polymers exhibit conductivities up to 0.01 S/cm. Mechanism, kinetics and versatility of the photopolymerization of phenyl azide and a series of substituted phenyl azides is discussed.

1. Introduction

The discovery that polymers like polyacetylene and polypyrrole show an appreciable electrical conductivity after doping [1,2] opened the alluring prospect of conducting elastomers. Materials with the ideal combination of high conductivity and the well-known superb ease in processibility of polymers could lead to unprecedented applications. Although a high conductivity in polymers can be obtained, the materials do not exhibit the usual polymer properties wished for, they are infusable and insoluble instead.

Polymer chemists started to encounter these intrinsic properties of the stiff-chain conjugated polymers by chemical modifications. Elegant synthetic routes have been developed, in which non-conjugated precursor polymers are chosen as intermediates [3,4]. These precursors are soluble in organic solvents and are easy to process by either spin-coating or precipitation. Processing being completed, the conjugation is induced by a thermally activated chemical reaction and conductivity is obtained by subsequent doping. The soluble precursor route is successfully used in the synthesis of polyacetylene [3] and poly-p-phenylene

J. L. Brédas and R. R. Chance (eds.), Conjugated Polymeric Materials:
Opportunities in Electronics, Optoelectronics, and Molecular Electronics, 115–131.

vinylidene [4]. Another breakthrough has been achieved by synthesizing conjugated polymers that are substituted with alkyl side chains [5]. These polymers are, due to the large conformational freedom of the side-chains, soluble in various organic solvents. As examples we mention substituted polythiophenes and poly-p-phenylene vinylidenes. Finally, water-soluble conductors have been introduced with the self-doped polymers [6,7]. Aqueous solutions of polypyrroles substituted with alkylsulfonate side-chains are even stable in a rather heavily doped state. The solubility in water may be attributed to the polyelectrolyte character of these polymers [7].

A completely different approach to arrive at processible conducting polymers is found in advanced deposition techniques of the polymers concurrently with their synthesis. In this approach the intractibility of the conducting polymers is circumvented by making use of new polymerization techniques, new technology, and even new polymers, all leading to useful sheets, films, pattern-wise films etc. Most of the work is carried out with the stable highly-conducting polypyrroles. Large sheets of polypyrrole, suitable for use in batteries, are made by a continuous electrochemical polymerization of pyrrole on a rotating anode [8]. In other studies an image-wise growth of polypyrrole on specific substrates is obtained. Silicon can be transformed into an electrode for pyrrole polymerization upon illumination [9]. In this case the photoconductivity of silicon is used and the illumination can be performed area selective. Recently, another image-wise polymerization is obtained by the photosensitized polymerization of pyrrole [10]. In this case a catalyst for oxidative pyrrole polymerization is generated at the surface of the substrate only at the illuminated areas.

A very neat approach is disclosed by Salaneck et al. in which polypyrrole doped with $FeCl_3$ is deposited by Chemical Vapour Deposition techniques. In a controlled way, polypyrrole is grown on a variety of substrates using a dry vacuum processing technique [11].

In this paper we present a new photodeposition technique of a new polyheterocycle, poly-1,2-azepine. This polymer is made from phenyl azides in a photopolymerization process. A preliminary communication on this subject has been published [12].

Despite numerous reports in the literature concerning both gas-phase polymerization and photopolymerization only a few examples are known in which a gaseous monomer is polymerized photochemically. These examples are limited by the lack of suitable monomers and are therefore limited to (meth)acrylates and some uncontrolled polymerizations of aromatic compounds as phenols, anilines etc.

1.1 Phenyl azide photochemistry

The photochemistry of phenyl azides (1) has been the subject of intensive research since its discovery by Wolff in 1912 [13]. Despite numerous reports on the formation of large amounts of intractable tars as main product in this photochemistry [13-20], only a few contradictory notes have been made concerning the structure of this tar and its possible mechanism of formation [15,16,19]. In most photoreactions of phenyl azides studied both singlet (**2**) and triplet (**3**) phenyl nitrenes act as intermediate [21-24] (Fig. 1).

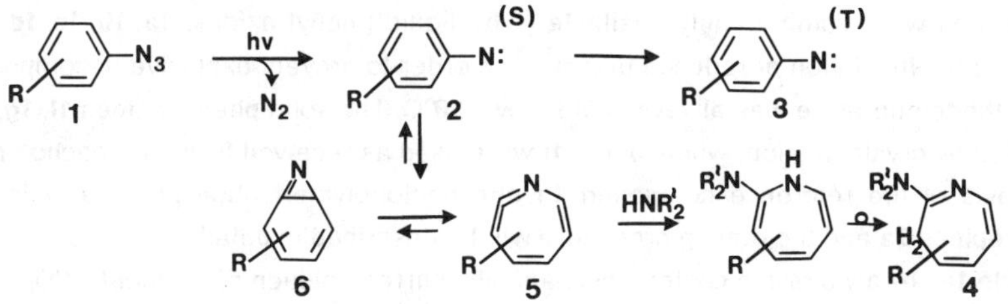

Figure 1. The photochemistry of phenyl azide

Upon irradiation of phenylazide **1** with UV-light nitrogen is eliminated and a singlet nitrene **2** is formed as the initial product. The next step involves either ringenlargement to azacycloheptatetraene **5**, possibly via bicyclic azirine **6** or, by intersystem crossing, a transition to triplet phenyl nitrene **3**. The latter can dimerize to azobenzene or abstract hydrogen atoms to form aniline. The most important chemistry of phenyl azide is performed via azacycloheptatetraene **5**. In the presence of nucleophiles, including primary and secondary amines, **5** is trapped furnishing 2-substituted-3H-azepines **4**. The latter is formed after a 1,3-H shift of the 2-substituted-1H-azepine. Comprehensive chemical and spectroscopic investigations have revealed that azacycloheptatetraene is the lowest energy isomer of singlet phenylnitrene. The estimated lifetime of this intermediate is established to be approximately 1 ms in inert media [21,25]. For further details of the photochemistry of phenyl azide the reader is referred to review articles [13].

In the present paper we report our experimental evidence for the formation of poly-1,2-azepines as the main polymeric product in the photochemistry of phenyl azides. The ease of oxidation of these polymers is investigated in view of the formation of conducting polymers.

2.Experimental

Apparatus: IR spectra were recorded on a Bruker (AFS 45) FT-IR spectrophotometer, using pressed pellets of KBr. NMR spectra were recorded at 80.13 MHz (^{1}H) and 20.13 MHz (^{13}C) on a Bruker (WP 80 SY) spectrometer or at 50.13 MHz (^{13}C) on a Nicolet (NT200). UV spectra were recorded with a Varian Superscan-3 spectrophotometer. GPC analysis is performed with an array of 10^3, 10^4, 10^5 μm Styragel columns with a 0.05 M LiBr in DMF solution as eluent. Low molecular weight amines and emeraldines are used as reference. Film thickness was measured with a Dektak 3030. Conductivities were measured using a four-probe technique. Nonlinearities (d_{33}) were measured using a Nd:YAG laser at 1064 nm. The Corona poling at circa 1 MW/cm was performed at roomtemperature in the standard way.

Materials: The (substituted) phenyl azides **1a - 1i** were synthesized according to known procedures [26]. The starting materials were either the corresponding anilines or phenyl hydrazines, and were commercially available. The liquid phenyl azides, **1a**, **1b**, **1c**, **1d**, and **1e** were purified by distillation at reduced pressure. In order to prevent explosive decomposition of the azides the temperature was always kept below $100°C$. The solid phenyl azides **1f**, **1g**, and **1h** were purified by crystallization, while **1i** and **1j** were used as received from the reaction mixture. The synthesis of the reference compound **1a**, the photopolymerization process which yields poly-1,2-azepines **7a** and the doping procedure will be described in detail.

Phenyl azide 1a. To a water-ice cooled, mechanically stirred solution of 0.1 mol $NaNO_2$ in 110 ml H_2O was added dropwise a solution of 0.1 mol aniline in 150 ml 2N HCl. The mixture was stirred for 2 h at 0-5 $°C$. A solution of 0.1 mol NaN_3 in 110 ml H_2O was added dropwise, while N_2 evolution occurred. The reaction mixture was warmed to room temperature and stirring was continued for an additional 18 h. The product was worked up by extraction with 600 ml ether. The ether extract was washed twice with 200 ml 0.1 N H_2SO_4, 200 ml 2 % $NaHCO_3$ and 200 ml H_2O, respectively. From the dried ($MgSO_4$) extract a crude mixture was obtained by evaporating the solvent in vacuo at room temperature. The residue was purified by distillation at 22 $°C$ with 3-4 mm Hg, which yielded a colourless liquid (yield 75 %). IR (neat) $2120 cm^{-1}$; ^{1}H- NMR (80 MHz, DMSO)δ 6.55-7.50 (m,5 H)

Poly-1,2-azepine 7a. Phenyl azide is polymerized in a specially designed reaction vessel. A glass cilinder is capped with a fused silica window using a screw-cap. Two inlets with vacuum- taps are introduced to control the atmosphere. The height of the cilinder varied between 10 and 60 mm. To polymerize phenyl azides with low vapour pressures, like **1h** and **1i**, a reaction vessel has been used in which the distance between bottom and fused silica window is minimized to 10 mm. Irradiation is performed with either a deep UV lamp (OAI) with an intensity of 8 mW/cm^2 at the fused silica window or with a high pressure mercury arc (SP-500) with an intensity of approximately 12 mW/cm^2 at the fused silica window. Using the SP-500 lamp the temperature of the reaction mixture raised to circa $40°C$. Upon irradiation of 200 mg of pure phenyl azide in the reaction vessel with an inner height of 10 mm and in an atmosphere of nitrogen for 5.5 hours with a light intensity of 12 W/cm^2 (SP-500) a polymer film of approximately 3-6 μm has been formed onto the fused silica substrate. When the residue of phenyl azide is dissolved in toluene and refluxed for 5 days, the same polymer is precipitated from the solution. The polymer is collected by filtration, washed with diethylether and dried in vacuo. The yield of poly-1,2-azepine is 15%. Aniline is isolated as minor byproduct from the filtrate. Poly-1,2-azepine:^{1}H-NMR (DMSO-d_6) δ6.2-8.2 ppm (br. m.); ^{13}C-NMR (DMSO-d_6) δ 129, 119, 115 ppm (all br. s.); IR: 3050, 1660, 1600, 1500, 1240, 750, 690 cm^{-1}; UV (thin film): 202, 245, 320 (w. sh.).

Doped poly-1,2-azepine 7a. Conductivity measurements have been applied to films grown on specially designed substrates. Four Pd-electrodes are sputtered on a fused silica plate, in such a way that the distance from electrode to electrode is 1 cm. Using a mask (diameter 12 mm) poly-1,2-azepine is grown only on a part of the substrate (diameter 25 mm) connecting and covering parts of all electrodes. Finally the polymer is doped by either iodine (30 min. at 80°C) or arsenic pentafluoride (20 h. at room temperature, 400 Torr). Four-probe DC electrical conductivities were measured at field strengths varying between 0.1 and 10 V/cm. In all cases no dependence of conductivity on field strength was detected. The electronic character of the electrical conductivity was ascertained by verifying that a long exposure to the DC field did not change the conductivity significantly.

3.Results and discussions

3.1 *Photopolymerization phenyl azide*

Irradiation of neat phenyl azide with a 500-Watt high pressure mercury arc at room temperature in an atmosphere of nitrogen, followed by heating at 110°C in toluene, leads to the precipitation of insoluble dark-brown poly-1,2-azepine in 15 % yield. The same polymer, with only a small amount of low molecular weight impurities (< 5 %), is obtained as a thin film on the fused silica window of the reaction vessel, when gaseous phenyl azide is irradiated. In this case the polymer is slightly soluble in polar solvents like DMSO and DMF. During the photopolymerization the temperature of the reaction vessel increases slightly up to 40°C, due to heat emitted from the lamp. When substituted phenyl azides (**1b-1i**) are used as starting materials, a similar deposition of polymeric material is observed. The difference in rate of polymerization is discussed in the paragraph on film formation. Structure elucidation of the polymers is performed on the materials obtained as thin films.

We propose the formation of a (substituted) poly-1,2-azepine **7** by the polymerization of azacycloheptatetraene **5**, which in turn is formed from phenyl azide via singlet phenyl nitrene (Fig. 2)

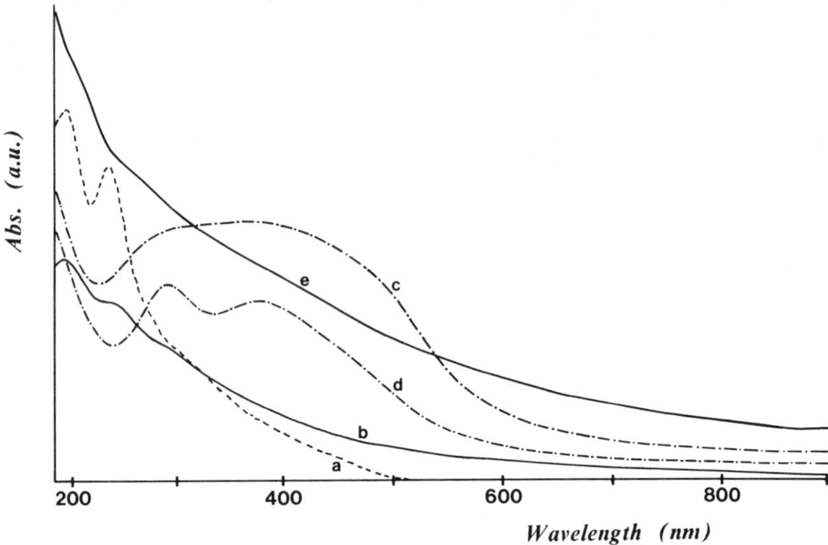

Figure 2. The photopolymerization of phenyl azide

The virgin poly-1,2-azepines, being antiaromatic poly(ketene)aminals, are easily oxidized when exposed to air. The UV-vis absorption spectrum of a thin film of poly-1,2-azepine **7a** as grown in nitrogen shows absorptions typically for azacycloheptatriene units (Fig. 3). Upon exposure of the polymer to air, the bands are broadened, while a long tail into the near infrared appears, characteristic of the presence of undefined defect levels in the band gap of the virgin polymer. Changes observed in the UV-vis spectrum when the film is exposed to air depend on film thickness, and is most pronounced in ultra-thin films. Apparently, the oxidation is diffusion controlled and limited to the boundaries.

Figure 3, The UV-VIS spectra of ultrathin (< 0.1 μm) films of **7a**. (a) the virgin polymer, (b) after exposure to air, (c) directly after exposure to I$_2$ at 80 °C for 30 min. (d) the sample of c after 30 min.; upon heating for 1 hr. at 80°C spectrum (b) returns. (e) after exposure to AsF$_5$ at room temperature at 400 torr for 20 hr.

These slightly oxidized poly-1,2-azepines are the main product in the photolysis of phenyl azide, as found in most studies published to date. We have performed most of our analyses on these, inadvertently, slightly oxidized polymers.

Table 1 Characteristic data of (substituted) poly-1,2-azepines.

Polymer	GPC[a]	UV[b]	IR[c]
7a	30.0	202, 245, 320(w.sh)	750, 690
7b	32.0[d]	200, 240(w), 340(sh)	845
7c	33.0[d]	n.a.	835
7d	30.5	202,280	780, 730
7e	30.2	n.a.	815
7f	30.5	n.a.	850(w)
7g	30.5	195, 240, 320(w.sh)	810, 740, 670(all w)
emeraldine[e]	30.5		

(a) GPC with DMF/0.05M LiBr, retention time of peak is given. (b) UV-Vis spectra of ultrathin films on fused silica. (c) IR out-of-plane vibrations in the region below 900 cm^{-1}. (d) These Rt's are overshadowed by hydrophobic interactions as proven with the corresponding azides and anilines. (e) Emeraldine and polyanilines have been studied in great detail[27].

The most significant data on the spectroscopic analyses are given in Table 1. Independent of the monomer chosen, a low average molecular weight for the polymers is found with GPC ($\overline{DP}$ about equal to emeraldine, i.e. 8-10) The GPC analysis was performed using 0.05 M LiBr / DMF as solvent. The use of LiBr is a prerequisite to eliminate aggregation of the polymers [28]. These aggregations yield bimodal GPC curves with a high apparent molecular weight. Low molecular weight amines and emeraldine have been used as standards. With respect to polystyrene standards the resulting $\overline{DP}$'s are higher by a factor of ten.

The out-of-plane C-H resonances in the IR fingerprint region indicate that no aromatic substitution of the phenyl azides occurs, thus excluding the formation of substituted polyanilines [27]. Additional evidence for ring expansion during polymerization is found from ^{1}H- and ^{13}C-NMR spectroscopy, although this analysis is hampered by the presence of paramagnetic species. The NMR spectra show very broad and low sensitivity resonances for the atoms of the azepine ring mainly by paramagnetic broadening. Powder ESR spectra of poly-1,2-azepine show a single isotropic Gaussian signal with a g factor of 2.0035 and a peak-to-peak line width of 6 G.

Two conclusions can be drawn when comparing polymerizations carried out in nitrogen and in air. Firstly, the rate of polymer formation is identical in both cases, indicating that *singlet* phenyl nitrene and not *triplet* phenyl nitrene is the intermediate, as already suggested by Platz [19]. Secondly, the polymerization performed in air is accompanied by simultaneous photooxidation of the polymer, furnishing carbonyl containing polymers (strong IR absorption at

1750 cm^{-1}), presumably by oxidation of the ketene-aminal group. A clear demonstration of this difference is given by the IR spectra of the polymers of 2,6-dimethoxyphenyl azide **7d** (Fig.4) , one formed in nitrogen and the other in air atmosphere.

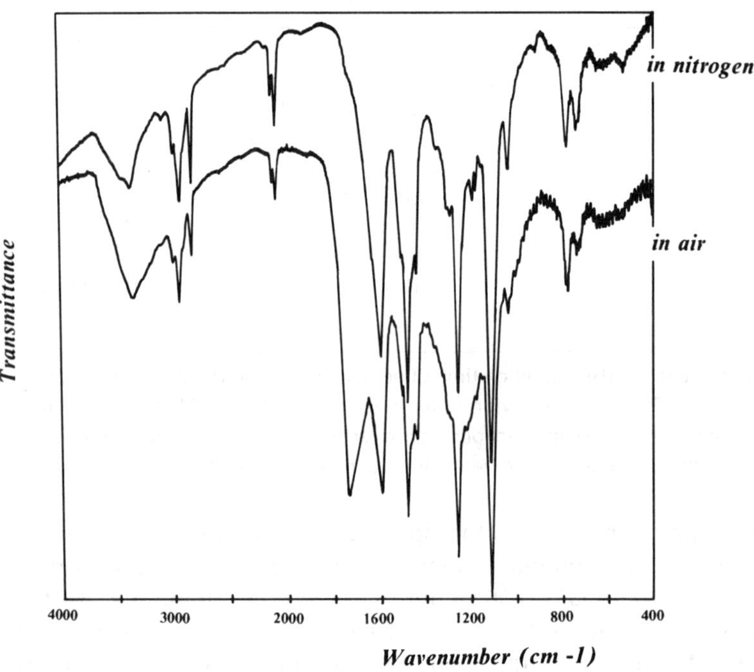

Figure 4. IR spectra of poly-1,2-azepine from 2,6-dimethoxyphenyl azide **1d** as formed in nitrogen and in air atmosphere.

For the electron-rich alkoxy substituted phenylazides **1c**, **1d**, **1e**, **1i** and especially **1f** the photooxidation accompanying the polymerization is most pronounced. The increased reactivity versus photooxidation by alkoxy substitution is well-known [29].

The proposed reaction mechanism is given in figure 5. The first product in the photochemistry of phenyl azide is singlet phenyl nitrene **2**, which is in equilibrium with azacycloheptatetraene **5**. On the otherhand there is always the formation of nucleophiles via triplet phenyl nitrene. The formation of aniline is well founded and we isolated aniline in the residue. We propose that this nucleophile adds to **5** furnishing e.g. 3-anilino-1H-azepine **8**. The initiation step of the polymerization is followed by the addition of a new, photochemically generated, azacycloheptatetraene in the propagation step. Termination is expected by e.g. 1,3-H shift of **9** furnishing a 3-H azepine as end-group, or by the addition of the N-H to other species like CO_2.

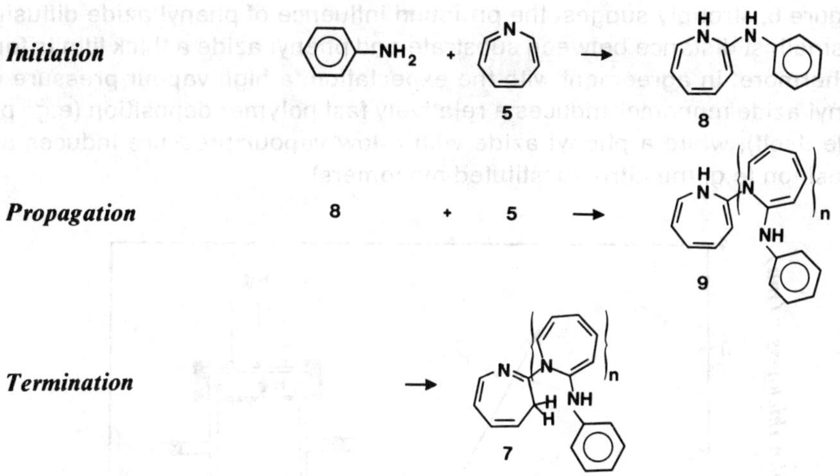

Initiation

Propagation

Termination

Figure 5. Proposed mechanism of polymerization of phenyl azide.

This polymerization mechanism is in agreement with the definition of a photopolymerization process. Although the term photopolymerization often has been used in stead of photoinitiated polymerization, it should be strictly reserved for polymerizations, in which every propagation step involves a new photochemical or -physical event.

3.2 Film formation

Thin films of poly-1,2-azepines can be grown on a variety of substrates, as long as they are transparent for UV-light. We have successfully used sheets of fused silica, polycarbonate, poly(methyl)methacrylate and PVDF. A strong adhesion of the polymer to the substrate is found. Probably, the polymer is covalently attached to the substrate.

The rate of film formation is a function of light intensity and phenyl azide concentration. Vapour pressure and diffusion coefficient of the (substituted) phenyl azides are determining monomer concentration near the substrate. Light intensity, absorption coefficient and quantum yield of azacycloheptatetraene formation are determining the concentration of reactive species and, hence, of polymer growth. Several experiments have been conducted to study the rate of film formation and to establish the rate-determining factor (i.e. diffusion of phenyl azide or decrease of light intensity due to absorption of growing polymer layer). With a constant light intensity of 12 mW/cm^2 and a constant irradiation time of 2 h., the film thickness has been measured as a function of distance between the fused silica substrate and the liquid phenyl azide. The data, as given in figure 6, strongly suggest the profound influence of phenyl azide diffusion. At the smallest distance between substrate and phenyl azide a thick film is formed. Furthermore, in agreement with the expectation, a high vapour pressure of the phenyl azide monomer induces a relatively fast polymer deposition (e.g. phenyl azide itself), while a phenyl azide with a low vapour pressure induces a slow deposition (e.g. the nitro-substituted monomers).

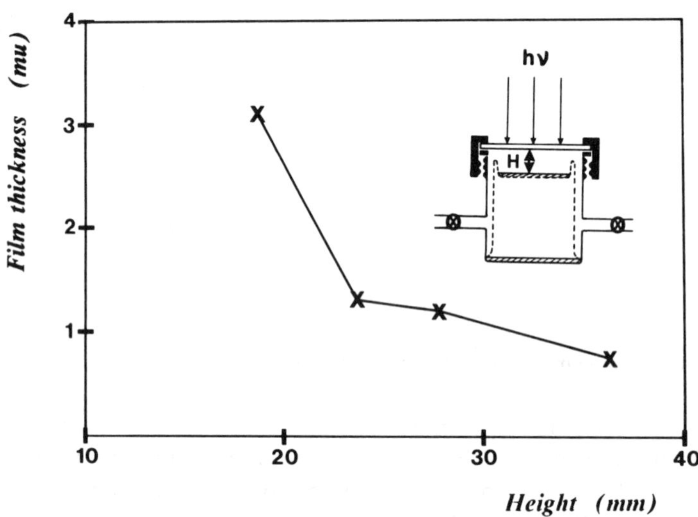

Figure 6. The thickness of polymer **7a** deposited as a function of distance between substrate and liquid **1a**. In all experiments the irradiation was performed for 2 h. at an intensity of 12 mW/cm².

Since the irradiation is performed through the growing polymer layer, the absorption of the active radiation by this film should limit its thickness. Typically, the thickness is between 1 and 5 μm. Upon extended irradiation with a 500-Watt high pressure mercury arc a maximum thickness of 10 μm has been reached. The decrease in rate of polymer growth during deposition is always observed, however, whether this decrease in rate is raised primarily by absorption or diffusion is not well understood.

When the irradiation is performed through a photomask in a reaction vessel as depicted in figure 7 , the polymer is grown in high-resolution patterns (figure 8).

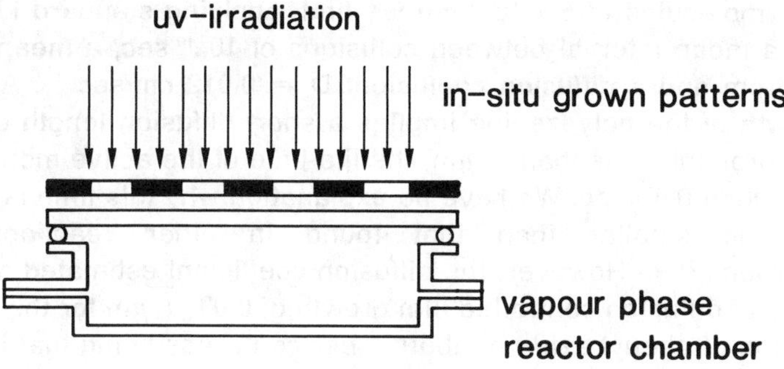

Figure 7. The reaction vessel for polymerization using a photomask.

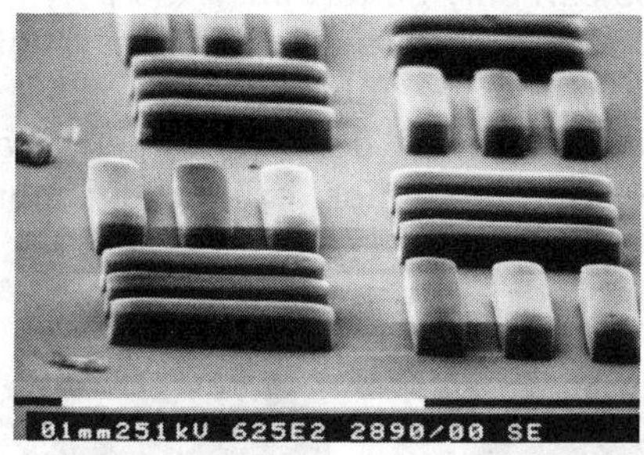

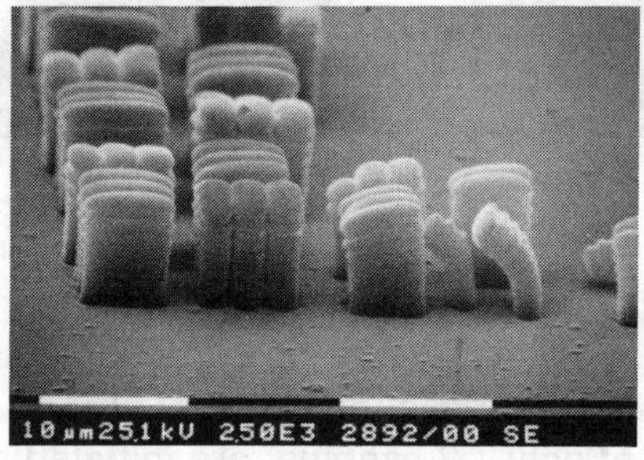

Figure 8. SEM photographs of poly-1,2-azepine **7a** as grown pattern-wise on fused silica with so-called contact illumination. The mask used exhibits equal lines and spacers

Several conclusions can be drawn from the formation of these high-resolution patterns. Firstly, the polymerization occurs unambiguously at the gasphase - solid phase interface. Secondly, since no diffusion of the reactive monomer azacycloheptatetraene to unirradiated areas is observed, we propose a very short life-time of the monomer. The mean free path of the active specimen will be determined by collisions with nitrogen atoms, as it is present as a minor component only. Assuming a reasonable cross section diameter for collision

126

with the nitrogen molecules of 5×10^{-8} cm we find, applying standard kinetic theory of gases, a mean interval between collisions of 10^{-10} sec, a mean free path of 2.7×10^{-6} cm and a diffusion coefficient $D = 0.012$ cm^{2}sec^{-1}. As the pattern-wise growth of the polyazepine implies a short diffusion length of the active monomer, probably less than 1 μm, the life-time of the active monomer should be shorter than 0.4 μsec. We have no explanation why this time is 3 orders of magnitude smaller than that found in other reactions of azacycloheptatetraene [21,25]. However, the diffusion coefficient estimated above leads to a velocity of diffusion controlled film growth of $0.03\sqrt{t}$ μm for the saturation concentration of phenyl azide of about 1 per cent. Assuming that in the experiment of figure 6 all phenyl azide molecules reaching the surface are indeed photochemically activated, we calculate a diffusion controlled film thickness of 2.5 μm, in full agreement with the experimental value. Finally, the shape of the structures grown indicates that the polymer acts as an optical waveguide. Namely, the steep structures can only be formed if the polymer guides the light to the end of the structure, which presupposes that light scattering and diffraction are negligible. Only then the light will remain concentrated at an end face of the structure that more or less duplicates the initial mask, and new azacycloheptatetraene **5** is generated almost exclusively near to that interface.

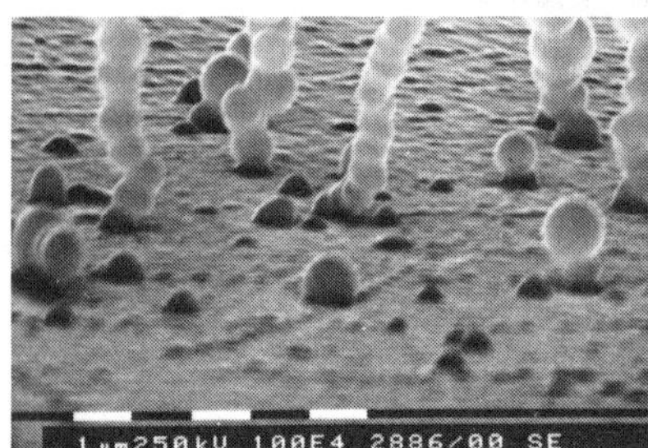
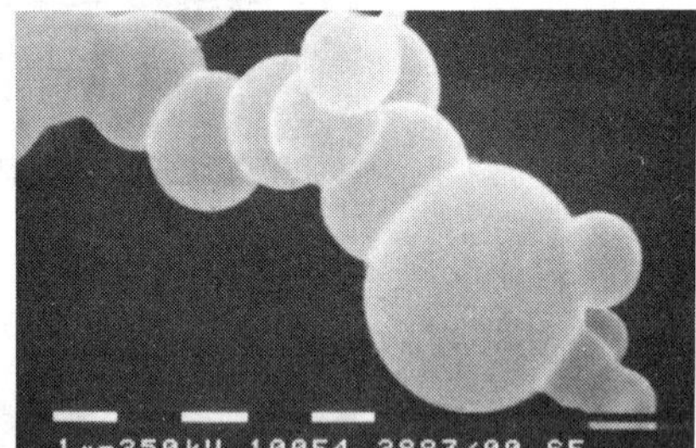

Figure 9. The uncontrolled formation of strings of polymer beads.

Occasionally we observed a very fast formation of strings of polymer beads. These polymer beads, also of poly-1,2-azepine, are initiated at the surface. The high growing velocity suggests that in this case the mass transport in the gas phase is no longer determined by diffusion as in the case of regular film growth, but that vortices are important. The conditions for their occurrence is as yet unknown. An example of the bead strings of poly-1,2-azepine is shown in figure 9.

The polymers films made from all phenyl azides studied are amorphous as measured by X-ray analysis.

3.3 *Doping the poly − 1,2 − azepines*

The ease of oxidation of **7** prompted us to investigate the possible formation of conducting polymers [1]. A large structural similarity exists between poly-1,2-azepine and the polyaniline free base [27]. Both polymers consist of a $\{C_6H_5N\}$ repeating unit that exhibits short delocalized structures connected by nitrogen. In view of this similarity, we have studied in detail the possible doping by oxidation and protonation, as is used successfully on polyaniline [27]. However, the conductivity of the virgin and of an acid- treated polymer **7** never exceeded 10^{-10} S/cm.

Doping the polymer films with either I_2 or AsF_5 was more successful and specific conductivities up to 10^{-2} S/cm are found. The I_2 doping is reversible due to evaporation of the dopant, as can be concluded from both UV-vis spectroscopy (fig. 3) and conductivity measurements. In an open cell the conductivity drops to values below 10^{-10} S/cm in 30 minutes. When the conductivity is measured in a closed cell directly after doping, we observe a more gradual increase in resistivity. Due to the closed cell construction, the loss of iodine is limited and the conductivity decreased only from 2.10^{-3} S/cm to 2.10^{-6} S/cm. In a vacuum oven at 80°C practically all I_2 can be evaporated again. In the case of AsF_5, the conductivities are more stable with respect to the evaporation of the dopant. In a nitrogen atmosphere conductivities of 10^{-3} S/cm are observed directly after doping and no significant decrease is found after storage for several days. However, in air the conductivity drops pretty fast to values below 10^{-10} S/cm.

UV-Vis spectra of thin doped samples (fig.3) show that the absorption due to the charge carriers is confined to rather low wavelengths in comparison with the absorptions in the near IR of the polarons in standard conjugated polymers. This reflects the only limited conjugation in the doped poly-1,2-azepines (see below).

Most of the substituted poly-1,2-azepines can be doped with I_2 and the conductivities measured are all between 10^{-3} and 10^{-5} S/cm. The small differences in conductivity observed for the different substituted polymers are mainly attributed to sample morphologies and not to substituent effects. UV-vis spectroscopy, however, showed that polymer **7f** is clearly a much better Charge-Transfer donor for I_2 than **7g**, but this is not reflected in a higher conductivity. These results show again that conductivity can be obtained from precursors lacking delocalization, as shown before[30]. Moreover, semiconducting properties can be observed in doped polymers with a high concentration of charge carriers. Those carriers can hop from one site to another. In these cases no transfer along an extended conjugated chain is required. The latter is, however, a prerequisite for high conductivities in the range of 10^3 to 10^5 S/cm.

The formation of conducting poly-1,2-azepines suggests that the charged species formed are stabilized by conjugation or even aromaticity. The species to be expected are radical cations and dications of azepines. Delocalization of

128

these species and 6-π electron aromaticity of the dication is evident both from theoretical studies and from experiments[31] . However, extension of the conjugation beyond a pair of rings is excluded owing to steric hindrance between the individual azepine rings (ortho-substituted aromatics). This explains both the relatively high energy of the polaron absorption band and the moderate conductivity. Semi-emperical SCF-LCAO calculations have been conducted on model compounds for the poly-1,2-azepines. The calculations on both 1H-azepine and 1'H-1,2'-diazepine (a dimer) in the neutral, monocation, and dication state showed that for 1H-azepine the dication is aromatic, while the diazepine-dication is a delocalized planar structure with the charge divided on both rings [31].

In order to arrive at planar polymers based on 1,4-substitution, we are currently studying the photopolymerization of bisazide 1j (figure 10). A hypothetical structure of the polymer network expected is given.

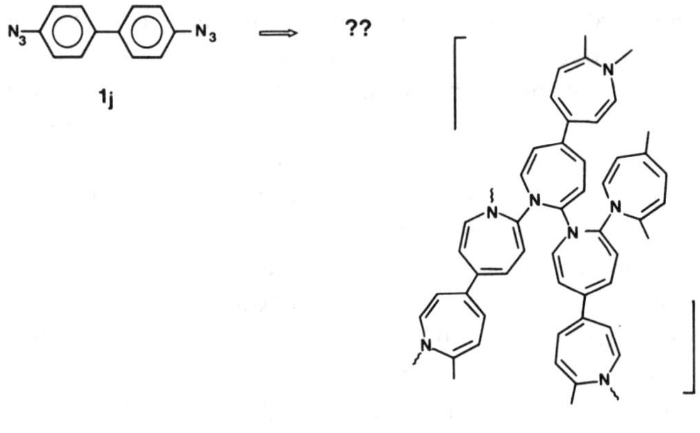

Figure 10. A hypothetical structure of polymer network formed from biphenyl bisazide **1j**.

3.4 Nonlinear optical properties

The possibility of introducing both electron withdrawing and electron donating substitutents to the phenyl azides and therefore to the poly-1,2-azepines prompted us to study second order nonlinearities $\chi^{(2)}$ of these polymers. Large $\chi^{(2)}$ nonlinearities are generally observed when amorphous polymers with charge-transfer units are poled using a strong DC electric field. The polar polymers obtained are investigated with respect to their possible applications in opto-electronics and frequency doubling.

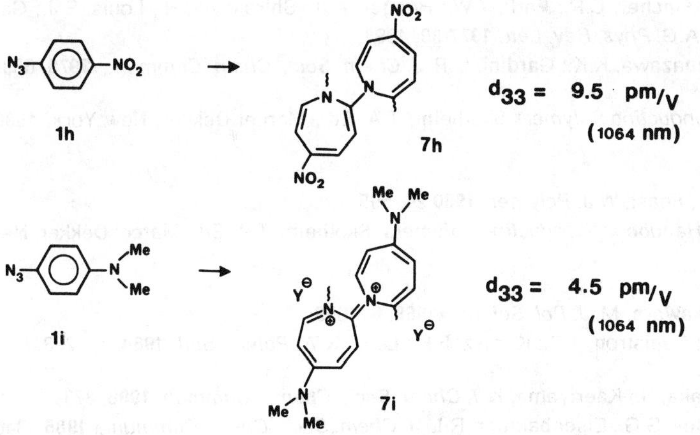

Figure 11. Nonlinear optical polymers with their d_{33} values at 1064 nm.

Poly-5-nitro-1,2-azepine **7h** should be regarded as a charge-transfer polymer in the neutral state. Air oxidation of **7h** is expected to be small, due to the electron withdrawing properties of the nitro group. Using Corona poling an, unstable, d_{33} of 9.5 pm/V is observed at 1064 nm. Poly-5-dimethylamino-1,2-azepine, on the other hand, should be regarded as a charge-transfer polymer in its oxidized state. The latter is expected to be formed reasonably easy due to the electron donating properties of the dimethylamino group. For **7i** we observed similarly an, unstable, d_{33} of 4.5 pm/V at 1064 nm. Both nonlinearities observed are rather small and unstable due to molecular relaxation. Moreover, residual absorption in the visible range of the spectrum (note the long tailing of the absorption bands in fig. 3) makes these polymers of minor importance for nonlinear optical applications. These results show, however, that both neutral and oxidized state of the poly-1,2-azepine are present.

4. Acknowledgement
The authors would like to thank G.L.J.A. Rikken for measuring the optical non-linearities.

5. References

1. Shirakawa, H.; Louis, E.J.; MacDiarmid, A.G.; Chiang, C.K.; Heeger, A.J. *J. Chem. Soc., Chem. Commun.* **1977**, 578;

130

Chiang, C.K.; Fincher, C.R.; Park, Y.W.; Heeger, A.J.; Shirakawa, H.; Louis, E.J.; Gau, S.C.;
MacDiarmid, A.G. *Phys. Rev. Lett.* **1977** 39, 1098;
Diaz, A.F.; Kanazawa, K.K.; Gardini, G.P. *J. Chem. Soc., Chem. Commun.*, **1979**, 653.

2. *Handbook Conducting Polymers* Skotheim, T.A. Ed., Marcel Dekker, New York, 1986, Vol 1
    and 2.

3. Edwards, J.H.; Feast, W.J. *Polymer,* **1980** 21, 595;
    Feast W.J. in *Handbook Conducting Polymers* Skotheim, T.A. Ed., Marcel Dekker, New York,
    1986, p1.

4. Kanbe, M.; Okawara, M. *J. Pol. Sci. A1,* **1968**, 6, 1058;
    Gagnon, D.R.; Capistron, J.O.; Karasz, F.E.; Lenz, R.W. *Polym. Bull.* **1984**, 12, 293.

5. Sato, M.; Tanaka, S.; Kaeriyama, K *J. Chem. Soc., Chem., Commun.* **1986**, 873;
    Jen, K.Y.; Miller, G.G.; Elsenbaumer, R.L. *J. Chem. Soc., Chem. Commun.,* **1986** 1346;
    Elsenbaumer, R.L.; Jen, K.Y.; Oboodi, R. *Synth. Met.,* **1986**, 15, 169;
    Havinga, E.E.; van Horssen, L.W. *Makromol. Chem., Macromol. Symp.,* **1989**, 24, 67.

6. Patil, A.O.; Ikenoue, Y.; Wudl, F.; Heeger, A.J. *J. Am. Chem. Soc.* **1987**, 109, 1858;
    Patil, A.O.; Ikenoue, Y.; Basescu, N.; Colaneri, N.; Chen, J.; Wudl, F. Heeger, A.J.,
    Synth. Met., **1987** 20, 151;
    Ikenoue, Y.; Chiang, J.; Patil, A.O.; Wudl, F.; Heeger, A.J. *J. Am. Chem. Soc.,* **1988**, 110, 2983.

7. Havinga, E.E.; van Horssen, L.W.; ten Hoeve, W.; Wynberg, H.; Meijer E.W.
    Polymer Bull. **1987**, 18, 277;
    Havinga, E.E.; ten Hoeve, W.; Meijer, E.W.; Wynberg, H. *Chem. Materials* **1989**, in press.

8. Naegele, D.; Bittihn, R. *Solid State Ionics,* **1988**, 28-30, 983.

9. Okano, M.; Itoh, K.; Fujishima, A.; Honda, K, *Chem. Lett.,* **1986**, 469.

10. Kern, J.-M.; Sauvage, J.-P., *J. Chem. Soc., Chem. Commun,* **1989**, 657.
    Sigawa, H.; Shimidzu, T.; Honda, K. *J. Chem. Soc., Chem. Commun.* **1989**, 132.

11. Mohammadi, A.; Hasan, M.-A.; Liedberg, B.; Lundstrom, I.; Salaneck, W.R. *Synth. Met.* **1986**,
    14, 189.

12. Meijer, E.W.; Nijhuis, S.; van Vroonhoven, F.C.B.M., *J. Am. Chem. Soc.,* **1988**, 110, 7209.

13. *Azides and Nitrenes, Reactivity and Utility* ; Scriven, E.F.V.,Ed.; Academic Press: New York,
    1984.

14. Horner, L.; Christmann, A.; Gross, A. *Chem.Ber.* **1963** , 96 , 399.

15. Reiser, A.; Leyshon, L.J. *J.Am.Chem.Soc.* **1971** , 93 , 4051.

16. Abramovitch, R.A.; Challand, S.R., Scriven, E.F.V. *J.Am.Chem.Soc.* **1972** , 94 , 1374.

17. Waddell, W.H.; Go, C.L. *J.Am.Chem.Soc.* **1982** , 104 , 5804.

18. Leyva, E.; Platz, M.S.; Persy, G.; Wirz J. *J.Am.Chem.Soc.* **1986** , 108 , 3783.

19. Leyva, E.; Young, M.J.T.; Platz, M.S. *J.Am.Chem.Soc.* **1986** , 108 , 8307.

20. Schrock, A.K.; Schuster, G.B. *J.Am.Chem.Soc.* **1984** , 106 , 5228.

21. Chapman, O.L.; Le Roux, J.P. *J.Am.Chem.Soc.* **1978** , 100 , 282.

22. Chapman, O.L.; Sheridan, R.S.; Le Roux J.P. *J.Am.Chem.Soc.* **1978** , 100 , 6245.

23. Chapman, O.L.; Sheridan, R.S. *J.Am.Chem.Soc.* **1979** , 101 , 3690.

24. Chapman, O.L.; Sheridan, R.S.; Le Roux, J.P. *Recl.Trav.Chim.Pays − Bas* **1979** , 98 , 334.

25. Shields, C.J.; Chrisope, D.R.; Schuster, G.B.; Dixon, A.J.; Poliakoff, M.; Turner, J.J. *J.Am.Chem.Soc.* **1987** , 109 , 4723.

26. Lindsay, R.O.; Allen, C.F.H. *Org. Synth.* **1955**, 3, 710;
Smith, P.A.S.; Boyer, J.H. *Org. Synth.* **1963**, 4, 75.

27. See for polyanilines: Green, A.G.; Woodhead, A.E. *J.Chem.Soc.* **1910** , 101 , 1117 and Macdiarmid, A.G.; Chiang, J.-C.; Huang, W.; Humphrey, B.D.; Somasiri, N.L.D. *Mol.Cryst.Liq.Cryst.* **1985** , 125 , 309.

28. Azuma, C.; Dias, M.; Mano, E.B. *Macromol. Chem., Macromol. Symp.* **1986**, 2, 169.

29. Schaap, A.P.; Zaklika, K.A., in *Singlet Oxygen* , Wasserman, H.H.; Murray, R.W., Eds., Academic Press, New York, 1970, p. 287.

30. Swager, T.M.; Grubbs, R.H. *J. Am. Chem. Soc.* **1987**, 109, 894 and references therein.

31. Staring, E.G.J.; Meijer, E.W., manuscript in preparation.

POLYHETEROARYLMETHINES, SYNTHESES AND PHYSICAL PROPERTIES

R. BECKER, G. BLÖCHL, H. BRÄUNLING
WACKER-CHEMIE GMBH
POSTFACH 12 60
D-8263 BURGHAUSEN/OBB.
W-GERMANY

ABSTRACT. Selfcondensation of 2-pyrrolealdehyde with phosphoroxichloride gives a product with a polypyrrylmethine like structure. Structure and reactivity of the methinegroup are investigated by ^{15}N-CPMAS and ^{13}C-CPMAS technique.

Quantum chemical calculations on polyheteroarylmethines according to structure I predicted very interesting electrical and optical properties for these compounds[1,2]. Though several syntheses, leading to products containing substructures of I have been published[3-8] no synthetic way is known which produces undoped, low bandgap material I.

$$\left[-\left(\underset{X}{\left\langle\!\!\!\right\rangle}\right)_n \underset{R}{C}=\left(\underset{Y}{\left\langle\!\!\!\right\rangle}\right)_m \underset{R}{C}- \right]_x \qquad I$$

Condensation products of pyrrolealdehyde

The known condensation[4] of pyrrolealdehyde with $POCl_3$ as condensating agent according to equation 1 gives II which can be regarded as a 100% doped polypyrrylmethine.

133

J. L. Brédas and R. R. Chance (eds.), Conjugated Polymeric Materials:
Opportunities in Electronics, Optoelectronics, and Molecular Electronics, 133–139.
© 1990 *Kluwer Academic Publishers. Printed in the Netherlands.*

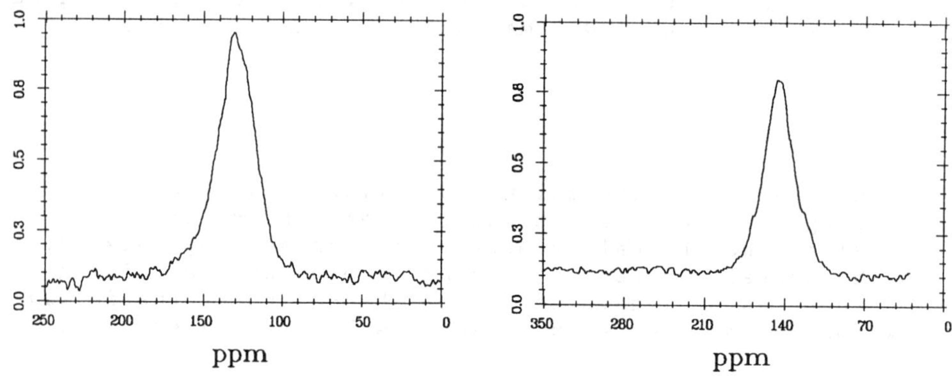

It contains one electron less per pyrrole ring than I and as
shown in eq 2 reduction should lead to the undoped material
III. The ^{13}C-CPMAS-spectrum of II (fig. 1) shows no signals
for sp^3 C-atoms and only a small signal for carbonyl C-atoms
between 150 - 180 ppm. ^{15}N-CPMAS-spectrum[9] (fig. 2) and the
elementary analysis are in accordance with the proposed
structure[9]. Since the methine -C- atom was expected to be
the most reactive site a 99 % methine- ^{13}C-labeled sample of
II was used for the reduction experiments.

Fig 1: ^{13}C-CPMAS-Spectrum
     of II

Fig 2: ^{15}N-CPMAS-Spectum
     of II

The ^{13}C-CPMAS-spectrum (fig. 3) of this labeled compound in
which the ring - C -atoms with natural ^{13}C- content are
quasi suppressed shows that the methine -C- atom exists
nearly quantitatively in the sp^2-form. No carbonyl-signal
between 150 - 180 ppm, is observed

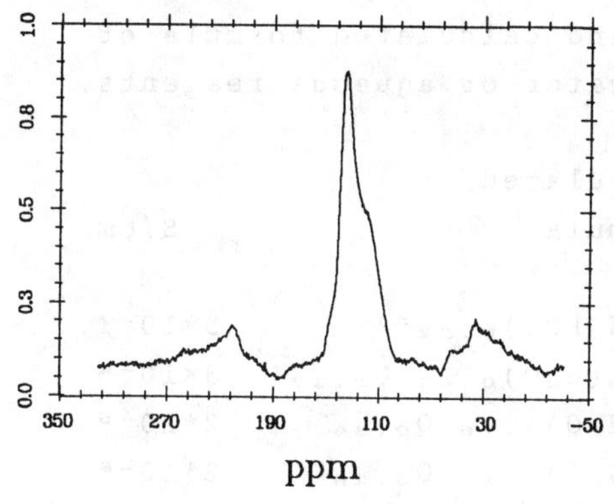

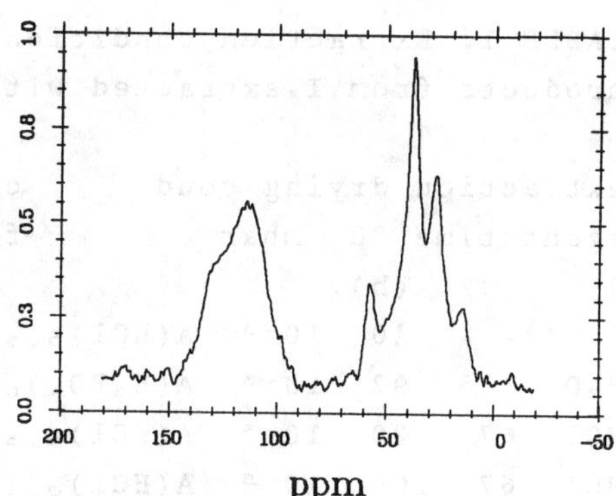

Fig 3: ^{13}C-CPMAS-Spectrum of II, 99% ^{13}C-labeled at the methine group

Fig 4: ^{13}C-CPMAS- Spectrum of ^{13}C labeled II after reduction with NaBH$_4$

Only small signals for sp^3-C - atoms between 20 - 50 ppm occur. A considerable amount of branching on tripyrrylmethane units can be excluded too.
The reduction with NaBH$_4$ however changes the ^{13}C-spectrum dramatically (fig. 4). This reaction did not lead to the desired II but attacked the methine-group, which is hydrogenated forming different sp^3-centers. The electro-chemical reduction was also unsuccessful. After a short time the electrical current decreased. Presumably an insulating layer was formed by electroreduction of the methine - group exhibiting a material with a similar structure as obtained by the reduction with NaBH$_4$.

Chemical behavior and electrical conductivity

Chloride and phosphate can be extracted from II with water or exchanged with aqueous solutions of acids and salts. While chloride is extracted with water within 1.5 hours nearly quantitatively the mobility of phosphate is lower. Treatment with aqueous ammonia eliminates both phosphate and chloride. Concentrated hydrochloric acid extracts phosphate under retention of the original chloride content. Table 1 shows the results of the extraction experiments.

TABLE 1. Extraction conditions and calculated formula of products from I extracted with water or aqueous reagents.

| extraction | | drying cond. | | calculated | |
agent	time (h)	°C	mbar	formula	S/cm
[a] $-$	$-$	20	10^{-3}	$A(HCl)_{0.54}(H_3PO_4)_{0.52}$[b]	$5*10^{-7}$
H_2O	1.5	97	10^{-3}	$A(H_3PO_4)_{0.38}(H_2O)_{0.45}\,O_{0.17}$	$3*10^{-4}$
HCl	67	20	10^{-3}	$A(HCl)_{0.53}(H_2O)_{0.76}\,O_{0.34}$	$2*10^{-6}$
HCl	67	100	10^{-5}	$A(HCl)_{0.10}(H_2O)_{0.16}\,O_{0.24}$	$2*10^{-8}$
NH_3	3	97	10^{-3}	$A(NH_3)_{0.2}(H_3PO_4)_{0.06}(H_2O)_{0.16}O_{0.6}$	$< 1*10^{-10}$

a) stored in air

b) $A = C_5H_3N$

In all cases when I is exposed to water the elementary analysis shows an uptake of water, which is difficult to eliminate and is partly retained even at 100 °C and

2×10^{-5} mbar.

The sites, where H_2O is bound to the molecule are unknown. Fig. 5 shows the ^{13}C-CPMAS-spectrum of the sample treated with concentrated aqueous HCl dried at 10-3 mbar. Though the elementary analysis shows that the product contains 0.76 molecules of water and 0.24 atoms of oxygen per ring, no signal for carbonyl or hydroxymethylgroups can be detected in the NMR-spectrum. Since heating to 100 °C at 2*10-5 mbar eliminates a great part of water and chlorine we assume that water is bound only weakly to the polymer. Evaporation, of hydrogenchloride under high vacuum at 100 °C is in accordance with the proposed structure of hydrochlorides of pyridine-like bases. Treatment of II with aqueous ammonia should give the base with the formula IV. The ^{15}NMR-spectrum however (fig. 6) shows only a content of unprotonated basic nitrogen below 30 %. On the other hand the ^{13}C-spectrum of the ammonia treated sample (fig. 7) shows an enhanced amount of carbonyl- or imino- C - atoms compared with the starting material. Since the signals between 150 and 180 ppm occured after the treatment with ammonia it is obvious, that II reacted with ammonia and the low amount of basic N found in the ^{15}N-NMR is not representative for the original compound II. The additional content of nitrogen, found in the elementary analyses is consistent with the addition of ammonia to the polymer

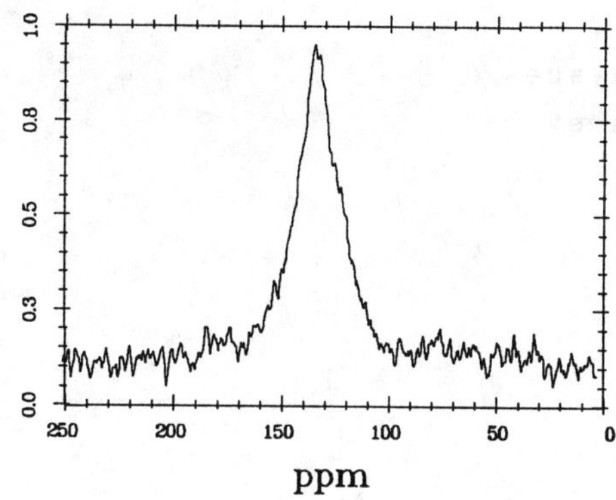

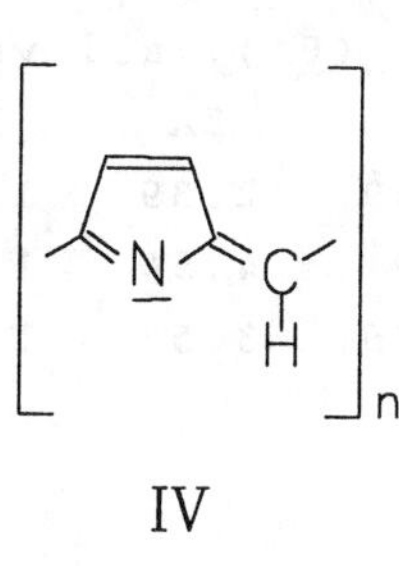

Fig. 5: ^{13}C_CPMAS-Spectrum of II after
extraction with conc. HCl

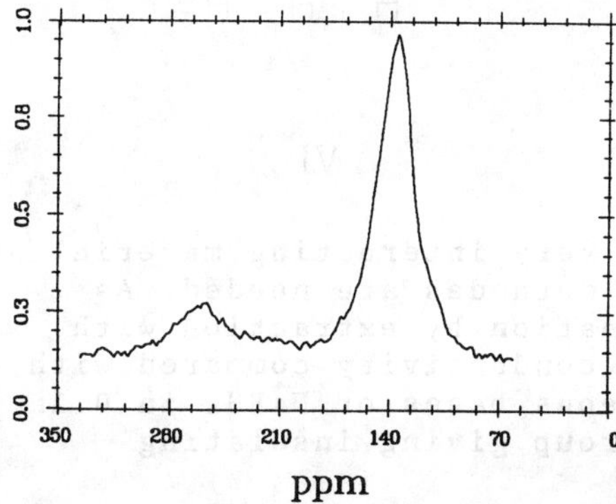

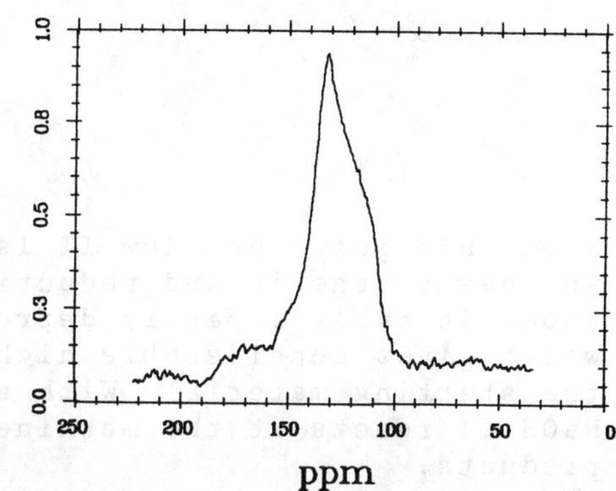

Fig. 6: ^{15}N CPMAS-Spectrum   Fig. 7: ^{13}C-CPMAS-Spectrum
of II after extraction with
aq. ammonia

The structures IIa, III and IV can be considered as one
system. Doping of III leads to structure units like II and
deprotonation of II would give structure units like IV. The
base IV should have a band gap comparable with polyacetylene
as calculated by Bredas and coworkers [10]. Table 2 shows the
calculated values.

Tab 2: Calculated values of III[1]
       and IV[10] compared with polyace-
       tylene (PA), all values in eV

Comp.	IP	EA	Eg
III	3.49	2.39	1.10
IV	6.0	4.54	1.47
PA	4.76	3.52	1.44

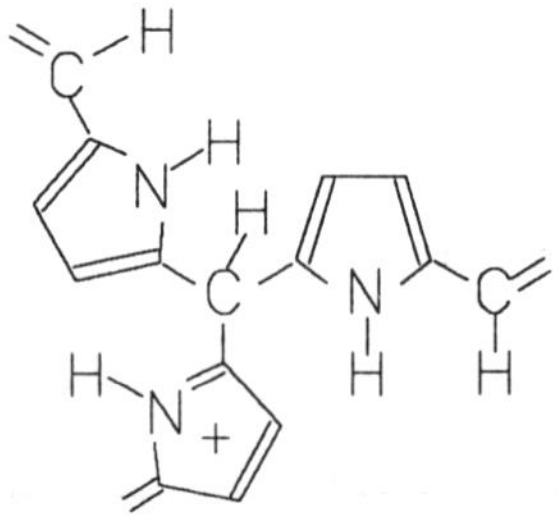

V

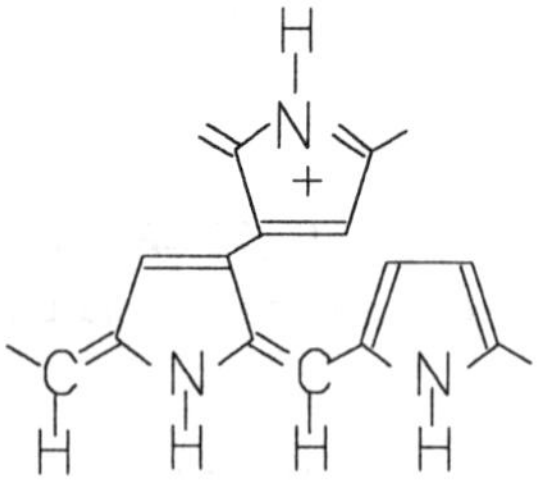

VI

From this point of view II is a very interesting material
and new synthetic and reduction methodes are needed. As
shown in table 1 partly deprotonation by extraction with
water gives considerable higher conductivity compared with
the starting material. With aqueous bases or $NaBH_4$ in 0.1n
NaOH II reacts at the methine group giving insulating
products.
The conductivity between 10^{-4} and 10^{-6} S/cm is still low
compared with other known cond. polymers. Since now the NMR
- spectra exclude significant amounts of sp^3 - carbon atoms
and crosslinking over tripyrrylmethine carbon atoms (V) as
well, we conclude that a low molcular weight is the reason
for the low conductivity. Not excluded till now can be
structures of formula VI, which also would act as crosslinking units, but should not lower the conductivity.
The optical properties which are less sensitive to the chain
length than conductivity are found as expected for such
compounds.
The NIR - spectra show a band gap value of about 1 eV.
II is photoelectrical sensitiv, the spectral dependence of
the photosensitivity essentially follows the absorption
spectrum. The quantum yield is still far below 1%. In ESR
experiments a Pauli spin concentration of $1*10^{19}$ spins/g [6]
is found, a value also found in light doped polypyrrole and
polythiophene.

Acknowledgment:
This work is supported by the German Ministry of Research and Technology (BMFT). We thank Prof J. L. Bredas for calculations and helpful discussions, Prof M. Hanack Tübingen for ^{13}C-, Prof. H. - H. Limbach, Freiburg for ^{15}N- CPMAS spectra and Prof H. Meier, Bamberg for photoelectrical measurements.

LITERATURE

1) J. M. Toussaint, B. Thomas, J. M. Andre and J. L. Bredas, Synth. Metals 28, (1989), C 205 - 210

2) M. Kertesz and Y. S. Lee, J. Phys. Chem. 1987, 91,2690

3) R. Jira and H. Bräunling, Synth. Metals 17 (1987) 691

4) H. Bräunling and R. Becker, E. Appl 285 021

5) R. Becker, G. Blöchl, H. Bräunling, Ger. Appl. 3829753

6) R. Becker, G. Blöchl, H. Bräunling, IWEPP 89 submitted for publication

7) S. A. Jehnecke, Nature 322, 345 1986

8) A. O. Patil and F. Wudl, Macromolecules 1988, 21,540

9) B. Wehrle, H - H Limbach and H. Bräunling, submitted for publication

10 J. L. Bredas, private communication

CONFORMATION OF CONJUGATED POLYMERS IN GOOD SOLVENT

J.P. AIME*
EXXON Research and Engineering Company, Clinton Township
Route 22 East, Annandale, New Jersey 08801.

1. Introduction

One of the most interesting recent advances in the field of conjugated
polymers has been the synthesis of such macromolecules soluble in
common organic solvents (1,2,3). There are many reasons for studying
conjugated polymers in solution. A simple one is that basic
information such as molecular weight can be obtain since it is
possible to discriminate a single chain from an aggregate (several
so-called "solutions" are in fact suspensions of aggregates).

Furthermore solution of conjugated polymers has also stimulated a
more fundamental research. For example polydiacetylenes exhibit a
color transition by varying the temperature or the solvent quality. An
accurate understanding of the phase transition is of the main
importance in order to control polymer processing from solution. Also
solvatochromism and slight color change with the temperature are
currently observed and suggested a large amount of work about the
effect of the conformational disorder upon the electronic
distribution.

Conjugated polymers in solution differ from the current saturated
polymers not only from their conjugated backbone structure, but also
because side-groups are needed to improve the solubilities. The
solutions give a very good opportunity for studying the relevant
parameters which control the chain conformation.

We are mainly interested with a description of the local correlation
along the chain. Intuitively, we can expect that a measurement of a
length describing the local rigidity will be related to the optical
properties of the polymer. Here we report small angle neutrons
scattering (SANS) experiments on solutions of poly-butylthiophene
(p-BT) and three polydiacetylenes with side-groups formula
$(CH_2)_nOCONHCH_2C_4H_9$ with n=3 or 4 and $(CH_2)_4OSO_2C_6H_4CH_3$ respectively
for poly-3BCMU, poly-4BCMU and poly-PTS12. Also we will present
results obtained by light scattering on saturated polymers. The
comparison between these different polymers will provide us the
possibility to underline the origin of the local stiffness.

*J. L. Brédas and R. R. Chance (eds.), Conjugated Polymeric Materials:
Opportunities in Electronics, Optoelectronics, and Molecular Electronics*, 141–148.

2.Structural study of conjugated polymers in solution.

In many scattering experiments (Light, Xray or neutrons) we measure the Fourier transform $S(q)$ of the monomer density autocorrelation function $\langle\rho(r)\rho(0)\rangle$, q is the scattering wave vector.(With an incident wavelength λ and a scattering angle θ we have $q=4\pi \lambda^{-1} \sin\theta/2$). The autocorrelation function tends to the average density $\langle\rho\rangle^2$ for large r, and the structural information is contained in a finite region where $\langle\rho(r)\rho(0)\rangle$ deviates from the final value. The quantity of interest is therefore the Fourier transform of the density fluctuation. In the present study we are concerned with the measurement of the statistical length b which characterizes the correlation at relative short scale: it takes typically length b to reverse the chain's orientation. Values of b lie between 20 and 600 Å, small angle neutron scattering studies (SANS) are very convenient for this range.

TABLE 1. Conformation of conjugated
polymers in good solvent.

polymer	Solvent	Le (A)	b (A)
PTS12	nitrobenzene	25	300
P3BCMU	DMF	27	310
P4BCMU	Toluene	26	310
p-BT	THF	11	55

In table I are reported the statistical lengths measured on the different conjugated polymers. The experimental SANS results are in complete agreement with the Porod-Kratky or worm-like-chain model (4,5) describing the macromolecule as a continuous line with fluctuation in curvature. The series of polydiacetylenes exhibits a rather stiff behavior, b=300 or 310 Å while the poly-butylthiophene is more flexible b=55 Å. At this stage, we note that backbone conjugated structure does not necessarily imply a large local rigidity.

SANS results provide also information about the conformation of the side groups with respect to the backbone, that is the extension Le of the substituents in the solvent. It is of the main importance since we know that for having soluble conjugated polymers we need to make a substitution upon the monomer unit. The results are reported in column 3 (table 1). The side groups are almost fully extended in the solvent. The latter result can be understood in the following way: - the side groups can be considered as short chains swollen in the solvent, - the steric repulsion between neighboring monomer unit reduce the space available for each side group.

Since conjugated polymers can be described as a macromolecules with statistical lengths, we expect a temperature dependence following the relation: $b = aT^{-1}$ (6), where a is a constant which characterize the stiffness of the macromolecules. The SANS experiment was performed with PTS12 samples in two good solvents: DMF and nitrobenzene. PTS12 samples were chosen, because no hydrogen bonds are involved like in poly-nBCMU. Consequently only substituents and conjugated backbone contribute to the chain conformation. The asymptotic behavior of the structure factor for a worm-like-chain is given by (7)

$$\lim_{L,q\to\infty} \frac{L}{b} P(q) = \frac{\pi}{qb} + \frac{4}{3}(\frac{1}{qb})^2 \qquad (1)$$

where L is the total contour length and $P(q)=S(q)/S(0)$ is normalized (4). At small enough scale, the chain behaves like a stiff rod. That is, the larger the statistical length, the larger the extension of the plateau towards the small scattering wave vectors q in a qg(q) plot (see figure 1). For a value of b equal to 300 Å at 25°C, the T^{-1} law predicts a value of b equal to 255 Å at 75°C neglecting that the macromolecule is not infinitely flexible at infinite temperature.

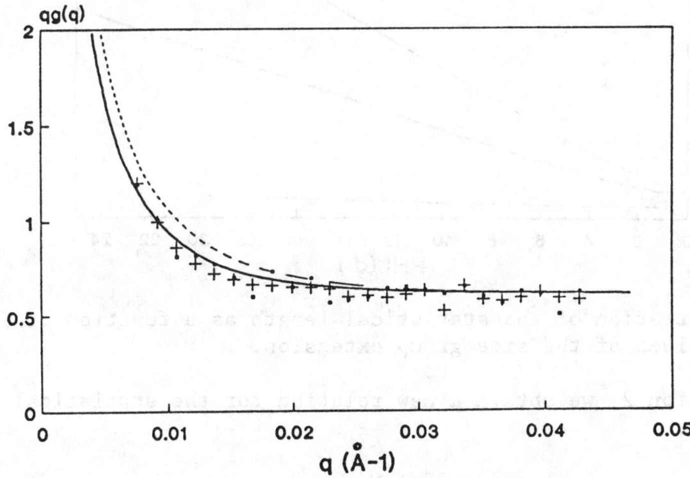

Figure 1: Scattering functions plotted as qg(g) vs q in absolute values for PTS12 in nitrobenzene at two different temperatures: T=25°C: • , T=75°C:+ . Computed structure factors are also shwon for b=300 Å (continuous line) and b=255 Å (dashed line).

144

Figure 1 shows an identical structure factor at 25°C and 75°C. The thermal behavior of the chain conformation of PTS12 does not follow the one expected for a worm-like chain. Up to now, the best way for understanding the absence of variation of the statistical length with the temperature is to take into account a rotation between monomer units. Using the relation given in reference 6:

$$<I_e> = \frac{1}{\phi} I_1 \int_0^{\phi} \sqrt{1-k^2 \cos^2(\beta)} \; d\beta \quad (2)$$

Equation 2 gives the average effective moment of inertia governing the fluctuation in curvature of the macromolecule. ϕ is the rotation between neighboring monomer units along the contour length.

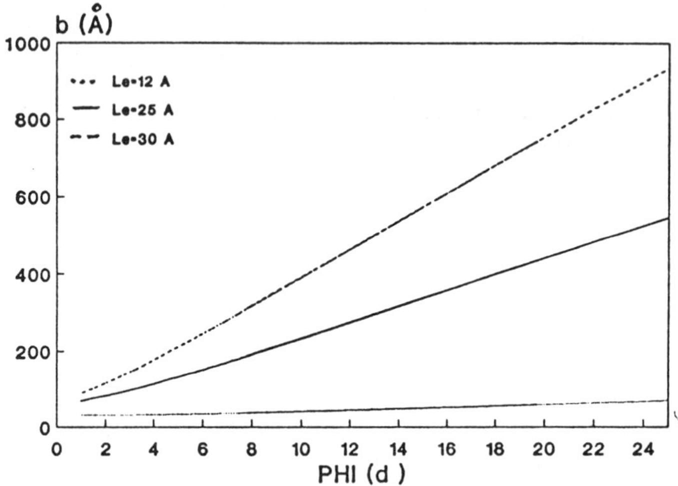

Figure 2: Variation of the statistical length as a function of ϕ for different values of the side group extension.

Using equation 2, we obtain a new relation for the statistical length (6) given by:

$$b = \frac{2E<I_e>}{T} \quad (3)$$

where E is the elastic modulus.
Equations 2 and 3 give an implicit dependence of the angular deviation between monomer units upon b. An increase of ϕ will lead to an increase of the average effective moment of inertia of the ribbon and therefore to an increase of the statistical length (see figure 2).

This result means that the two types of fluctuation which occur along the chain have an opposite effect upon the value of the statistical length (6).

3. Conformation of polymer with side group.

In order to precise the effect of the side group upon the chain conformation, we report briefly light scattering results obtain on saturated polymers (8). We have chosen polymers as simple as possible: poly-octene and poly-decene. They differ only by the number of CH_2 units attached at the backbone: 5 and 7 respectively. Using the relation given the radius of gyration for chain with statistical length (9):

$$<R^2> = \frac{1+2U}{1+U} \frac{L_w b}{6} \left[1 - \frac{1+U}{1+2U} \frac{3b}{2L_w} \left(1 - \frac{b}{L_w}\right) \right.$$

$$\left. - \frac{(1+U)^2}{1+2U} \frac{3b^3}{4L_w^3} \left(1 - \left(1 + \frac{U}{1+U} \frac{2L_w}{b}\right)^{-1/u}\right) \right] \quad (4)$$

where $U = M_w/M_n - 1$ and L_w is the measured contour length.
  The equation 4 gives the radius of gyration as a function of the molecular weight (contour length), the polydispersity and the statistical length.

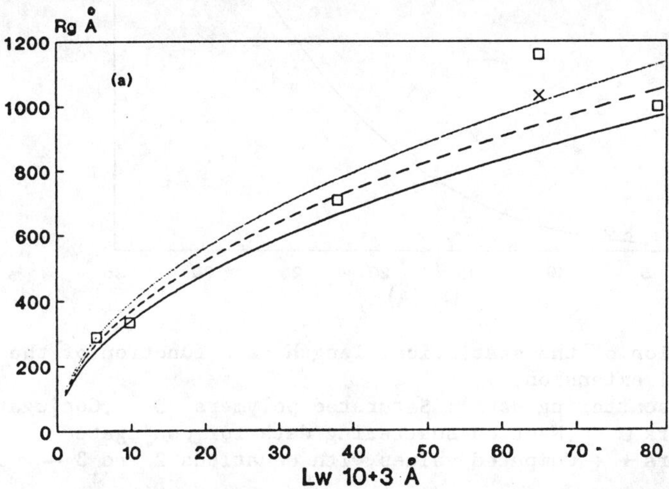

146

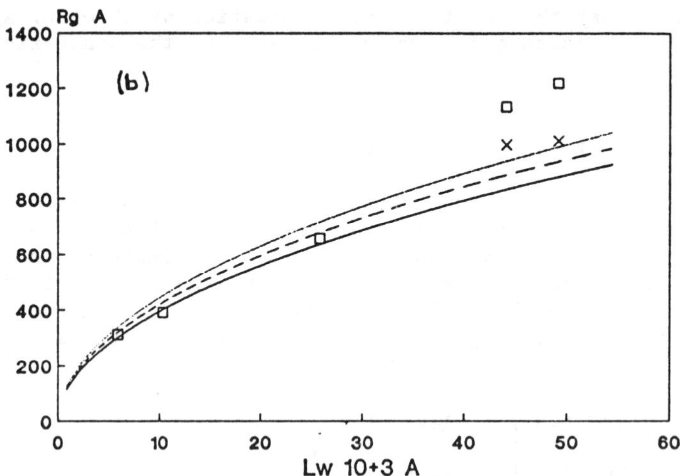

Figure 3: Variation of the Radius of Gyration as a function of Mw,
using the relation 5 and Mw/Mn=1.35
a- Polyoctene □ ; x Radius of gyration for Mw/Mn = 1.35 (see
Text) b = 55 ____; b = 65 __ ; b = 75 - -.
b- Polydecene □ ; x Radius of gyration for Mw/Mn = 1.35
b = 75 ____; b = 85 __ ; b = 95 - -.

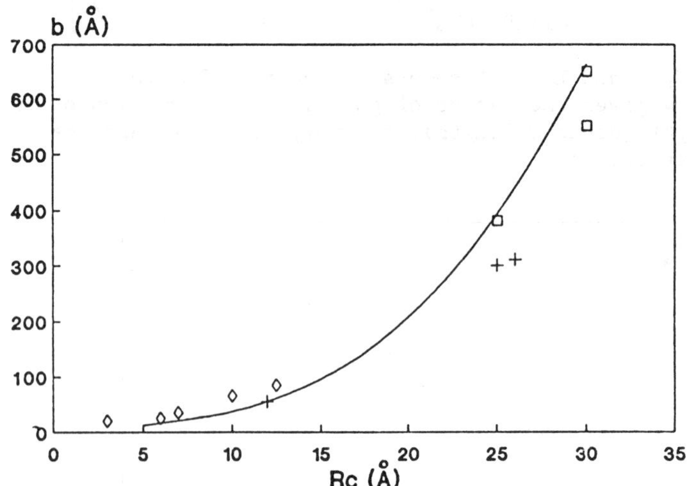

Figure 4: Variation of the statistical length as a function of the
lateral extension.
Light scattering data : Saturated polymers ◊   ,Conjugated
polymers □   ;Neutron scattering data for conjugated
polymers + ; Computed values with equations 2 and 3 ─   .

Results are reported in the figures 3. The statistical lengths deduced are 65 Å for poly-octene and 85 Å for poly-decene. These values are much larger than the ones obtain with the standart linear saturated polymer: typically b is between 20 and 30 Å. Also the local rigidity depends unambiguously of the extension of the side groups since the poly-decene is more rigid than the poly-octene.

Figure 4 summarizes the statistical length measured for various saturated and conjugated polymers. The highest values of b (550 and 660 Å) correspond to the polydiacetylenes with R - $(CH_2)_9OCOCH(CH_3)_2$, $(CH_2)_9OCOCH_2C_6H_5$ and $(CH_2)_9OCOCH_2$(1-naphthyl) (10). Using equations 2 and 3, we report values of b computed with ϕ equal to 18 degrees and E equal to 10^{10} dyn cm^{-2}. The general evolution is surprinsigly well reproduced meaning that the internal structures are almost the same. That is, the average torsion between monomer units (at least for the different polydiacetylenes) are identical in spite of the large change in the magnitude of the statistical length.

4. Conclusion.

The extension of the side group and their mutual orientation appear to be the key parameters for understanding the chain conformation. It means also that the measured statistical length is not directly related to the electronic distribution or conjugation length. In other words the conjugation length will be related to the torsion (11). The latter remark is well supported with the experimental result given by Plachetta and Schulz (10). They observe a maximum of absorption similar to the other polydiacetylenes while the statistical length is about twice larger: 550 or 660 Å instead of 300 Å.

Acknowledgements:

I wish to thank several colleagues involved in that work: F. Bargain, M. Schott in G.P.S, M. Rawiso at Orsay, S. Ramakrishnan, R.R. Chance and M. W. Kim at Exxon. Institut Laue Langevin (Grenoble, France) and Laboratoire Leon Brillouin (Saclay, France) are thanked for time allocation for SANS experiments.

* Permanent address:
 Groupe de Physique des Solides, Universite Paris VII, 2 Place Jussieu, Paris 75251, France.

References

[1] Patel G.N., Chance R.R., and Witt J.D., J. Chem. Phys. 70 (1979) 4387.

[2] Elsenbaumer R.L., Jen K.Y., and Oboodi R., Synth. Met. 15 (1986) 164.

[3] Hotta S., Rughooputh S., Heeger A.J., and Wudl F., Macromolecules
     20, (1987) 212.
[4] Rawiso M., Aime J.P., Fave J.L., Schott M., Muller M.A.,
     Schmidt M., Baumgartl H. and Wegner G., J. Phys. Paris, 49
     (1988) 861.
[5] Aime J.P., Bargain F., Schott M., Eckhardt H., Elsenbaumer R.L.
     and G.G. Miller, Phys. Rev. Let., 62 (1989) 55.
[6] Aime J.P. and Bargain F., Europhys. Lett., 9 (1989) 35.
[7] Des Cloizeaux J., Macromolecules 6, (1973) 403.
[8] Aime J.P., Ramakrishnan S., Chance R.R., Kim M.W., submit to J.
     de Physique.
[9] Oberthur R.C., Makromol. Chem. 179 (1978) 2693.
[10] Plachetta C. and Schulz R.C., Makromol. Chem., Rapid
     commun., 3 (1982) 815.
[11] Rossi G., Chance R.R., Silbey R., J. Chem. Phys. in Press.

ELECTRONIC STRUCTURE OF PROCESSABLE CONDUCTING POLYMERS

R. LAZZARONI[*], M. LÖGDLUND, S. STAFSTRÖM, W.R. SALANECK,
Department of Physics, Linköping University, S-58183, Linköping, Sweden

D.D.C. BRADLEY, R.H. FRIEND
Cavendish Laboratory, Cambridge University, Cambridge CB3 OHE, UK

N. SATO
Department of Chemistry, Tokyo University, Meguro, Tokyo 153, Japan

E. ORTI, J.L. BREDAS
*Service de Chimie des Matériaux Nouveaux, Université de Mons,
B-7000, Mons, Belgium*

ABSTRACT. The electronic structure of poly-3-hexylthiophene is studied in the solid state with photoelectron spectroscopy, as the polymer is gradually doped from $NOPF_6$. The evolution of the core level binding energies is related to the modification of the electron density on the conjugated backbone, due to the creation of polaron and bipolaron defects. Upon doping, valence spectra show a shift in the Fermi level of the system, and at saturation doping a finite density of states is observed at the Fermi level. In the case of poly-para-phenylene vinylene, the experimental density of states of the neutral polymer is related to the results of Valence Effecttive Hamiltonian calculations. Comparison with gas phase data on styrene leads to an estimation of the polarization energy.

1. Introduction

Processability and solubility are of prime importance in the field of conducting polymers to develop these materials for wide-scale applications. These properties are also very interesting for more fundamental purposes, because they allow for the preparation of clean and well-characterized samples, thereby greatly improving the quality of spectroscopic data. For instance, probing the density of valence states with Ultraviolet Photoelectron Spectroscopy (UPS) cannot be carried out properly on electrochemically-prepared polyheterocycles, due to the

---

* Permanent address: Service de Chimie des Matériaux Nouveaux, Université de Mons,Belgium

149

J. L. Brédas and R. R. Chance (eds.), Conjugated Polymeric Materials:
Opportunities in Electronics, Optoelectronics, and Molecular Electronics, 149–162.
© 1990 *Kluwer Academic Publishers. Printed in the Netherlands.*

surface inhomogeneity of the films, or on air-exposed samples, due to surface contamination.

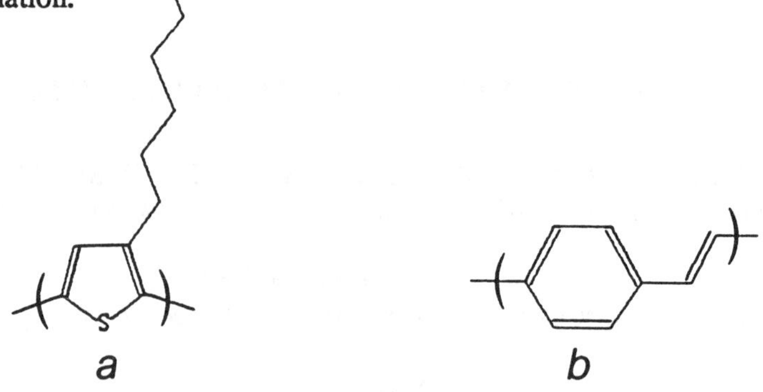

Figure 1: The structure of poly-3-hexylthiophene (a) and poly-para-phenylene vinylene (b).

In this work, we use the processability properties [1-4] of poly-3-hexylthiophene (P3HT; Fig. 1a) and poly-para-phenylene vinylene (PPV; Fig. 1b) to prepare very clean and very thin films, which are then studied with photoelectron spectroscopy. In the former case, we follow (i) the evolution of the charge density on the chain with core level spectroscopy (XPS), and (ii) the modification of the density of states near the Fermi level as the polymer is gradually doped, and we relate the observed changes to the existence of charged defects (polarons, bipolarons and their lattices) characteristic of non-degenerate ground state systems. The experimental results are compared with the predictions of recent theoretical calculations [5] performed with the Valence Effective Hamiltonian (VEH) and Modified Neglect of Diatomic Overlap (MNDO) techniques.

In the case of PPV, the density of valence states in the neutral form is interpreted in relation with the VEH-obtained spectrum. The experimental curve of the polymer is also compared to gas phase data on styrene, which can be considered as a "model monomeric unit" for PPV.

2.   Experimental

P3HT is prepared chemically as reported earlier [6]; molecular weight measurements show that the average chain length is around 50 monomer units. The polymer is dissolved in spectral-grade chloroform at a concentration of 1 mg/ml. From this solution, thin films, i.e. with a thickness less than 2000 Å, are prepared by a slow dip-coating process on aluminum-coated, optically-flat silicon substrates, following the method described previously [7]. The samples are immediately introduced into the spectrometer. Through this step, air exposure does not exceed a few minutes. The samples are then pumped down to 10^{-6} Torr, transferred into a preparation chamber ($P < 10^{-10}$ Torr), and heated for a few minutes to 180°C, in order to clean the surface from hydrocarbon impurities and the remaining solvent. We have observed that this step is of prime importance to obtain distinct and reproducible UPS spectra.

Doping with NOPF$_6$ is performed by cooling down the sample to just below room temperature, and moving it back into a special chamber, where doping can be carried out under a flowing dry nitrogen atmosphere. A precise amount of freshly-prepared doping solution (0.03 ml at 10^{-2} M) is applied onto the film surface, the sample being tilted 30º from the horizontal. It is important to note that P3HT is not soluble in acetonitrile, so the doping process does not significantly affect the quality of the films. The excess solution is collected in a groove-like receptacle at the lower end of the sample holder, apart from the polymer film. This operation can be repeated several times to increase gradually the doping level of P3HT.

The sample surface is finally rinsed with pure acetonitrile to wash away any excess NOPF$_6$ salt. The chamber is then sealed, the nitrogen flow is stopped and the film is dried at a pressure of 10^{-6} Torr before moving the sample into the analysis chamber at 10^{-10} Torr. We have found that this procedure is very effective in keeping the surface clean from atmospheric contamination. For instance, in a separate control experiment, no oxygen is detected on the surface of a sputter-cleaned copper substrate after a 5-minute exposure to the N$_2$ flow [8].

PPV is synthesized via the soluble sulfonium precursor route [3,4]. A 200 Å-thick precursor film is spin-cast on aluminum-coated silicon substrates, and annealed at 300°C under N$_2$ atmosphere for 10 hours to yield the conjugated polymer. After transfer into the spectrometer, the PPV samples are shortly heated to 180°C to remove hydrocarbon contamination from the surface.

XPS measurements are carried out with an unmonochromatized Mg Kα source, whose resolution results in a full-width at half maximum of 0.9 eV for the Au4f$_{7/2}$ line. This line is also used to calibrate the binding energy scale at 83.8 eV. The standard error on the measurements is ±0.1 eV. UPS spectra are recorded with a monochromatized HeI ($h\nu = 21.2$ eV) or HeII ($h\nu = 40.8$ eV) radiation. P3HT samples are kept at - 40°C during all the measurements; as shown earlier [7],

152

conformational defects along the P3HT chain occur only at higher temperatures. Hence, in our case, the chains can be assumed to be completely planar.

3. Results and Discussion

3.1   DOPING OF POLY-3-HEXYLTHIOPHENE

3.1.1. Estimation of the doping level. The amount of dopant in the polymer is determined by measuring the intensities of the core level lines of the counterion (PF_6^-). The intensity ratio of these lines to those characteristic of the polymer, when weighted by cross-section factors, gives a fair estimate of the doping level. In this work, we used the P_{2p}/S_{2p} ratio; the standard error is around 10% in atomic ratio. We focussed our investigations upon three domains of doping level: below 5% (low doping), around 35% (high doping) to allow for comparison with previous studies [9], and around 45% (saturation doping). At this stage, further exposure to the doping agent does not induce additional changes in the spectra.

3.1.2. Core level lines of the polymer. In order to assess the effect of the charge transfer on the electron density of the polymer, the usual procedure would be to measure the chemical shift of the polymer lines (here, C1s and S2p) upon doping. However, the raw data are not meaningful in that regard, because the doping also induces other effects, such as a shift in the reference energy. Photoelectron spectra in the solid state are referred to the Fermi level, but the vacuum level is the chemically significant reference. For a given material, however, the energy difference between the vacuum and the Fermi level is a constant value, the work function Φ. This relation cannot be applied when a shift in the Fermi level (ΔE_F) occurs as a consequence of a chemical modification of the sample, e.g. in the intercalation of graphite [10] or in the doping of conducting polymers [11]. In these cases, the core level binding energies must be corrected to take ΔE_F into account before any interpretation.

The shift of the Fermi level can be determined from the UPS measurements; the total width of the spectrum, i.e., the distance between the Fermi energy and the cut-off of the secondary electrons (E_{CO}), gives a direct estimation of the work function, according to the following formula:

$$(E_{CO} - E_F) + \Phi = h\nu$$

Here, we use the photon energy of the HeI radiation (21.2 eV). The change in the work function, $\Delta\Phi$, which is measured, is equal to the shift of the Fermi level (ΔE_F).

After this correction, it is possible to discuss the evolution of the core level binding energies upon doping (Fig. 2). In the undoped state, the C1s level appears at 285.6 eV; as expected, it consists of only one peak, containing the contributions of the ring and of the aliphatic side-chain.

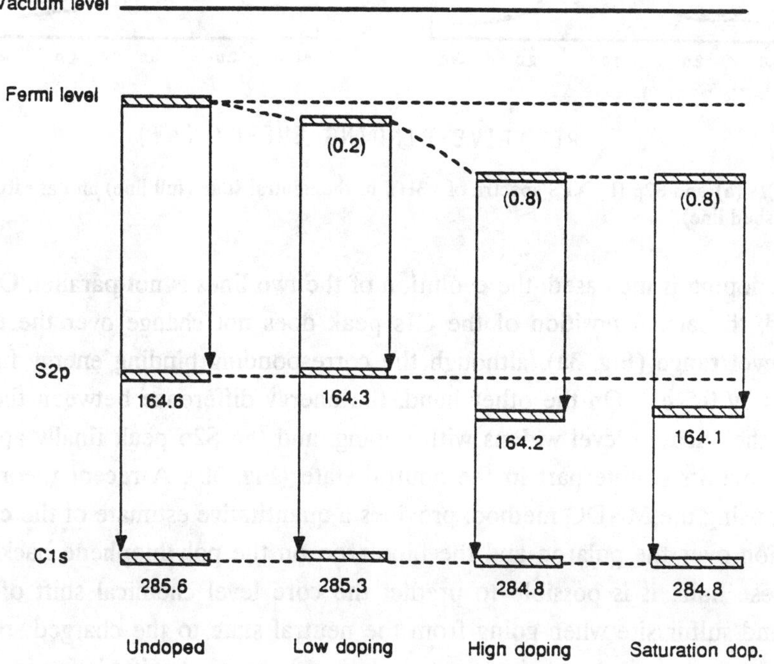

Figure 2: Evolution of the S2p and C1s binding energies (eV) in P3HT upon doping. The values are referred to the Fermi level; the shift of the Fermi level relative to the undoped state is indicated in parentheses.

The S2p line is located at 164.6 eV. At low doping levels, although the measured binding energies decrease, the position of both lines relative to the vacuum level remains constant. This is probably due to the fact that the contribution of the positively-charged units is too weak to be detected, relative to that of the still neutral regions.

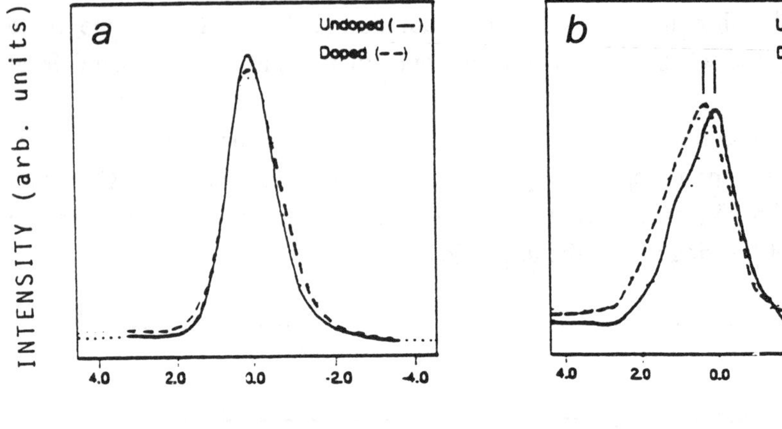

Figure 3: C1s (a) and S2p (b) XPS spectra of P3HT in the neutral state (full line) and at saturation doping (dashed line).

As the doping is increased, the evolution of the two lines is not parallel. On the one hand, the actual position of the C1s peak does not change over the whole doping level range (Fig. 3a), although the corresponding binding energy further decreases by 0.5 eV. On the other hand, the energy difference between the S2p line and the vacuum level widens with doping, and the S2p peak finally appears 0.3 eV above its counterpart in the neutral state (Fig. 3b). A recent theoretical work [5], using the MNDO method, provides a quantitative estimate of the charge distribution over the polaron and the bipolaron on the polythiophene backbone. From these data, it is possible to predict the core level chemical shift of each carbon and sulfur site when going from the neutral state to the charged species. The link between calculated charges and observed chemical shifts is found in the pioneering work of the Uppsala group [12]. From measurements on model molecules containing carbon, nitrogen, oxygen, or sulfur, shifts were shown to be linearly related to the charges. The MNDO-calculated relative values then give directly the chemical shift of the carbon and sulfur atoms of the conjugated backbone, in the polaron or the bipolaron state, relative to the neutral state. We use the theoretical data obtained on polythiophene for the study of P3HT, since the electronic structure of the π system is not affected by the presence of the alkyl groups. This has been clearly demonstrated by optical absorption and photoelectron spectroscopies as well as theoretical calculations [6,7,13].

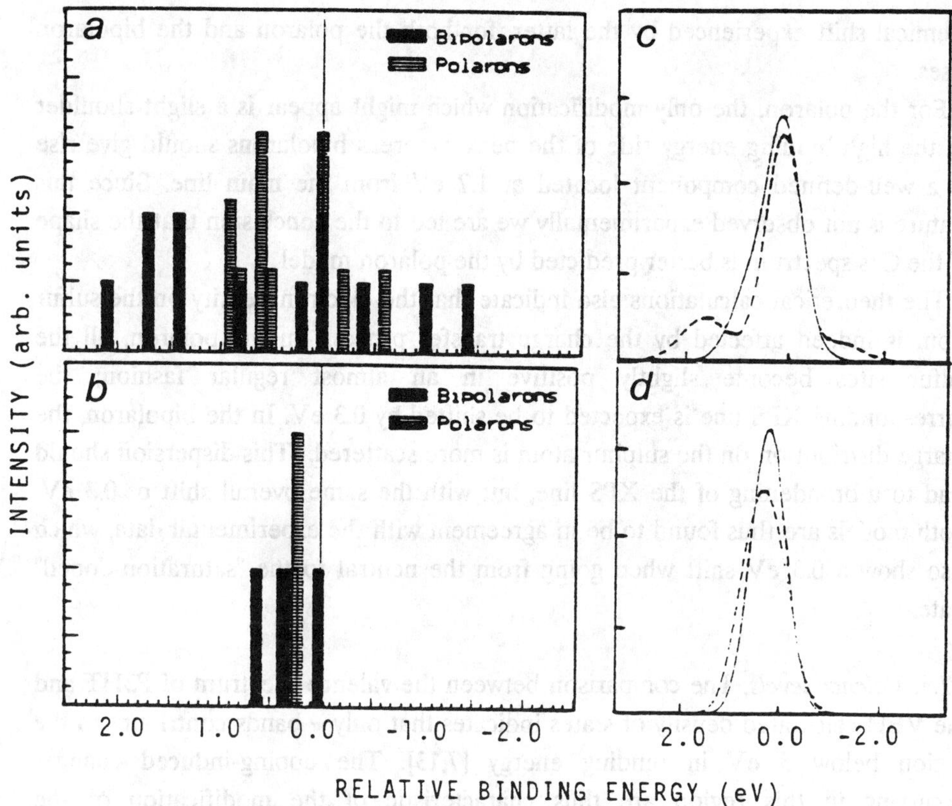

Figure 4: Calculated chemical shifts for the C1s(a) and S2p (b) lines in polythiophene; convoluted C1s (c) and S2p 3/2 (d) spectra for the polaron (full line) and bipolaron (dashed line) configurations. Binding energies are relative to the neutral chain

The predicted shifts are plotted in Figure 4 (a,b) for a model chain consisting only of polarons, or bipolarons. The calculated spectra (Fig. 4c,d), which are obtained by convoluting the data with Gaussian functions (full width at half maximum: 0.8 eV), are to be compared with the XPS measurements at saturation doping. For the C1s line, the formation of polarons is expected to induce a 0.6 eV-shift of the contribution of the backbone atoms towards higher binding energies, whereas the bipolaron-containing chain should yield a more broadened spectrum, with the most intense features also shifted towards higher binding energies. However, the component due to the atoms of the alkyl groups, which do not undergo any polarization, is not expected to move; because of the large number of sp^3 carbons and the distribution of the contributions of the sp^2 carbons, the former completely dominate the spectrum and overcast the small chemical shift experienced by the latter, for both the polaron and the bipolaron cases.

For the polaron, the only modification which might appear is a slight shoulder on the high binding energy side of the peak, whereas bipolarons should give rise to a well-defined component located at 1.7 eV from the main line. Since this feature is not observed experimentally we are led to the conclusion that the shape of the C1s spectrum is better predicted by the polaron model.

The theoretical calculations also indicate that the electron density on the sulfur atom is indeed affected by the charge transfer process. In the polaron, all the sulfur sites become slightly positive in an almost regular fashion; the corresponding XPS line is expected to be shifted by 0.3 eV. In the bipolaron, the charge distribution on the sulphur atom is more scattered. This dispersion should lead to a broadening of the XPS line, but with the same overall shift of 0.3 eV. Both models are thus found to be in agreement with the experimental data, which also show a 0.3 eV shift when going from the neutral to the "saturation-doped" state.

3.1.3. Valence levels. The comparison between the valence spectrum of P3HT and the VEH calculated density of states indicates that only π bands contribute to the region below 5 eV in binding energy [7,13]. The doping-induced changes occurring in this region are thus characteristic of the modification of the π-electron system. The main peak appearing in this region (Fig. 5a, lower curve) is located at 3.5 eV in the neutral polymer. It corresponds to π states strongly localized on separate thiophene rings. Upon doping, it slightly shifts to larger binding energies, in agreement with the results of the VEH calculations. This is due to the fact that the charge transfer drives the ring geometry towards a more quinoid structure. As a consequence, the bonding interaction between the two β carbons of the same ring is enhanced, thereby stabilizing the corresponding electronic level [14].

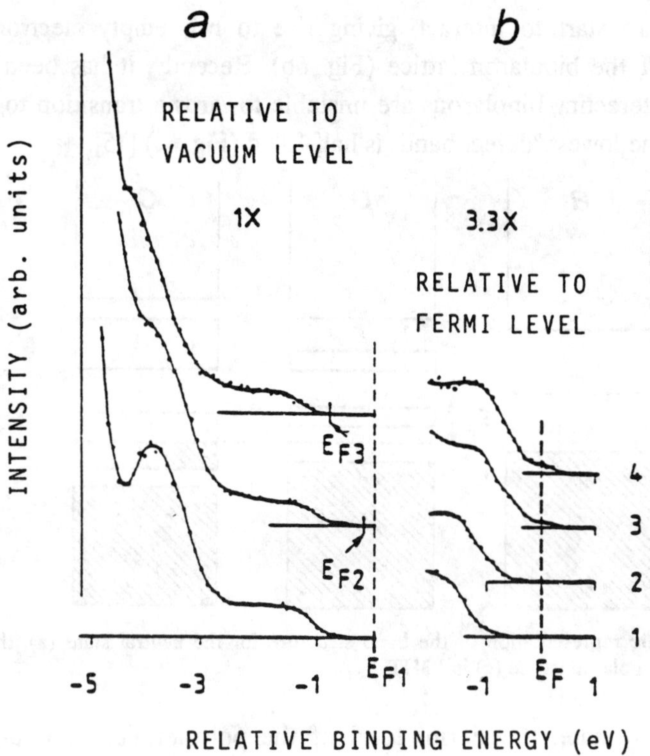

Figure 5: UPS spectra of the highest occupied π states aligned to the vacuum level (a) and to the Fermi level (b), for undoped (1), lightly doped (2), highly-doped (3), and saturation-doped (4) P3HT.

If the spectra are aligned relative to the vacuum level, it also appears clearly that the Fermi level shifts towards the valence band edge upon doping. A shift of 0.2 eV is already observed at low doping levels; it increases to reach its maximum value (0.8 eV) at high doping. This effect is of prime importance to interpret the results of core level spectroscopy, as discussed above. More fundamentally, it is the consequence of the creation of charged defects on the conjugated backbone. The first ionization of neutral chains leads to the formation of polarons; as a consequence, two new electronic states appear within the bandgap, the lowest lying one being singly-occupied. As the dopant concentration is increased, the polarons recombine into doubly-charged bipolarons. Finally, at high doping levels, bipolarons start to interact, giving rise to two empty electronic bands, characteristic of the bipolaron lattice (Fig. 6b). Recently, it has been proposed that strongly-interacting bipolarons are unstable towards a transition to a polaron lattice, where the lowest "defect band" is half-filled (Fig. 6c) [15].

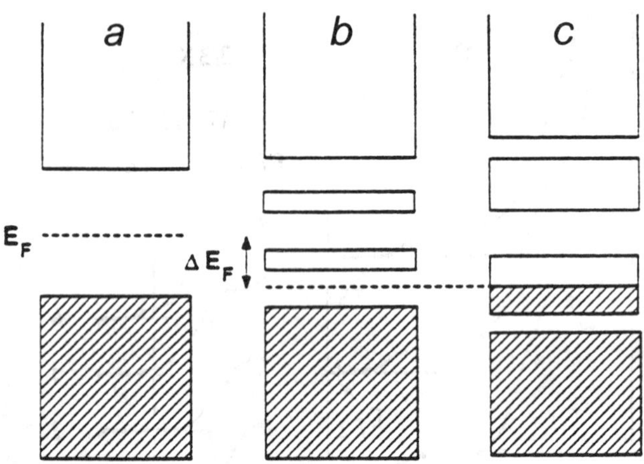

Figure 6: Schematic representation of the band structure for the neutral state (a), the bipolaron lattice (b), and the polaron lattice (c) in P3HT.

In the neutral state, the Fermi level of the polymer lies at mid-gap, as is generally the case in conventional undoped semiconductors. The VEH-calculated shift of the Fermi level, when going to high doping, is 0.7 eV, for either the bipolaron or the polaron lattice, in excellent agreement with the experimental value of 0.8 eV. According to a recent study [16], the movement of the Fermi energy towards the valence band should be a gradual, rather than step-like, process. Thus, the value of 0.2 eV observed at "low doping levels" appears quite reasonable.

The major difference between the bipolaron and the polaron lattices consists in the existence of a finite density of states at the Fermi level in the latter case. A careful investigation of that part of the spectrum could then allow to distinguish between the two configurations. This is indeed the case. A density of states at E_F is detected at saturation doping (Fig. 5b, upper curve), in agreement with the existence of the polaron lattice. In the neutral state, the band edge is located around 1 eV, as expected for a 2 eV-bandgap semiconductor. At low doping levels, the edge gets closer to the Fermi level, due to the shift described above. Optical spectra have shown the typical features of polarons in this range of dopant concentration (<5%) [17]. These states could in principle be observed with photoelectron spectroscopy a few tenths of eV above the valence band, but their number is probably too low to provide a significant signal. At high doping, the Fermi level is very close to the band edge, as predicted for the bipolaron lattice, and a very weak density of states may have already appeared at E_F. However, this feature is unambiguously detected only at saturation doping, suggesting that the bipolaron configuration can survive up to high doping levels ($\approx 35\%$).

3.2. POLY-PARA-PHENYLENE VINYLENE

The HeI and HeII UPS spectra of PPV are shown in the upper part of Figure 7. The HeI radiation is more sensitive to the low-lying π states (compare the intensity of the lowest binding energy feature) whereas HeII allows to probe high binding energy levels ($E_b > 15$ eV). The differences in the intensities of the bands between the two spectra are related to the dependence of the photoemission cross section on the energy of the incoming photon. The theoretical VEH curve is obtained by convoluting the calculated density of states with 0.7 eV-wide Gaussian functions. The input geometry for the polymer chain is built from the results of a MNDO optimization of the vinyl group, and assuming C-C bond lengths of 1.40 Å and bond angles of 120° for the phenyl ring. Since ab initio methods are known to give too wide a valence band and do not take into account the polarization energy in the solid state, the VEH scale has been compressed by a factor 0.785 and shifted to lower binding energies to align the first features in the experimental and the theoretical spectra. This value of the scaling factor has been shown to be of general application for electron-containing systems with either a π-electron or a σ-electron backbone [18,19].

It appears that the theoretical curve nicely agrees with the UPS data, so that all the major experimental features can be assigned. The lowest lying peak, located at 6.5 eV from the vacuum level, corresponds to a π band largely delocalized over the conjugated backbone. The second peak, which appears as a shoulder on the UPS spectra, is due to π states localized on the phenyl rings. C2p-based σ bands give rise to high-intensity peaks in the 5-10 eV region.

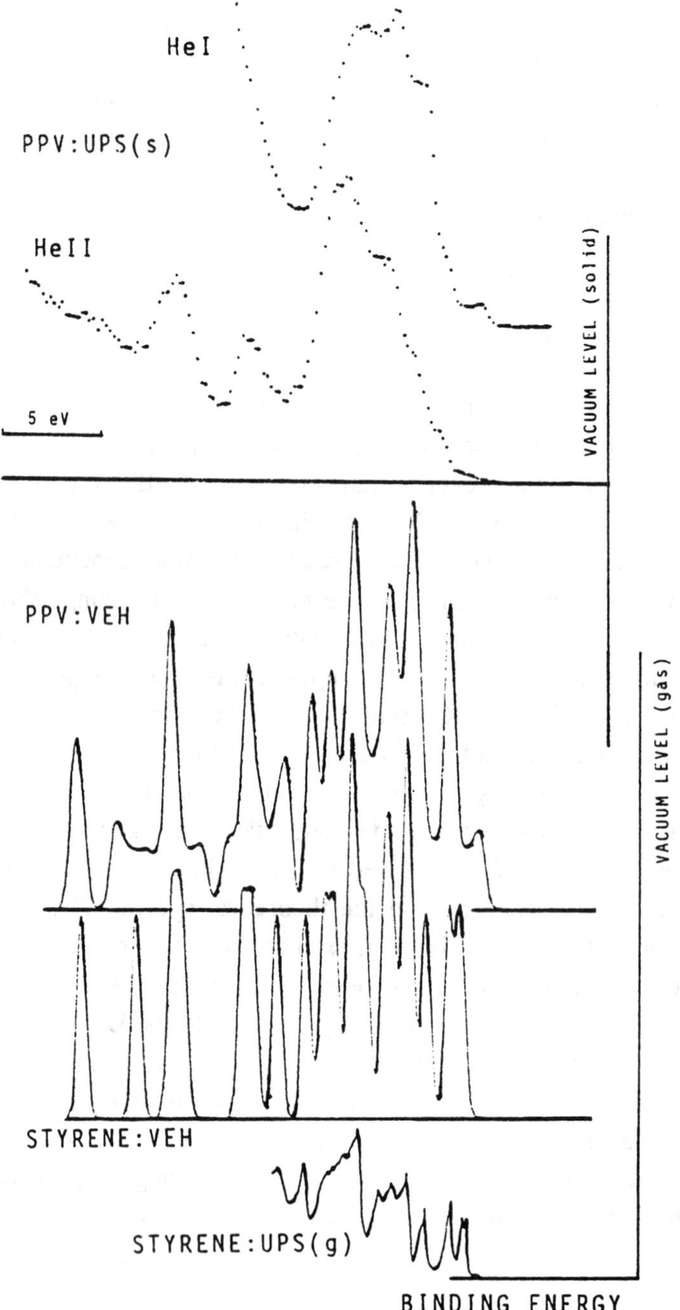

Figure 7: Upper spectra: HeI and HeII spectra of PPV. Center curves: VEH-calculated density of states of PPV and styrene. Bottom spectrum: HeII spectrum of styrene.

Finally, the last two peaks, observed only with HeII, correspond to orbitals with large C2s contributions [20]. A recently-published XPS valence spectrum [21] shows two additional peaks located at 24 eV and 30 eV from the Fermi level, i.e., in a region which cannot be probed easily with UPS. It should be noted that, when compared to our compressed VEH curve, the highest-lying XPS band appears far up from the upper edge of the theoretical density of states. In our opinion, this discrepancy is too large to be explained by a drawback of the theoretical method.

The lower part of Figure 7 displays the VEH and UPS curves for styrene, a model molecule for the PPV monomeric unit. Again, the compressed theoretical spectrum and the experimental results are in excellent agreement. As expected when comparing the VEH density of states of the "monomer" and the polymer, the lowest lying peak of PPV, which results from the π band delocalized over the chain, has no counterparts in styrene. To align the gas phase UPS spectrum of styrene [22] with the other spectra, it is necessary to shift the binding energy scale. As a consequence, the vacuum level, which is the reference energy for gas phase measurements, does not correspond anymore to the position determined from the solid state spectra of PPV. The observed difference (Figure 7) can be explained in terms of polarization energy in the solid state. The 1.5 eV value we find here is in the range typical of π-electron-containing molecules [23]. It is noteworthy that the same value is found when comparing ethylbenzene in the gas phase and polystyrene [24]. It thus seems that the presence of the π conjugated backbone in PPV has no major influence on this phenomenon.

4. Conclusions

As shown in this paper, solubility and processability of conjugated polymers represent a major advance for fundamental studies of the electronic structure of these systems. In the case of P3HT, the evolution of the core level binding energies can be related to the appearence of charged defects (polarons, bipolarons) upon doping. We also observe a doping-induced shift of the Fermi level towards the valence band edge, in quantitative agreement with the predictions of VEH calculations. At saturation doping, a finite density of states appears at the Fermi level; this result is consistent with the existence of a polaron lattice in highly-doped polythiophenes.

In the case of PPV, the good agreement between the VEH-calculated density of states and the UPS spectra allows for a detailed assignment of the experimental bands. Comparison with the results on styrene in the gas phase points out the presence of a highly delocalized π band in the polymer, and provides an estimation of the polarization energy in the solid state.

5. Acknowledgments

This work is supported by the Swedish Board for Technical Development (STU), the Swedish Natural Sciences Research Council (NFR), and the Nordic Fund for Industrial Research (NI).

6. References

[1] M. Sato, S. Tanaka, K. Kaeriyama; J. Chem. Soc. Chem. Commun. 873 (1986)

[2] K.Y. Jen, G.G. Miller, R.L. Elsenbaumer; J. Chem. Soc. Chem. Commun. 1346 (1986)

[3] D.D.C. Bradley, R.H. Friend, H. Lindengerger, S. Roth; Polymer 27, 1709 (1987)

[4] D.R. Gagnon, F.E. Karasz, E.L. Thomas, R.W. Lenz; Synth. Met. 20, 85 (1987)

[5] S. Stafström, J. L. Brédas; Phys. Rev. B. 38, 4180 (1988)

[6] O. Inganäs, W. R. Salaneck, J. E. Österholm, J. Laakso; Synth. Met. 22, 395 (1988)

[7] W. R. Salaneck, O. Inganäs, B. Thémans, I. O. Nilsson, B. Sjögren, J. E. Österholm, J. L. Brédas, S. Svensson; J. Chem. Phys. 89, 4613 (1988)

[8] M. Lögdlund, R. Lazzaroni, W. R. Salaneck, S. Stafström, J. O. Nilsson, X. Shuang, J. E. Österholm, J. L. Brédas; in Proceedings of the 3rd International Winterschool on Electronic Properties of Polymers, in press

[9] Y. Jugnet, G. Tourillon, T. Minh Duc; Phys. Rev. Lett. 56, 1862 (1986)

[10] G. K. Wertheim, P. M. Th. Van Attekum, G. Basu; Solid State Commun. 33, 1127 (1980)

[11] W. R. Salaneck, R. Erlandsson, J. Prejza, I. Lundström, O. Inganäs; Synth. Met. 5, 125 (1983)

[12] K. Siegbahn, C.Nordling, G. Johansson, J. Hedman, P. F. Heden, K. Hamrin, U. Gelius, T. Bergmark, L. U. Werme, R. Manne, Y. Bayer; in *ESCA Applied to Free Molecules* (North-Holland, Amsterdam, 1969)

[13] B. Thémans, J. M. André, J. L. Brédas, Synth. Met. 21, 149 (1987)

[14] J. L. Brédas, B. Thémans, J. G. Fripiat, J. M. André, R. R. Chance; Phys. Rev. B 29, 6761 (1984)

[15] S. Kivelson, A.J. Heeger; Synth. Met. 17, 183 (1987)

[16] A. Saxena, J. D. Gunton; Synth. Met. 15, 23 (1986)

[17] G. Harbeke, E. Meier, W. Kobel, M. Egli, H. Kiess, E. Tosatti; Solid State Commun. 55, 419 (1985)

[18] J.L. Brédas; in *Handbook of Conducting Polymers*, T. Skotheim Ed. (Dekker, New York, 1986) pp. 859-913

[19] J.L. Brédas, G.B. Street; J. Chem. Phys. 82, 3284 (1985)

[20] A more detailed analysis of the valence spectrum of PPV will be published elsewhere.

[21] J. Obrzut, M.J. Obrzut, F.E. Karasz; Synth. Met. 29, E109 (1989)

[22] D.W. Turner, C. Baker, A.D. Baker, C.R. Brundle; in *Molecular Photoelectron Spectroscopy* (Wiley-Interscience, London, 1970)

[23] N. Sato, K. Seki, H. Inokushi; J. Chem. Soc. Farad. Trans. II 77, 1621 (1981)

[24] W.R. Salaneck; in *Characterization of Molecular Structure of Polymers by Photon, Electron, and Ion Probes*, T.J. Fabish, D. Dwight, H.R.Thomas Ed's (Am. Chem. Soc., Washington, 1981) pp. 121-149

BRIDGED MIXED VALENCE PHTHALOCYANINATO-METAL COMPOUNDS

A. HIRSCH, M. HANACK*
Institut für Organische Chemie
Auf der Morgenstelle 18
D-7400 Tübingen
West Germany

ABSTRACT. Bridged mixed valence phthalocyaninato-metal dimers, trimers and polymers (M = Fe, Co) with the central metal in the oxidation state of +II and +III and cyanide and pyrazine as bridging ligands have been synthesized. The synthesis has been carried out at either by coupling of PcM(L)CN (M = Fe(III), Co(III); L = pyz, py, t-bupy) with $PcFe(NH_3)_2$ and substituting the weak coordinated ammonia or by thermal decomposition of PcFe(pyz)CN. The infrared, Mößbauer and electrical properties of these compounds have been investigated.

1. Introduction

Macrocyclic transition metal complexes linked by linear organic bridging ligands exhibit semiconducting properties (Fig. 1). In this way a large variety of polymeric compounds has been investigated [1].

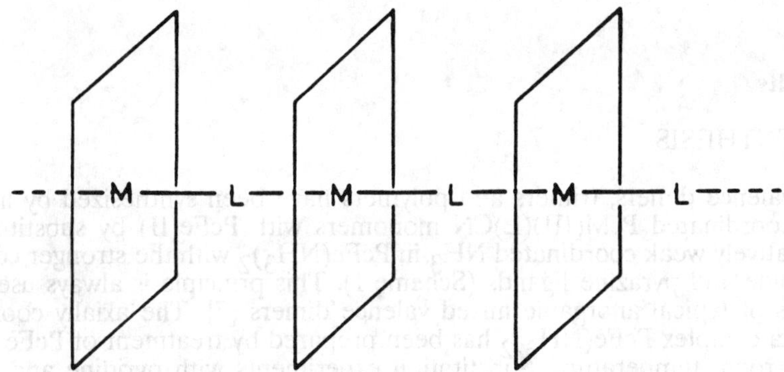

Figure 1. Bridged macrocyclic metal complexes

By oxidativ doping the electrical conductivity can be increased by several orders of magnitude. Herein the doping process occurs mainly at the macrocycle whereas the oxidation state of the central metal e.g. Fe(II) is not changed as shown by

J. L. Brédas and R. R. Chance (eds.), Conjugated Polymeric Materials:
Opportunities in Electronics, Optoelectronics, and Molecular Electronics, 163–169.
© 1990 *Kluwer Academic Publishers. Printed in the Netherlands.*

Mößbauer spectroscopy [2]. By using, e.g. cyanide or tetrazine as bridging ligands, semiconducting materials can be obtained without further oxidative doping [3,4]. The high conductivity of, e.g. $[PcFe(tz)]_n$, can be related to a charge transfer from the central Fe(II) to tetrazine [5]. The mixed valence polymer, shown in Fig. 2, contains the central metal atoms (Fe, Co) in the formal oxidation states of $+II$ and $+III$.

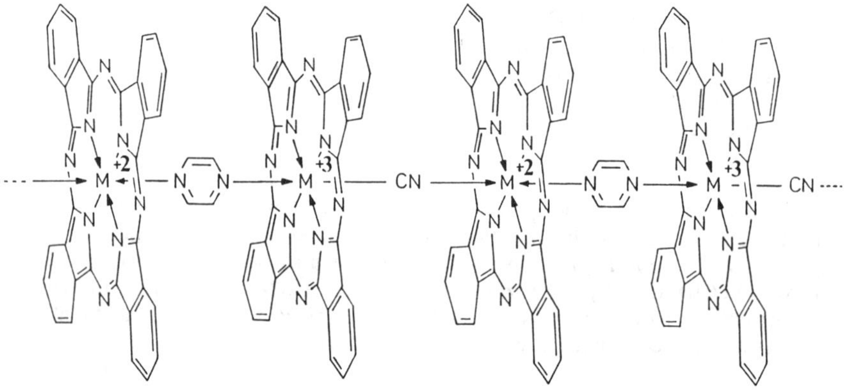

Figure 2. Mixed valence polymer

The use of a neutral as well as a negatively charged bridging ligand (pyz and CN^-) leads to a system with an overall charge neutrality. A structurally similar type of mixed valence compounds is the "Creutz-Taube ion", where one electron is delocalized across the pyrazine ligand onto the entire molecule [6].

2. Results

2.1. SYNTHESIS

Mixed valence dimers, trimers and polymers have been synthesized by linking of axially coordinated PcM(III)(L)CN monomers with PcFe(II) by substituting the comparatively weak coordinated NH_3 in $PcFe(NH_3)_2$ with the stronger coordinating cyanide and pyrazine ligands (Scheme 1). This principle is always used in the synthesis of typical anorganic mixed valence dimers [7]. The axially coordinated ammonia complex $PcFe(NH_3)_2$ has been prepared by treatment of PcFe in liquid NH_3 at room temperature. Substitution experiments with pyridine and pyrazine confirmed the expected coordination behavior. Reaction of $PcFe(NH_3)_2$ with stoichiometric amounts of pyrazine and pyridine leads to $[PcFe(pyz)]_n$ and $PcFe(py)_2$ respectively, whereas substitution of pyridine and pyrazine by NH_3 in the corresponding compounds is not possible even if an excess of ammonia is used.

Scheme 1.

Synthesis of bridged mixed valence phthalocyaninato metal compounds

Dimer and trimer:

$PcFe(L)CN + PcFe(NH_3)_2 \longrightarrow PcFe(L)CNPcFe(NH_3) + NH_3$

$\underline{1}$: L = pyridine

$2\,PcM(L)CN + PcFe(NH_3)_2 \longrightarrow PcM(L)CNPcFePc\,M(L)CN + 2\,NH_3$

$\underline{2}$: M = Fe, L = py
$\underline{3}$: M = Fe, L = t-bupy

$\underline{4}$: M = Co, L = py
$\underline{5}$: M = Co, L = t-bupy

Polymer:

$PcM(pyz)CN + PcFe(NH_3)_2 \longrightarrow [pyzPcMCNPcFe]_n + 2\,NH_3$

$\underline{6}$: M = Co

The three valent monomers $PcM(L)CN$ (M = Fe, Co) have been synthesized by degradation reactions of the polymers $[PcMCN]_n$ with the bases pyridine, t-butyl-pyridine and pyrazine.

The mixed valence compounds $\underline{1}$ - $\underline{6}$ have been prepared at 80°C in toluene or ethanol as solvent. The reaction time is 72 h. For M(II), M(III) = Fe a polymeric solid is also formed by thermal decomposition. Treatment of $PcFe(pyz)CN$ in a solvent (toluene, ethanol) leads to partial homolytic cleavage of CN radicals from this monomer. The resulting very reactive pentacoordinated -pyzPcFe(II) species serves as an intermediate in the formation of bridged mixed valence complexes. Such an intermediate can react with the pyrazine site as well as the cyanide site of $PcFe(pyz)CN$. The content of Fe(II) and Fe(III) is dependent on the reaction time and temperature. The general formula of such a system of solids is $[PcFe(III)_xPcFe(II)_{(1-x)}(pyz)_y(CN)_x]_n$ ($\underline{7}$) where $0 \le x \le 1$ and $0.5 \le y \le 1$.

2.2. IR SPECTROSCOPY AND CONDUCTIVITIES

Table 1 shows the cyanide valence frequencies and the room temperature conductivites of the mixed valence compounds $\underline{1}$ - $\underline{6}$ and the corresponding monomers $PcM(L)CN$. The formation of a cyano bridge leads to an increase of the CN valence frequency. This can be explained by a decrease of electron density of the antibonding p-orbital due to coordination to a second metal atom. An increase of the bond order and therefore of the valence frequency is also observed at the polymerisation of monomeric $K[PcM(III)(CN)_2]$ compounds (M = Fe, Co, Rh, Cr, Mn) to the bridged $[PcM(III)CN]_n$ systems [4].

Table 1. IR data and conductivities

Compound	$\nu_{CN}[cm^{-1}]$	$\sigma_{RT}[S/cm]$
PcFe(py)CN 1 2	2131 2136 2135	1×10^{-10} 3×10^{-5} 2×10^{-5}
PcFe(t-bupy)CN 3	2132 2135	6×10^{-5} 1×10^{-4}
PcCo(py)CN 4	2146 2150	3×10^{-12} 3×10^{-8}
PcCo(t-bupy)CN 5	2143 2151	2×10^{-13} 5×10^{-10}
PcCo(pyz)CN 6	2148 2152	2×10^{-11} 1×10^{-6}

Also the electrical conductivities are increasing if a bridged mixed valence compound is formed. The highest conductivities are reached if only Fe is used as central metal. In the mixed metal complexes 4, 5 and 6 charge delocalisation is less favourable.

2.3 ^{57}FE-MÖSSBAUER SPECTROSCOPY

The ^{57}Fe-Mößbauer data of the compounds synthesized are given in table 2. The data for the isomer shift and quadrupole splitting of the Fe(II) centers lie in the typical range for low spin hexacoordinated phthalocyaninato- iron complexes [8,9]. This points to a similar coordination behaviour of the PcM(L)CN moiety and a N-base like pyridine. The isomer shift of the Fe(II) center in compound 3 is comparatively small which can be related to a high decrease of electron density of the 3d-orbitals.

Table 2. ^{57}Fe-Mößbauer data

Compound	T[K]	ΔE_Q[mm/s]	IS[mm/s][a]
1[b] M(II)	293	1.79	0.26
M(III)		147	-0.05
2[b] M(II)	293	2.01	0.24
M(III)		1.61	0.09
3[b] M(II)	293	1.98	0.11
M(III)		1.22	0.03
4[c] M(II)	77	1.93	0.33
5[c] M(II)	293	2.01	0.23
6[c] M(II)	293	2.00	0.26
PcFe(py)CN[c]	293	1.57	0.04
PcFe(t-bupy)CN[c]	293	1.57	0.04
PcFe(pyz)CN[c]	293	1.60	0.05
[PcFe(pyz)]$_n$[c]	293	2.00	0.25

a) Relative to metalic iron.
b) Two quadrupole doublets.
c) One quadrupole doublet.

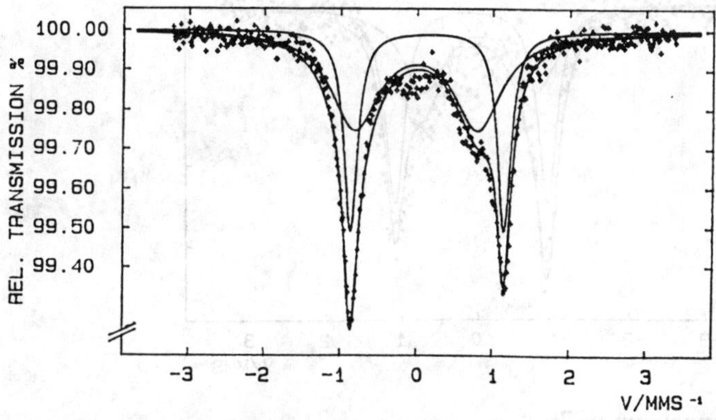

Figure 3. Mößbauer spectrum of **2**

2.4. THE SYSTEM $[PcFe(III)_xPcFe(II)_{(1-x)}(pyz)_y(CN)_x]_n$

The heating of PcFe(III)(pyz)CN leads to a system of solids containing Fe(II) and Fe(III) centers bridged with pyrazine and cyanide. The content of Fe(II) and Fe(III) is dependent on the reaction temperature. Table 3 shows the Mößbauer data and the conductivities of such solids.

Table 3. Mößbauer data and conductivites of $[PcFe(III)_xPcFe(II)_{(1-x)}(pyz)_y(CN)_x]_n$ compounds

Compound	center	ΔE_Q[mm/s]	IS[mm/s]	σ_{RT}[S/cm]
$\underline{7a}$[a]	Fe(II) Fe(III)	2.00 1.32	0.11 0.04	5×10^{-5}
$\underline{7b}$[b]	Fe(II) Fe(III)	2.01 1.47	0.19 0.07	3×10^{-5}
$\underline{7c}$[c]	Fe(II) Fe(III)[d]	1.99	0.24	9×10^{-6}

a) 50°C, toluene
b) 70°C, toluene, ethanol
c) 80°C, toluene, ethanol
d) not detectable

The IR spectrum of $\underline{7c}$ is identical with that of $[PcFe(pyz)]_n$ (x = 0), (y = 1).

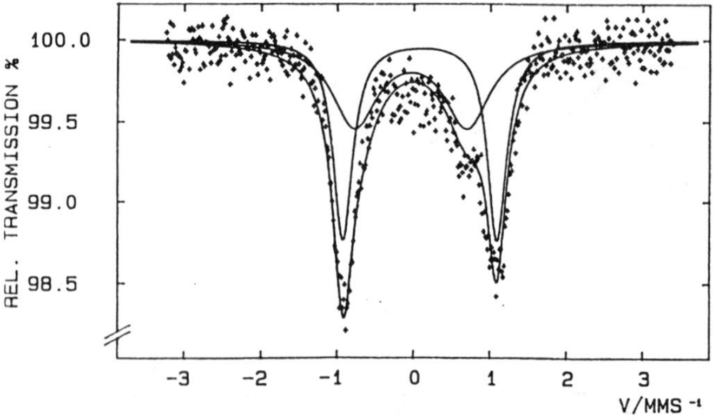

Figure 4. Mößbauer spectrum of $\underline{7b}$

If PcFe(pyz)CN is heated at 80°C in toluene/ethanol 1:1 over a period of 5 days a solid is formed (7c) which exhibits nearly the same physical properties as [PcFe(pyz)]$_n$. There is no coordinated cyanide present and all of the iron has been reduced to Fe(II). In this case x is 0 and y is 1. If beside the proposed mechanism of the formation of these compounds a preceding dimerisation of PcFe(pyz)CN to NCPcFepyzPcFeCN takes place, y in addition can show values ranging from 0.5 to 1. Otherwise y is always 1. The cleavage of a CN radical from this dimer also leads to a reactive pentacoordinated intermediate able to form mixed valence complexes together with PcFe(pyz)CN. There is no direct evidence to date in the formation of such a dimer. The solids 7a and 7b show Fe(II) as well as Fe(III). The isomer shifts of the Fe(II) centers are comparatively small which points to a low electron density of the corresponding 3d orbitals. That can be caused by a charge delocalisation to the polymer chain. The pressed powder conductivities of 7a - 7c lie in the same range than those of the dimer and trimers containing only Fe as central metal.

3. References

[1] M. Hanack et al. in T.J. Skotheim, Handbook of Conducting Polymers, Marcel Dekker, New York 1985.
[2] M. Hanack, A. Leverenz, Synth. Met. 22, 9 (1987).
[3] M. Hanack, S. Deger, A. Lange, Coord. Chem. Rev. 83, 115 (1988).
[4] A. Datz, J. Metz, O. Schneider, M. Hanack, Synth. Met. 9, 31 (1984).
[5] M. Hanack, A. Lange, M. Rein, R. Behnisch, G. Renz, A. Leverenz, Synth. Met. 29, F1 (1989).
[6] C. Creutz, H. Taube, J. Am. Chem. Soc. 91, 3988 (1969).
[7] C. Creutz, Prog. Inorg. Chem. 30, 1 (1983).
[8] B.N. Diel, T. Inabe, N.K. Jaggi, J.W. Lyding, O. Schneider, M. Hanack, C.R. Kannewurf, T.J. Marks, L.H. Schwarz, J. Am. Chem. Soc. 106, 3207 (1984).
[9] F. Calderazzo, S. Frediani, B.R. James, G. Pampaloni, K.J. Remier, J.R. Sams, A.M. Serra, D. Vitali, Inorg. Chem. 21, 2302 (1982).

POLYANILINE PROCESSED FROM SULFURIC ACID AND IN SOLUTION IN SULFURIC ACID: ELECTRICAL, OPTICAL AND MAGNETIC PROPERTIES

Y. Cao, P. Smith and A.J. Heeger
Institute for Polymers and Organic Solids
University of California, Santa Barbara 93106

ABSTRACT. Both the salt and the base forms of polyaniline can be completely dissolved in concentrated sulfuric acid at room temperature, with polymer concentrations ranging from extremely dilute to more than 20% (w/W). This solubility opens the way to processing pure, partially crystalline polyaniline or composites of polyaniline with other commercial polymers into fibers and films, etc. We summarize the structural data and the electrical, optical and magnetic properties of polyaniline (in its various forms) processed from sulfuric acid, and the optical and magnetic properties of the emeraldine salt in solution in sulfuric acid.

1. Introduction

Although there has been considerable progress toward the development of soluble conducting polymers (such as substituted polythiophenes and substituted polyparaphenylene-vinylene), polyaniline in sulfuric acid is the first example of a stable, concentrated, solution of a conjugated polymer in which the material is converted by the solvent to the conducting form and from which the polymer can be processed directly into the metallic salt with no need for subsequent doping.[1]

Polyaniline has been investigated extensively since the beginning of this century.[2] There has ben renewed interest in this polymer, in recent years, as a conducting polymer.[3] Among conducting polymers, polyaniline is unique in that its electronic structure and electrical properties can be reversibly controlled both by charge transfer doping (to vary the oxidation state of main chain) and by protonation.[4] The wide range of associated electrical, electrochemical and optical properties, coupled with good stability, make polyaniline attractive as an electronic material for potential use in a variety of applications.

Throughout the extensive literature, polyaniline has been generally categorized as an intractable material. Recently, however, two groups[1,5] have reported methods to dissolve and process polyaniline without changing of the structure of the polymer (in N-methylpyrrolidinone[5] or in concentrated sulfuric[1] and other strong acids[1]). In contrast to alternative methods for achieving solubility through preparation of substituted[6] polyaniline or through the synthesis of graft or block copolymers,[7] the resulting films[1,5] and fibers[1] are highly conductive after processing. The use of

171

J. L. Brédas and R. R. Chance (eds.), Conjugated Polymeric Materials:
Opportunities in Electronics, Optoelectronics, and Molecular Electronics, 171–193.
© 1990 *Kluwer Academic Publishers. Printed in the Netherlands.*

concentrated acids[1] has specific advantages[8] in that both the salt and the base form of polyaniline can be completely dissolved at room temperature, with polymer concentrations ranging from extremely dilute to more than 20% (w/w), in concentrated protonic acids such as H_2SO_4, CH_3SO_3H, and CF_3SO_3H.

The solubility of this conducting polymer opens the way to processing pure, partially crystalline polyaniline or composites of polyaniline with the other commercial polymers into fibers and films, etc. In addition, this solubility enables extensive characterization of polyaniline as a macromolecular system (e.g. viscosity in solution as a probe of molecular weight, etc.) and of polyaniline as a conducting polymer (e.g. optical studies of spin-cast films as a probe of electronic structure of the salt or base forms). The latter is the subject of this paper.

The structure of the semiconducting emeraldine base form of polyaniline is

$$[(1A)(2A)]_n \tag{1}$$

where

$$(1A) = (B\text{---}NH\text{---}B\text{---}NH\text{---}) \tag{2a}$$

and

$$(2A) = (B\text{---}N=Q=N\text{---}). \tag{2b}$$

In the above, B denotes a C_6H_4 ring in the benzenoid form and Q denotes a C_6H_4 ring in the quinoid form. The existence of one quinoid ring (out of four) in the emeraldine base has been well established in the literature.[9] The emeraldine base can be fully reduced to leucoemeraldine, $(1A)_n$, a large bandgap insulator with π-π^* transition at nearly 4 eV. Upon protonation of $[(1A)(2A)]_n$ to the emeraldine salt, there is a structural change (with no change in the number of electrons) leading to a half-filled band and a metallic state (described as a polaronic metal) of the form[10-13]

$$[1S]^{\cdot}(A^-)_n = [B\text{---}NH\text{---}B\text{---}NH^+\text{---}]^{\cdot}_n(A^-)_n \tag{3}$$

where A^- is the counter ion (e.g. ClO_4^-, Cl^-, HSO_4^- etc.), and the $[\ ]^{\cdot}$ denotes one unpaired electron per formula unit. Further oxidation of $[1S]^{\cdot}(A^-)_n$ is expected to yield the fully-oxidized copolymer with alternating quinoid and benzonoid monomer units:

$$(B\text{---}NH^+=Q=NH^+\text{---})_n \tag{4}$$

with an H-atom bonded to each nitrogen, and with one counterion for each charge on the chain. Since the $(-N^+=Q=N^+-)$ unit is a charged bipolaron, the polymer structure in (4) is that of a charged bipolaron lattice with a two-fold degenerate ground state. Note that (4) can in principle be obtained directly from $(1A)_n$ by charge transfer doping.

In the following sections, we briefly review recent progress in the structure and properties of the various forms of polyaniline with emphasis on the use of sulfuric acid as a means of characterizing and processing the polymer.

2. Structure[14]

Polyaniline (emeraldine salt) can be recovered in *partially crystalline* form from solutions in sulfuric acid[5] This is demonstrated by the wide-angle X-ray diffraction pattern of such a polyaniline film, shown as the upper solid curve in Figure 1a. Essentially identical data were obtained by X-ray scans from the film and powder samples. In addition, we found
that the as-synthesized emeraldine salts, $[1S]^{+}(Cl^{-})_n$ (dotted curve) and $[1S]^{+}(ClO_4^{-})_n$ (lower solid curve), are partially crystalline. The data indicate that the $[1S]^{+}(HSO_4^{-})_n$ samples and $[1S]^{+}(Cl^{-})_n$ sample are more highly crystalline than the $[1S]^{+}(ClO_4^{-})_n$ sample.

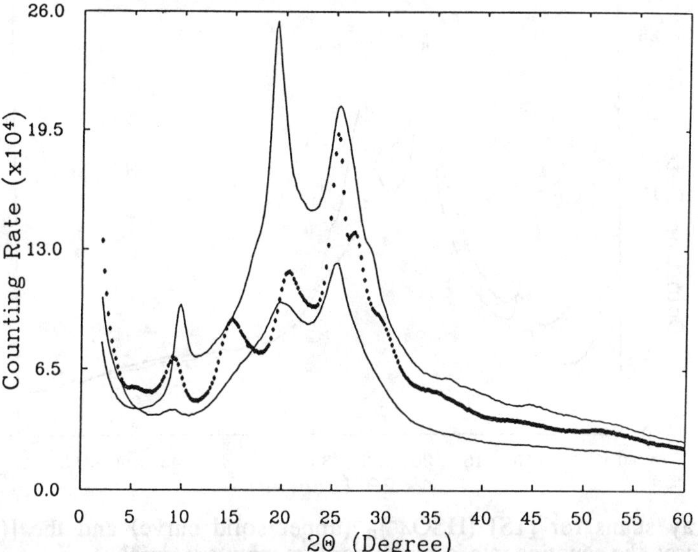

Fig. 1a. X-ray scans for $[1S]^{+}(HSO_4^{-})_n$ (upper solid curve), $[1S]^{+}(Cl^{-})_n$ (dotted curve), and $[1S]^{+}(ClO_4^{-})_n$ (lower solid curve). For the upper curve, the counting rate is counts per five minute internal; for the other two curves, the counting rate is counts per four minute interval.

The intensity of the various reflections and their clear definition above the background amorphous scattering indicate that the $[1S]^{+}(A^{-})_n$ materials have significant crystallinity. For the $[1S]^{+}(HSO_4^{-})_n$ and $[1S]^{+}(Cl^{-})_n$ samples, as many as eight reflections can be identified. The relatively narrow widths of the diffraction peaks (full width at half maximum as narrow as approximately 1°) indicate relatively long range structural coherence.

Figure 1b compares the corresponding wide angle X-ray scans for the $[1S]^{+}$ material (as precipitated from sulfuric acid) and the emeraldine base prepared by carefully compensating the same material by exposure to aqueous ammonia solution. Although compensation reduces the crystallinity, the resulting emeraldine base shows X-ray reflections which clearly indicate structural order.

Attempts to establish the details of the interchain packing and crystal structure through comparison of the calculated structure factors with the data in Figures 1 and 2 are currently underway. However, some more general aspects of the structure can be inferred directly from the data. The coherence lengths parallel and perpendicular to the chains (based on the narrowest lines at $2\theta \approx 10^\circ$ and 19°, respectively) are comparable. From the Scherrer formula,

$$\zeta \approx 2\pi\lambda/\delta(2\theta)$$

where $\delta(2\theta)$ is the full width at half-maximum, we find $\zeta \approx 10$ nm.

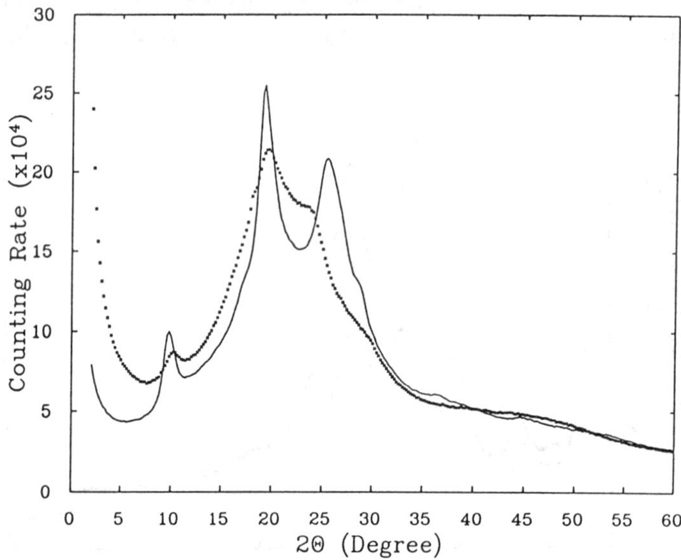

Fig. 1b. X-ray scans for $[1S]^\cdot(HSO_4^-)_n$ (upper solid curve) and the $[(1A)(2A)]_n$ emeraldine base; the counting rate is counts per five minute internal.

The similarities between the molecular structure of the emeraldine base and those of poly(p-phenylene sulfide),[15a] poly(p-phenylene oxide),[15b] and even poly(p-phenylene terephthalamide)[15c] suggest that a similar interchain packing might be expected. Indeed, the two strongest equatorial peaks in Figure 2 (at $2\theta=19.5^\circ$ and $2\theta=22.8^\circ$) correspond to the similar peaks observed in all three materials.

3. Spectroscopy

By utilizing the partial solubility of polyaniline in dimethylformamide (DMF), dimethylsulfoxide (DMSO) and tetrahydofuran (THF) and the more complete solubility in N-methylpyrrolidinone (NMP), films of chemically polymerized polyaniline have been prepared for spectroscopic studies.[11] With DMF, DMSO and THF, it is necessary to remove insoluble materials in order to cast films from solution. Since we have recently found that this soluble part has low molecular weight and usually consists of only about

20% of the total mass,[16] there is a need to extend the spectroscopic studies to evaluate the effect of molecular weight on the principal absorption features.

Spectroscopic measurements have provided information on the electronic structure of the insulating emeraldine base form of polyaniline[17-19] and the evolution of the electronic band structure as a function of the degree of protonation. On the basis of magnetic susceptibility experiments,[10-13,20] it was proposed that the electronic structure changed from a semiconductor in the emeraldine base (the structure given in equation 1) to a metal in the protonated emeraldine salt (the structure given in equation 3) as a result of the formation of a polaron band resulting from an internal redox reaction[17,21] described as proton-induced spin unpairing.[20] Upon protonation, the absorption at 2 eV in the emeraldine base polymer disappears and two new absorptions appear, centered at 1.5 and 2.9 eV. Using the results of band structure calculations,[17] the latter have been assigned[17] to excitations from the highest and second-highest occupied energy bands to the partially filled polaron band. The long tail extending into the infrared has been assigned to intraband free carrier excitations.[12,13]

Recently, the results of photoinduced absorption studies (i.e. excitation spectroscopy) of the emeraldine base polymer have been reported.[22-24] Photoinduced absorptions were observed[23,24] at 0.9eV, 1.4eV and 3.0 eV, accompanied by photo-induced bleaching at 1.8eV and at energies above 3.5 eV. The 1.4 and 3.0 eV photoinduced absorptions have been interpreted, by analogy with the absorption spectrum of the emeraldine salt, to the photoproduction of polarons in emeraldine base.[24] Stafstrom et al[25] concluded that positive and negative polarons as well as positive bipolarons account very well for the 1.4 and 3.0 eV photoinduced absorptions. No specific explanation has been proposed for the 0.9 eV photoinduced peak.

We have recently published the results of a comprehensive spectroscopic study of the emeraldine salt of polyaniline in dilute solution in concentrated sulfuric acid and in the form of thin films spin-cast from sulfuric acid solutions. Three spectral features are observed, at 1 eV (with a tail extending deeper into the infrared), at 1.5 eV, and at 3 eV. In the films, the relative intensities of the 1eV and 1.5 eV absorption bands were found to be strongly dependent on both the molecular weight and the protonation level; the 1.0 eV absorption is strongest in the fully protonated emeraldine salt with the highest molecular weight. We also present a method for preparation and stabilization of the fully oxidized polyaniline charged bipolaron lattice form of polyaniline (the structure in equation 4) and give the first spectroscopic characterization of this novel alternating copolymer. We analyze the available data in comparison with the results of band structure calculations[17,18] to provide the basis for an understanding of the electronic structure of the four principal forms of polyaniline: the fully reduced leucoemeraldine, $(1A)_n$; the partially oxidized emeraldine base, $[(1A)(2A)]_n$; the oxidized and fully protonated emeraldine salt, $[1S]^{\cdot} (A^-)_n$; and the fully oxidized bipolaron lattice, $(-B-NH^+=Q=NH^+-)_n$.

3.1 EXPERIMENTAL METHODS AND TECHNIQUES

The polyaniline samples used for the spectroscopic studies were synthesized according the method described in detail previously.[16] The oxidizing agent was prepared by dissolving 23.0 g (0.11 mole) of ammonium persulfate (Aldrich) into 250 ml 1.5M aqueous HCl solution. This was slowly added (while stirring vigorously) to 250 ml of aqueous HCl solution of the same molarity that contained 20 ml (0.22 mole) of aniline. During the entire 3 hour period of addition, the temperature of the polymerizing mixture was carefully maintained at 0°C (to within 1°C). After the oxidant was added, the reaction mixture was left stirring at 0°C for an additional hour. The precipitated polyaniline was recovered from the polymerization vessel, filtered, and then washed with distilled water until the washing liquid was completely colorless. In order to remove oligomers and other organic by-products, the precipitate was washed with several portions of methanol until the methanol solution was colorless. Finally, the material was washed twice with ethyl ether and subsequently dried at room temperature for 48 hours in dynamic vacuum, until constant mass was reached. The pristine polyaniline salt was converted to the base form by treatment with 3% aqueous NH_4OH solution for two hours, followed by washing with distilled water, methanol and ethyl ether. Polyaniline synthesized in this way has been thoroughly characterized in terms of viscosity (as a measure of molecular weight), structure and magnetic properties, as well as electrical conductivity.

The emeraldine salt films were spin-cast from a viscous 3% (w/w) sulfuric acid solution of as-polymerized polyaniline onto sapphire substrates at spin rates of $1.5 - 2.0 \times 10^3$ cpm during a period of 2-3 minutes. After spinning, the film on the substrate was left in air for 1 hour in order to complete the precipitation of the polyaniline. The resulting thin film was then washed several times in distilled water to remove any residue of free sulfuric acid and subsequently dried in dynamic vacuum. Such polyaniline emeraldine salt films appear homogeneous, and they adhere tightly to the substrate. Typical film thicknesses were 5000Å as determined by Dektak measurements. In order to convert a film to the base form, it was treated with 3% aqueous NH_4OH solution, followed by washing with distilled water and drying in dynamic vacuum. Fully reduced leucoemeraldine films with structure $(1A)_n$ were prepared by dipping an emeraldine base film into phenylhydrazine for several hours under nitrogen followed by thorough washing with deoxygenated acetone.

Polyaniline films with different degrees of protonation were prepared by dipping the films (on the substrate) for 30 minutes into a large excess of previously prepared HCl solution, the pH of which was preset and determined by a Fisher Model 955 pH meter. The films were again dried in dynamic vacuum. Since we found that the spectra did not change after 30 minutes in the HCl solutions, this is sufficient time for the thin film to come to equilibrium with the solution.

Solution spectra were obtained from samples prepared by diluting a homogeneous 3% solution of polyaniline in sulfuric acid by 97% H_2SO_4 to appropriate concentrations. Unless otherwise stated, the solution spectra were obtained at the dilute concentrations of 10^{-4} mole/liter.

The inherent viscosity of the polyaniline base (in concentrated sulfuric acid) was used as an indicator of the molecular weight. Measurements of the inherent viscosity of the polyaniline emeraldine base (obtained by exposure to aqueous NH_3 solutions and subsequently washed) were carried out using an Ubbelohde viscometer at 25°C using 0.1% w/w polymer solutions in H_2SO_4. The polyaniline batch synthesized for this study had an inherent viscosity of 1.0 dl/g at 25°C (0.1 wt% of the emeraldine base in H_2SO_4). The material was fractionated for studies of the molecular weight dependence. Samples with $\eta_{in}=0.18$ dL/g were made from the low molecular weight fraction extracted from the same batch; i.e. that fraction which was soluble in THF. Samples with $\eta_{in}=1.6$ dL/g sample were made from that fraction which was insoluble in THF (after the elimination of

THF-soluble part by extraction). Note that the fraction which was insoluble in THF was soluble in concentrated sulfuric acid. Thus, the fraction which is insoluble in common organic solvents is not crosslinked, but simply has a relatively high molecular weight.

The fraction of the sample which gives η_{in}=1.6 dL/g has a molecular weight (as estimated from the Mark-Houwink relations[26]) in the range from about 15,000 (rigid chain limit) to about 60,000 (flexible chain limit). Based on these estimates of the molecular weight, the concentration (10^{-4}mole/liter) used for the solution spectroscopic studies is dilute; i.e. involving negligible interchain interaction.

Electronic absorption spectra (from $\approx$0.3 eV to $\approx$ 6.0 eV) were recorded with a Perkin-Elmer Lamda 9 UV/VIS/IR spectrophotometer. Mid-IR spectra were obtained with a Perkin-Elmer 1330 Infrared Spectrophotometer.

3.2. EXPERIMENTAL RESULTS

3.2.1 *Molecular weight (viscosity).* The absorption spectra of polyaniline films cast from sulfuric acid are strongly dependent on the molecular weight (viscosity). Figure 2 shows spectra of three thin films spin-cast from sulfuric acid solution and subsequently treated by 0.5M HCl solution to achieve full protonation. To study the effect of molecular weight on the absorption spectrum, we used emeraldine base samples fractionated from the same preparation batch (polymerization and compensation) as described in the previous section. This procedure prevents any uncertainty in the viscosity value and avoids slight variations in the preparation procedure.

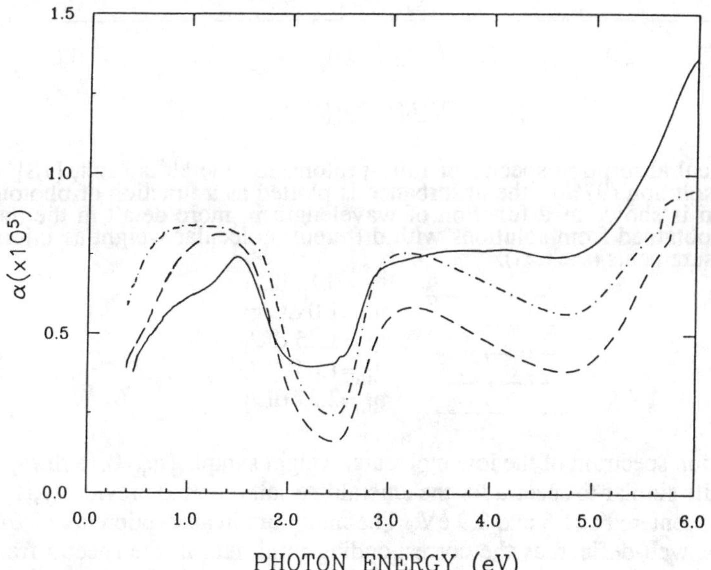

Fig. 2. Optical absorption spectra of films of the fully protonated emeraldine salt $[1\ddot{S}]^{+}(A^{-})_n$, spin-cast from sulfuric acid; the absorption coefficient is plotted as a function of photon energy. The three spectra were obtained from films with different molecular weights as inferred from viscosity measurements (see text).

———————— η_{in}=0.18 dL/g   — — — η_{in}=1.0 dL/g   — — — — — η_{in}=1.6 dL/g

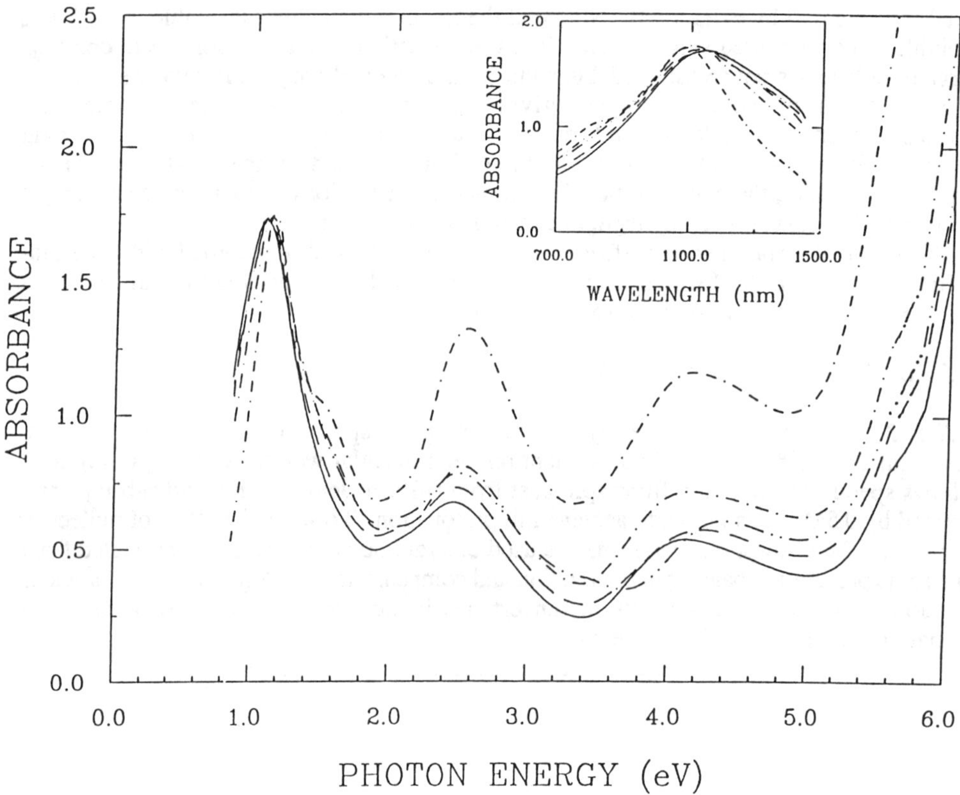

Fig. 3.   Optical absorption spectra of fully protonated emeraldine salt, $[1S]^{\cdot}(A^{-})_n$, in sulfuric acid solution (97%); the absorbance is plotted as a function of photon energy. The IR region is shown as a function of wavelength in more detail in the inset. The spectra were obtained from solutions with different molecular weight as inferred from viscosity measurements (see text).

η_{in}=0.18 dL/g
η_{in}=1.0 dL/gπ
η_{in}=1.45 dL/g
η_{in}=1.65 dL/g
η_{in}=2.27 dL/g

The absorption spectrum of the low molecular weight sample (η_{in}=0.18 dL/g) shown in Figure 2 is quite similar to spectra for the emeraldine salt reported previously;[11,17] there are two peaks, centered at 1.5 and 2.9 eV. The minimum in absorption near 2.0-2.4 eV is not as deep or well-defined as the corresponding minimum in the spectra from higher molecular weight samples, implying excess absorption near 2 eV; in addition, there is residual interband absorption near 4 eV. Since both of these features are characteristic of the emeraldine base, the spectrum of the low molecular weight material implies incomplete conversion to the emeraldine salt. The fact that these features remained after protonation for the η_{in}=0.18 dL/g sample suggests that the low molecular weight (protonated) material is not fully metallic, consistent with the low electrical conductivity of this sample (after precipitation from sulfuric acid, we find $\sigma= 4$ S/cm for $\eta_{in}= 0.18$ dL/g compared with $\sigma=33$ S/cm for $\eta_{in}=1.6$ dl/g). In comparison with previously reported spectra, an important difference in the spectrum in Figure 1 for the low molecular weight sample is the clear shoulder at 1.0 eV.

Figure 2 shows that upon increasing the molecular weight through $\eta_{in}=1.0$ dL/g to $\eta_{in}=1.6$ dL/g, the intensity of the 1 eV feature increases until the infrared absorption becomes a broad band with a long tail extending into the deep infrared, consistent with the metallic character of these samples. In addition, for the highest molecular weight sample, there is a single asymmetric absorption band which peaks at 3.0-3.2 eV with no residual interband absorption near 4 eV from the emeraldine base.

Since the films used in earlier studies[11,17] were cast from DMF or DMSO, it is not surprizing that only the 1.5 eV peak had been reported. The fraction (~20%) of polyaniline emeraldine base which is soluble in these solvents has the same viscosity (and hence approximately the same low molecular weight) when redissolved in sulfuric acid as the THF-soluble emeraldine base used to obtain the solid curve on Figure 1. Moreover, unlike the homogeneous films obtained by spin-casting from sulfuric acid, in our experience the thin films cast from organic solvent solution are more granular, leading to decreased spectral resolution due to stronger scattering.

In Figure 3, we show the <u>solution</u> spectra of samples prepared with different viscosity. We present the data as a function of wavelength to emphasize the subtle changes found in near infrared region. Comparing Figures 2 and 3, one finds that the two near infrared features appear at almost the same positions, 1.0 and 1.5 eV, respectively, while their relative intensities are inverse in the solid and solution spectra. In the dilute solution spectra, the intensity of 1.0 eV feature was always much higher than that of the 1.5 eV absorption; in fact, the 1.5 eV absorption is completely absent in the high molecular weight material, while the 1.0 eV absorption (with a long tail extending farther into the infrared) completely dominates. This result implies that the 1.0 eV absorption band arises from an <u>intrachain excitation,</u> while the 1.5 eV transition is due to <u>interchain</u> charge separation. Note that the 3.0 eV peak in the solid film spectra was red shifted to about 2.5 eV in the solution spectra; i.e. opposite to the usual solid state/solution effect.

3.2.2. *Degree of protonation.*

Figure 4 shows a series of absorption spectra of polyemeraldine spin-cast from solutions with $\eta_{in}=1.6$ dL/g, and subsequently equilibrated with aqueous HCl solutions of different pH. Chiang et al[27] have calibrated the relationship between the pH of the solution in which emeraldine base samples were equilibrated and the fractional protonation; the lower pH gives the higher degree of protonation. We demonstrated previously that treatment of emeraldine base with aqueous solution of pH=1 resulted in essentially complete protonation of the emeraldine salt.[16] As can be seen from Figure 4, at pH=5 (where the protonation begins), the band at 2 eV shifted slightly to lower energy and the intensity decreased, while a new shoulder appeared at about 1.4 eV, indicative of partial protonation. The simultaneous existence of the 2 eV and 1.4 eV bands suggests phase segregation of fully protonated and unprotonated domains. At pH=3, although the protonation level is still very low (the protonated fraction[25] is less 0.05) the 2 eV absorption has disappeared completely, while in the near-infrared region only a single, almost symmetric peak is observed centered at about 1.4 eV. Further increase of protonation level (by equilibration in pH as low as

180

0.15) leads to an important red shift and broadening of this peak. Noting the existence of the two features in this spectral region (as described in previous section), the data indicate that the intensity of the lower energy (intrachain) feature increases with increasing level of protonation, qualitatively consistent with an increase in the number of free carriers. The complete absence of the well-defined absorption at 4 eV (characteristic of the π-π^* transitions in leucoemeraldine and in the emeraldine base) is also consistent with metallic behavior; the oscillator strength has shifted into the infrared.

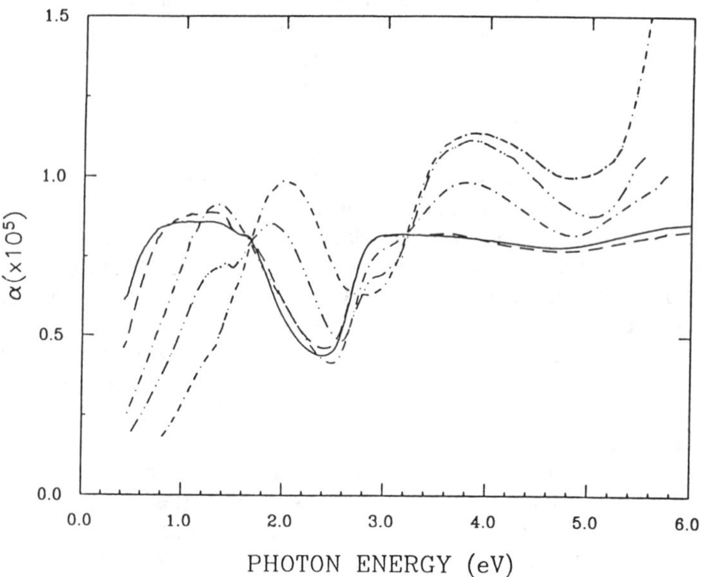

Fig. 4. Optical absorption spectra of polyaniline film spin-cast from sulfuric acid and subsequently exposed to media with different pH to control the protonation level. The absorption coefficient is plotted as a function of photon energy. The film was made from relatively high molecular weight material with η_{in}=1.6 dL/g.

_ _ . _ _ . _ _	pH>10
_ _ .. _ _ .. _ _	pH=5
_ _ . _ _ . _ _	pH=3
_ _ _ _ _	pH=2
_________	pH=0.15

3.2.3. _Leucoemeraldine and the Emeraldine Base._ The spectra of leucoemeraldine, $(1A)_n$, and the emeraldine base, $[(1A)(2A)]_n$, have been extensively studied elsewhere.[28,29] We include the spectra of these two important forms of polyaniline obtained from films spin-cast from sulfuric acid (see Figure 5) in order that they can be directly and quantitatively compared with those of the emeraldine salt and the bipolaron lattice. Leucoemeraldine is an insulator with a large band gap. The π-π^* transition shows an onset at approximately 3 eV with a peak at 3.7-3.8 eV. The emeraldine base shows two principal absorption bands (see Figure 5), with maxima at 2 eV and at 3.9 eV, respectively.

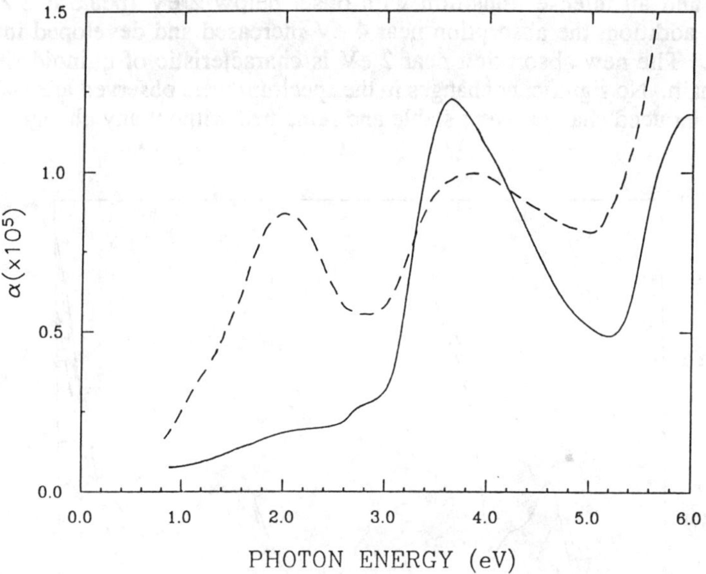

Fig. 5 Optical absorption spectra of spin-cast films of leucoemeraldine and of the emeraldine base.

    ——————— $(1A)_n$

    — — — $[(1A)(2A)]_n$

We found that when fully reduced leucoemeraldine film is exposed to air, the 2 eV peak appeared quickly, within the first several hours. The intensity continued to increase for the first two days, saturated during the fourth day, and remained almost constant during several more days of exposure. The relative intensity of the 2 eV to 4 eV absorption peaks of this auto-oxidized and air-stabilized polyaniline was much smaller than that of the polyemeraldine base. Comparing with the spectra of the various oxidation states of octaaniline,[30] we conclude that the degree of oxidation for this sample is near to y=0.25 (corresponding protoemeraldine).

3.2.4. *Fully-oxidized polyaniline in sulfuric acid.* Asturias et al[31] noted that overoxidized polyaniline will gradually degrade toward the emeraldine structure (in 80% acetic acid solution) due to the hydrolysis of -C=N- double bonds). We find, however, that fully oxidized polyaniline is stable in concentrated sulfuric acid (97%).

Figure 6 shows the spectral changes that occur as a result of the overoxidation of polyaniline in sulfuric acid. After taking the initial spectrum, a small drop of a solution of 2.5 g of $(NH_4)_2S_2O_8$ oxidant dissolved in 100 g of 97% sulfuric acid was added into a quartz cell containing a 3×10^{-4} mole/liter solution of polyaniline in concentrated sulfuric acid. After vigorously shaking the quartz cell for one minute, the absorption spectrum was recorded; the spectrum was then recorded again at 15 minutes and at 10 hours after addition of oxidant. As can be seen from Figure 6, the spectrum changed almost immediately after addition of oxidant; the characteristic features of the emeraldine salt disappeared, and an intense transition with onset below 2 eV (peak at 2.2 eV) was observed. In addition, the absorption near 4 eV increased and developed into a well-defined peak. The new absorption near 2 eV is characteristic of quinoid rings in the polyaniline chain. No significant changes in the spectrum were observed after 40 minutes; the oxidation induced changes were stable and remained without any change at least for several days.

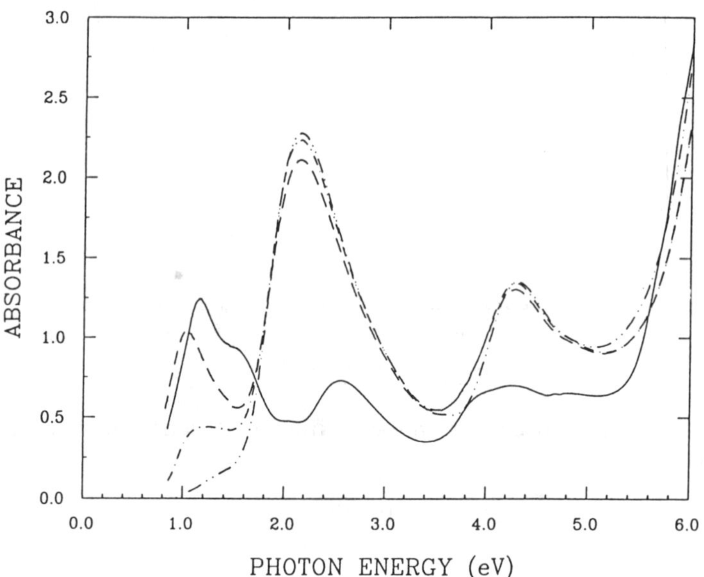

Fig.6. Solution optical absorption spectra showing the formation of the fully oxidized bipolaron lattice form of polyaniline.

—————————	initial emeraldine base
— — — —	1 minute after addition of ammonium persulfate
— — — —	5 minutes after addition of ammonium persulfate
—— ——	10 hours after addition of ammonium persulfate

That this procedure does indeed yield the fully oxidized charged bipolaron lattice was demonstrated by nuclear magnetic resonance (NMR). ^{13}C NMR spectra of the poly(emeraldine) salt before and after oxidation in sulfuric acid solution are shown in Figure 7. For the emeraldine salt in solution (Fig. 7a), we observe only two lines: a more intense line at 129.7 ppm attributable to the proton-bonded carbons in the benzenoid rings and a weaker line at 144.7 ppm attributable to the nitrogen-bonded carbons in the benzenoid rings. The fact that the peaks corresponding to carbons in quinoid rings completely disappear (in comparison with the spectra obtained for the emeraldine base and salt in the solid state[9a,32]) demonstrates that all benzene rings become equivalent in sulfuric acid solution. After subsequent oxidation (as described in the previous paragraph), two additional features appear: one line at around 136 ppm corresponding to proton-bonded carbons in quinoid rings, and a second line at 161.5 ppm corresponding to the nitrogen-bonded carbons in quinoid rings. The downfield shift to 161.5 ppm in from

157 ppm in the neutral emeraldine base[9d,33] suggests that the fully oxidized polyaniline [i.-e. after treatment with $(NH_4)_2S_2O_8$] in sulfuric acid must be protonated.

Comparing the positions and relative intensities of the 2 eV and 4 eV peaks in Figure 6 with the data obtained from the fully-oxidized tetramer[9b] and octamer[30] suggests a formal positive charge on each nitrogen in the protonated structure in sulfuric acid solution (i.e. a doubly charged bipolaron associated with each quinoid unit). These results imply that the polyaniline was fully-oxidized to the copolymer with alternating quinoid and benzonoid monomer units with the structure given in (4). The absence of any sign of free carrier absorption in the infrared indicates complete conversion of the polaronic metal to the doubly charged bipolaron lattice after oxidation of the protonated emeraldine salt.

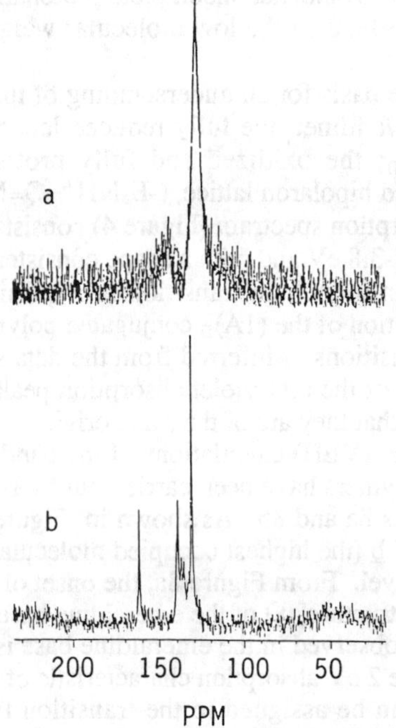

Fig. 7. ^{13}C NMR spectra of the emeraldine salt before (7a) and after (7b) oxidation in sulfuric acid

The stabilization of the bipolaron lattice structure provides an opportunity for studies of this new degenerate ground state polymer, at least in solution. However, although the bipolaron lattice form appears to be stable in concentrated sulfuric acid, hydrolysis occurs when the solution is exposed to moisture from the atmosphere. As a result, attempts toward recovering (4) in the pure solid state have thus far been unsuccessful. Efforts in this direction are continuing in our laboratory.

3.3. DISCUSSION OF SPECTROSCOPIC DATA

The molecular weight dependence of the $[1S]^{\cdot}(A^-)_n$ absorption spectrum demonstrated in Figure 1 is important, for it emphasizes the need for long, uninterrupted chains in order to obtain the intrinsic electronic structure. Short chains (low molecular weight) appear to affect the electronic structure in two ways. First, they lead to localization of the carriers as inferred from the loss of the infrared free carrier absorption (at 1 eV and below in Figure 1). This is consistent with the inferences obtained from magnetic susceptibility studies;[34] only in well-ordered partially crystalline samples does one clearly observe the temperature independent Pauli spin susceptibility expected for delocalized Fermions. Second, the residual 2 eV and 4 eV absorption present in the low molecular weight samples (after exposure to pH=0.15) indicate incomplete protonation. This may be due to the relative importance of end effects in the low molecular weight oligomers which are soluble in THF and DMF.

The available data provide the basis for an understanding of the electronic structure of the four principal forms of polyaniline: the fully reduced leucoemeraldine, $(1A)_n$; the emeraldine base, $[(1A)(2A)]_n$; the oxidized and fully protonated emeraldine salt, $[1S]^{\cdot}(A^-)_n$; and the fully oxidized bipolaron lattice, $(-B-NH^+=Q=NH^+-)_n$.

For leucoemeraldine, the absorption spectrum (Figure 4) consists of a single asymmetric absorption with maximum at 3.7-3.8 eV and with a shape consistent with that expected for the joint density of states for an intrachain transition in a quasi-one dimensional band structure. This is the π-π^* transition of the $(1A)_n$ conjugated polymer. For the emeraldine base structure there are two transitions as inferred from the data shown in Figure 4. The close agreement of the energies of the ultraviolet absorption peaks for the two materials, $(1A)_n$ and $[(1A)(2A)]_n$ implies that they are of the same origin.

Valence Effective Hamiltonian (VEH) calculations of the band structures of the $(1A)_n$ and $[(1A)(2A)]_n$ conjugated polymers have been carried out by Boudreaux et al;[18] their results are reproduced in Figures 8a and 8b. As shown in Figure 8a, the π-π^* transition in leucoemeraldine is from band $\underline{b}$ (the highest occupied molecular orbital or HOMO), to the flat band above the Fermi level. From Figure 8a, the onset of absorption is estimated as about 4 eV. The VEH band structure[18] of the emeraldine base is reproduced in Figure 8b. The high energy transition observed in the emeraldine base is from band $\underline{b}$ to the flat band 4 eV higher in energy. The 2 eV absorption characteristic of the existence of quinoid rings in the emeraldine base can be assigned to the transition from band $\underline{b}$ to band $\underline{b'}$; however, the observed energy ($\approx$2 eV) is higher than that inferred from the calculations (approximately 0.8 eV). These assignments of the optical transitions are in agreement with the results of photoconductivity measurements which indicate the photoproduction of free carriers for both 2 eV and 4 eV pumping.

From the data of Figures 1 and 2, we conclude that the absorption spectrum of partially crystalline films of the high molecular weight and fully protonated emeraldine salt, i.e. $[1S]^{\cdot}(A^-)_n$, has three important spectral features;

> (i) an <u>intra</u>chain absorption band at 1 eV (with a long tail into the deep infrared)
> (ii) an <u>intra</u>chain absorption band at 3 eV (with onset at approximately 2.5 eV),

and

> (iii) an <u>inter</u>chain absorption at 1.5 eV.

The 1 eV absorption appears to be dominated by the free carrier intraband absorption (the Drude-like tail extending into the deep infrared). Although there may be a weak interband contribution (with a finite energy gap) as well, this is uncertain because of the strength of the metallic free carrier absorption.

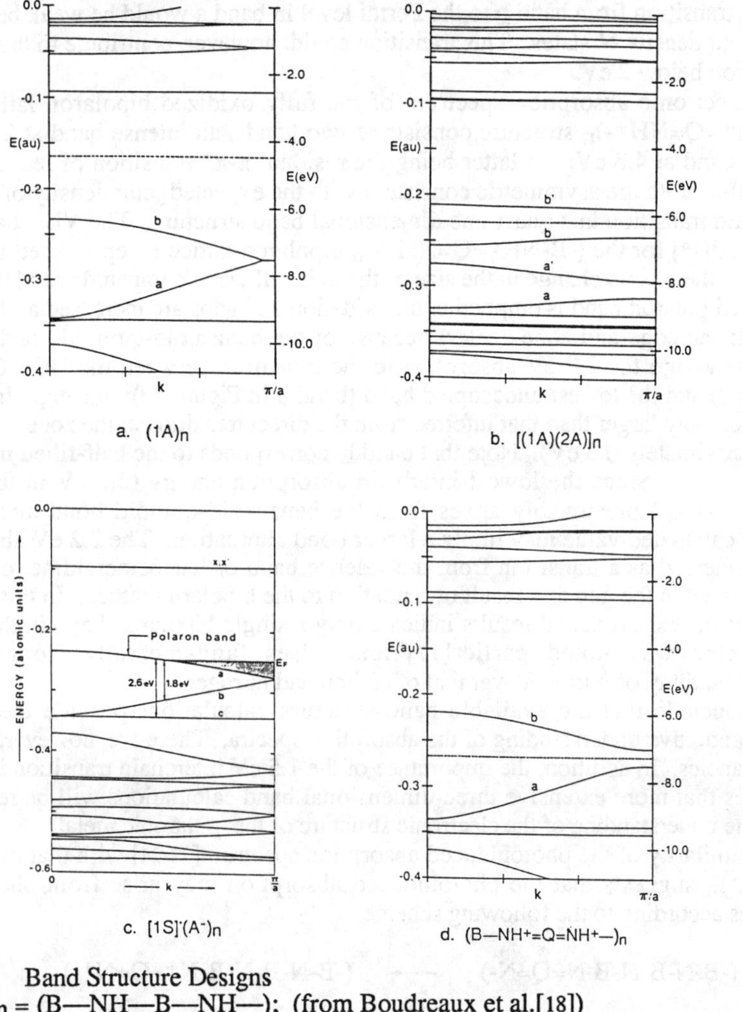

Fig. 8.   Band Structure Designs
   a. $(1A)_n = (B—NH—B—NH—)$; (from Boudreaux et al.[18])
   b. $[(1A)(2A)]_n = [(B—NH—B—NH—)(B—N=Q=N—)]_n$; (from Boudreaux et al.[18])
   c. $[1S]^{\cdot}_n = [B—NH—B—NH^+—]^{\cdot}_n$; (from Stafstrom et al[17])
   d. $(B—NH^+=Q=NH^+—)_n$, (from Boudreaux et al[18])

These features of the $[1S]^{\cdot}(A^-)_n$ spectrum can also be compared with the results of electronic band structure calculations. The single chain VEH band structure for the $[1S]^{\cdot}(A^-)_n$ polaron metal[17] is reproduced in Figure 8c. Since the highest occupied band (band $\underline{a}$) is half-filled, the intraband Drude-like contribution is an obvious feature to be expected for the $[1S]^{+}(A^{-})_n$ polaronic metal. The 3 eV intrachain absorption must then arise from vertical transitions from the Fermi level in band $\underline{a}$ to the higher energy flat bands, which in the VEH band structure are about 4 eV above E_F. Although allowed, the vertical transition from band $\underline{b}$ to the Fermi level in band $\underline{a}$ would be weak because of the small joint density of states. This transition could, however, contribute to the background absorption below 2 eV.

The electronic absorption spectrum of the fully oxidized bipolaron lattice with the $(-B-NH^+=Q=NH^+-)_n$ structure consists of two bands, an intense band at 2.2 eV and a weaker band at 4.3 eV; the latter being the residual π-π^* transition of leucoemeraldine, $(-B-NH)_n$. Both are asymmetric consistent with the expected joint density of states for an intrachain transition in a quasi-one dimensional band structure. The VEH band structure calculated[18] for the $(-B-NH^+=Q=NH^+-)_n$ bipolaron lattice is reproduced in Figure 8d. Although there is no change in the size of the unit cell when compared with $[1S]^{+}(A^{-})_n$, the half-filled polaron band is emptied upon oxidation and gaps are increased at the symmetry points (zone edge and zone center) because of the benzenoid-quinoid bond alternation. Thus we assign the 2.2 eV absorption to the transition between the HOMO (band $\underline{a}$ in Figure 8d) and the lowest unoccupied band (band $\underline{b}$ in Figure 8d); the experimental value is considerably larger than that inferred from the direct transition at the zone edge in Figure 8d (approximately 0.5 eV). Note that band $\underline{b}$ corresponds to the half-filled metallic band in $[1S]^{+}(A^{-})_n$. Since the lowest interband absorption energy (2.2 eV in the bipolaron semiconductor) presumably arises from the benzenoid-quinoid bond alternation, the smaller calculated value may imply a larger bond alternation. The 2.2 eV absorption can also be viewed as a transition from the valence band of leucoemeraldine to a bipolaron band formed in the gap as a result of oxidation to the bipolaron lattice. In this context, we note that the experimental results indicate only a single bipolaron band in the gap. This asymmetry was noted earlier[17] and arises fundamentally from the higher electronegativity of nitrogen over that of carbon and nitrogen.

We conclude that the available band structure calculations provide the basis for a semiquantitative understanding of the absorption spectra. There are, however, quantitative discrepancies. In addition, the importance of the 1.5 eV interchain transition in $[1S]^{+}(A^{-})_n$ indicates that more extensive three-dimensional band calculations will be required for a complete understanding of the electronic structure of the polaronic metal.

The similarity of the photoinduced absorption spectrum[22-24] with that of the metallic $[1S]^{+}(A^{-})_n$ suggests that the photoinduced absorption may arise from photogenerated polarons according to the following scheme:

$$(-B-N-B-N-B-N=Q=N-) \quad \xrightarrow{\hbar\omega} \quad (-B-N-B-N-B-N^+=Q=N^+-)$$

$$\downarrow$$

$$(-B-N-B-N^{+\cdot}-B-N-B-N^{+\cdot}-)$$

Although consistent with the photoinduced infrared active vibrational (IRAV) mode spectrum, this scheme must be verified by photoinduced electron spin resonance measurements, since the mechanism implies the photogeneration of unpaired spins via charge induced spin unpairing (i.e. directly analogous to the proton induced spin unpairing).

4. Magnetic Susceptibility[34b]

The discovery[5] that when in solution in concentrated sulfuric acid, polyaniline is in the protonated form, $[B-NH-B-NH-]^{+\cdot}{}_n$, and that it is recovered as the partially crystalline salt, $[B-NH-B-NH-]^{+\cdot}{}_n(HSO_4{}^-)_n$, from solutions in sulfuric acid (by precipitation in water or methanol) has opened the way to a more complete characterization of the polymer and to studies directed toward the determination of the intrinsic properties of the ordered material. For example, the temperature independence of the spin susceptibility[34] of the more highly ordered crystalline material above 125K is consistent with the Pauli spin susceptibility expected for a metal with a density of states at the Fermi level estimated as 1 state per eV per formula unit (two rings).

The solubility in concentrated sulfuric acid allows the investigation of the novel state of metallic chains of polyaniline in solution in sulfuric acid. Since the material should be fully protonated when in solution in sulfuric acid, this would be the first example of a metallic polymer as a "liquid metal". On the other hand, when in solution the interchain interactions are clearly much weaker than in the solid state. As a result, the electronic structure is expected to be more nearly one-dimensional, and the possibility of a Peierls' transition with the formation of an energy gap in the excitation spectrum might be anticipated.[35,36] In this case, the metallic chains would distort to form a superlattice structure with an associated energy gap in the excitation spectrum. The existence of such a gap should be evident from the magnitude of the magnetic susceptibility. Alternatively, unless the $[B-NH-B-NH-]^{+\cdot}{}_n$ chains are rigid rods when in solution, disorder would be expected to play a significant role (and perhaps to even suppress the Peierls' instability).[37]

Solutions of emeraldine salt were prepared by dissolving an appropriate amount of the polymer in 97% sulfuric acid.[5,16] The mixture was left overnight at room temperature under moderate stirring, to yield a homogeneous solution. Part of the solution was transferred into an ESR tube under nitrogen and sealed under vacuum. Using this method, solutions of emeraldine salt in sulfuric acid with concentrations 3 wt%, 5 wt%, 7 wt% and 10 wt% were prepared.

The magnetic susceptibility was determined by means of ESR measurements (9.5 Ghz). The magnitude of χ was obtained by comparison of the results of double integration of the signal from the sample and from a calibrated (National Bureau of Standards) ruby sample as reference. Since the reference is in the cavity with the sample, this method automatically compensates for changes in Q arising from loss due to the ionic conductivity of the sulfuric acid. The measured susceptibility for all the different concentration solutions was in the range of $(2\pm1)\times10^{-6}$ emu/mole (2 rings) at room temperature. Since the ESR signal originates from the polyaniline in solution, the magnitude of χ is given in units appropriate to the amount of emeraldine salt present in the solution. Because of the low Q of the cavity (due to ionic conductivity of the sulfuric acid) and the small number of unpaired spins, we were unable to detect the ESR signal at concentrations less than 3 wt%.

Figure 9 shows the temperature (T) dependence of the susceptibility in the case of the solution with 10 wt% polyaniline in sulfuric acid. The temperature dependence of the peak-to-peak linewidth (ΔHpp) of the ESR absorption line is presented in Figure 10. All the solutions which were studied (with different polymer concentrations) exhibited the same dependences of χ and ΔHpp versus T as those shown in Figure 9 and Figure 10. The magnetic susceptibility of the $[\text{B-NH-B-NH-}]^{+\cdot}_n(\text{X}^-)_n$ salt obtained by precipitation from sulfuric acid (and subsequently treated with HCl to ensure complete protonation) is also shown for comparison in Figure 9.

As the temperature is lowered (Figure 9), the susceptibility drops from 1.5×10^{-6} at room temperature to 3×10^{-7} emu/mole (2 rings) below 200K, at which point χ is reduced from that of the solid $[\text{B-NH-B-NH}^+\text{-}]^{\cdot}_n(\text{X}^-)_n$ salt by more than a factor of 200. The linewidth of the ESR signal in the frozen glass is approximately an order of magnitude greater than that of the solid $[\text{B-NH-B-NH}^+\text{-}]^{\cdot}_n(\text{X}^-)_n$ salt. It is temperature independent below 200 K, and begins to narrow above 200 K.

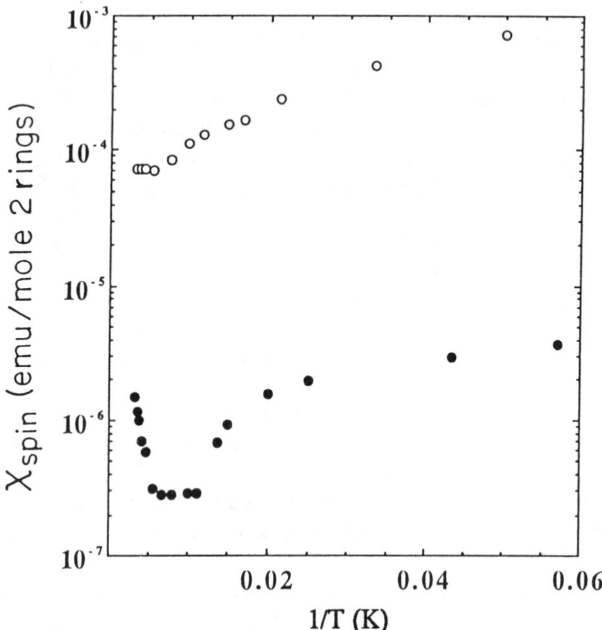

Fig. 9.   Magnetic susceptibility(χ) versus temperature (T); black points, emeraldine salt in 10% (w/w) solution in acid; open circles, solid emeraldine salt.

As noted above, because of the reduced interchain electron transfer interaction for conjugated chains in solution, the electronic structure is expected to be more nearly one-dimensional. Thus, the possibility of a Peierls' transition with the formation of an energy gap in the excitation spectrum might be anticipated. The susceptibility results shown in Figure 9 suggest that this is indeed the case for polyaniline in solution in sulfuric acid. Such a large decrease in χ cannot be accounted for in the context of the half-filled band expected for $[\text{B-NH-B-NH-}]^{+\cdot}{}_n$. To reduce χ by such a large factor would require an increase in the band width by the same factor; i.e. to a band width of more than 200 eV!

The small density of states, ~0.004 states per eV per formula unit (two rings), indicates the formation of an energy gap at the Fermi surface, consistent with a Peierls' transition. This residual value may arise from a sum of two contributions:

(i)   A small fraction of the polyaniline is not in solution, but is suspended as fine particles;
(ii) States in the gap as a result of disorder in the $[\text{B-NH-B-NH-}]^{+\cdot}{}_n$ chains.

The small measured density of states sets an upper limit on the sum of these two contributions.

The magnetic susceptibility data imply, therefore, that polyaniline in sulfuric acid forms a true solution rather than a suspension of fine particles; at least about 99.6% of the polymer is in molecular solution (and that estimate is a lower limit). This is confirmed by the measured linewidth (Figure 10). There is no indication of even a small fraction of the material with a linewidth of about 1-2 Gauss; i.e. that of the solid $[\text{B-NH-B-NH-}]^{+\cdot}{}_n(\text{X}^-)_n$ salt. Note that this conclusion is valid at low temperatures (e.g. 100K) implying that even in the glass of frozen sulfuric acid, the $[\text{B-NH-B-NH-}]^{+\cdot}{}_n$ chains are molecularly dispersed and in matrix isolation.

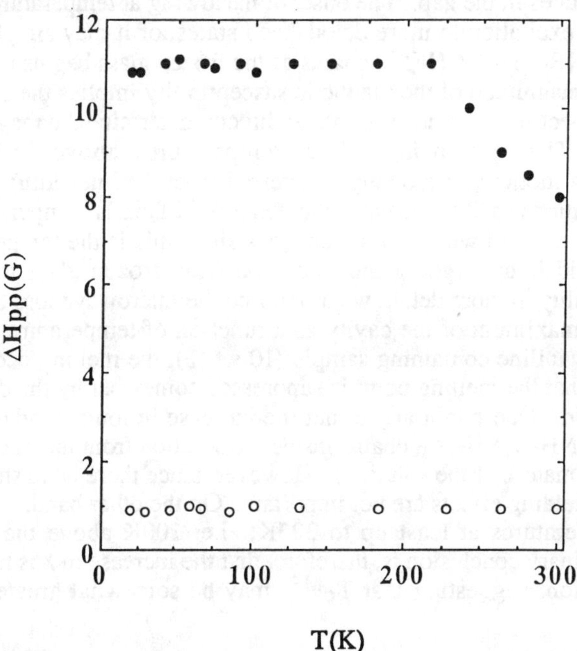

Fig. 10. ESR line width ΔHpp versus the temperature T; black points, emeraldine salt in 10% (w/w) solution in acid; open circles, solid emeraldine salt.

Unless the $[\text{B-NH-B-NH-}]^{+\cdot}{}_n$ chains are rigid rods when in solution, disorder would be expected to play a significant role and perhaps to even suppress the Peierls' instability. It is known, however, that for conjugated chains in solution, the conformation is determined by a competition between the increased entropy associated with the many degrees of freedom of a coiled or worm-like chain and the increased electronic energy associated with the breaking of π-bonds.[38] Moreover, even if the insulating conjugated chain is coiled or worm-like, the introduction of free carriers leads to the self-consistent formation of straight-chain segments in the vicinity of the carriers, i.e. conformons.[39] Thus, for the $[\text{B-NH-B-NH-}]^{+\cdot}_n$ chains, which have one carrier per formula unit, one is perhaps not surprized to find evidence of a rod-like conformation. That this must indeed be the case is implied by the small value of the residual susceptibility for chains in matrix isolation in the frozen glass. The weak disorder is not sufficient to suppress the Peierl's transition, nor is it sufficient to yield a density of localized states within the energy gap greater than about 0.004 states per eV per formula unit (two rings). Moreover, the increase in $\chi(T)$ at low temperatures can be used as an indication of disorder-induced localized states. Even this contribution is extremely small compared with that observed in the partially crystalline material cast from sulfuric acid. We conclude that in solution, the $[\text{B-NH-B-NH-}]^{+\cdot}_n$ chains have a relatively long persistence length (and a correspondingly long localization length) implying that the chain conformation is probably rod-like.

The order of magnitude increase in the ESR linewidth of the $[\text{B-NH-B-NH-}]^{+\cdot}_n$ chains in the frozen matrix is consistent with the interpretation that the residual susceptibility arises from localized states in the gap. The onset of narrowing at temperatures above 200 K may imply thermal excitation to more delocalized states, or it may simply reflect the onset of motion of the $[\text{B-NH-B-NH-}]^{+\cdot}_n$ chains as the frozen glass begins to melt.

Although the small magnitude of the magnetic susceptibility implies the existence of a gap in the excitation spectrum, it is not possible to directly determine the magnitude of the gap from the data. The up-turn in $\chi(T)$ at temperatures above 250K (and the corresponding onset of motional narrowing as inferred from the linewidth) may indicate the onset of 1d fluctuations as T approaches the mean field Peierls' temperature (T_c^{MF}). However, one must be careful with such an analysis since this is the temperature range where the sulfuric acid is undergoing the transition from frozen glass to liquid. To investigate this possibility in more detail, we monitored the microwave loss by measuring the full width at half maximum of the cavity as a function of temperature. The results imply that for the polyaniline containing sample (10 wt %), the melting occurs between 220K and 230K, and that the melting point is suppressed somewhat by the dissolution of the polymer in the acid. One might argue that the increase in ionic conductivity could partially shield the $[\text{B-NH-B-NH-}]^{+\cdot}_n$ chains inside the solution from the microwave field (and thereby underestimate χ of the solution). However, since there is no step in χ at the melting point, such shielding effects are not important. On the other hand, the increase in χ continues for temperatures at least up to 325K; i.e. 100K above the solid-liquid transition. Our preliminary conclusion is, therefore, that the increase in χ is not an artifact of the melting transition, suggesting that T_c^{MF} may be somewhat greater than room temperature.

5. Summary and Conclusion

The results of structural, spectroscopic and magnetic studies of the various forms of polyaniline carried out recently at UC-Santa Barbara have been briefly reviewed. Although polyaniline can be prepared in partially crystalline form, the molecular weight dependence of the $[1S]^{\cdot}(A^-)_n$ absorption spectrum is particularly important, for it demonstrates the need for long, uninterrupted chains in order to obtain the intrinsic electronic structure.

From the spectroscopic data, we conclude that high molecular weight and fully protonated emeraldine salt, $[1S]^{\cdot}(A^-)_n$, has three important spectral features; an intrachain absorption band at 1 eV (with a long tail into the deep infrared), an intrachain absorption band at 3 eV (with onset at approximately 2.5 eV), and an interchain absorption at 1.5 eV. The 1 eV absorption appears to be dominated by the free carrier "metallic" intraband absorption.

We have re-analyzed the available data in comparison with the results of band structure calculations to provide the basis for an understanding of the electronic structure of the four principal forms of polyaniline: the fully reduced leucoemeraldine, $(1A)_n$; the partially oxidized emeraldine base, $[(1A)(2A)]_n$; the oxidized and fully protonated emeraldine salt, $[1S]^{\cdot}(A^-)_n$; and the fully oxidized bipolaron lattice, $(-B-NH^+=Q=NH^+-)_n$.

Studies of the magnetic susceptibility of the emeraldine salt in the solid state and in solution in sulfuric acid have demonstrated that in sulfuric acid, the $[B-NH-B-NH-]^{+\cdot}{}_n$ chains are molecularly dispersed in a genuine solution (as opposed to a suspension of fine particles). The indications of weak disorder imply that the chains have a relatively long persistence length (and a correspondingly long localization length) suggesting that the chain conformation is probably rod-like.

The magnetic susceptibility of protonated $[B-NH-B-NH-]^{+\cdot}{}_n$ chains in solution was found to be decreased with respect to χ in the solid state by more than two orders of magnitude. This decrease was interpreted as indicative of a Peierls' transition; in solution the metallic chains apparently distort to form a superlattice which leads to the opening of a characteristic energy gap in the excitation spectrum.

Acknowledgment: The spectroscopic studies were supported by the Office of Naval research under N00014-83-K-0450; the synthesis and solution processing were supported by the National Science Foundation through an MRG grant (NSF DMR-87-03399); X-ray measurements and analysis were supported by the Office of Naval Research, and the ESR studies were supported by the National Science Foundation (NSF DMR85-21392).

References

1.   Andreatta, A., Cao, Y., Chiang, J.-C., Smith, P. and Heeger, A.J. (1988) Synth. Met., 26, 383.

2a.   Green, A.G. and Woodhead, A.E. (1910) J. Chem. Soc., 97, 2388 ; (1912) 101, 1117.

2b.   Wilstetter, R. and Dorogi, S. (1909) Ber., 42, 2147 and (1909) 42, 4118.

3.   MacDiarmid, A.G., Chiang, J.-C., Halpern, M., Mu, S.-L., Somasori, N.L.D., Wu,W. and Yaniger, S.I. (1985) Mol. Cryst. Liq. Cryst., 121, 173.

4.   Salaneck, W.R., Lundstrom, I., Huang, W.-S. and MacDiarmid, A.G. (1986) Synth. Met., 13, 291.

5.   Angelopoulos, M., Asturias, G.E., Ermer, S.P., Ray, A., Scherr, E.M., MacDiarmid, A.G., Akhtar, M., Kiss, Z. and Epstein, A.J. (1988) Mol. Cryst. Liq. Cryst., 160, 151.

6.   Wang, L., Jing, X. and Wang, F. (1989) Synth. Met., 29, E363.

7.   Li, S., Dong, H. and Cao, Y. (1989) Synth. Met., 29, E329.

8.   A special case is the early report of solubility of undoped polyphenylenesulphide in a mixture of SbF_3 and SbF_5; see Frommer, J., Elsenbaumer, R.E. and Chance, R.R. (1984) ACS Symp. Ser., 242, 44. However, the resulting solution is unstable. Moreover, after casting, the resulting films are no longer reversibly soluble in the same solvent, implying either a change in structure or significant crosslinking (or both).

9a.   Hjertberg, T., Salaneck, W.R., Lundstrom, I., Somasiri, N.L.D., MacDiarmid, A.G. (1985) J. Polym. Sci., Polym. Lett. Ed., 23, 503.

9b.   Cao, Y., Li, S., Xue, Z. and Guo, D. (1986) Synth. Met., 16, 305.

9c.   Ray, A., Asturias, G.E., Kershner, D.L., Richter, A.F., MacDiarmid, A.G. and Epstein, A.J. (1989) Synth. Met., 29, E141.

9d.   Kaplan, S., Conwell, E.M., Richter, A.F. and MacDiarmid, A.G. (1988) J. Am. Chem. Soc., 110, 7647; *ibid* (1989) Macromolecules, 22, 1669.

9e.   Richter, A.F., Ray, A., Ramanathan, K.V., Manohar, S.K., Furst, G.T., Opella, S.J., MacDiarmid, A.G. and Epstein, A. J. (1989) Synth. Met., 29, E243.

10a. MacDiarmid, A.G., Chiang, J.-C., Richter, A.F. and Epstein, A.J. (1987) Synth. Met., 18, 285.

10b. MacDiarmid, A.G., Chiang, J.-C., Huang, W.-S., Humphrey, B.D. and Somarisi, N.L.D. (1985) Mol. Cryst. Liq. Cryst., 121, 181.

11.   Epstein, A.J., Ginder, J.M., Zuo, F., Bigelow, R.W., Woo, H.-S., Tanner, D.B., Richter, A.F., Huang, W.S. and MacDiarmid, A.G. (1987) Synth. Met., 18, 303.

12.   Ginder, J.M., Richter, A.F., MacDiarmid, A.G. and Epstein, A.J. (1987) Solid State Comm., 63, 97.

13.   Epstein, A.J., Ginder, J.M., Zuo, F., Woo, H.-S., Tanner, D.B., Richter, A.F., Angelopoulos, M., Huang, W.-S. and MacDiarmid, A.G. (1987) Synth. Met., 21, 63.

14.   Moon, Y., Cao, Y., Smith, P. and Heeger, A.J. (1989) Polymer Commun., 30, 196.

15a. Tabor, B.J., Magre, E.P. and Boon, J. (1971) Eur. Polym. J., 7, 1127.

15b. Boon, J. and Magre, E.P. (1969) Makromol. Chem., 126, 130.

15c. Northold, M.G. (1974) Eur. Polym. J., 10, 799.

16.   Cao, Y., Andretta, A., Heeger, A.J. and Smith, P., Polymer (in press).

17.   Stafstrom, S., Brédas, J.L., Epstein, A.J., Woo, H.S., Tanner, D.B., Huang, W.S. and MacDiarmid, A.G. (1987) Phys. Rev. Lett., 59, 1464.

18.   Boudreaux, D.S., Chance, R.R., Wolf, J.F., Shacklette, L.W., Brédas, J.L., Themans, B., Andre, J.M., Silbey, R. (1986) J.Chem. Phys., 85, 4584.

19.   Conwell, E.M., Duke, C.B., Paton, A., Jeyadev, L. (1988) J.Chem. Phys., 88, 331.

20.   Wudl, F., Angus, R.O., Lu, F.L., Allemand, P.M., Vachon, D.J., Nowak, M., Liu, Z.X and Heeger, A.J. (1987) J. Amer. Chem. Soc., 109, 3677.

21.   Wnek, G. (1986) Synth. Met., 15, 213.

22.   Epstein, A.J., Ginder, J.M., Roe, M.G., Gustafson, T.L., Angelopoulos, M. and MacDiarmid, A.G. (1988) Mat. Res. Soc. Symp., 109, 317.

23.   Roe, G., Ginder, J.M., Wigen, P.E., Epstein, A.J., Angelopoulos, M., MacDiarmid, A.G. (1988) Phys. Rev. Lett., 60, 2798.

24.   Kim, Y.H., Foster, C., Chiang, J., Heeger, A.J. (1988) Synth. Met. 26, 49; (1989) Synth. Met., 29, E29.

25.   Stafstrom, S., Sjogen, B. and Bredas, J.L. Synth. Met. (in press; Proceedings of ICSM '88).

26a   Baird, D.G. and Smith, J.K. (1978) J. Polym. Sci., Polym. Chem. Ed., 16, 61.
26b.   Kamide, K. and Yukio, M. (1978) Kobunshi Ronbunshi, 35, 467.
27.   Chiang, J.-C. and MacDiarmid, A.G. (1986) Synth. Met., 13, 193.
28a.   McManus, P.M., Cushman, R.J. and Yang, S.C. (1987) J. Phys. Chem., 91, 744.
28b.   Genies, E.M. and Laporski, M. (1987) J. Electrochem., 220, 67.
29.   Phillips, S.D., Yu, G., Cao, Y. and Heeger, A.J. (1989) Phys. Rev. B, 39, 10702.
30.   Lu, F.-L., Wudl, F., Nowak, M. and Heeger, A.J. (1986) J. Amer. Chem. Soc., 108, 8311.
31.   Asturias, G.E., MacDiarmid, A.G. and Epstein, A.J. (1989) Synth. Met., 29, E157.
32.   Devreux, F., Bidan, G., Sayed, A.A. and Tsintavis, C. (1985) J. Physique, 46, 1595.
33.   Menardo, C., Nechtschein, M., Rousseau, A., Travers, T.P. and Hany, P., Synth. Met. (in press).
34a.   Fite, C., Cao, Y. and Heeger, A.J., Sol. State Commun. (in press).
34b.   Fite, C., Cao, Y. and Heeger, A.J., Phys. Rev. Lett. (submitted)
35.   Peierls, R.E. (1955) "Quantum Theory of Solids," Oxford University Press, London , p. 108.
36.   Devrees, J.T., Evrard, R.P. and Van Doren, V.E. (eds.) (1978), "Highly Conducting One Dimensional Solids," Plenum, New York.
37.   Mele, E.J. and Rice, M.J. (1981) Phys. Rev. B, 23, 5397.
38.   Spiegel, D.R., Pincus, P.A. and Heeger, A.J. (1988) Polymer Commun., 29, 264.
39.   Pincus, P., Rossi, G. and Cates, M. E. (1987) Europhys. Lett., 4, 41.

POLYANILINE VERSUS POLYACETYLENE, OR, RINGS VERSUS BONDS AND THE ROLES OF BARRIERS AND CRYSTALLINITY

A.J. EPSTEIN
Department of Physics and
Department of Chemistry
The Ohio State University
4108 Smith Lab, 174 West 18th Avenue
Columbus, Ohio 43210-1106

and

A.G. MACDIARMID
Department of Chemistry
S. 34th Street and Spruce Streets
University of Pennsylvania
Philadelphia, Pennsylvania 19104-6323

Abstract. Wide-spread study of polyacetylene for the past decade has demonstrated that the energy gap of *trans*-polyacetylene is due to the electron-phonon interaction with some contribution from the Coulomb interaction. Injected charge carriers in soliton and polaron states have modest masses, of order m_e. The time dynamics for the relaxation of photoinduced charge indicate that there are no very long-lived states. The leucoemeraldine and emeraldine forms of polyaniline differs substantially from polyacetylene. Their large energy gaps have origin in the electronic structure of C_6 rings. Additional injected charges form very long-lived polaron states, the masses of which are 50-100 m_e. The dramatic difference in the behavior of polyaniline with respect to polyacetylene reflects the important role of ring-torsion angles. Highly doped New-polyacetylene has conductivity σ in excess of 10^4 S/cm. Combined conductivity, thermopower, susceptibility and magneto-resistance studies show that the charge motion is three-dimensional and barrier limited. In contrast, heavily doped polyaniline (emeraldine salt) has conductivities up to 10^2 S/cm. Extensive transport studies show that though the charge motion is three-dimensional, the "metallic" regions are sufficiently small such that charging energy limited tunneling dominates the measured resistance. The ability to derivatize the polyaniline system enables the control of the separation between the polyaniline chains. Systematic study of this has shown the important role of localization of charge as the chains are separated. Recent advances in synthesis and processing of polyaniline including oriented films, fibers, and even a self-doped derivative have opened up new opportunities for basic studies and applications.

1. Barriers and Three-Dimensional Conductivity In Polyacetylene

The electronic structure and charge defect states of *trans*-polyacetylene has been studied extensively [1, 2]. The important role of the electron-phonon interaction leading to a bond alternating state and a Peierls-gap in the electronic structure has been well-developed. In the absence of Coulomb repulsion, the charge defect states in *trans*-polyacetylene are neutral and charged solitons which have energy levels are at mid-gap. The ability of the neutral solitons (with spin) to diffuse along the chain is

195

J. L. Brédas and R. R. Chance (eds.), Conjugated Polymeric Materials:
Opportunities in Electronics, Optoelectronics, and Molecular Electronics, 195–205.
© 1990 *Kluwer Academic Publishers. Printed in the Netherlands.*

dependent on the density of defects in the chain. For polymers such as *cis*-polyacetylene, polythiophene, and polypyrrole which have "external" gap in addition to the Peierls gap, the most stable excitations are polarons and bipolarons. Polymers such as polyacetylene, polythiophene, and polypyrrole are essentially charge conjugation symmetric with the valence conduction band being composed of carbon $2p_z$ orbitals. As a result, the polaron states formed have electronic energy levels symmetrically in the gap at $\pm \omega_0$.

Heavily doped Shirakawa polyacetylene has a conductivity of order 50-1000 S/cm [3]. The conductivity of heavily iodine doped Shirakawa polyacetylene is at the lower end of this range. Extensive temperature dependent conductivity, thermopower, susceptibility, optical and structural studies had lead to the suggestion that although there was a finite density of states at the Fermi level for this system, the charge transport was via hopping [4].

Preparation of New-polyacetylene has opened up questions concerning the origin of the high conductivities in conducting polymers [5]. When heavily doped with iodine, New-polyacetylene has $\sigma(295K)$ of 10^4-10^5 S/cm. The conductivity decreases upon lowering the temperature, varying as $\sigma = \sigma_0 \exp[-(T_1/(T_0+T))]$, i.e., barrier limited transport [6-8]. For New-$(CH)_x$ use of this formula lead to the suggestion that with aging, the islands seem to increase in size while the barriers themselves became larger, suggesting clustering of defects with time. In contrast, for New-$(CH)_x$ produced with an additional reducing agent, ARA-$(CH)_x$, the distance between barriers (islands) decreases with aging, suggesting new interruptions in the chains [8]. Optical absorption, photoinduced infrared absorption, magnetic susceptibility and conductivity studies of undoped and iodine doped New-and ARA-$(CH)_x$ were earlier reported [8]. Electronic structure of the undoped and highly conducting iodine doped New-and ARA-$(CH)_x$ are essentially identical to that of conventional Shirakawa $(CH)_x$ [8].

We have extended [9] these experiments to look at the detail of the microscopic electronic structure of doped N-$(CH)_x$ and doped S-$(CH)_x$ utilizing magnetic susceptibility, which measures the density of states, $N(E_F)$, thermoelectric power, which measures the first derivative with respect to energy of the density of states, and magnetoresistance, which measures the second derivative of the density of states with respect to energy, all evaluated at the Fermi energy, E_F, Equations 1-3:

$$\chi^P = \mu_B^{\ 2}\, N(E_F) \qquad\qquad (1)$$

$$S = \left(\frac{\pi^2}{3}\right)\left(\frac{k_B^{\ 2}T}{e}\right)\left(\frac{d\ln\sigma(E)}{dE}\right)_{E=E_F} \propto \left(\frac{d\ln N(E)}{dE}\right)_{E=E_F} \qquad (2)$$

$$\left(\frac{\Delta\rho(H)}{\rho}\right) = -\left(\frac{1}{2\sigma}\right)\left(\mu_B H\right)^2\left(\frac{d^2\sigma(E)}{dE^2}\right)_{E=E_F} \qquad (3)$$

It is noted that the contribution to the magnetoresistance from variation of N(E) with E near E_F is usually negligible unless E_F is near a sharp feature in N(E) [10].

Figure 1 represents the density of states at Fermi energy as a function of iodine doping for both S-$(CH)_x$ [11] and N-$(CH)_x$ [9] obtained utilizing $\chi^P = \mu_B^2 N(E_F)$. At high doping levels, χ^P is nearly identical for N-$(CH)_x$ and S-$(CH)_x$, implying $N(E_F) \approx$ 0.08 states/eV-C for both systems. The difference in magnetic behavior of N-$(CH)_x$ and S-$(CH)_x$ at low doping levels likely reflects inhomogeneous doping of the more

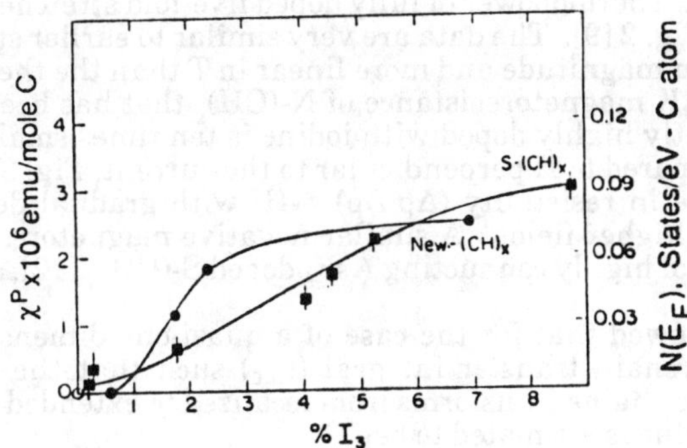

Figure 1– Pauli magnetic susceptibility and N(E$_F$) of S–(CH)$_x$[19] and N–(CH)$_x$[9] as a function of iodine doping. At high doping levels, y>0.06, the density of states at the Fermi energy is comparable for the two materials.

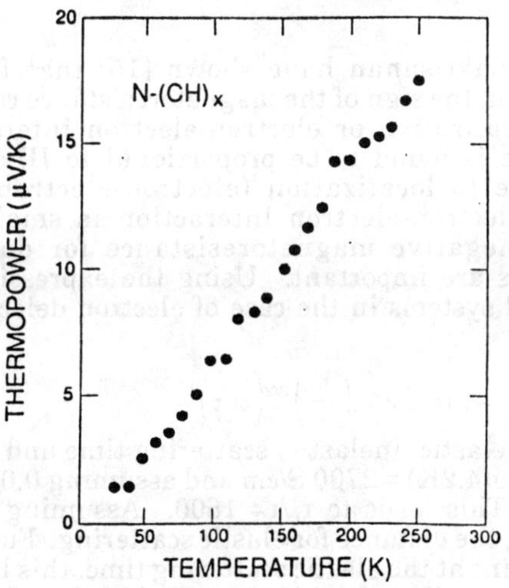

Figure 2– Thermopower of 500% stretched N–(CH)$_x$ film vs temperature measured parallel to stretch (chain) axis [9].

dense N-(CH)$_x$. Thermopower of fully doped five-fold stretched N-(CH)$_x$ is linear with temperature, Fig. 2 [9]. The data are very similar to earlier studied AsF$_5$ doped (CH)$_x$ but are lower in magnitude and more linear in T than the thermopower of I$_2$ doped S-(CH)$_x$. The 4.2K magnetoresistance of N-(CH)$_x$ that has been 500 percent stretched and subsequently highly doped with iodine is ten times smaller for H parallel to the current as compared to H perpendicular to the current, Fig. 3 [9]. In small fields, the relative change in resistivity ($\Delta\rho$ / ρ) $\propto$-H^2 with gradual deviation from quadratic dependence at higher fields. A similar negative magnetoresistance was reported in an early study of highly conducting AsF$_5$ doped S-(CH)$_x$ [12].

Firsov showed that for the case of a quasi-one-dimensional system there is a threshold interchain transfer integral ($t_{\perp c}$) such that the electronic system of a bundle of single chains transforms from localized to extended states for $t_\perp > t_{\perp c}$ [13]. This critical value is estimated to be

$$t_{\perp c} \approx \left(0.3\hbar/2 \right)\left(\tau \tau_2 /2 \right)^{-1/2} \tag{4}$$

which depends not only on τ_2 (backward scattering relaxation time) but also on τ_1 (forward scattering relaxation time). Here τ is the total relaxation time. The influence of τ_1 could be understood on qualitative grounds because it leads to an additional disorder which should enhance the tendency toward localization. Weak localization may be suppressed as a result of the dephasing effects of scattering. Kivelson and Heeger [14] performed a similar analysis though disregarding forward scattering altogether. It is noted that if $\tau_1 = \tau_2 = \tau/2$ then $t_{\perp c} = \sim 0.3\hbar \tau^{-1}$, very similar to 0.5$\hbar$ τ^{-1} obtained by Kivelson and Heeger. It is also noted that Kivelson and Heeger introduced a more stringent requirement if the charge transfer between chains is incoherent.

Lee and Ramakrishnan have shown [15] that for a system near a metal-insulator transition, the sign of the magnetoresistance could distinguish between the importance of localization or electron-electron interactions. At low fields the magnetoresistance is found to be proportional to H^2 with negative (positive) to contributions due to localization (electron-electron interaction), though the contribution of electron-electron interaction is smaller by a factor of $(k_Fl)^{-2}$. Observation of negative magnetoresistance for doped (CH)$_x$ indicates that localization effects are important. Using the expression from Kawabata [16] for three-dimensional systems in the case of electron delocalization by magnetic fields ($\omega_c \tau \ll 1$)

$$\Delta\rho/\rho = -\left(1/12\sqrt{3} \right)\left(\frac{\tau_\varepsilon}{\tau} \right)^{3/2}\left(\omega_c \tau \right)^2 \tag{5}$$

where τ (τ_e) is the elastic (inelastic) scattering time and $\omega_c = (eH/mc)$ is the cyclotron frequency. Using $\sigma(4.2K) = 2700$ S/cm and assuming 0.08 charge carriers per carbon, $\tau = 2.7 \times 10^{-15}$ sec. This leads to $\tau_e/\tau \simeq 1600$. Assuming mean free path L=$v_F\tau$, we obtain L=25-30Å, the distance for elastic scattering. Further assuming the electrons jump between chains at the elastic scattering time, this leads to a crude estimate that the electrons visit a radius of $\sim\sqrt{1600}=40$ chains and travel ~2000Å in the chain direction between inelastic scatterings. It is noted that this is smaller than the number of chains in a typical fiber of polyacetylene [17]. These results strongly suggest that the difference between N-(CH)$_x$ and S-(CH)$_x$ is increasing delocalization and that electrons *do* visit a significant number of chains between inelastic scatterings. Higher conductivities would, within this model, result from both reduced intrachain defects and increased interchain coupling.

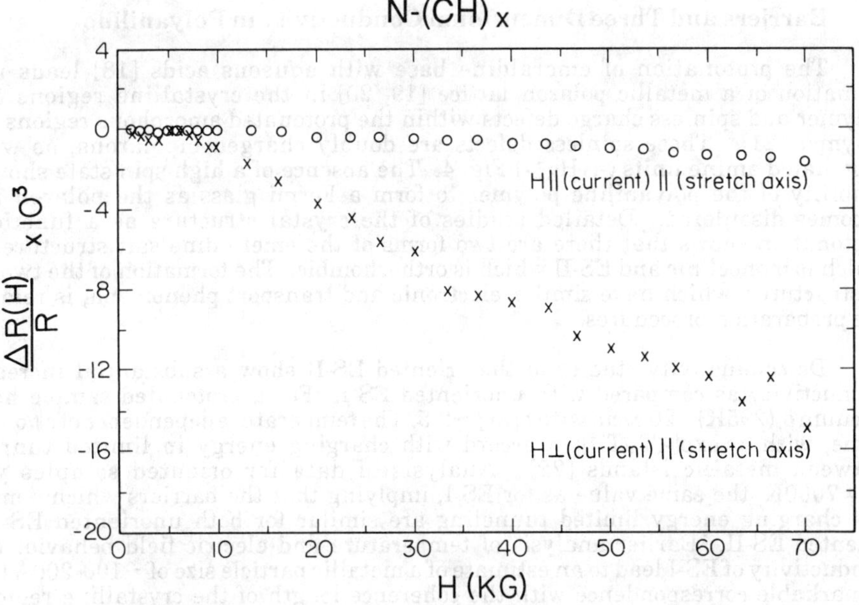

Figure 3–   Magnetoresistance of stretched N-(CH)$_x$ film vs magnetic field
for two orientations of magnetic field. Current is along the
stretch axis in both cases [9].

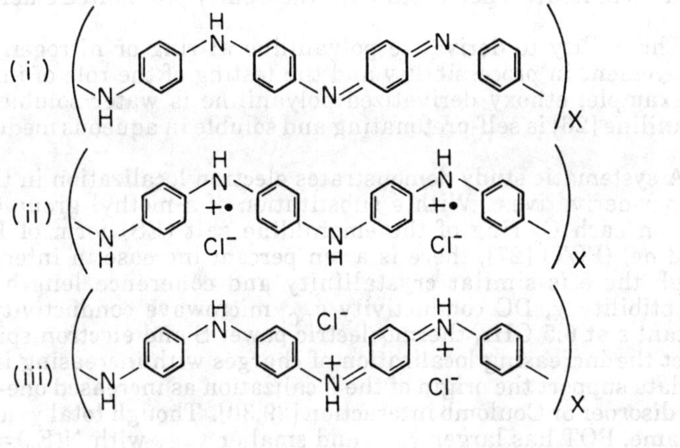

Figure 4–   Schematic illustration of ideal repeat unit for (i) emeraldine
base, (ii) polaron form of emeraldine salt, (iii) bipolaron
form of emeraldine salt.

2. Barriers and Three Dimensional Conductivity in Polyaniline

The protonation of emeraldine base with aqueous acids [18] leads to the formation of a metallic polaron lattice [19, 20] in the crystalline regions of the polymer and spinless charge defects within the protonated amorphous regions of the polymer [21]. These spinless defects are doubly charged bipolarons, as well as protonated amine units ($-NH_2^+-$) Fig. 4. The absence of a high spin state shows the inability of the polyaniline polymer to form a Fermi glass as the polaron metal becomes disordered. Detailed studies of the crystal structure as a function on protonation shows that there are two forms of the emeraldine salt structure, ES-I which is monoclinic and ES-II which is orthorhombic. The formation of the two types of structures, which have similar electronic and transport phenomena, is related to the preparation procedures.

Dc conductivity studies of the oriented ES-II show a substantial increase in conductivity as compared with unoriented ES-I. For a protonated sample held in vacuum $\sigma_\parallel$(295K)~20 S/cm with $\sigma_\parallel/\sigma_\perp$~2.5. The temperature dependence of $\sigma_\parallel/\sigma_\perp$ is the same, with $\sigma \propto \exp[-(T_0/T)^{\frac{1}{2}}]$ in accord with charging energy in limited tunneling between metallic islands [22]. Analysis of data for oriented samples yields $T_0 = 7000K$, the same value as for ES-I, implying that the barriers which dominate the charging energy limited tunneling are similar for both unoriented ES-I and oriented ES-II. Earlier analysis of temperature and electric field behavior of the conductivity of ES-I lead to an estimate of a metallic particle size of ~100-200Å [22] in remarkable correspondence with the coherence length of the crystalline regions in the protonated ES-I and ES-II [21]. Application of contactless microwave conductivity and dielectric measurements and infrared spectroscopy lead to the conclusion that the intrinsic conductivity within the metallic islands is substantially higher than the measured microscopic conductivity [23], while audio frequency conductivity studies suggest hopping of charge among positively charged polaron and bipolaron or neutral defect states in the lightly protonated emeraldine polymer [24].

The ability to derivatize polyaniline at ring or nitrogen positions enables an improvement in processibility and the testing of the role of interchain interaction. For example, ethoxy derivatized polyaniline is water soluble [25] and sulfonated polyaniline [26] is self-protonating and soluble in aqueous media.

A systematic study demonstrates electron localization in the polyaniline (PAN) polymer derivatives. With a substitution of a methyl group (CH_3) for a hydrogen atom on each C_6 ring of the emeraldine salt (ES) form of PAN to form poly(o-toluidine) (POT) [27], there is a ten percent increase in interchain separation [28] though there is similar crystallinity and coherence length [21, 28]. Magnetic susceptibility χ, DC conductivity σ_{DC}, microwave conductivity σ_{MW}, and dielectric constant ε at 6.5 GHz, thermoelectric power S and electron spin resonance together reflect the increasing localization of charges with increasing interchain separation. The data support the origin of the localization as increased one-dimensionally rather than disorder or Coulomb interaction [29,30]. Though total χ at room temperature is the same, POT has larger χ_{Curie} and smaller χ_{Pauli} with $N(E_F) = 1.76$ states/ev-2 ring calculated from $\chi_{Pauli} = \mu_B^2 N(E_F)$ compared to 3.5 for emeraldine salt (ES) [19]. This implies greater localization of the electrons [31]. The larger EPR linewidth for POT similarly suggests less motional narrowing [32] and hence more localization.

The increased localization in POT is further reflected in much smaller ε of POT ($x \equiv [Cl]/[N] = 0.5 \equiv$ POT-ES) at 6.5 GHz (Fig. 5) as compared with ε of ES [33]. Also the σ_{MW}, (Fig. 5) of POT-ES is much smaller than that of ES, while an order of

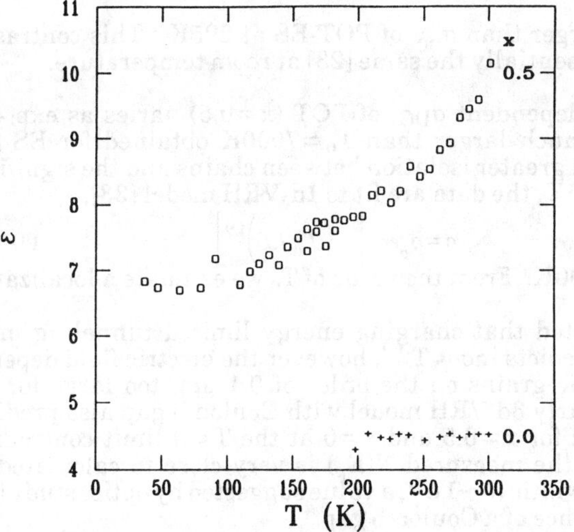

Figure 5—   T-dependence of dielectric constant of POT at (x=0.5)(□) and ES(+)[33].

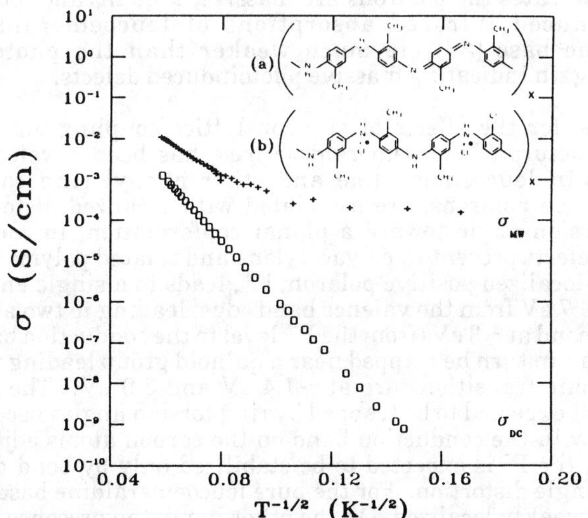

Figure 6—   T-dependence of σ_{DC} (□) and σ_{MW}(6.5 GH$_z$) (+) for POT(x=0.5) [33].   The σ_{DC} data shown were obtained for a pressed pellet of POT powder that was doped in 1 M HCl; σ_{MW} data were for NMP-cast film.   Inset is structue of (a) POT base (b) POT HCl salt.

magnitude larger than σ_{DC} of POT-ES at 295K. This contrasts with ES where σ_{MW} and σ_{DC} are essentially the same [23] at room temperature.

The T-dependent σ_{DC} of POT (x = 0.5) varies as $\exp[-(T_0/T)^{1/2}]$ (Fig. 6) with $T_0 \sim 30000K$ much larger than $T_0 = 7000K$ obtained for ES [22]. Given a crystal structure with greater isolation between chains and the significant difference of σ_{MW} and σ_{DC} at 295K, the data are fit to 1d-VRH model [33],

$$\sigma = \sigma_o \, exp\left| -\left(T_0/T\right)^{1/2}\right| \qquad (6)$$

with $T_0 = 37000K$. From the value of T_0 we estimate a localization of roughly 7.4Å.

It is noted that charging energy limited tunneling model [34] for granular metals also predicts $\ln\sigma \propto T^{-1/2}$, however the electric field dependence of σ implies the size of metallic grains on the order of 0.1 µm, too large for the application of the model. Similarly 3d VRH model with Coulomb gap also predicts $T^{-1/2}$-law, but finite difference of ε for x = 0.5 and x = 0 at the T = 0 limit contradicts application of this model. Also, the measured $N(E_F)$ is very close to calculated $N(E_F) = 4/\pi aW$ where $a \sim 10$Å, bandwidth $W \sim 1$ eV, a value suggested by optics study [27], again inconsistent with the presence of a Coulomb gap.

Photoinduced infrared absorption studies [35] show that the photoinduced infrared modes are much weaker in intensity than the photoinduced electronic transitions indicates the polarons are massive, >60 m_e and long lived. Similarly, the photoinduced infrared absorptions of leucoemeraldine base [36] and pernigraniline base [37] are much weaker than the photoinduced electronic transitions, again indicating massive photoinduced defects.

A model for the effects of electron lattice coupling via ring rotation on the electronic structure of leucoemeraldine base has been developed [38]. The charge defect states in leucoemeraldine and other phenyl ring containing polymers, particularly hole polarons, are associated with localized distortions in the ground state ring torsion angle toward a planar conformation, in contrast with the bond alternation defects present in polyacetylene and related polymers. It has been shown that this self-localized positive polaron, P^+, leads to a single energy level in the gap split off by ~ 0.7 eV from the valence band edge, leading to two absorptions, one at 0.7 eV and the second at ~ 3 eV (from the P^+ level to the conduction band). It is anticipated that the P^+ can in turn be trapped near a quinoid group leading to a formation of $P^+{}_Q$ whose electronic transitions are at ~ 1.4 eV and 3.0 eV. The negative polaron, in contrast, is not expected to be trapped by ring torsion angles because of the absence of charge density in the conduction band on the carbon atoms adjacent to the nitrogen sites. Hence, the P^- is expected to be stabilized only by bond distortion and not by ring torsion angle distortion. For the pure leucoemeraldine base the P^- is anticipated to all be only weakly localized. On the other hand, the presence of a quinoid or imine group in the polymer chain will act as a trap for the P^- to form a $P^-{}_Q$ with the energy level of which is expected to be approximately in the middle of the π to π^* gap.

Additional relevant excitations include the formation of a charge transfer exciton which has negative charge centered on a quinoid with $+e$ distributed on the phenyl groups on either side of the quinoid. The "exciton" is expected to exist in excited states, EX^* and also in long lived metastable states due to metastable ring conformations, $EX^\dagger$. Figure 7 summarizes a proposed decay scheme [36] for electrons and holes photoexcited from the ground state for pure leucoemeraldine base (with the presence of a few quinoid groups) and also for emeraldine base. The formation of trap

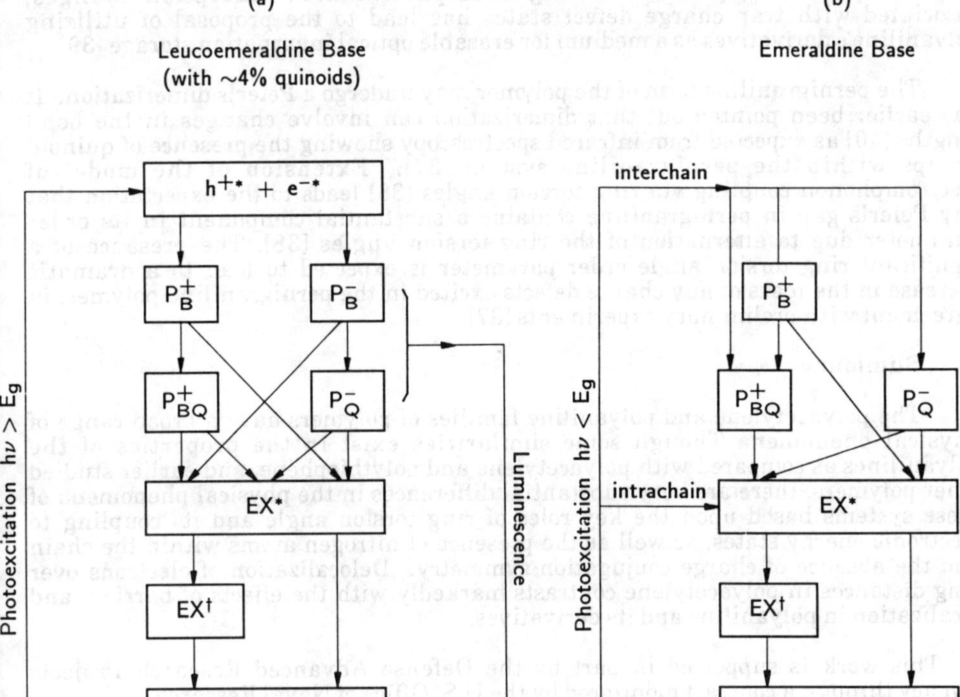

Figure 7–   Schematic representation of a model for the photoproduction and time dependent relaxations of defect states in (a) leucoemeraldine base for above the $\pi - \pi^*$ gap photoexcitation and (b) emeraldine base for photoexcitation in the "exciton" band.

charge at P^+_Q and P^-_Q is reflected in the long life times observed in the photoinduced spectroscopies. The presence of long lived photoinduced absorption changes, associated with trap charge defect states has lead to the proposal of utilizing polyaniline's derivatives as a medium for erasable optical information storage [39].

The pernigraniline form of the polymer may undergo a Peierls dimerization. It has earlier been pointed out that dimerization can involve changes in the bond lengths [40] as expected from infrared spectroscopy showing the presence of quinoid groups within the pernigraniline system [37]. Extension of the model of electron/phonon coupling via ring torsion angles [38] leads to the expectation that any Peierls gap in pernigraniline contains a substantial component in its order parameter due to alternation of the ring torsion angles [38]. The presence of a significant ring torsion angle order parameter is expected to lead to a dramatic increase in the mass of any charge defects excited in the pernigraniline polymer, in agreement with preliminary experiments [37].

3.   Summary

The polyacetylene and polyaniline families of polymers have a broad range of physical phenomena Though some similarities exist in the properties of the polyanilines as compared with polyacetylene and polythiophene, and earlier studied other polymers, there are very substantial differences in the physical phenomena of these systems based upon the key roles of ring torsion angle and its coupling to electronic energy states, as well as the presence of nitrogen atoms within the chain and the absence of charge conjugation symmetry. Delocalization of electrons over long distances in polyacetylene contrasts markedly with the effects of barriers and localization in polyaniline and its derivatives.

This work is supported in part by the Defense Advanced Research Projects Agency through a contract monitored by the U.S. Office of Naval Research.

References

1.   For reviews on recent progress, see Proc. Int. Conf. Synth. Met., Santa Fe, NM, USA (1988), Synth. Met. **28** (1989).
2.   See, for example, T. Skotheim (ed.), Handbook of Conducting Polymers, Vols. 1 and 2, Dekker, New York (1986).
3.   C.K. Chiang, C.R. Fincher, Jr., Y.W. Park, A.J. Heeger, H. Shirakawa, S.C. Gau, and A.G. MacDiarmid, Phys. Rev. Lett. **39**, 1098 (1977).
4.   A.J. Epstein, H. Rommelmann, R. Bigelow, H.W. Gibson, D.M. Hoffman, and D.B. Tanner, Phys. Rev. Lett. **50**, 1866 (1983).
5.   H. Naarmann and N. Theophilou, Synth. Met. **22**, 1 (1987).
6.   Th. Schwimmel, W. Reiss, J. Gimeiner, G. Denninger, M. Schwoerer, H. Naarmann, and N. Theophilou, Solid State Commun. **65**, 1311 (1988).
7.   N. Basescu, Z.-X. Liu, D. Moses, A.J. Heeger, H. Naarmann, and N. Theophilou, Nature **327**, 403 (1987).
8.   N. Theophilou, D.B. Swanson, A.G. MacDiarmid, A. Chakraborty, H.H.S. Javadi, R.P. McCall, S.P. Treat, F. Zuo, and A.J. Epstein, Synth. Met. **28**, D35 (1989).
9.   H.H.S. Javadi, A. Chakraborty, C. Li, N. Theophilou, D.B. Swanson, A.G. MacDiarmid, and A.J. Epstein, submitted.
10.   M.I. Katnelson and A.S. Scherbakov, J. Phys. C **19**, 5173 (1986).

11. A.J. Epstein, H. Rommelmann, M.A. Druy, A.J. Heeger and A.G. MacDiarmid, Solid State Commun. **38**, 683 (1981).
12. J.F. Kwak, T.C. Clarke, R.L. Greene, and G.B. Street, Solid State Commun. **31**, 355 (1979).
13. Yu.A. Firsov, in "Localization and Metal-Insulator Transition", H. Fritzche and D. Adler (Editors), Plenum Press, New York) (1985).
14. S. Kivelson and A.J. Heeger, Synth. Met. **22**, 371 (1988).
15. P.A. Lee and T.V. Ramakrishnan, Rev. Mod. Phys. **57**, 287 (1985).
16. A. Kawabata, Solid State Commun. **34**, 431 (1980).
17. A.J. Epstein, H. Rommelmann, R. Fernquist, H.W. Gibson, M.A. Druy, and T. Woerner, Polymer **23**, 1211 (1982).
18. A.G. MacDiarmid and A.J. Epstein, J. Chem. Soc., Faraday Trans (1989), in press.
19. J.M. Ginder, A.F. Richter, A.G. MacDiarmid, and A.J. Epstein, Solid State Commun. **63**, 97 (1987).
20. S. Stafstrom, J.L. Bredas, A.J. Epstein, H.S. Woo, D.B. Tanner, W.S. Huang, and A.G. MacDiarmid, Phys. Rev. Lett. **59**, 1464 (1987).
21. M.E. Jozefowicz, R. Laversanne, H.H.S. Javadi, A.J. Epstein, J.P. Pouget, X. Tang, A.G. MacDiarmid, Phys. Rev. B **39**, 12,958 (1989).
22. F. Zuo, M. Angelopoulos, A.G. MacDiarmid, A.J. Epstein, Phys. Rev. B **36**, 3475 (1987).
23. H.H.S. Javadi, K.R. Cromack, A.G. MacDiarmid, A.J. Epstein, Phys. Rev. B **39**, 3579 (1989).
24. F. Zuo, M. Angelopoulos, A.G. MacDiarmid, A.J. Epstein, Phys. Rev. B **39**, 3570 (1989).
25. S.K. Manohar, A.G. MacDiarmid, A.J. Epstein, Bull. Am. Phys. Soc. **34**, 583, (1989) and to be published.
26. J. Yue, A.G. MacDiarmid, and A.J. Epstein, to be published.
27. Y. Wei, W.W. Focke, G.E. Wnek, A. Ray, and A.G. MacDiarmid, J. Phys. Chem. **93**, 495 (1989).
28. M. Jozefowicz, et al., to be published.
29. N.F. Mott and E. Davis, Electron Proc. Non-cryst. Mat. (Clarendon Press, Oxford, 1979).
30. P.W. Anderson, Phys. Rev. **109**, 1492 (1958); Rev. Mod. Phys. **50**, 470 (1978).
31. M. Milovanovic, S. Sachdev and R.N. Bhatt, Phys. Rev. Lett. **63**, 82 (1989).
32. R. Kubo and K. Tomita, J. Phys. Soc. Japan **9**, 888 (1954).
33. Z. Wang, H.H.S. Javadi, A. Ray, A.G. MacDiarmid, and A.J. Epstein, submitted.
34. P. Sheng and B. Abeles, Phys. Rev. Lett **28**, 34 (1972).
35. R.P. McCall, J.M. Ginder, M.G. Roe, G.E. Asturias, E.M. Scherr, A.G. MacDiarmid, and A.J. Epstein, Phys. Rev. B **39**, 10174 (1989).
36. R.P. McCall, J.M. Ginder, J. M. Leng, H.J. Ye, S.K. Manohar, J.G. Masters, G.E. Asturias, A.G. MacDiarmid, and A.J. Epstein, Phys. Rev. B (1990), in press.
37. J.M. Ginder, R.P. McCall, J.M. Leng, H.J. Ye, S.K. Manohar, Y. Sun, and A.G.MacDiarmid, to be published.
38. J.M. Ginder, A.J. Epstein, and A.G. MacDiarmid, Solid State Commun. **72**, 987 (1989).
39. R.P. McCall, J.M. Ginder, and A.J. Epstein, to be published.
40. M.C. dos Santos and J.L. Bredas, Phys. Rev. Lett. **62**, 2499 (1989); see also M.C. dos Santos and J.L. Bredas, Synth. Met. **29**, E321 (1989).

LINEAR POLYENES: THE INTERPLAY BETWEEN ELECTRONIC STRUCTURE, GEOMETRIC STRUCTURE, AND NONLINEAR OPTICAL PROPERTIES

J.M. TOUSSAINT, F. MEYERS, AND J.L. BREDAS
Service de Chimie des Matériaux Nouveaux
Département des Matériaux et Procédés
Université de Mons
avenue Maistriau 21
B-7000 MONS (BELGIUM)

ABSTRACT. This contribution deals with theoretical investigations of the electronic and optical properties of two series of all-trans linear polyenes. In the first case, the focus is on unsubstituted polyenes and the geometry relaxation processes occurring in the first singlet, one-photon optically allowed excited state. Calculations are performed at various levels of sophistication, from a simple Su-Schrieffer-Heeger (Hückel-like) Hamiltonian up to Pariser-Parr-Pople single CI and Restricted Hartree-Fock ab initio Hamiltonians. It is found that explicit consideration of the electron-lattice coupling is essential in order to obtain a coherent evolution of the $1B_u$ state geometry relaxation, in going from short polyenes to long polyenes and polyacetylene. The relevance of our results in terms of the nonlinear optical properties of these compounds is pointed out. In the second case, the electronic structure and second-order polarizability, β, are calculated at the ab initio level for two series of novel push-pull polyene molecules, benzodithiapolyenals and dithiolylidenepolyenals. The benzodithiapolyenal molecules have recently been reported to present among the largest $\mu.\beta$ values ever measured. The theoretical results allow for an in-depth understanding of the properties of these systems.

1. Introduction

Since the discovery that polyacetylene can be made highly electrically conductive through doping with electron donors or acceptors [1], extensive theoretical and experimental works have been devoted to this polymer. These have for example led recently to the synthesis of a new type of polyacetylene which exhibits conductivities after doping as high as that of copper [2,3].

Most of the peculiar electrical, optical, and magnetic properties appearing upon doping of polyacetylene have been rationalized by invoking the formation of nonlinear elementary excitations of soliton-type [4,5]. Solitons in all-trans polyacetylene are topological kinks that produce a reversal of the bond alternation pattern (see Figure 1). It takes about seven to ten bonds (i.e., 15 to 20 carbon atoms) for the bond alternation reversal to be complete. The presence of solitons on a polyacetylene chain also affects the polymer electronic structure: localized electronic states associated to solitons appear around midgap and thus produce novel subgap optical transitions that can be detected on doping [5].

J. L. Brédas and R. R. Chance (eds.), Conjugated Polymeric Materials:
Opportunities in Electronics, Optoelectronics, and Molecular Electronics, 207–219.
© 1990 *Kluwer Academic Publishers. Printed in the Netherlands.*

Figure 1. Reversal of the bond alternation pattern induced by the presence of a soliton along a polyacetylene chain (a neutral soliton, i.e. a radical defect, is illustrated here).

Following the pioneering work of Ducuing and co-workers [6], it has been realized that conjugated polymers inherently possess very high nonlinear optical responses [7,8]. It has been recently pointed out that the influence of electron-lattice coupling and e.g. of the presence of nonlinear elementary excitations such as solitons could also be significant for the nonlinear optical properties [9-13]. In all-trans polyacetylene, photoexcitation across the gap into the $1B_u$ state produces electron-hole pairs that are found to decay very rapidly (in 10^{-13} sec.) into pairs of separated, positively and negatively charged, solitons [14-16]. Such an evolution is in agreement with the early theoretical predictions of Su and Schrieffer [17]. It is important to note that this process results in an efficient charge separation mechanism, a feature essential to provide large optical nonlinearities [18]. Furthermore, in the off-resonance regime, instantons (i.e., virtual soliton-antisoliton pairs) [19] have been suggested to yield enhanced optical nonlinearities with respect to a purely rigid lattice situation [8-10].

In short polyenes, spectroscopic studies indicate that strong geometry relaxations also take place in the first B_u one-photon optically-allowed excited state [20,21]. These geometry relaxations are traditionally modeled theoretically using Bond Order/Bond Length (BOBL) relationships [22]. In this framework, the geometry relaxations follow exactly the wavefunction characteristics in the excited state. They are calculated to become smaller as the chain length increases and tend to be negligible for chains containing over 15-20 carbon atoms [23,24]. Such an evolution is, however, inconsistent with the situation appearing in very long chains, i.e., polyacetylene, where the first B_u state strongly relaxes to produce a pair of charged solitons, as mentioned above [5,14,15].

Therefore, in this paper, we review the calculations we have performed in order to remove the inconsistency between the traditional understanding of the $1B_u$ relaxation process in short polyenes and the perception of the situation prevailing in all-trans polyacetylene and provide a coherent picture of the evolution between short and long polyenes. First, we investigate the relaxation process in the $1B_u$ excited state of polyene chains by means of the methodology of Su, Schrieffer, and Heeger [4], as adapted by Brédas et al. [25]. We make use of a Hückel Hamiltonian with bond-length dependent transfer integrals and σ-bond compressibility. We study chains ranging from 10 to 58 carbon atoms in order to understand the evolution of the relaxed geometry as a function of chain length. Second, calculations are carried out at a much higher level of sophistication (combined Restricted Hartree-Fock ab initio / Pariser-Parr-Pople Configuration Interaction level) for three polyene molecules: hexatriene (C_6H_8), decapentaene ($C_{10}H_{12}$), and tetradecaheptaene ($C_{14}H_{16}$). We are in this way able to assess our results on a firmer ground.

Finally, we report on ab initio calculations of the electronic structure and second-order polarizability of a series of push-pull polyene molecules, based on benzodithiapolyenals and dithiolylidenepolyenals. Our interest in these compounds is that they have recently been shown to present among the largest $\mu.\beta$ values ever measured [26] and are thus suited for second-order nonlinear optical applications.

2. Geometry Relaxation in the First B_u Excited State of Linear Polyenes

2.1. SU-SCHRIEFFER-HEEGER HAMILTONIAN APPROACH

In order to be in a position, on the one hand, to examine the evolution in going from short to long polyenes and, on the other hand, to make a significant comparison to polyacetylene, we have chosen to work first at the Su-Schrieffer-Heeger Hamiltonian level. This corresponds to a Hückel technique with bond-length dependent transfer integrals and σ-bond compressibility [4].

The parameters originally optimized for polyacetylene by Su, Schrieffer, and Heeger [4] are applied to polyene chains ranging in size from 10 to 58 carbon atoms [11]. These parameters lead to: (i) a degree of bond length alternation of 0.14 Å in the ground state (single bond equal to 1.47 Å and double bond equal to 1.33 Å), (ii) a (somewhat underestimated) bandgap of 1.4 eV in trans-polyacetylene, and (iii) a creation energy of 0.92 eV for a pair of totally separated solitons [11].

We search the configuration space to find the geometry providing the lowest $1B_u$ excited state total energy. Two types of situation are mostly investigated:
(i) We allow for geometries similar to those obtained with traditional BOBL relationships in the framework of Pariser-Parr-Pople (PPP) calculations including configuration interaction [27]. The amplitude and the extent of the relaxation are both optimized.
(ii) We look for geometries resulting in the formation of a soliton pair (two-soliton geometries). Both the distance between the solitons and the soliton widths are optimized.

The results corresponding to the two-soliton geometries are given in Table 1 for chains containing from 10 to 58 carbon atoms.

In the longest polyene chain we investigated here (58 carbon atoms), results are identical to those obtained for the infinite polyacetylene chain [4,25]. The $1B_u$ excited-state geometry relaxes to form a pair of solitons (with a creation energy of 0.922 eV), which lowers the $1B_u$ excited state total energy by 38% relative to the vertical excitation energy. The solitons are calculated to be optimally located on sites 19 and 40, i.e., they are as distant from one another as from the chain ends. Each soliton is found to extend over about 15 carbons, as in trans-polyacetylene [4].

As the chain length decreases, the total energy lowering due to the relaxation of the $1B_u$ excited state becomes smaller. However, even in the case of decapentaene, this energy lowering due to a soliton pair formation still amounts to 23% of the vertical transition energy.

Very importantly, the relaxed excitation energies obtained for excited-state geometries calculated on the basis of BOBL relationships are 0.2-0.3 eV larger than in the two-soliton formation situation. The major result of these Hückel-like calculations is thus to find that the relaxation effects in the $1B_u$ excited state of polyene molecules are qualitatively similar to those in trans-polyacetylene. Even in the shortest chains considered here, the $1B_u$ excited state is found to relax optimally in such a way as to produce a pair of solitons. In order to be accommodated in shorter chains, the solitons shrink in size: their width decreases from $2l=15$ for trans-polyacetylene, to $2l=11$ for chains containing between 30 and 50 carbon atoms, $2l=7$ for chains between 20 and 30 carbon atoms, and $2l=3$ in chains containing less than 20 carbons. In all cases, the solitons tend to be centered on locations separating them equally from one another and from the chain ends. The evolution between short and long polyene chains is thus found to be fully coherent.

TABLE 1. Evolution as a function of chain length n of: (i) the vertical excitation energy (in eV) to the first B_u excited state, E_{vert}; (ii) the corresponding relaxed excitation energy (in eV), E_{rel}; (iii) the relative total energy lowering (in %) due to the relaxation of the $1B_u$ excited state; (iv) the optimal soliton width, $2l$ (in number of sites); (v) the optimized site locations of the soliton defects, n_1 and n_2; and (vi) the HOMO-LUMO separation (in eV) in the relaxed geometry. The calculations are performed in the framework of the Su-Schrieffer-Heeger Hamiltonian [11].

n	E_{vert}	E_{rel}	%	$2l$	n_1,n_2	HOMO-LUMO
58	1.477	0.922	37.6	15	19,40	0.14
46	1.515	0.939	38.0	11	15,32	0.20
38	1.557	0.983	36.9	11	11,28	0.20
26	1.682	1.105	34.3	7	9,18	0.30
22	1.760	1.178	33.1	7	7,16	0.55
18	1.876	1.340	28.6	3	5,14	0.70
14	2.064	1.541	25.3	3	5,10	0.98
10	2.400	1.853	22.8	3	3,8	1.16

It is important to stress that the two-soliton formation in the $1B_u$ excited state of decapentaene appears to agree better with experimental data than does the situation where the $1B_u$ excited state geometry is described in terms of BOBL relationships. In the former case, the average bond-length variations between the $1B_u$ excited state and the ground state are calculated to be 0.070 and 0.087 Å for the double and single bonds, respectively. In the latter case, the corresponding values are 0.043 and 0.047 Å [24]. These results are to be compared with the experimental estimates of Granville et al. [20] which provide average variations of 0.085 and 0.081 Å for double and single bonds, respectively. Furthermore, the relaxation energy experimentally measured in the $1B_u$ state of decapentaene (at 77 K) is on the order 0.8 eV [20]. Such a relaxation represents 20% of the vertical excitation energy [20]. Within the two-soliton configuration, the relaxation is calculated to be about 23% of the vertical transition energy, while it is only 10-12% in the BOBL-derived $1B_u$ excited state [24].

The overall picture obtained within the simple Su-Schrieffer-Heeger Hamiltonian is that there exists a smooth evolution of the relaxation of the $1B_u$ excited-state geometry in going from short polyenes to polyacetylene. This evolution is consistent with the strong geometry relaxations experimentally observed in short polyenes (e.g. decapentaene) as well as polyacetylene. In all cases, the relaxation is such as to produce the formation of a soliton-antisoliton pair on the chain. Note that a relaxation tending to produce a two-soliton geometry is not unlike the ionic valence-bond pictures used to describe the $1B_u$ excited state [27].

It is however important to address the question of the $1B_u$ excited-state relaxation using a more sophisticated theoretical approach, in particular taking explicitly into account the effects of electron-electron interactions and electron correlation. The calculations have therefore been extended on short polyenes, using a combined PPP Configuration Interaction and Hartree-Fock ab initio approach.

2.2 PPP-CI/RHF ab initio APPROACH

It is usually the case that PPP Hamiltonians are suitably parameterized to reproduce excited state transition energies. However, they do not provide reliable geometry optimizations in terms of total energy differences. On the other hand, if ab initio techniques (which can afford good total energy differences) can be used quite easily to optimize ground-state geometries, they soon prove to be too costly when applied to optimize excited-state geometries of systems containing over one hundred basis functions.

Therefore, we have sought to combine these two techniques in order to be able to perform calculations on relatively large molecules and to assess more reliably the trends in the first B_u excited state geometry [12]. We have thus considered: (i) geometry optimizations at the Restricted Hartree-Fock (RHF) 3-21G split valence basis set level [28], which allow for a correct sketch of the ground-state potential energy curve; and (ii) PPP Configuration Interaction (CI) calculations including single excitations from the ground-state reference framework and based on the various 3-21G optimized geometries. The PPP program and parameters we have used can be found in Ref. 29. The parameters are chosen in such a way as to provide a degree of bond-length alternation in the ground state almost identical to that calculated at the RHF 3-21G level, see below. Although it does not predict a correct ordering of the excited states of polyene molecules, a single CI approach is known to describe adequately the first B_u excited state [22,23,27]. Combining the ground state total energies with the PPP vertical transition energies, leads to the relative excited-state total energies corresponding to various geometrical situations.

We have considered three linear polyene molecules: hexa-1,3,5-triene (C_6H_8), deca-1,3,5,7,9-pentaene ($C_{10}H_{12}$), and tetradeca-1,3,5,7,9,11,13-heptaene ($C_{14}H_{16}$). The RHF ab initio calculations are performed in the ground state for three different geometry situations [12]:

(i) In the first case, we carry out a full (i.e., all bond lengths and valence angles) 3-21G optimization of the ground state geometry assuming coplanar conformations.

(ii) In the second case, we consider the geometry corresponding to the relaxed $1B_u$ excited-state, obtained using BOBL relationships with the PPP Hamiltonian [24]. All bond angles are optimized at the ab initio 3-21G level, while the carbon-carbon bond lengths are kept at their PPP B_u-state values and the carbon-hydrogen bond lengths are set at their optimal ab initio ground-state values.

(iii) In the third case, we partly optimize the geometry corresponding to the formation of a pair of solitons, as obtained at the Hückel/SSH Hamiltonian level described previously. Here, we force the two carbon-carbon bonds surrounding the soliton centers to be of equal lengths. The values of these bond lengths and those of the bonds towards the ends of the molecules are optimized. The carbon-carbon bonds located between the solitons are chosen to be identical to the single and double bonds appearing in the 3-21G optimized ground-sate geometry. All the bond angles are fully optimized, while the carbon-hydrogen bonds are fixed at the ab initio ground-state optimal values.

In Table 2, we present the ground-state geometries of hexatriene, decapentaene, and tetradecaheptaene, as optimized at the RHF ab initio 3-21G level. The decapentaene PPP values [24] are also indicated. The PPP values are obtained by using the following bond order (l_{pq})/ bond length (R_{pq}, in Å) relationship: $R_{pq} = 1.51 - 0.19\, l_{pq}$

TABLE 2.   Optimized RHF/3-21G carbon-carbon bond lengths (in Å) for the ground state of hexatriene, decapentaene, and tetradecaheptaene. The numbers between parentheses for decapentaene refer to the PPP Hamiltonian-optimized values [24]. The atoms are labeled starting from one end of the molecule.

	C_6H_8	$C_{10}H_{12}$	$C_{14}H_{16}$
R(C1-C2)	1.322	1.322 (1.325)	1.322
R(C2-C3)	1.462	1.461 (1.465)	1.461
R(C3-C4)	1.327	1.329 (1.330)	1.329
R(C4-C5)		1.456 (1.463)	1.455
R(C5-C6)		1.330 (1.332)	1.331
R(C6-C7)			1.454
R(C7-C8)			1.331

The evolution of the degree of bond-length alternation along the polyene chains obtained on the basis of BOBL relationships are depicted in Figure 2. It is clearly observed that in this framework the geometry relaxation in the first singlet B_u excited state strongly decreases with increasing chain length. In hexatriene, all bonds are significantly affected by the excitation; in the middle of the molecule, the sign of bond dimerization reverses. This is only slightly the case in decapentaene, where the two central bonds are almost equal and the bond alternation for the outer bonds reaches +0.09 Å. In tetradecaheptaene, however, the sign of the bond alternation remains positive all along the molecule, indicating no single bond-double bond character reversal. (Note that by convention, we consider the sign of the bond alternation to be positive when it is the same as in the ground state).

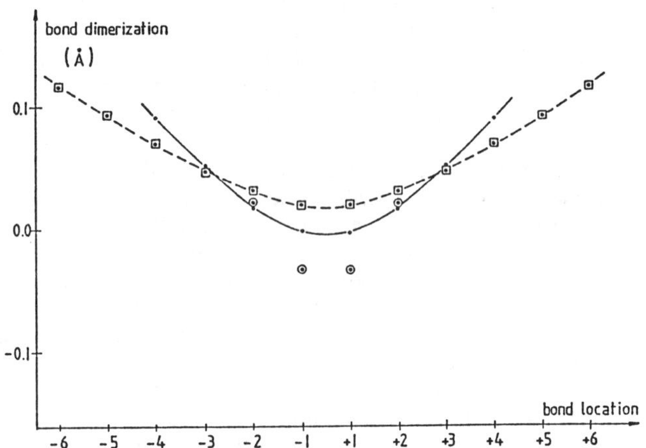

Figure 2.   Illustration of the RHF ab initio evolution of the bond dimerization value (in Å) along the molecules of hexatriene (double circles), decapentaene (closed circles, solid line), and tetradecaheptaene (squares, dashed line), in the BOBL-derived relaxed geometry for the first B_u excited state. The sign of the dimerization is by convention taken to be positive if it is the same as in the ground state. The bonds are labeled starting from the central bond of the molecule.

In Figure 3, we present the results obtained when considering the possibility of formation of a soliton-antisoliton pair as the $1B_u$ state relaxes. A comparison of Figures 2 and 3 illustrates that the two-soliton geometry leads to bond length modifications relative to the BOBL-derived geometries, which are smaller in the hexatriene case but much larger in the longer chains. We note that for hexatriene, it is actually rather misleading to speak in terms of the formation of a two-soliton configuration since the soliton and the antisoliton are localized on adjacent sites. Hexatriene is therefore too short a chain to provide for a complete bond alternation reversal in the middle of the molecule when considering a two-soliton configuration. Such a reversal, however, clearly occurs in decapentaene and tetradecaheptaene. In decapentaene, the average bond-length modification with respect to the optimal ground-state geometry is 0.073 Å. We stress again that, relative to the 0.08 Å experimental estimate of Granville et al. for the relaxed B_u state [20], this value is in much better agreement than that provided by BOBL relationships (on the order of 0.04 Å, i.e. twice smaller than the experimental value).

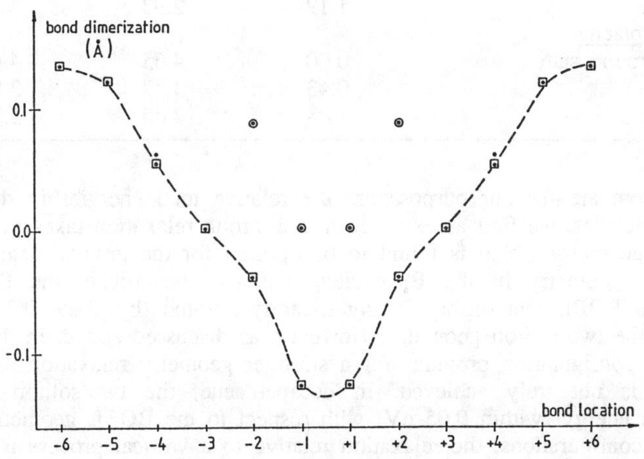

Figure 3. Illustration of the RHF ab initio evolution of the bond dimerization value (in Å) along the molecules of hexatriene (double circles), decapentaene (closed circles and dashed line), and tetradecaheptaene (squares, dashed line), in the two-soliton relaxed geometry for the first B_u excited state. The sign of the dimerization is by convention taken to be positive if it is the same as in the ground state. The bonds are labeled starting from the central bond of the molecule.

In Table 3, we list for the three kinds of investigated geometries: (i) the relative total energies in the ground state, as obtained at the Hartree-Fock ab initio 3-21G level; (ii) the vertical transition energies to the first optically-allowed singlet B_u state, as calculated at the PPP-CI level; and (iii) the relative total energies in the $1B_u$ excited state, obtained by simply summing the first two terms.

TABLE 3. Comparison of the energies involved in hexatriene, decapentaene, and tetradecaheptaene for the three geometry situations investigated in this work: (A) relative RHF ab initio total energies in the ground state (the fully optimized RHF ab initio 3-21G value being taken as reference); (B) PPP-CI vertical transition energies to the first singlet B_u state; (C) relative total energies in the first singlet B_u state (calculated by summing the first two terms). All energies are given in eV.

	A	B	C
hexatriene			
ab initio ground state	0.00	5.86	5.86
BOBL	0.63	4.14	4.77
two-soliton	0.27	4.83	5.10
decapentaene			
ab initio ground state	0.00	5.03	5.03
BOBL	0.51	3.56	4.07
two-soliton	1.19	2.93	4.12
tetradecaheptaene			
ab initio ground state	0.00	4.65	4.65
BOBL	0.43	3.37	3.80
two-soliton	1.08	2.65	3.73

What we need to compare for our purpose are the relative total energies in the excited state. For all three molecules, we find as expected that a strong relaxation takes place in the excited state, i.e. the geometry which is found to be optimal for the ground state does not constitute the optimal geometry in the B_u excited state. In hexatriene, the B_u relaxed geometry derived from BOBL relationships is significantly favored (by about 0.3 eV) over that corresponding to the two-soliton geometry. However, as discussed above, in the case of hexatriene, the BOBL configuration provides for a stronger geometry relaxation and a two-soliton configuration is not truly achieved. In decapentaene, the two-soliton geometry becomes very close in energy (within 0.05 eV) with respect to the BOBL geometry for the excited state. In both configurations, the relaxation relative to a vertical process is found to be on the order of 0.9 eV, a value in excellent agreement with the 0.84 eV value which is experimentally measured [20]. Importantly, in the case of tetradecaheptaene, the two-soliton geometry corresponds to the most stable situation for the excited-state, being about 0.07 eV lower in energy than the BOBL geometry. The gain in stability of the two-soliton configuration with respect to the BOBL configuration as chain length increases is fully consistent with the experimental observation on the photogeneration of soliton pairs in the first $1B_u$ excited state of polyacetylene. In long chains, a BOBL-derived configuration would in contrast lead to negligible relaxation.

An interesting feature is uncovered when: (i) observing in Figures 2 and 3 the evolution in bond length alternation for decapentaene and tetradecaheptaene; and (ii) comparing it to the geometry relaxation process due to a photogenerated electron-hole pair in polyacetylene (see Figure 2a of Ref. 19). From this observation, we suggest that the configuration given by BOBL relationships corresponds to the early stage of the process of electron-hole separation into a pair of solitons whereas the two-soliton configuration relates to the final stage of the separation process. This separation is actually induced through the explicit coupling of the electronic structure to the lattice, an ingredient which is absent from the BOBL approach where the geometry has to follow the wavefunction characteristics. It is worth pointing out that in the two-soliton configuration, the phase of the bond alternation pattern between the locations of the soliton and the antisoliton is <u>opposite</u> to that of the wavefunction bonding-antibonding pattern.

The results obtained from the combined RHF ab initio/PPP-CI approach thus confirm that in linear polyenes containing at least ten carbon atoms, the electron-lattice coupling plays a significant role in the relaxation of the first B_u excited state. We stress that the fast relaxation of the $1B_u$ state related to the decay of a photoinduced electron-hole pair into a pair of charged solitons has been invoked to be an important factor in the large nonlinear optical response of polyacetylene [8-10,19]. Such a decay indeed provides major shifts in oscillator strengths and a very effective charge separation mechanism, which are essential to large hyperpolarizabilities [6,7,18].

Our work thus suggests that going beyond frozen geometry models of the polarizabilities and hyperpolarizabilities by incorporating electron-lattice coupling could also prove to be essential to describe properly the optical nonlinearities in short and intermediate-sized polyenes [30,31], as well as in other oligomers where electron-lattice coupling effects are known to be important (polypyrroles, polythiophenes,...).

We note that the two-soliton geometry relaxation leads to the appearance of new electronic states in the gap, see Figure 4 [11]. This feature leads to important shifts in oscillator strengths. It would be therefore most interesting to carry out photoinduced absorption experiments in the subpicosecond regime. Such measurements indeed constitute an ideal means to probe the energies and fast time evolution of the new optical absorptions resulting from such a relaxation in linear polyenes.

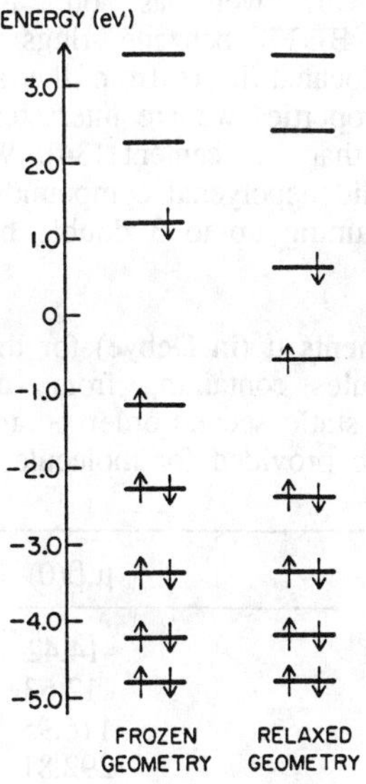

Figure 4. Evolution of the one-electron energy levels in the first B_u excited state of decapentaene, in going from the frozen ground-state geometry (left) to the relaxed excited state geometry (right).

3. Electronic Structure and Second-Order Polarizability of Benzodithiapolyenals

Recently, Lehn and co-workers [32] have succeeded in synthesizing two series of push-pull polyene molecules containing up to 8 double bonds in the polyene segment: benzodithiapolyenals (BDTP's) and dimethylanilinopolyenals (DMAP's), see Figure 5. Through EFISHG (Electric-Field Induced Second-Harmonic Generation) measurements, Barzoukas et al. [26] have shown that the longest of these compounds possess among the largest $\mu.\beta$ values that have ever been reported, on the order of 7000-9000 x 10^{-48} esu at 1.34 μm, leading to static values of about 3000 x 10^{-48} esu. The value of $\mu.\beta$ as a function of the length of the polyene segment is experimentally observed to evolve as $n^{2.4}$, where n is the number of double bonds in the polyene segment [26]. Interestingly, no saturation effect is obtained even after eight double bonds.

A simple model, detailed in Ref. 33, suggests that μ should evolve as $n^{0.5}$, in which case β would almost have a square dependence on the number of double bonds. We have tried to check on that dependence by performing RHF ab initio 3-21G calculations of the electronic structure and second-order polarizability β, on the benzodithiapolyenal series. We chose to focus first our attention on the BDTP compounds because we wished to rationalize the lower $\mu.\beta$ values (relative to the dimethylanilinopolyenal series) measured for the shortest chains, in particular that containing only one double bond in the polyenic segment. The $\mu\beta(0)$ is indeed $\approx$20 x 10^{-48} esu for the BDTP compound with one double bond in the polyenic segment, to be compared to $\mu\beta(0)$ of$\approx$200 x 10^{-48} esu for the corresponding DMAP molecule.

Full geometry optimizations have been carried out at the 3-21G level on the BDTP molecules containing up to four double bonds as well as on a series of dithiolylidenepolyenal (DTLP) molecules where the BDTP benzene rings are simply replaced by a double bond (connecting the carbons located in α from the sulfurs), see Figure 5. It was indeed observed that the molecular properties we are interested in (dipole moment, hyperpolarizability) are almost unaffected by that replacement [34]. We therefore decided to pursue the theoretical works on the dithiolylidenepolyenal compounds since they allow us to perform β calculations on molecules containing up to 3 double bonds in the polyene segment.

TABLE 4. RHF ab initio 3-21G dipole moments μ (in Debye) for the optimized geometries of dithiolylidenepolyenal molecules containing from (n=) 0 to 4 double bonds in the polyenic segment. The static second-order polarizabilities β (in 10^{-30} esu) and $\mu.\beta$ terms (in 10^{-48} esu) are provided for molecules with up to three double bonds.

	μ	$\beta(o)$	$\mu.\beta(0)$
n=0	4.04	-3.57	-14.42
n=1	5.12	2.47	12.63
n=2	5.52	21.18	116.95
n=3	5.79	50.61	292.81
n=4	6.39	-	-

The ground state dipole moments for the dithiolylidenepolyenals are given in Table 4. We observe that the dipole moment actually evolves very little with the length of the polyene segment. This can be understood on the basis that the charges induced by the donor (dithiolylidene group) on the one hand, and the acceptor (carbonyl) group on the other hand, are calculated to remain very localized and do not spread over the polyene segment [34]. The μ evolution is found to go as $n^{0.1}$

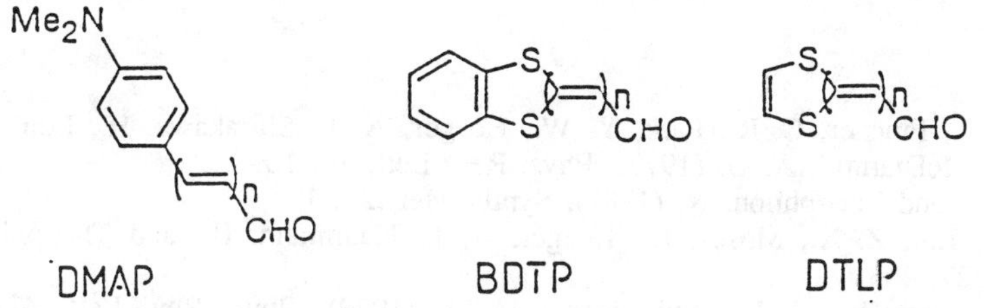

Figure 5. Geometric structures of the dimethylanilinopolyenal (DMAP), benzodithiapolyenal (BDTP), and dithiolylidenepolyenal (DTLP) molecules.

In Table 4, we also present the 3-21G ab initio static second-order polarizabilities, $\beta(0)$, as well as the $\mu.\beta(0)$ terms. Two important remarks can be made.

First, the theoretical $\mu.\beta(0)$ are in excellent agreement with the values that can be interpolated from the EFISHG measurements of Barzoukas et al. (see Figure 2 of Ref. 26). This lends much confidence in our theoretical approach. The DTLP $\mu.\beta(0)$ calculated values evolve as $n^{2.6}$, in excellent agreement to the experimental data on BDTP's.

Second, we observe that the β value switches sign in going from the n=0 to the n=1 DTLP molecule. As will be detailed elsewhere [34], the origin of this effect is due to the fact that in the n=0 compound, the most intense optical transition involves the promotion of an electron from the HOMO molecular orbital to the (LUMO+1) molecular orbital which is localized on the dithiolylidene part of the molecule. For n=1, the HOMO-LUMO transition becomes the most intense, the LUMO being, on the contrary, localized on the polyenal part of the molecule. As the length of the polyenic segment increases, the (LUMO+1) MO remains at the same energy level; on the other hand, the LUMO keeps stabilizing and accounts more and more predominantly for the charge transfer in the excited state. On the basis of this simple two-level analysis, the change in sign of β can be easily rationalized.

Furthermore, this effect also explains the low $\mu.\beta$ values in the BDTP and DTLP molecules with small polyenic segments, relative to the corresponding DMAP molecules. In the latter case, the benzene rings are located <u>along</u> the charge-transfer path, thereby effectively increasing the conjugated length of the polyenic segment and providing a greater separation of charges upon excitation. In the former case, the benzene ring of the BDTP molecules or the double bond that replaces it in the DTLP molecules, are located in a direction <u>opposite</u> to the charge-transfer path, thereby counteracting an efficient charge transfer in the smaller molecules.

4. Acknowledgements

The authors acknowledge stimulating discussions with A.J. Heeger, B.E. Kohler, R. Silbey, and J. Zyss. The collaboration with UCSB is supported by a joint grant from the US National Science Foundation and the Belgian National Fund for Scientific Research, FNRS (NSF-INT-8912268 / FNRS-1.5.089.90F). The authors are indebted to F. Van De Ven for his efficient technical assistance. They thank FNRS for the use of the KUL Supercomputer Facility; FNRS, IBM-Belgium, and FNDP-Namur for the use of the SCF Facility; and the University of Mons CCI Computing Center.

5. References

[1] Chiang, C. K., Fincher, C. R., Park, Y. W., Heeger, A. J., Shirakawa, H., Louis, E. J., Gau, S. C., and McDiarmid, A. G. (1977), Phys. Rev. Lett. 39, 1098.

[2] Naarmann, H. and Theophilou, N. (1987), Synth. Met. 22, 1.

[3] Basescu, N., Liu, Z. X., Moses, D., Heeger, A. J., Naarmann, H., and Theophilou, N. (1987), Nature 327, 403.

[4] Su, W. P., Schrieffer, J. R., and Heeger, A. J. (1979), Phys. Rev. Lett. 42, 1698; (1980), Phys. Rev. B 22, 2099; Rice, M. J. (1979), Phys. Lett. A 71, 152.

[5] Heeger, A. J., Kivelson, S., Schrieffer, J. R., and Su, W. P. (1988), Rev. Mod. Phys. 60, 781.

[6] Sauteret, C., Hermann, J. P., Frey, R., Pradere, F., Ducuing, J., Chance, R. R., Baughman, R. H. (1976), Phys. Rev. Lett. 36, 956; Hermann, J. P. and Ducuing, J. (1974), J. Appl. Phys. 45, 5100.

[7] Nonlinear Optical Properties of Polymers, A. J. Heeger, J. Orenstein, and D. R. Ulrich (eds.), Materials Research Society Symposium Proceedings, Vol. 109 (1988).

[8] Nonlinear Optical Properties of Organic Molecules and Crystals, D. S. Chemla and J. Zyss (eds.), Academic, New York (1987).

[9] Sinclair, M., Moses, D., Akagi, K., and Heeger, A. J., in Ref. 7, p. 205; Heeger, A. J., in this volume.

[10] Sinclair, M., Moses, D., Akagi, K., and Heeger, A. J. (1988), Phys. Rev. B 38, 10724; Sinclair, M., Moses, D., McBranch, D., Heeger, A. J., Yu, J., and Su, W. P. (1989), Synth. Met. 28, D655.

[11] Brédas, J. L. and Heeger, A. J. (1989), Chem. Phys. Lett. 154, 56.

[12] Brédas, J. L. and Toussaint, J. M., J. Chem. Phys., in press.

[13] Brédas, J. L. (1989), in Electronic Properties of Polymers and Related Compounds, H. Kuzmany, M. Mehring, and S. Roth (eds.), Springer Series in Solid State Sciences, Springer Verlag, Vol. 91, 204.

[14] Sinclair, M., Moses, D., and Heeger, A. J. (1986), Solid State Commun. 57, 343.

[15] Rothberg, L., Jedju, T. M., Etemad, S., and Baker, G. L. (1986), Phys. Rev. Lett. 57, 3229; (1987), Phys. Rev. B 36, 7529.

[16] Roth, S. and Bleier, H. (1987), Adv. Phys. 36, 385.

[17] Su, W. P. and Schrieffer, J. R. (1980), Proc. Natl. Acad. Sci. U.S.A. 77, 5626.

[18] Heflin, J. R., Wong, K. Y., Zamani-Khamiri, O., and Garito, A. F. (1988), Phys. Rev. B 38, 1573.

[19] Sinclair, M., Moses, D., McBranch, D., Heeger, A. J., Yu, J., and Su, W. P. (1989), Phys. Scripta T27, 144.

[20] Granville, M. F., Kohler, B. E., and Snow, J. B. (1981), J. Chem. Phys. 75, 3765; Kohler, B. E., Spiglanin, T. A., Hemley, R. J., and Karplus, M. (1984), J. Chem. Phys. 80, 23.

[21] Kohler, B. E. and Spiglanin, T. A. (1984), J. Chem. Phys. 80, 5465; Aoyagi, M., Ohmine, I., and Kohler, B. E., J. Phys. Chem., in press.

[22] Salem, L. (1966) Molecular Orbital Theory of Conjugated Systems Benjamin, New York.

[23] Schulten, K., Ohmine, I., and Karplus, M. (1976), J. Chem. Phys. 64, 4422; Tavan, P.

and Schulten, K. (1986), J. Chem. Phys. 85, 6602.

[24] Brédas, J. L., Dory, M., and André, J. M. (1985), J. Chem. Phys. 83, 5242.

[25] Brédas, J. L., Chance, R. R., and Silbey, R. (1982), Phys. Rev. B 26, 5843.

[26] Barzoukas, M., Blanchard-Desce, M., Josse, D., Lehn, J. M., and Zyss, J. (1989), Chem. Phys. 133, 323.

[27] Hudson, B. S., Kohler, B. E., and Schulten, K. (1982), in Excited States, E. C. Lim, Academic Press, New York, Vol. 6, p. 1.

[28] Binkley, J. S., Pople, J. A., and Hehre, W. J. (1980), J. Am. Chem. Soc. 102, 939.

[29] Greenwood, H. H. (1972) Computing Methods in Quantum Organic Chemistry, Wiley-Interscience, London, pp.152-155 and pp. 186-203.

[30] de Melo, C. P., and Silbey, R. (1988), J. Chem. Phys. 88, 2567.

[31] Soos, Z. G., and Ramasesha, S. (1989), J. Chem. Phys. 90, 1067.

[32] Blanchard-Desce, M., Ledoux, I., Lehn, J. M., Malthete, J., and Zyss, J. (1988), J. Chem. Soc. Chem. Commun., 736.

[33] Dulcic, A., Flytzanis, C., Tang, C. L., Pepin, D., Fetizon, M., and Hoppiliard, Y. (1981), J. Chem. Phys. 74, 1559.

[34] Meyers, F., Zyss, J., and Brédas, J. L., to be published.

SEMICONDUCTOR DEVICE PHYSICS IN CONJUGATED POLYMERS

J H BURROUGHES, C A JONES, R A LAWRENCE and R H FRIEND
Cavendish Laboratory
Madingley Road
Cambridge CB3 OHE   U.K.

ABSTRACT. Conjugated polymers which can be processed to form thin, coherent films can be used as the active layers in semiconductor device structures. We have used the Durham precursor route to polyacetylene to fabricate a range of unipolar devices including Schottky barrier diodes, MIS diodes and MISFET's. Although carrier mobilities are low, limited by thermally-activated transport between chains, these devices work well, and we find that these structures are remarkably free of surface states and bulk defect states with energy levels within the gap. The fundamental excitation of the trans-polyacetylene chain is the self-localised soliton-like kink defect in the bond alternation pattern along the chain; the soliton has associated with it an energy state at mid-gap of non-bonding p_z character. We have been particularly concerned to demonstrate the formation of solitons from charges injected into the polyacetylene layer in these device structures, and have measured the changes in the optical properties that accompany soliton formation. For the MIS structures working in accumulation mode we find the 'mid-gap' transition from soliton level to the band edge at energies ranging between 0.55 eV (characteristic of the bulk) for polyacetylene on polymeric insulator layers to 0.8 eV when formed on silicon dioxide. We discuss this spread in energies in terms of the different surface structures at these different interfaces. We also show data for the characteristic vibrational excitations of the soliton, including both the IR-active translation modes and the Raman active amplitude modes.

1.   Introduction

Conjugated polymers are polymers in which the structure of the chain is defined by σ bonds, but the π-electron system delocalized along the chain can support excited electronic states. It is this combination of the properties of the σ and π electrons which allows these polymers to survive in a wide range of oxidised and reduced states, and gives rise to the wide range of interests in them, ranging from their use as electrochemical insertion electrodes, as high-conductivity/low-density metals, as materials for non-linear optics, and as semiconductors. There are basic issues common to most of these topics, principally the ways in which charge can be stored on the polymer chain, and the ways by which it can be

J. L. Brédas and R. R. Chance (eds.), Conjugated Polymeric Materials:
Opportunities in Electronics, Optoelectronics, and Molecular Electronics, 221–245.
© 1990 *Kluwer Academic Publishers. Printed in the Netherlands.*

moved both along and between polymer chains. Associated with the anisotropic properties of these materials are excitations of the chain which are strongly coupled to the local bond order and which show therefore a strongly non-linear response. Our interests are in the semiconductor properties of these polymers, and in this paper we show how the formation of self-localised excitations of the polymer chain determines the operation of a variety of semiconductor device structures.

1.1   EXCITED STATES IN POLYACETYLENE, SOLITONS

The π-electron system of the trans isomer of polyacetylene can be considered as an example of a one-dimensional, half-filled electron band, and bond dimerisation as the result of a Peierls distortion of the regular chain. Within the Peierls electron-phonon coupling model for the chain dimerisation and $\pi - \pi^*$ band gap, the gap ($2\Delta_0$) is determined by the dimensionless electron phonon coupling constant, λ and a cut-off energy, E_c comparable to the bandwidth, 4t, through the relation [1]

$$\Delta_0 = 2E_c \exp\{-1/2\lambda\} \tag{1}$$

Fixing to experimental values for the gap, of 1.7 eV and the bandwidth, of 12 eV, the value of λ is required to be of order 0.2, rather larger than can be attributed to electron-phonon coupling alone [2], and it is widely held that the enhanced bond dimerisation is due to the effects of the on-site Coulomb repulsion energy, U_0. This is shown for example in the work on Baeriswyl and Maki [3], in which, for U_0 small (<4t), bond dimerisation is enhanced. It is possible to choose realistic values of λ (= 0.1) and $U_0/4t$ (= 0.6-0.7) and to obtain a satisfactory value for the energy gap.

The modelling of polyacetylene in its excited states is usually performed using a Hückel theory, with the various parameters (bandwidth, electron-phonon coupling etc.) chosen empirically. This approach has been remarkably successful, and we shall follow it in the present work for the modelling of the effects of disorder on the chain. The fundamental excitations of the half-filled band Peierls distorted chain are known to be phase kinks, or solitons, in the pattern of the bond alternation. This was shown for polyacetylene [4,5] to take the form of the bond alternation defects shown in figure 1. An important insight into the nature of these excitations from the work of Su et al [4] is that the bond alternation defect is not localised at a single carbon site, as indicated for convenience in the schematic representation in figure 1, but is spread over some 10 to 15 carbon sites. This delocalisation is crucial to the energetics of the stabilisation of the soliton, and is clearly demonstrated experimentally. For this situation, it is possible to use a continuum model for the polyacetylene chain, within which various simple analytic results are found. Thus, the gap parameter, Δ varies through the soliton as

$$\Delta(x) = \Delta_0 \tanh(x/\xi) \tag{2}$$

with the soliton half-width, $\xi = v_F/\Delta_0$ where v_F is the Fermi velocity [1]. The dynamical mass, M_S of the state described by equation 2 is given by

$$M_S = \frac{4\Delta_o^3}{3\pi\lambda v_F^2 \Omega_o^2}$$

(3)

where Ω_o is an averaged bare phonon frequency [6]. Δ_o and hence ξ and M_S depend very sensitively upon the value of λ (through equation 1), and we show later that much of the variation in behaviour due to differing types of disorder in the polyacetylene used in the MIS and MISFET structures can be modelled by allowing variation in the effective electron-phonon coupling constant, λ. The soliton has associated with it an energy level which within one-electron theory lies at mid-gap and is of p_z non-bonding character. This is as shown in figure 1, where for the injection of negative charge, this state is doubly occupied. The corresponding positively charged state is unoccupied.

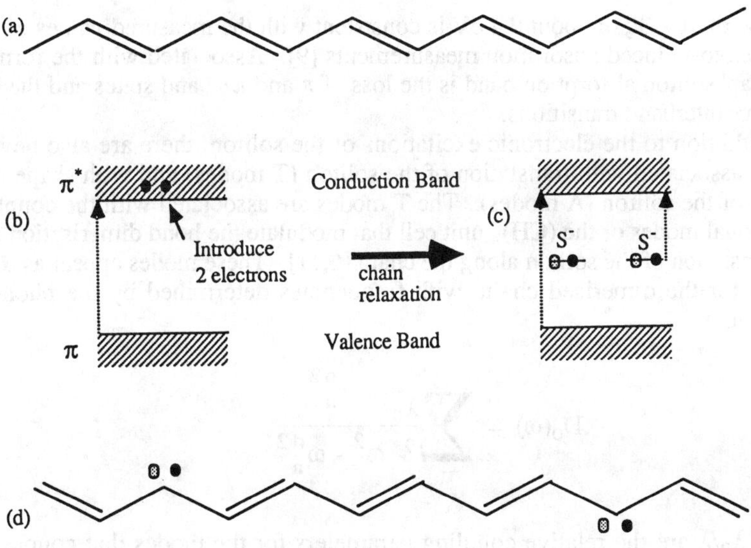

Figure 1.   Schematic representation of the formation of a pair of solitons on a polyacetylene chain. (a) shows the undistorted chain and (b) shows the associated band scheme. The interband π-π^* optical transition is shown as a dotted arrow. If two electrons are introduced onto the chain, initially into the conduction band as shown in (b), the chain relaxes to the form shown in (d), with a reversed sense of bond alternation in the centre of the chain separated by two solitons. Associated with the two solitons are non-bonding π states in the gap, created from one (doubly occupied) valence and one (empty) conduction band state. These two states are doubly occupied, as shown in (c), and each carries a negative charge. New optical transitions from the soliton level to the conduction band are indicated with a dotted arrow; the oscillator strength for the interband transition is weakened through the loss of the band states. A complementary picture holds for positive charge, which is accommodated as unoccupied solitons levels.

The formation of solitons through the addition of charge to the polymer chain modifies the optical properties, with new optical transitions allowed between the mid-gap soliton levels and the band edges, as indicated for the negative soliton to conduction band transition in figure 1. The dipole matrix element for this transition is large, and is enhanced above that for the π-π^* interband transition by a factor of $\pi^2\xi/a$, where a is the average carbon-carbon spacing along the chain [7,8] The effect of Coulomb interactions, characterised through the on-site Coulomb repulsion, U_0 is to modify the energies of the electronic transitions between band states and the localised soliton states [9,10]. Transitions involving charged solitons are reduced in energy, and modelling the effects of U as a perturbation in the one-electron model [9],

$$\Delta E = \Delta_0 - \frac{aU_0}{3\xi} = \Delta_0\left(1 - \frac{aU_0}{3\hbar v_F}\right) \tag{4}$$

A value of $aU_0/3\xi$ of about 0.4 eV is consistent with the measured values of ΔE obtained from photo-induced absorption measurements [9]. Associated with the formation of the 'mid-gap' soliton absorption band is the loss of π and π^* band states and the bleaching of the π-π^* interband transitions.

In addition to the electronic excitations of the soliton, there are also new vibrational modes associated with translation of the soliton (T modes) and with shape or amplitude modes of the soliton (A modes). The T modes are associated with the coupling of those vibrational modes of the $(CH)_x$ unit cell that modulate the bond dimerisation amplitude to the translation of the soliton along the chain [6,11]. These modes appear as Raman-active modes for the dimerised chain, with frequencies determined by the phonon response function,

$$D_0(\omega) = \sum_n \frac{\lambda_n}{\lambda} \frac{\omega_n^{o\,2}}{\omega^2 - \omega_n^{o\,2}} \tag{5}$$

where λ_n/λ are the relative coupling parameters for the modes that couple to the bond dimerisation (n = 1 to 3 for polyacetylene) and ω_n^o are the bare phonon frequencies for these modes. $D_0(\omega)$ determines the frequencies of the Raman modes of the dimerised chain, through $D_0(\omega) = -1/(1-2\lambda)$ where λ is the effective electron-phonon coupling constant. It determines the frequencies of the T modes of the soliton, through $D_0(\omega) = -1/(1-\alpha)$ where α is a parameter that characterises the pinning of the soliton due to disorder etc., and finally, it determines the frequencies of the Raman-active A modes of the soliton, through $D_0(\omega) = -1/(1-2K\lambda)$, where K is calculated to be 0.723 [12-14].

1.2 SEMICONDUCTOR DEVICE PHYSICS

There are various ways of introducing excitations into a semiconductor. Chemical doping and photoexcitation of electron-hole pairs are both well studied routes for the conjugated polymers, but the construction of semiconductor device structures such as the

MIS (Metal Insulator Semiconductor) device, offers the third method, through the injection of charge to form regions of space charge density (as in the depletion layer) or surface charge density (as in accumulation or inversion layers in the MIS device).

Devices such as the Schottky barrier diode and the MIS and MISFET structures depend for their operation on the control of charge density within the semiconductor through the external gate voltage. In the MIS and MISFET devices, the gate is separated from the active semiconductor by a thin insulator layer with a high dielectric strength, and the applied gate voltage is used to create a very high electric field at the interface between the semiconductor and insulator. This can be used to bend energy bands in the semiconductor towards or away from the Fermi energy. We follow the example of the p-type semiconductor applicable to the polyacetylene as used in our work. With a large negative gate voltage, the valence band is pulled upwards, and if it crosses the Fermi energy, holes are created in the surface layer of the semiconductor. This surface charge layer is known as the accumulation layer. As the gate voltage is reduced and the accumulation layer is removed, a depletion layer is formed in which the mobile charges associated with the extrinsic dopants (acceptors) are pushed away from the interface. If the gate voltage is further raised, then it may be possible to lower the conduction band towards the Fermi level, and allow occupation of conduction band states at the interface, to form a charge inversion layer of n-type carriers. The MIS structure is particularly versatile, and for our purposes has the special advantage that in the accumulation and inversion modes, the charge injected is present without the countercharge from a dopant species, and we can hope to make measurements on carriers which are not perturbed by the presence of dopants between the chains.

There are stringent conditions that must be met before a semiconductor can be used in this type of device. Firstly, the semiconductor must be free of defects that have associated energy levels within the semiconductor gap. Charge densities for extrinsic doping, for example, are chosen to be low (in our work in the range 10^{16} to 10^{17} cm^{-3}) in order that widths of depletion regimes are of the correct size, of order 0.1 to 1 μm. Polymers do of course possess many defects from regular crystalline order, and for the polyacetylene that we use, prepared by the Durham precursor route, we have many chain bends or twists that interrupt the straight chain sequences. However, we can expect quite generally that the degree of conjugation at a bend or twist will be lowered, and we can expect that the associated π and π^* states will move deep into the bands.

Secondly, the interface between the insulator and the semiconductor must be free of surface states, caused for example by dangling bonds at the semiconductor surface. The maximum surface charge density that can be achieved in the accumulation or inversion layer in the MIS device is fixed by the dielectric strength, E_b of the insulator, as $n_{max} = \varepsilon_0 \varepsilon_r E_b / q$ where ε_r is the relative dielectric constant. With silicon dioxide as the insulator, this gives of order 2×10^{13} electrons/cm^2; this is a low value in comparison with the 10^{15} or so carbon atoms/cm^2 at the polyacetylene surface. Polymers do not, of course, have unsatisfied bonds at the surface, since bonds are only formed along the polymer chain, and we can expect that if the structure can be fabricated without inclusion of contaminants in the interfacial region, the device should be free of surface states. A likely source of contamination is from included oxygen, and it is probably always necessary to process polymeric semiconductor devices in the absence of oxygen.

226

Thirdly, fabrication of semiconductor devices requires control of the concentration of the extrinsic charge carriers, typically in the concentration range 10^{15} to 10^{18} charges/cm^2, with the further requirement that the dopant species be immobile at the operating temperature of the device. This poses a major challenge for the use of conjugated polymers in these applications because the chemical and electrochemical methods commonly used to 'dope' conjugated polymers, in which doping is achieved by diffusion of dopants through the solid polymer at room temperature, are quite inappropriate here. Apart from the problem of dopant mobility, it is very difficult to control doping levels at the very low dopant concentrations required (10^{-6} molar %), and at such low levels it is very unlikely that the doping would be homogeneous. In the case of the Durham route polyacetylene that we use in this work, the polymer as made is extrinsically doped with p-type carriers in a convenient concentration we are able to use it without further treatment. However, for other polymers it will be necessary to develop new methods for doping in the fashion required here.

The picture described above for the polymer, with the formation of self-localised states with energy levels at mid-gap to accommodate the injected charge calls for a very different description of the device operation to that used for inorganic semiconductors. Specifically, we expect that charge storage in the polymer will be in soliton states created in sufficient number to store all the injected charge. We expect to see the Fermi energy pinned close to the centre of the gap, and to see the density of soliton states vary with external field. Our principal motivation in carrying out this study of the device physics of polyacetylene was to find out how the relaxation of injected charges to form charged solitons controls the device operation and distinguishes this class of polymeric semiconductor devices from traditional structures. Besides the range of electrical measurements that characterise the device performance, we have carried out an extensive range of optical measurements, since the clearest evidence for the formation of solitons in polyacetylene is through the appearance of additional optical absorption below the band gap, associated with the vibrational and electronic excitations of the solitons. We expect, therefore, to see changes in the optical properties of the polyacetylene in these devices; depletion should reduce the 'mid-gap' optical absorption whereas accumulation or inversion layer formation should introduce 'mid-gap' states onto previously undistorted chains, and increase the mid-gap optical absorption.

2. Polymer Processing and Device Fabrication

Despite the considerable progress in the past decade that has been made in the understanding of the electronic properties of conjugated polymers, there has been relatively little work reported on their use as the active component in semiconductor device structures. There are several reasons for this; the most important of which is that most conjugated polymers cannot be conveniently processed to the forms required in these devices. Most conjugated polymers are not readily soluble in easily-handled solvents, and are infusible. This has severely limited the scope for construction of devices [15,16]. Improvements in the processing of conjugated polymers, through electrochemical deposition during polymerisation, and the use of solution-processible poly(3-alkyl thiophenes) has allowed better device fabrication [17,18], and the use of model oligomers which can be deposited by vacuum sublimation has recently been reported [19]. There are also reports of field-induced conductivity measured in MISFET (Metal-Insulator-Semiconductor Field Effect Transistor) structures [20-23].

2.1   DURHAM POLYACETYLENE

A major advance in the control of the polymer processing is in the use of a solution-processible precursor polymer which can be converted to the conjugated polymer after processing. This was first demonstrated by Edwards and Feast [24-26] for the preparation of polyacetylene (the Durham route) and it is polyacetylene produced by this route that we have used in the present studies. We have made thin-film devices by spin-coating the precursor polymer, poly((5,6-bis(trifluoro-methyl)-bicyclo[2,2,2]octa-5,7-diene-2,3-diyl)-1,2-ethenediyl), in solution onto the required substrate, followed by heat treatment to convert to the polyacetylene, by elimination of hexafluoroorthoxylene [27-35]. The Durham precursor route is shown in figure 2; the precursor polymer is soluble in common solvents such as acetone and 2-butanone, and the heat treatment to achieve the thermal conversion from precursor to polyacetylene is achieved at temperatures of up to 100°C. The precursor polymer has very good film-forming properties, and it is straightforward to control the thickness of the resultant polyacetylene film over the range 100 Å to several μm.

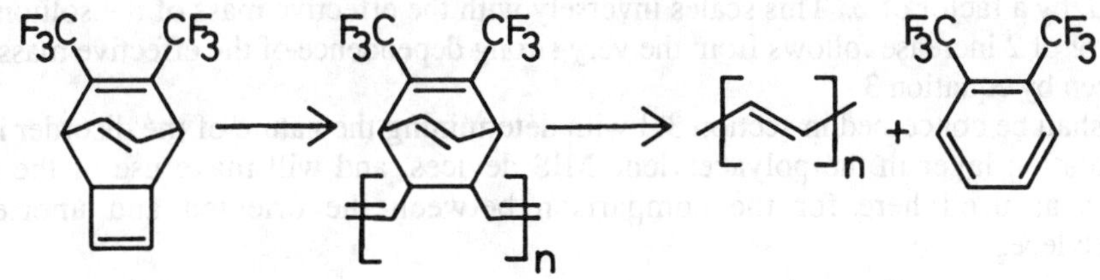

Figure 2.   Durham precursor route to polyacetylene

Polyacetylene produced by the Durham route can be obtained in a variety of morphologies. If prepared in thin-film form by spin-coating the precursor onto a substrate and thermally converting to polyacetylene in situ the polymer retains much of the disorder that was transferred from the precursor polymer in solution. Although polyacetylene, as a rigid-rod polymer, shows a strong tendency to crystallise, these films show coherence lengths as determined from X-ray measurements of typically of no more than 40 Å [36]. In contrast, if a free-standing film of the precursor polymer is stretched during the thermal conversion it is possible to obtain a very high degree of orientation and an increase in the crystallinity [37,38].

The electronic structure of polyacetylene is very sensitive to the presence of conformational defects on the chains, and there are significant differences between the properties of polyacetylene prepared as unoriented and as stretch-oriented films. We find that the band-gap in the unoriented material is raised, with the peak in the interband absorption lifted from 1.9 to 2.3 eV [28], and we find concomitant increases in the Raman-active vibrational modes. In spite of the short lengths of straight chain, however, we find that the electronic excitations of the chain still have the character of the solitons discussed in section 1, though the increased π-π^* gap affects these considerably [30-32].

We can model the effect of the conformational chain defects in the unoriented samples in a consistent way by using a value of the effective electron-phonon coupling constant, λ, that is some 8% higher than for the oriented samples. The use λ of as the principal parameter affected by disorder was shown successfully to account for the dispersion of the resonant Raman scattering in Shirakawa polyacetylene [11]; as applied here to unoriented Durham polyacetylene, we require that the chain defects do not impose a preferred sense of bond alternation on the chains, so that it is possible to propagate a soliton through the chain defect.

The 8% increase in the value of λ is measured directly from Raman scattering, using the product rule due to Horovitz [11]. This increase in λ accounts for the increase in gap, $2\Delta_0$ from 1.9 to 2.3 eV through equation 1. The energy of the transition between soliton and the band edge, as seen in measurements of photoinduced absorption for charged solitons, is raised from 0.45 to 0.55 eV. The scaling of this energy with Δ_0 is shown in equation 3. The IR-active T modes of the soliton are found at 520, 1287 and 1365 cm^{-1} in the oriented samples, and at 625, 1287 and 1373 cm^{-1} in the oriented samples. The lowest of these modes is the Goldstone mode that should appear at zero frequency in the absence of pinning. In addition to the increase in frequency, which indicates that the pinning parameter, α is larger in the unoriented samples, the intensity of the T modes (which can be conveniently ratioed to that of the electronic soliton to band transition) is sharply reduced, by a factor of 2. This scales inversely with the effective mass of the soliton, and the factor of 2 increase follows from the very strong dependence of the effective mass with gap given by equation 3.

We shall be concerned in section 3.3 with determining the nature of the disorder in the accumulation layer in the polyacetylene MIS devices, and will make use of the same analysis as used here for the comparison between the oriented and unoriented polyacetylene.

2.2   DEVICE FABRICATION

Device fabrication requires the successive deposition of a series of layers of metal, polymer and insulators, with appropriate patterning in the case of the MISFET for the formation of the source and drain areas. We use simple metal evaporation, of gold as an ohmic contact and, typically, aluminium as a blocking contact. Thus, to fabricate a Schottky barrier diode, we deposit a bottom layer of gold onto the substrate, coat this with the polyacetylene layer, and put the low work function metal such as aluminium on top on this.

For the MIS structures where we require an insulator layer, we have made extensive use of silicon processing techniques to provide both gate, as a heavily n-doped layer, and insulator, as silicon dioxide grown on the silicon substrate. This structure provides an MIS device when coated with polyacetylene and capped with an evaporated thin gold top contact. For the MISFET structure it is convenient to put the source and drain contacts onto the insulator layer, and put the polyacetylene layer down on top as the last stage in the construction. In the example of the MISFET structure is shown in figure 3, the source and drain contacts are of poly n-silicon; we have also made structures with gold.

MIS structures have also been fabricated with a variety of organic insulator layers. As an example, we have made devices with 200 Å of gold on a silica substrate, a

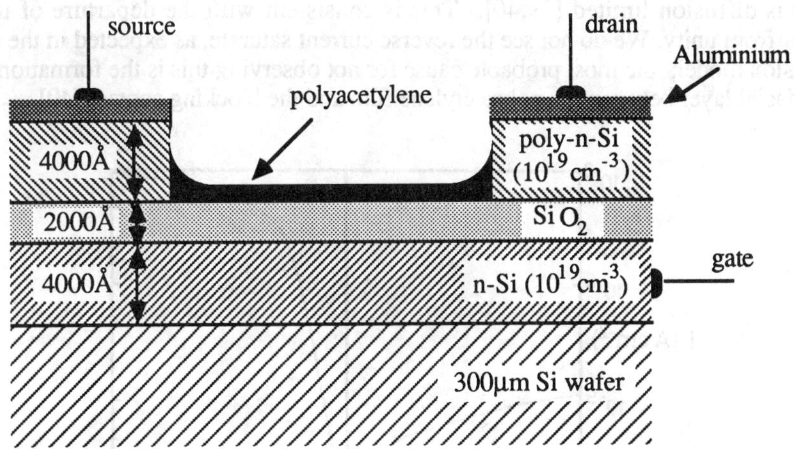

Figure 3.   Schematic diagram for a polyacetylene MISFET structure. Dimensions shown are to scale, except the channel width (20 μm) and length (1.5 m)

polyacetylene layer about 200 Å in thickness, followed with a spin-coated layer of PMMA (poly methyl methacrylate) 1500 Å in thickness, and capped with a thin (200 Å) gold top contact.

We perform all fabrication steps involving the polyacetylene inside a glove-box which is maintained with oxygen and water levels of less than 5 ppm. The glove box contains, therefore, the spin-coater, temperature-controlled ovens, and the metals evaporator.

3.   Characterisation of Devices

3.1   SCHOTTKY-BARRIER DIODES

3.1.1 *Electrical characterisation*   The characteristics of the Schottky diodes constructed as above provide important information about the polyacetylene. The current density versus voltage characteristics for Schottky diodes constructed as above are shown in figure 4, in which |log(I)| is plotted versus bias voltage for both forward and reverse biases. The ratio of forward to reverse current reaches a value of typically 5×10^5 at bias voltages of around 1.5 V. The usual parameterisation for the variation of current density with bias voltage is $J = J_{Sat.}\exp\{qV/nkT\}$ for $V>3kT/q$, and where n is the ideality factor, which for a 'perfect' diode is equal to 1. As seen in figure 4, we find a value for n of about 1.3 at low forward biases. At higher forward biases the current is limited by the bulk resistance. In the thermionic emission model for conduction across the junction, the saturation current density, $J_{Sat.}$ is given by $J_{Sat} = A^*T^2\exp\{q\Psi/\kappa T\}$, with A^*, the effective Richardson constant, which for an effective electron mass of one equals 120 $AK^{-2}cm^{-2}$, and Ψ is the barrier height [39]. By extrapolating from the linear region to the log(J) axis in figure 4, we estimate that $J_{Sat}= 4.4 \times 10^{-13}$ A cm^{-1}, and hence $\Psi=1.28$ eV. The applicability of the thermionic emission model here is questionable; it is more likely that the forward current flow is diffusion limited [39,40]. This is consistent with the departure of the ideality factor from unity. We do not see the reverse current saturate, as expected in the thermionic emission model; the most probable cause for not observing this is the formation of a small interfacial layer between the polyacetylene film and the blocking contact [40].

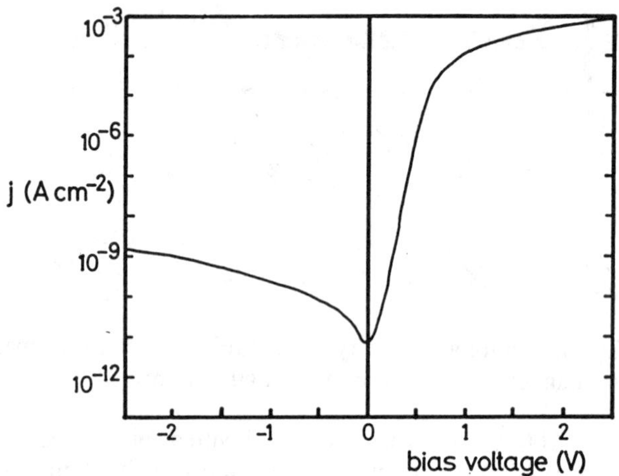

Figure 4.   Modulus of current versus voltage for a Schottky diode.   Sample area is 0.636cm^2.

The variation of the Schottky diode capacitance with reverse bias voltage provides information about the density of charged acceptors in the depletion layer. In the simple case of the abrupt junction (which should be appropriate here), the measured differential capacitance, C, is expressed as

$$\frac{1}{C^2} = \frac{1}{A^2}\left(\frac{2(\phi+V)}{\varepsilon_r\varepsilon_0 q N_a}\right) \tag{6}$$

where N_A is the acceptor concentration and ϕ is the built-in potential [39]. Figure 5 shows the C(V) characteristics for a polyacetylene-aluminium device. It can be seen that the relation in equation 6 is obeyed well for reverse bias, and we obtain values of $N_A = 9 \times 10^{15}$ cm^{-3} and $\phi = 0.94$ V. We thus find that the electrical characteristics of the Schottky barrier between aluminium and polyacetylene indicate that the device is operating in the conventional way, with movement of the depletion layer boundary with applied bias voltage. This is an important result, indicating clearly that the integrated density of states in the vicinity of the Fermi energy is finite. A very different picture, for example, holds for the case of Schottky diodes formed with amorphous silicon-hydrogen alloys, in which the Fermi energy lies in a region of band-tailing into the gap, so that the space charge density in the depletion region varies strongly with distance from the metal junction [41].

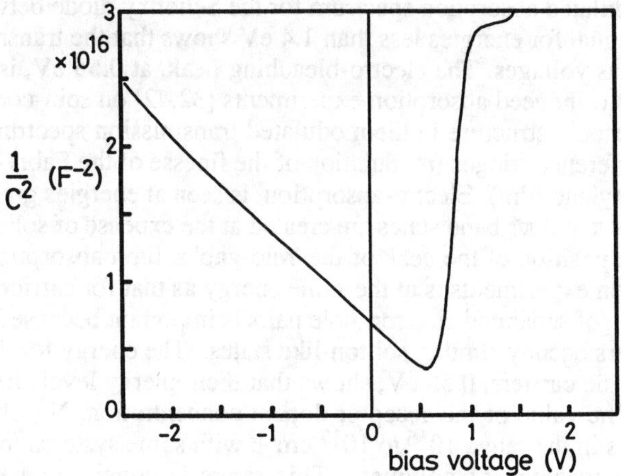

Figure 5. $1/C^2$ versus bias voltage for a Schottky diode.   Sample area = 0.636 cm^2, $N_A = 9 \times 10^{15}$ cm^{-3}, $\phi = 0.94$ V.

3.1.2 *Electro-optic Properties*   The soliton physics of polyacetylene must of course give a very different description for the device operation from that for conventional semiconductors. We can show that the mobile, electronic charges that move away from the junction to form the depletion layer do have the characteristics of charged solitons by looking for changes in the optical absorption, particularly below the band edge. Figure 6

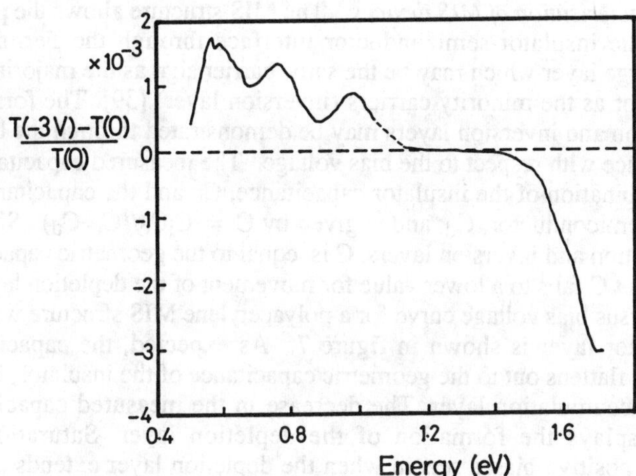

Figure 6.   Voltage-modulated optical transmission for a Schottky diode, [T(V) - T(0)] / T(0) versus photon energy (E$_p$), with V= -2V with respect to the blocking contact.

232

shows the voltage-modulated absorption spectrum for the Schottky diode between 0.4 and 1.8 eV. The positive signal for energies less than 1.4 eV shows that the transmission does increase for reverse bias voltages. The electro-bleaching peak, at 0.55 eV, is at the same energy as found in photo-induced absorption experiments [32,42] on spin-coated Durham polyacetylene. The periodic structure in the modulated transmission spectrum below the bandgap is due to interference fringes (modulation of the finesse of the Fabry-Perot etalon formed by the polyacetylene film). Electro-absorption is seen at energies greater than 1.4 eV, as expected if extra π and π^* band states are created at the expense of soliton states.

The finding that the position of the peak of the 'mid-gap' soliton absorption band seen in the electromodulation experiments is at the same energy as that for carriers introduced through photoexcitation of separated electron-hole pairs is important because it establishes that the extrinsic carriers occupy similar, soliton-like states. The energy for the electronic excitation of the extrinsic carriers, 0.55 eV, shows that their energy levels lie deep in the semiconductor gap. The value of the acceptor dopant concentration, N_A obtained from $1/C^2$ versus V plots lies in the range 10^{16} to 10^{17} cm^{-3}, with some systematic variation of this value from batch to batch of polymer. This range is consistent with the sign, magnitude and temperature variation of the thermopower [42]. We associate the value of N_A with the total number of soliton states present in the band of 'mid-gap' soliton states. The constant value for N_A obtained over a wide range of reverse bias voltages indicates that very few other states in the gap lie close to the soliton band. Though the Fermi energy must lie close to the soliton band, in other respects the operation of the polyacetylene-aluminium Schottky diode is similar to that for conventional devices.

3.2 MIS AND MISFET STRUCTURES

3.2.1 *Electrical characterisation of MIS devices*   The MIS structure allows the possibility of band bending at the insulator-semiconductor interface through the Fermi level to produce a surface charge layer which may be the same carrier sign as the majority carriers (accumulation layer) or as the minority carriers (inversion layer) [39]. The formation of accumulation, depletion and inversion layers may be demonstrated through the behaviour of the device capacitance with respect to the bias voltage. The measured capacitance, C, is that of the series combination of the insulator capacitance, C_i, and the capacitance of the active region of the semiconductor, C_d, and is given by $C = C_i C_d/(C_i+C_d)$. Since C_d is large for the accumulation and inversion layers, C is equal to the geometric capacitance of the insulating layer, but C falls to a lower value for movement of the depletion layer.

The capacitance versus bias voltage curve for a polyacetylene MIS structure with silicon dioxide as the insulator layer is shown in figure 7. As expected, the capacitance for negative gate voltages flattens out to the geometric capacitance of the insulator, indicating the formation of an accumulation layer. The decrease in the measured capacitance for positive voltages displays the formation of the depletion layer. Saturation of the capacitance for large positive biases sets in when the depletion layer extends across the polyacetylene film (estimated to be about 1200Å for this device). [We have not seen convincing evidence for inversion layer formation for the MIS structure at high positive biases; the usual problem for the 2-terminal device is that the mechanism for the generation of minority carriers is through the thermal generation of e-h pairs at the interface region, and transport of the majority carriers through the bulk of the semiconductor to the

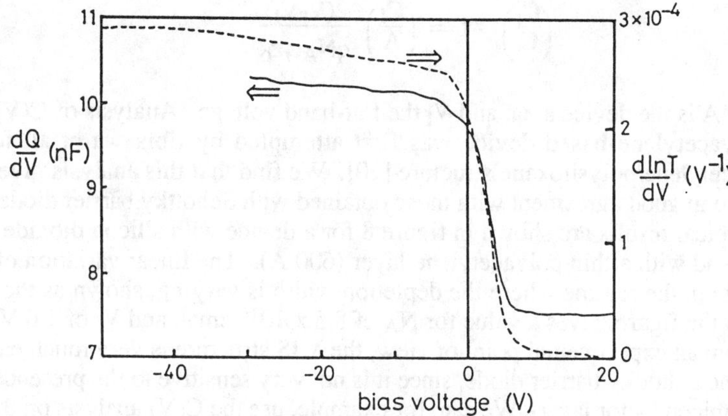

Figure 7. The differential optical transmission, $\partial \ln T/\partial V$ at 0.8 eV, and differential capacitance, $\partial Q/\partial V$ versus bias voltage for the MIS structure. Measurements were made at 500 Hz and with an a.c. modulation of 0.25 eV.

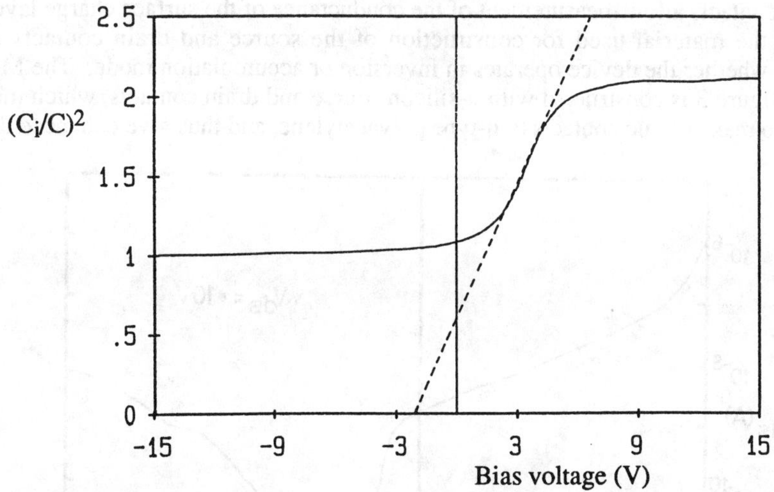

Figure 8. $(C_i/C)^2$ versus bias voltage for a thin polyacetylene film MIS structure (600 Å). $N_A = 8.5 \times 10^{16}$ cm^{-3}, $V_f = 1.6$ V.

back contact. This is usually limited by the thermal generation rate, which in the case of an anisotropic semiconductor such as polyacetylene is further impeded [31] by the difficulty of achieving charge separation for a system showing predominantly one-dimensional diffusion.]

The variation of C with bias voltage in the depletion regime has been modelled by Goetzberger and Nicollian [43] as

$$\left(\frac{C_i}{C}\right)^2 - 1 = \left(\frac{C_i}{A}\right)^2 \frac{2(V-V_f)}{qN_A\varepsilon_r\varepsilon_o} \qquad (7)$$

where A is the device area, and V_f the flat-band voltage. Analysis of C(V) in this way for a polyacetylene-based device was first attempted by Ebisawa et al, for a Shirakawa polyacetylene/polysiloxane structure [20]. We find that this analysis gives values for N_A that are in good agreement with those obtained with Schottky barrier diodes from equation 6. Typical results are shown in figure 8 for a device with silicon dioxide as the insulator layer and with a thin polyacetylene layer (600 Å). The linear variation of $1/C^2$ with bias voltage in the regime where the depletion width is varying, shown as the broken straight line in the figure, gives a value for N_A of 8.5 x 10^{16} cm^{-3}, and V_f of 1.6 V.

From an experimental point of view, the MIS structure is very much easier to fabricate than the Schottky barrier diode, since it is not very sensitive to the presence of pin-holes in the semiconductor layer. We can, for example, use the C(V) analysis on the MIS structure to obtain values of N_A for polymers which have shorter chain lengths than the Durham polyacetylene used here and which show poorer film-forming properties.

3.2.2 *MISFET structures* Demonstration that the charges at the polyacetylene/insulator interface in the MIS structure are mobile is made in a MISFET structure, in which source and drain contacts allow measurement of the conductance of the surface charge layer. The choice of the material used for construction of the source and drain contacts should determine whether the device operates in inversion or accumulation mode. The MISFET shown in figure 3 is constructed with n-silicon source and drain contacts, which might be expected to make ohmic contacts to n-type polyacetylene, and thus give enhanced channel

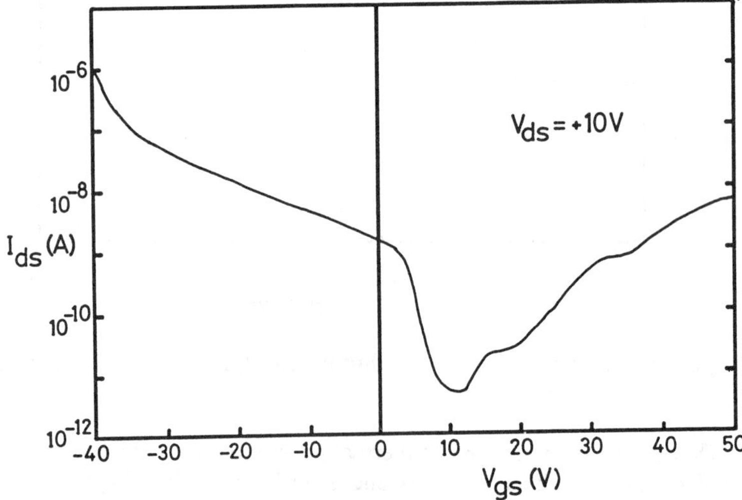

Figure 9. I_{DS} versus V_{GS} at constant V_{DS} (+10V) for a MISFET with poly n-silicon source and drain contacts, as shown in figure 3.

conductance from formation of an n-type inversion layer for positive gate voltages. Figure 9 shows the variation of I_{DS} with V_{GS} at constant V_{DS}. The channel conductance has a minimum at $V_{GS} = +10V$ (full depletion) and rises for gate voltages both more positive (inversion layer) and more negative (accumulation layer), with a maximum on/off ratio for this structure of 100,000 ($V_{GS} = -40V$ to $V_{GS} = +10V$). We see, therefore, the behaviour expected for positive bias, though the strong enhancement of the conductivity for negative bias indicates that the accumulation layer is still able to make good contact to the source and drain electrodes. It is unusual to find a device able to operate in both modes, and we consider that the n-silicon work function matches that of the mid-gap 'soliton' band in polyacetylene, and that in contrast to traditional semiconductors, the Fermi level does not move far from the mid-gap band with addition of either p or n carriers.

The MISFET structure provides a very convenient means of controlling the charge carrier concentration in the surface layer of the semiconductor. We have determined the carrier mobility from conductance measurements as a function of gate voltage, and find low values, typically 10^{-4} $cm^2V^{-1}s^{-1}$. Similar values for the mobility are estimated for photogenerated carriers, and the extrinsic carriers in the as-made polyacetylene [42]. We discuss mechanisms for charge transport further in 3.4.

3.2.3 *Electronic excitations of solitons* As discussed in section 1, charge is stored in polyacetylene in 'soliton' localised states, which have non-bonding p_z 'mid-gap' states associated with them. We expect to see, therefore, a modulation in the optical properties of the active semiconductor region of these devices with applied bias voltage. The devices described above have all been constructed so that it is possible to pass sub band-gap light through the structure and thus to observe the changes in optical properties as the gate voltage is varied. For the MIS and MISFET structures we expect a decrease in the device transmission below the band-edge as the device is driven towards accumulation, and new charged solitons are introduced onto the polyacetylene chains at the interface with the insulator. As with the experiments on the Schottky diodes, discussed above, the experiment is easily performed by modulating the gate voltage at a convenient frequency and monitoring with a lock-in amplifier the optical transmission signal that is modulated by the gate voltage.

The electromodulation spectrum for an MIS device with silicon dioxide as insulator is shown in figure 10, spectrally resolved between 0.4 and 1.1 eV, with the upper limit set by the onset of interband absorption of the silicon substrate. A decrease in the transmission through the device is observed with negative bias, with the formation of an absorption band, peaking at 0.8 eV. This energy is higher than the 0.55 eV found for photo-induced or extrinsic charge carriers, as discussed in section 3.1. We attribute the larger value here to increased disorder in the polymer chains at the interface with the insulator, as discussed in section 3.3. The electromodulation signal provides a very powerful means of characterising the device operation. For example, there is always a question about the participation of surface states at the semiconductor-insulator interface in charge storage. Here, however, we can compare directly the differential capacitance, $C = \partial Q/\partial V$ with the fractional change in optical transmission in the soliton absorption band, $\partial \ln T/\partial V$ (measured here at the peak at 0.8 eV). If all the charge at the polymer-insulator interface is stored in soliton-like states then $\partial \ln T/\partial V$ should scale with $\partial Q/\partial V$ as the bias voltage is varied. This is shown to be the case in figure 7, in which $\partial \ln T/\partial V$ and the

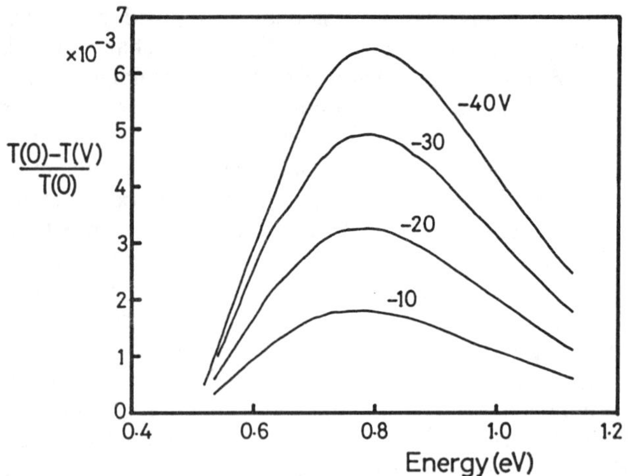

Figure 10. Voltage-modulated transmission for a MIS diode, with silicon dioxide as insulator, $[T(0) - T(V)] / T(0)$ versus the photon energy, for various values of V (all negative with respect to the gate)

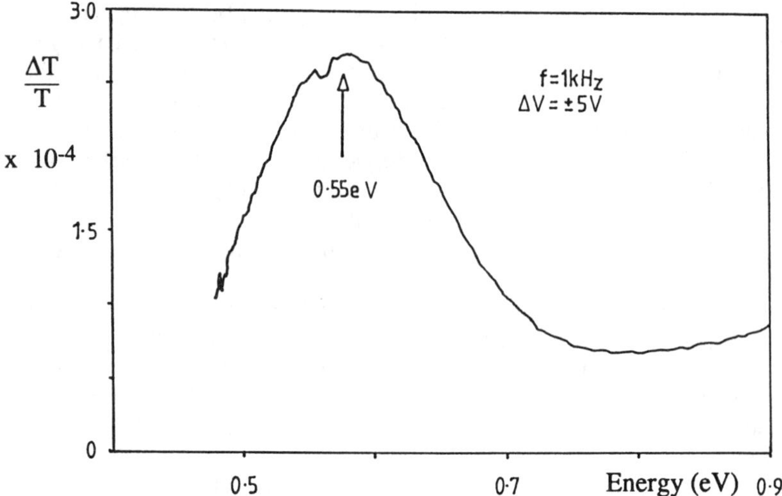

Figure 11. Voltage-modulated transmission for a MIS diode, with PMMA as the insulator. The peak in absorption at 0.55 eV is indicated.

differential capacitance are both plotted against bias voltage. Note that both switch from depletion to accumulation at the same voltage, and that the ratio of $\partial \ln T/\partial V$ to $\partial Q/\partial V$ remains constant in the accumulation regime. This ratio gives the optical cross-section per charge injected; we find the value at the peak in the 'mid-gap' absorption (0.8 eV) to be 1.2×10^{-15} cm^2. This value is a little larger than the value of 8×10^{-16} cm^2 determined from dopant-induced absorption [44]. We consider that the value of the optical cross-section does vary with the position of the 'mid-gap' absorption feature, as we show for the MIS device below fabricated with a polymer insulator layer, and we consider that this variation is significant.

The majority of our work on the MIS devices has been with devices based on silicon/silicon dioxide; these structures are easy to handle, and there is considerable advantage in using an insulator with a high dielectric strength since this controls the maximum charge density that can be obtained in the accumulation layer. However, we have also fabricated devices with a wide variety of organic insulator layers. We show in figure 11 the electromodulated optical transmission for the device fabricated with PMMA described in section 2.2 spectrally resolved between 0.4 and 0.9 eV. We find for this device that the peak in the soliton to band-edge transition is at 0.55 eV, and that the optical cross-section at the peak is 1.6×10^{-15} cm^2. As we discuss in section 3.3 we consider that these values are representative of carriers located on chains in the bulk of the material, and we attribute the differences in these quantities for the silicon dioxide insulator layer devices with extra disorder in the polyacetylene surface layer.

3.2.4 *Vibrational excitations of solitons* Besides the electronic transitions associated with the soliton state, there is extra IR activity due to new vibrational modes which couple to motion of the charged soliton along the polymer chain. These are seen in both doping [30] and photoexcitation experiments [32] in unoriented Durham polyacetylene. Figure 12 shows the results from an FTIR experiment on an MIS structure, showing the difference in transmission for the MIS device biased between 0 and -50V with respect to the gate [33,34]. Three peaks observable; an absorption above 4000 cm^{-1} (this is the low energy side of the electronic absorption, as in figure 10), and two sharp vibrational features at 1379 and 1281 cm^{-1}. We identify these as the two higher-frequency translation modes of the soliton. The value of 1379 cm^{-1} lies between that measured for photo-induced charges, 1373 cm^{-1} [32] and for dopant-induced charges, 1410 cm^{-1} [30], and provides information about the chain conformation and degree of conjugation. We note in particular that the intensity of this mode in relation to that of the electronic absorption band is much lower than found in photoinduced absorption measurements [32]. The oscillator strength scales inversely with effective mass of the charge carriers, and we estimate that the mass for the charges present in the accumulation layer is a factor of 3.6 higher than that of the photoexcited charge carriers in oriented Durham polyacetylene. We were not able to measure the electromodulation spectrum below 1000 cm^{-1} due to a very large absorption peak in the SiO$_2$, and we were thus unable to observe the lowest frequency, 'pinned' translational mode.

New Raman-active modes associated with the width modulation of the soliton are also expected. These have not been seen in doping or photoexcitation experiments because of the inherent experimental difficulties, but the modulation experiment that we can perform with the MIS structure is well suited for this measurement. We have carried out experiments using the MISFET structure of the type shown in figure 3, but with a very thin polyacetylene layer (100 Å) to ensure that the whole depth of the polymer film was accessible for Raman scattering [35]. Raman scattering off the top surface of the structure was measured with a double spectrometer, with a multi-diode array detector. Spectra were obtained from subtraction of a "background" scan from a "signal" scan. Here the "signal"

238

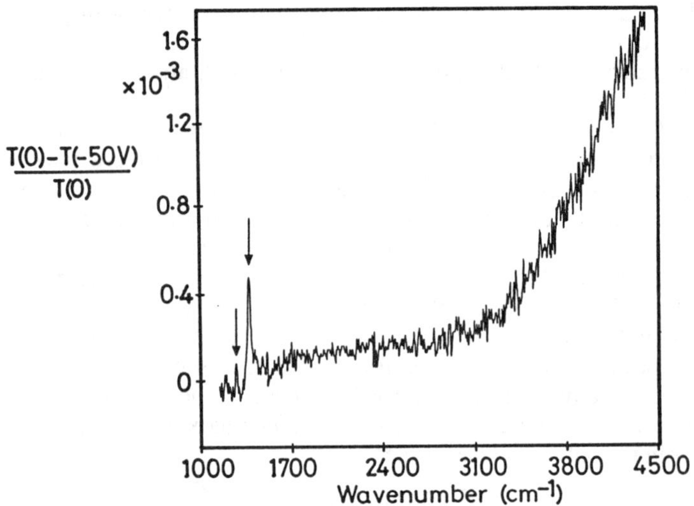

Figure 12. Voltage-modulated transmission for a MIS diode, [T(0) - T(-50V)] / T(0) in the spectral range 1000 and 5000 cm⁻¹.

was the measured scattering from a device with the gate at -30 V with respect to the source and drain, and the "background" was the same scattering signal with the gate set to 0 V. Typical measurement times were in the region of 16 hours. Under these conditions the charge density in the accumulation layer is equivalent to 2.21×10^{-4} charges/carbon atom for a 100 Å film. The spectral region from 900 cm⁻¹ to 1700 cm⁻¹ was studied at a range of laser wavelengths from 476.5 nm to 584.0 nm.

Two different classes of feature are observed in the difference spectrum of accumulation layers in Durham polyacetylene shown in figure 13. The main class of feature is negative and arises from the bleaching of the resonance-enhanced A_g modes that dominate the normal Raman spectrum of *trans*- polyacetylene. The two main modes are near 1100 cm⁻¹ and 1500 cm⁻¹ with the exact frequency a function of the excitation energy. The bleaching of these two main modes dominates the difference spectrum. The dispersion of the bleaching is the same as that of the normal spectrum, but the bleaching fraction (the ratio of the integrated bleaching signal to the normal signal) changes with excitation energy, being 7×10^{-3} at 476 nm, 6.9×10^{-3} at 488 nm, 7.7×10^{-3} at 514 nm and 1.6×10^{-2} at 584 nm. There are also small positive features to the low energy side of each of the bleaching features. These are much less intense, do not disperse with excitation energy, and occur at 1065 cm⁻¹ and 1455 cm⁻¹.

We attribute the positive features in the difference spectrum to the A modes of the soliton. The frequencies of these modes can be calculated using the methods discussed in section 1.1 [12-14]. We find that the modes, at 1065 cm⁻¹ and 1455 cm⁻¹, are at rather higher frequencies than the values of equal to 973 cm⁻¹ and 1414 cm⁻¹ calculated with a value of electron-phonon coupling of 0.200 appropriate to unstretched Durham polyacetylene [30,32]. There is clear evidence from the non-zero frequency of the Goldstone mode for the strong effect of pinning in this system - this could readily account for the increased frequency observed for the A mode. The presence of a defect on the

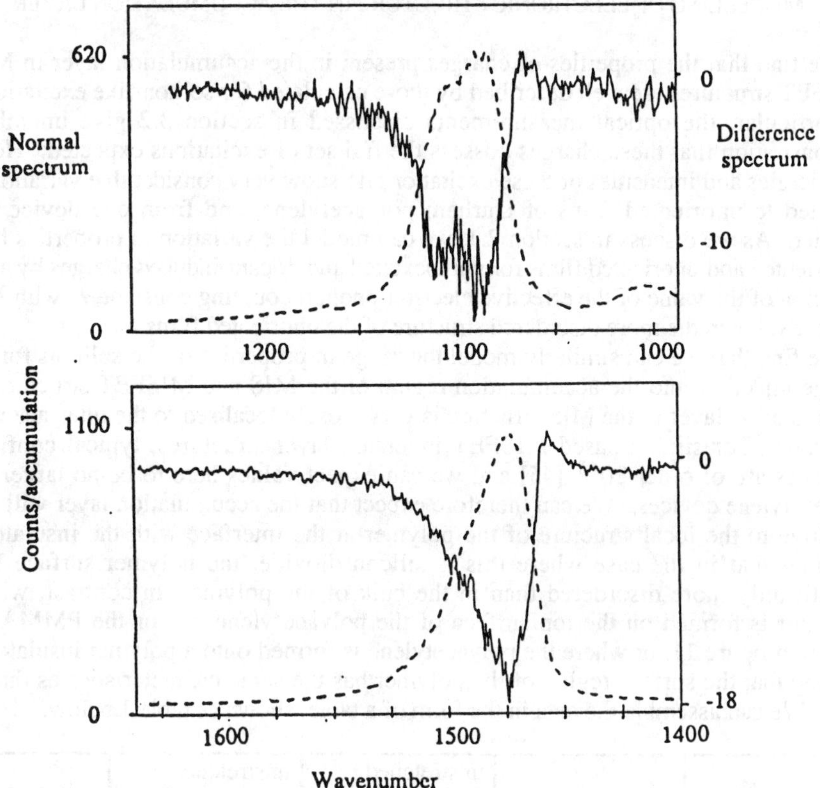

Figure 13. Difference spectra due to charge accumulation layer in a polyacetylene MISFET structure. The solid line is the difference signal; the broken line is the normal Raman spectrum. The excitation wavelength is 584 nm.

polyacetylene chain is also expected to bleach the normal Raman spectrum. Terai et al calculate [14] that this bleaching will extend over a range of 30 carbon atoms - that is, a 30% bleaching for a 1% soliton density. This large bleaching effect corresponds to the enhancement in the band to band optical absorption coefficient as calculated by Kivelson et al [8]. We find that excitation in the red gives a larger bleaching fraction than blue excitation. This suggests that the soliton defects prefer to reside on polyacetylene with a smaller band gap - the less disordered chains. Quantitatively, the areal soliton density was calculated simply from the field across the insulating layer to be 2.21×10^{-4} charges/carbon atom. Using the calculated enhancement factor of 30, this should give us a bleaching of 0.007 - which is as we observe in the blue, though below that observed in the red.

3.3   MODELLING OF ELECTRONIC STRUCTURE IN THE ACCUMULATION LAYER

We find that the properties of charges present in the accumulation layer in MIS and MISFET structures are well described by those calculated for soliton-like excitations, and in particular, the optical measurements discussed in section 3.2 give unambiguous demonstration that these charges possess the full set of excitations expected. However, the energies and intensities of these excitations do show very considerable variations from oriented to unoriented films of Durham polyacetylene, and from one device type to another. As we discuss in section 2.1, we can model the variation in properties between the oriented and unoriented films for photoexcited and dopant-induced charges by allowing variation of the value of the effective electron-phonon coupling constant, λ, with λ taking a larger value in the more disordered structure of the unoriented films.

We find that we can similarly model the range in properties of the solitons formed by charge injection into the accumulation region of the MIS and MISFET structures. The accumulation layer in the MIS structure is very closely localised to the interface with the insulator. For silicon-based MISFET inversion layer structures, typical confinement distances are of order 20 Å [45] and we can expect values here to be no larger for the polyacetylene devices. We can therefore expect that the accumulation layer will be very sensitive to the local structure of the polymer at the interface with the insulator. We consider that in the case where this is silicon dioxide, the polymer surface layer is significantly more disordered than in the bulk of the polymer. In contrast, where the insulator is formed on the top surface of the polyacetylene, as for the PMMA device shown in figure 11, or where the polyacetylene is formed onto a polymer insulator layer, we find that the surface region of the polymer has the same characteristics as that of the bulk. We can assemble the data in the form of a table, shown as table 1 below.

	stretched Durham	unstretched Durham, in bulk or at polymer interface	unstretched Durham at SiO_2 interface	scales as:
$\pi \rightarrow \pi^*$ peak	1.9 eV	2.3 eV	-	Δ
$\pi \rightarrow S^+$, $S^- \rightarrow \pi^*$	0.45 eV	0.55 eV	0.8 eV	Δ
$\dfrac{m^*}{m^*_{stretched}}$	1	2	3.6	Δ^3
σ at peak in $\pi \rightarrow S^+$, $S^- \rightarrow \pi^*$	-	$1.6 \times 10^{-15}\ cm^2$	$1.2 \times 10^{-15}\ cm^2$	$\dfrac{1}{\Delta}$

TABLE 1   Parameters obtained from absorption and photoinduced absorption measurements [30,32] and from electroabsorption measurements on MIS and MISFET structures.

The variation with gap parameter, Δ, as given by the formalism presented in section 1.1, is shown in the right-hand column. The peak in the interband absorption may be taken as 2Δ; from equation 4 we expect the energy of the 'mid-gap' soliton to band edge transition to scale as Δ, from equation 3 we expect the effective mass to scale as Δ^3, and the value of the optical cross-section for the mid-gap' soliton to band edge transition scales as $1/\Delta$ [7,8]. It may be seen that the data shown in the table are consistently modelled.

3.4 CHARGE TRANSPORT IN POLYACETYLENE

Charge transport in doped conjugated polymers has been extensively investigated but remains poorly understood. There was considerable early interest in the nature of the 'insulator-metal' transition achieved by chemical doping. It was observed that the conductivity reaches a metallic level at about 1% doping whereas the magnetic susceptibility remains low up to about 6% doping, at which level a transition to a Pauli-like behaviour is seen. There was considerable discussion as to whether the conduction processes in this intermediate regime were due to sliding of self-localised charged solitons along the polymer chain, but it is now clear that the transitions seen at around 6% doping are due at least in part to a phase transition between different orderings of the dopant ions around the chains [46].

In the limit of low charge carrier concentrations it is clear that carrier mobilities are low, and it is generally accepted that the rate-limiting process is that of charge transfer between chains. For the particular case of trans-polyacetylene, the topological character of the soliton raises some problems here, since it cannot move from one chain to another. There are two models developed for interchain charge transfer. The first of these is due to Kivelson [47] and relies on the presence of a substantial concentration of neutral 'soliton' states. The process of interchain charge transfer is then achieved by transfer of charge from a charged soliton on one chain to a neutral soliton on an adjacent chain. This process, termed 'intersoliton hopping' is clearly specific to polyacetylene, both because of the particular nature of the soliton excitation and because of the requirement for both charged and neutral solitons. The second model is one in which the charged solitons are present in pairs on a single chain, and constrained (by for example interchain coupling or finite chain lengths) to remain close to one another. The doubly charged excitation may then be treated as a bipolaron, and the process of interchain charge transport is then controlled by the rate at which these can hop (or possibly tunnel) between chains [42]. The 'bipolaron model' is implicit in the scheme shown in figure 1 for charge injection onto the polyacetylene chain. We consider that there is very strong evidence that this is the correct mechanism, for charge transport for carriers introduced through chemical doping, photoexcitation, and through charge injection.

The intersoliton model depends crucially on the presence of 'neutral solitons' which can be readily ionised and thus participate in the charge transport. Trans-polyacetylene, produced either by the Shirakawa route or by the Durham route, does contain a high concentration of spin defects which are detected in ESR experiments, and which can be shown to be due to the presence of non-bonding p_z states with wavefunctions spread over some 10 to 20 carbon sites. These spins appear during the thermal isomerisation from cis to trans, and are present typically in concentrations of 10^{18} - 10^{19} cm^{-3}. However, the bulk of the evidence gathered from photoexcitation experiments is that though they may be 'soliton-like' in the nature of their wavefunctions, they do not possess energy levels within the gap, and are not involved in the production of metastable photoexcited states [42,48,49]. Turning to the experiments we have performed with charge accumulation layers, we are able to achieve rather higher levels of charge injection than obtained in the photoexcitation experiments. At the higher gate voltages used in the data shown in figure 10, the surface charge density is up to 4×10^{12} cm^{-2} (at -40 V). As we discuss in section 3.3 we consider that the charge accumulation layer is strongly localised to the interface with the insulator, probably within 20 Å. The total areal concentration of the neutral spin defects within such a layer is a factor of 2 lower than the charge concentration (at $V_g = -40$ V), and we do not therefore consider that the charge injected into the polymer is stored on the spin defects. As previously discussed [42,49] we consider that the neutral spin defects are present on disordered regions of the chain, which possess a high π-π^* energy gap, and thus excluded from participation in the electronic processes that take place within the gap.

The 'bipolaron model' is generally applicable for conjugated polymers, whether or not they may in principle possess a degenerate ground state. As discussed by Chance et al [50], like-charged soliton pairs are analogous to bipolarons and are not topologically restrained from inter-chain hopping. The process by which they do this will involve an intermediate stage in which one of the two charges has transferred to an adjacent chain, and the instantaneous description is of two polarons on adjacent chains. If the second charge then follows the first, the bipolaron has moved from one chain to another, and has surmounted an energy barrier equal to the stabilization energy of the bipolaron (or soliton anti-soliton pair). Townsend and Friend [42] show that there is a well-defined activation energy for the mobility of photocarriers, of about 0.31 eV, and identify this as the bipolaron stabilisation energy. The general expression for the hop rate of a self-localized carrier is $R = \omega \exp\{-W_h/k_B\theta\}$ where W_h is the hop energy and ω is the attempt-to-hop frequency [51], and this is shown to be consistent with the room temperature mobilities observed if the average distance between hops is set at 2.1 nm.

For the case of photocarriers, Townsend and Friend [42] were able to identify from spectroscopic measurements that they are present as like-charged pairs of solitons; hence the description of the transport in terms of bipolaron motion along and between chains. For carriers introduced through chemical doping the situation is less clear because in general we have less spectroscopic information. It is interesting to compare the properties of as-made samples of Durham polyacetylene. These contain the usual level of spin 1/2 defects, as already discussed, and they also contain a low concentration of charges which are responsible for the DC conductivity. Measurements on Schottky barrier diodes formed between aluminium and the polyacetylene presented in section 3.1 do provide the necessary information to characterise the charges present. From the current/voltage and capacitance/voltage characteristics we know that the charge carriers are p-type and present in a concentration of about 10^{16} cm^{-3}, and that this does not vary strongly with temperature. From the differential optical transmission measurements through the depletion regime we also know that these p-type carriers show the same 'mid-gap' optical absorption signature that are found in photo-induced absorption for the photocarriers, with a peak in absorption at 0.55 eV. The mobility of the dark carriers is now easily fixed by the value of the dark conductivity. The room temperature carrier mobility inferred from the conductivity (3×10^{-8} S/cm) is about 2×10^{-5} cm^2/Vs, and it shows an activation energy of about 0.4 eV [42]. These values are rather similar to those we have established for the photocarriers, and we consider that the same model for transport is appropriate for both the extrinsic carriers and the photocarriers.

Turning to the carriers present in the accumulation and inversion layers in the MISFET structures, we have established that the mobilities are comparable to those of the extrinsic carriers in the as-made polyacetylene and to those of the photocarriers. The MISFET structure provides a very versatile system for the control of the carrier density and for the systematic measurement of the transport properties. However, as we have discussed in section 3.3, the accumulation or inversion layer in the MISFET structure is confined to the interface between the polymer and insulator, and is very sensitive to the surface structure of the polymer. It is clear therefore that the study of transport processes within the accumulation or inversion layers must be coordinated with the characterisation of the polymer structure at the interface with the insulator layer, and we consider that this is an important direction for future study.

4. Conclusions

The work which we have discussed in this chapter does demonstrate that it is relatively straightforward to produce semiconductor devices with good characteristics, once the processing of the polymer is controlled. We have been pleasantly surprised in the course of this work to find that our devices show little evidence for the role of defect states in the gap, either bulk or at the surface, and we consider that similar properties can be expected in a range of other polymer.

Semiconductor devices based on silicon are of course of major technological importance. The prospects for 'useful' polymer devices do not lie in any of the markets where silicon has a hold. We consider, however, that there are real possibilities for development of devices which exploit the physics novel to the polymers. Two areas satisfy this condition; first the possibilities for large area fabrication, and second, the exploitation of the large electro-optic effects in, for example, electro-optic modulators.

Acknowledgements

We thank British Petroleum plc for support for this work.

References

1. Takayama, H., Lin-Liu, Y. R. and Maki, K., (1980) Phys. Rev. **B21**, 2388

2. Baeriswyl, D., in "Electronic Properties of Polymers", ed. Mort, J., (Wiley, 1982)

3. Baeriswyl, D. and Maki, K,. (1985) Phys. Rev. **B31**, 6633

4. Su, W. P., Schrieffer, J. R. and Heeger, A. J., (1979) Phys. Rev. Lett. **42**, 1698; (1980) Phys. Rev. **B22**, 2099; Err., (1983) **B28**, 1138

5. Rice, M. J., (1979) Phys. Lett. **71A**, 152

6. Horovitz, B., (1982) Solid State Commun. **41**, 729

7. Suzuki, N., Ozaki, M., Etemad, S., Heeger, A. J. and MacDiarmid, A. J., (1980) Phys. Rev. Lett. **45**, 1209; erratum **45**, 1463

8. Kivelson, S., Lee, T-K., Lin-Liu, Y. R., Peschel, I., and Yu, L., (1982) Phys. Rev. B**25**, 4173

9. Vardeny, Z. and Tauc, J., (1985) Phys. Rev. Lett. **54**, 1844; (1986) Phys. Rev. Lett. **56**, 1510

10. Baeriswyl, D., Campbell, D. K. and Mazumdar, S., (1986) Phys. Rev. Lett. **56**, 1509

11. Ehrenfreund, E., Vardeny, Z., Brafman, O. and Horovitz, B., (1987) Phys. Rev. B**36**, 1535.

12. Hicks, J. C. and Blaisdell, G. A., (1985) Phys. Rev. B**32**, 919

13. Terai, A., Ono, Y. and Wada, Y., (1986) J. Phys. Soc. Japan **55**, 2889

14. Terai, A., Ono, Y. and Wada, Y., (1989) Synthetic Metals **28**, D353

15. Grant, P. M., Tani, T., Gill, W. D., Krounbi, M. and Clarke, T. C., (1980) J. Appl. Phys. **52**, 869

16. Kanicki, J., (1986) in 'Handbook on Conducting Polymers', Skotheim, T. J., Marcel Dekker, New York, pp. 544-660.

17. Garnier, F., and Horowitz, G., (1987) Synthetic Metals **18**, 693

18. Tomozawa, H., Braun, D., Phillips, S., Heeger, A. J., and Kroemer, H., (1987) Synthetic Metals, 1987, **22**, 63; ibid. (1989) **28**, 687

19. Garnier, F., Horovitz, G. and Fichou, D., (1989) Synthetic Metals **28**, 705

20. Ebisawa, E., Kurokawa, T., and Nara, S., (1983) J. Appl. Phys. **54**, 3255

21. Koezuka, H.,Tsumura, A. and Ando, T., (1987) Synthetic Metals **18**, 699

22. Koezuka, H., and Tsumara, A., (1989) Synthetic Metals **28**, 753

23. Assadi, A., Svensson, C., Willander, M. and Inganäs, O., (1988) Appl. Phys. Lett. **53**, 195

24. Edwards, J. H. and Feast, W. J., (1980) Polymer Commun. **21**, 595

25. Edwards, J. H.,Feast, W. J. and Bott, D. C., (1984) Polymer **25**, 395

26. Feast, W. J. and Winter, J. N., (1985) J. Chem. Soc. Chem. Comm. 202

27. Friend, R. H., Bott, D. C., Bradley, D. D. C., Chai, C. K., Feast, W. J., Foot, P. J. S.,Giles, J. R. M., Horton, M. E., Periera, C. M. and Townsend, P. D., (1985) Phil. Trans. Roy. Soc. Lond. A**314**, 37

28. Friend, R. H., Bradley, D. D. C., Pereira, C. M., Townsend, P. D., Bott, D. C. and Williams, K. P. J., (1986) Synthetic Metals **13**, 101

29. Bott, D. C., Brown, C. S., Chai, C. K., Walker, N. S., Feast, W. J., Foot, P. J. S., Calvert, P. D., Billingham, N. C. and Friend, R. H., (1986) Synthetic Metals **14**, 245

30. Friend, R. H., Bradley, D. D. C., Townsend P. D. and Bott, D. C., (1987) Synthetic Metals **17**, 267

31. Friend, R. H., Bradley D. D. C. and Townsend, P. D., (1987) J. Phys.D (Applied Physics) **20**, 1367

32. Friend, R. H., Schaffer, H. E., Heeger, A. J. and Bott, D. C.   (1987)   J. Phys. C (Solid State Physics) **20**, 6013

33. Burroughes, J. H., Jones, C. A. and Friend, R. H., (1988) Nature **335**, 137

34. Burroughes, J. H., Jones, C. A. and Friend, R. H., (1989) Synthetic Metals **28**, 735

35. Lawrence, R. A., Burroughes, J. H. and Friend, R. H., in "Electronic Properties of Conducting Polymers" ed. H. Kusmany et al, (Springer Series on Solid State Sciences, 1989 in press).

36. Brown, C. S., Vickers, M. E., Foot, P. J. S., Billingham, N. C. and Calvert, P. D., (1986) Polymer **27**, 1719

37. Kahlert, H. and Leising, G., (1985) Mol. Cryst. Liq. Cryst. **117**, 1

38. Sokolowski, M., Marseglai, E. A. and Friend, R. H., (1986) Polymer **27**, 1714

39. Sze, S. M., "Physics of Semiconductor Devices", 2nd Edition (Wiley-Interscience, New York, 1981).

40. Rhoderick, E. M. and Williams, R. H., "Metal-Semiconductor Contacts", 2nd Edition (Clarendon Press, Oxford, 1988)

41. Nemanich, R. J. and Thompson, M. J., in "Metal-Semiconductor Schottky Barrier Junctions and their Applications", ed. Sharma, B. L., (Plenum, New York 1984)

42. Townsend, P. D., and Friend, R. H., (1989) Synthetic Metals **28**, 735; Phys. Rev. **B40**, 3112

43. Goetzberger, A. and Nicollian, E. H., (1966) Appl. Phys. Lett. **9**, 12

44. Orenstein, J., in "Handbook of Conducting Polymers", Skotheim, T. J., ed.(Marcel Dekker, New York) (1986).

45. Ando, T., Fowler, A. B. and Stern, F., (1982) Rev. Mod. Phys. **54**, 437

46. Heeger, A. J., in "Handbook on Conducting Polymers", T. J. Skotheim, ed. (Marcel Dekker, New York) (1986), pp 729-756.

47. Kivelson, S., (1982) Phys. Rev. **B25**, 3798

48. Townsend, P. D. and Friend, R. H. (1987) J. Phys. C **20**, 4221

49. Colaneri, N. F., Friend, R. H., Schaffer, H. E. and Heeger, A. J., (1988) Phys. Rev. **B38**, 3960

50. Chance, R. R., Bredas, J-L. and Silbey, R., (1984) Phys. Rev. **B29**, 4491

51. Mott, N. F. "Conduction in Non-Crystalline Materials", (Clarendon Press, Oxford 1987).

CHEMISTRY AND PHYSICS OF MOLECULAR–BASED POLYMERS EXHIBITING A SPONTANEOUS MAGNETIZATION

Olivier KAHN

Laboratoire de Chimie Inorganique, URA 420

Université de Paris–Sud, 91405 Orsay

France

ABSTRACT. One of the main challenges in the field of the molecular materials is the design of molecular–based ferromagnets. Our basic strategy along this line consists of assembling ferrimagnetic chains within the crystal lattice in a ferromagnetic fashion. This can be achieved owing to the (almost) limitless flexibility of the molecular chemistry. The chains may be either regular or alternating. Examples of both situations are presented. $MnCu(pbaOH)(H_2O)_3$ with pbaOH=2–hydroxy–1,3–propylenebis(oxamato) is a regular chain compound ordering ferromagnetically at T_c=4.6 K, and $MnCu(obbz).1H_2O$ with obbz=oxamido–N,N'–bis(2–benzoato) is an alternating chain compound exhibiting a spontaneous magnetization below T_c=14 K. To get information on the mechanism of the magnetic ordering, a broad spectrum of physical techniques is utilized, including magnetic susceptibility and magnetization measurements, and EPR spectroscopy. The perspectives in this new field are briefly outlined.

1. Introduction

Several groups over the world are presently working on the design of molecular–based ferromagnets [1–6] and the first compounds of this kind have recently been reported. To the best of our knowledge, three families of compounds exhibiting potentially a spontaneous magnetization below a critical temperature T_c are currently investigated and in each family, one system, at least, has been found

J. L. Brédas and R. R. Chance (eds.), Conjugated Polymeric Materials:
Opportunities in Electronics, Optoelectronics, and Molecular Electronics, 247–261.
© 1990 *Kluwer Academic Publishers. Printed in the Netherlands.*

to order ferromagnetically. These families are : (i) the donor-acceptor compounds giving D^+A^- alternating stacks, in which both D^+ and A^- carry a local spin 1/2. So, $Fe(Me_5Cp)_2(TCNE)$ with Me_5Cp=pentamethylcyclopentadienyl and TCNE=tetracyanoethylene orders ferromagnetically at T_c=4.8 K [3,4], (ii) The Mn(II)-nitroxide chain compounds. For instance, $Mn(hfa)_2(NITMe)$ with hfa=hexafluoroacetylacetonato and NITMe=2-methyl-4,4,5,5-tetramethyl-4,5-dihydro-1H-imidazolyl-1-oxyl-3- oxide orders at T_c=7.8 K [5]; (iii) the Mn(II)-Cu(II) bimetallic polymers with bisbidentate extended ligands [6,7]. This family is currently studied in our group and we would like to review briefly the main results obtained so far.

2. Strategy

Our specific strategy to design molecular-based compounds exhibiting a spontaneous magnetization consists of assembling ordered bimetallic chains within the crystal lattice in a ferromagnetic fashion. If, as it is most often the case, the intrachain interactions between nearest neighbor metal ions A and B are antiferromagnetic, the chains are said to be ferrimagnetic and their ground state may be schematized as :

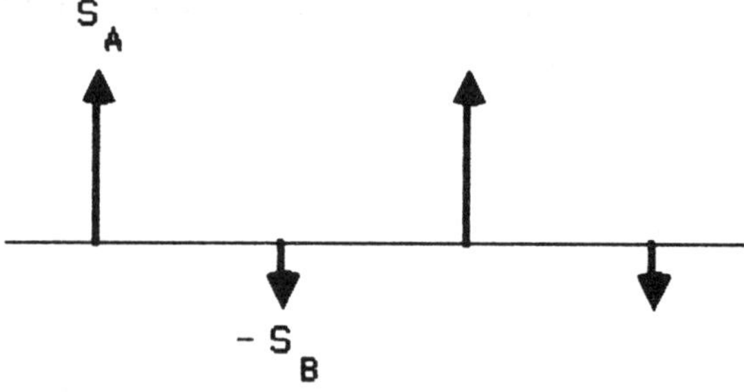

where S_A and S_B ($S_A \neq S_B$) are the local spins. In absence of any interchain interaction, the magnetic properties of such a system are quite characteristic. The $\chi_M T$ versus T plot, χ_M being the molar magnetic susceptibility per AB unit and T the temperature, exhibits a minimum. Upon cooling below the temperature of this minimum, $\chi_M T$ increases in a ferromagnetic-like fashion and diverges when T approaches zero [2,8,9]. This divergence may be considered as the onset of a magnetic ordering at 0 K. It is indeed well established that there is no magnetic ordering at a finite temperature for a purely one-dimensional system [10]. In fact, the chains cannot be perfectly isolated within the lattice. Either, they interact between them in an antiferromagnetic fashion and the divergence of $\chi_M T$ is stopped at a certain temperature close to the temperature of three-dimensional antiferromagnetic ordering [8,9,11], or they interact in a ferromagnetic fashion and below a critical temperature, the system behaves as a ferromagnet [6,7]. Our goal is to realize the conditions leading to this latter situation.

The main requirements of our strategy are the following : (i) $|S_A-S_B|$ must be as large as possible. The most favorable couple is Mn(II) and Cu(II) with 5/2 and 1/2 local spins, respectively; (ii) The intrachain antiferromagnetic interactions must be as large as possible. This condition is satisfied when using bridging groups known for their ability to transmit the electronic effects, like the conjugated bisbidentate ligands [12]; (iii) finally, the relative positions of the chains within the crystal lattice must favor the ferromagnetic situation instead of the antiferromagnetic one.

The first two requirements led us to synthesize bimetallic chain compounds according to one of the two schemes shown below, with Mn(II) cations reacting on copper(II) mononuclear anionic bricks :

The former scheme gives equally spaced bimetallic chains [6,11] of which examples are presented in the next Section and the latter alternating bimetallic chains [7,13] as those described in the following Section.

3. Regular bimetallic chains : MnCu(pba)(H$_2$O)$_3$.2H$_2$O and MnCu(pbaOH)(H$_2$O)$_3$.

The two compounds MnCu(pba)(H$_2$O)$_3$.2H$_2$O (**1**) and MnCu(pbaOH)(H$_2$O)$_3$ (**2**) with pba=1,3-propylenebis(oxamato) and pbaOH=2-hydroxy-1,3-propylenebis(oxamato) have very similar regular chain structures. The structure of **1** is shown in Figure 1.

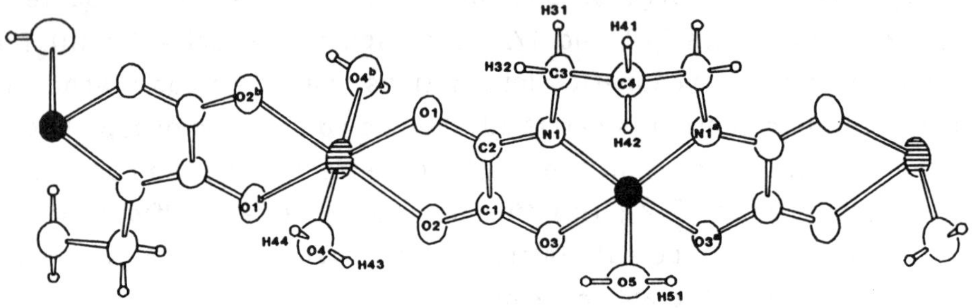

Figure 1. Structure of the bimetallic chain in **1**.

1 and **2** are obtained by reaction of Mn(II) ion on the copper(II) brick shown below :

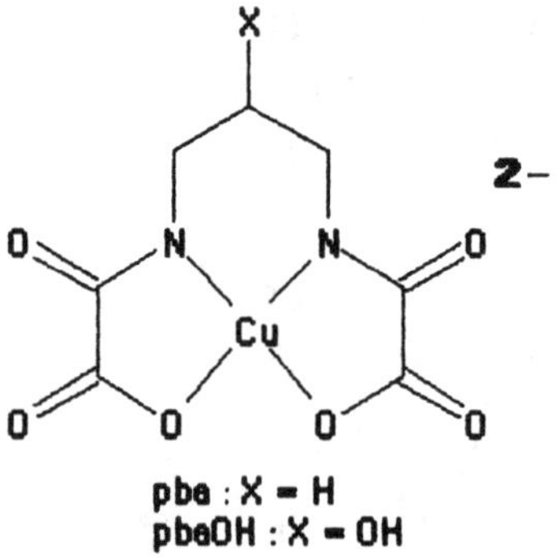

The relative positions of the chains within the crystal lattice are however different as shown in Figure 2.

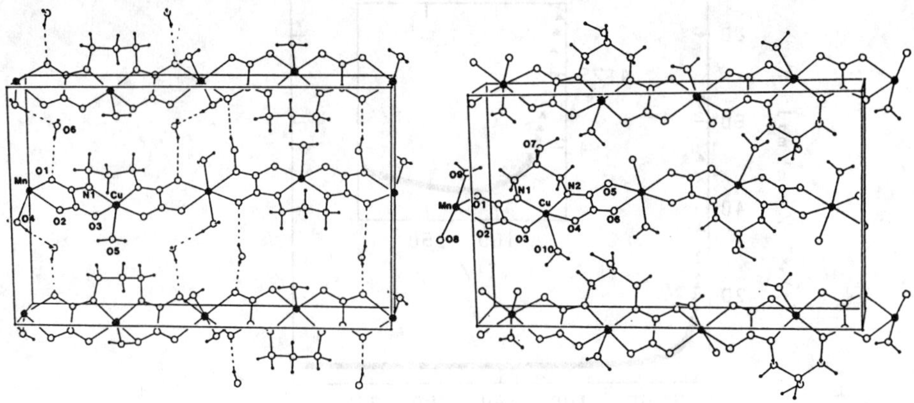

Figure 2. Perspective view of three neighboring chains in **1** (left) and **2** (right). The origin of the unit cell is the upper left-hand corner. The a-axis runs top to bottom and the b-axis left to right.

Both **1** and **2** crystallize in the orthorhombic system. The chains run along the b-axis. The shortest interchain separations along the a-axis involve metal ions of the same nature in **1** (Cu...Cu=6.545 $\mathring{A}$ and Mn...Mn=6.977 $\mathring{A}$) and of different nature in **2** (Cu...Mn=5.751 and 6.398 $\mathring{A}$). In a simplified fashion, we can say that in **2**, with respect to **1**, every other chain is displaced by slightly less than half of a repeat unit along b. In both compounds, the chains closest to one another are those related by a unit cell translation along the c-axis (Cu...Cu^c=Mn...Mn^c=5.2105 $\mathring{A}$ in **1** and 5.073 $\mathring{A}$ in **2**).

In the 30-300 K temperature range, **1** and **2** show the magnetic behavior characteristic of ferrimagnetic chains with the minimum of $\chi_M T$ around 115 K. On the other hand, upon cooling down below 30 K, $\chi_M T$ increases much faster for **2** than for **1** and below 5 K, the two compounds behave quite differently. **1** exhibits a maximum of $\chi_M T$ at 2.3 K and a maximum of χ_M at 2.2 K, due to the onset of a 3D antiferromagnetic ordering. In contrast, $\chi_M T$ for **2** diverges and becomes strongly field-dependent, which suggests that a ferromagnetic transition takes place (see Figure 3).

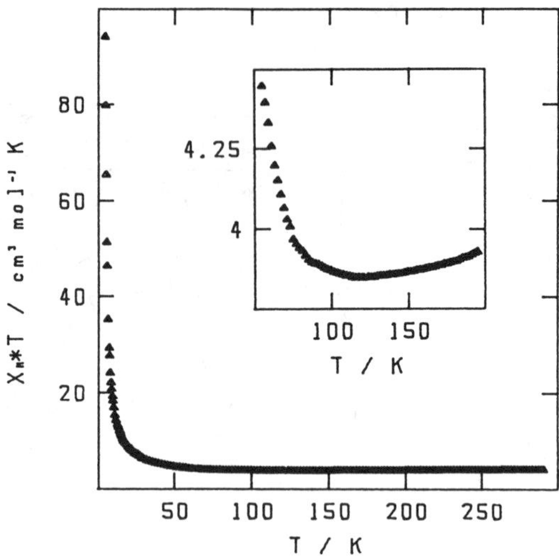

Figure 3 : Temperature dependence of $\chi_M T$ for **2**.

The temperature dependence of the magnetization M within a magnetic field of 3.10^{-2} G confirms that a 3D ferromagnetic transition occurs at T_c=4.6 K. Below this temperature, **2** exhibits a spontaneous magnetization and a hysteresis loop M=f(H) characteristic of a soft magnet.

To get informations on the mechanism of the 3D magnetic ordering and on the spin distribution in both compounds, we investigated the single crystal EPR spectra. The angular dependence of the linewidths is governed by the dipolar interactions between local spins. The EPR data also suggest that the main dipolar interaction must be parallel to the c-axis and this orients the local spins parallel to each other along this direction. It follows that the spin distributions should be as shown in Figure 4. The bc ferromagnetic planes alternate along the a-axis in an antiferromagnetic fashion in **1** and in a ferromagnetic fashion in **2**, leading to the ferromagnetic ordering observed at 4.6 K [14]. These differences of magnetic properties are related to the differences concerning the packing along the a-axis with Mn...Mn and Cu...Cu shortest separations in **1** and Mn...Cu shortest separations in **2**.

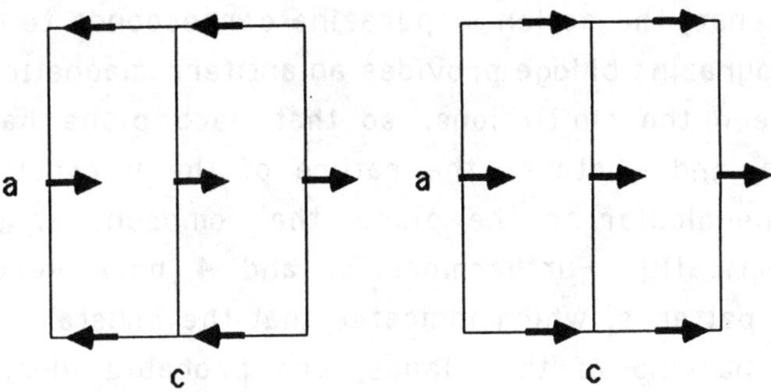

Figure 4. Schematic representation of the preferred spin orientation for **1** (left) and **2** (right) in the ac-plane. The arrows indicate the Mn(II) local spins; those of Cu(II) ions are oriented in the opposite directions. To make easier the comparison between the two spin distributions, the coordinates of the atoms for **2** have been translated by $x/a=-0.0310$ and $z/c=-0.2625$.

Both **1** and **2** react with pz=pyrazine in sealed tubes to give MnCu(pba)(pz) (**3**) and MnCu(pbaOH)(pz) (**4**), respectively. In order to respect the octahedral environment of Mn(II) ions, pz must bridge these ions, which leads to molecular planes as shown below :

254

In a certain sense, the action of pyrazine corresponds to molecular weaving. The pyrazine bridge provides an antiferromagnetic exchange pathway between the Mn(II) ions, so that each plane has a zero resulting spin, and whatever the nature of the interaction in the direction perpendicular to the plane, the compound should order antiferromagnetically. Furthermore, **3** and **4** have very similar powder X-ray patterns, which indicates that the crystal structures, including the packing of the planes, are probably identical. The magnetic data confirm these predictions. Both **3** and **4** exhibit the behavior characteristic of ferrimagnetic chains down to about 5 K, then order antiferromagnetically at 2.9 K, owing to the Mn...Mn interaction through pz.

The $MCu(pbaOH)(H_2O)_3.nH_2O$ componds with M = Fe, Co and Ni have also been investigated. All show the characteristic behavior of ferrimagnetic chains, i.e. the minimum of the $\chi_M T$ versus T plot. The Co(II) and Ni(II) derivatives present a 3D antiferromagnetic ordering, at 3.4 and 7 K, respectively. As far as the iron(II) derivative is concerned, the spins of the ferrimagnetic chains also tend to cancel within the crystal lattice, with however a small canting so that the resulting spin is not exactly zero. It follows that $FeCu(pbaOH)(H_2O)_3.3H_2O$ **(5)** presents a weak ferromagnetism with a small spontaneous magnetization below $T_c=10$ K. This behavior is revealed by the temperature dependence of the molar magnetization M shown in Figure 5. The field-cooled magnetization (FCM) is obtained by cooling within a field of 1 G. This FCM increases when the sample is cooled, then below 7 K saturates. When switching of the field at 2 K, a remnant magnetization is detected with vanishes at $T_c=10$ K. The zero-field-cooled magnetization (ZFCM) is obtained by cooling in zero-field, then warming within the field of 1 G. At any temperature below T_c, the ZFCM is smaller than the FCM, due to the fact that the temperature is too low for the domain walls to move freely. The ZFCM exhibits a maximum at T_c, as expected for a powder sample.

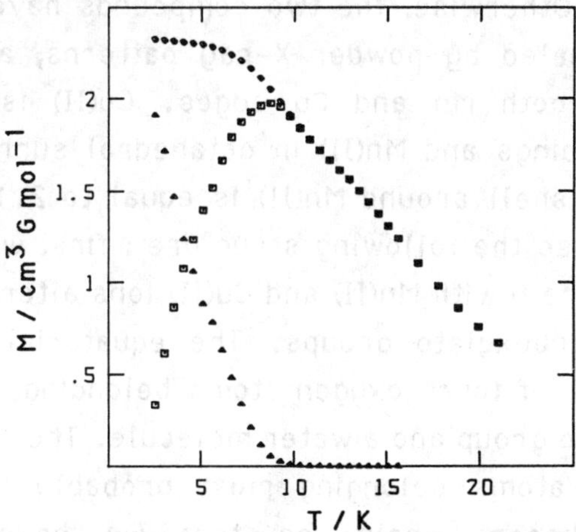

Figure 5. Temperature dependence of the magnetization M for **5** with a field of 1 G. ○, field-cooled magnetization; ▲, remnant magnetization; □, zero-field-cooled magnetization.

4. Alternating Bimetallic Chains : MnCu(obbz).5H$_2$O and MnCu(obbz).1H$_2$O.

The action of Mn(II) ion on the copper(II) brick noted [Cu(obbz)]$^{2-}$ shown below :

with obbz=oxamido-N,N'-bis(2'-benzoato) affords two compounds of formula MnCu(obbz).5H$_2$O (**6**) and MnCu(obbz).1H$_2$O (**7**). The four additional water molecules in **6** as compared to **7** are not coordinated

256

to the metal ions. Otherwise, the two compounds have very similar structures as revealed by powder X-ray patterns, and XANES and EXAFS spectra at both Mn and Cu edges. Cu(II) is in elongated tetragonal surroundings and Mn(II) in octahedral surroundings. The radius of the first shell around Mn(II) is equal to 2.16(3) $\mathring{A}$. Those data allow to propose the following structure : first we consider the chain shown in Figure 6 with Mn(II) and Cu(II) ions alternately bridged by oxamido and carboxylato groups. The equatorial plane around Mn(II) would consist of four oxygen atoms belonging to an oxamido group, a carboxylato group and a water molecule. The apical sites are two other oxygen atoms belonging most probably to carboxylato groups of an adjacent chain, so that we have a two- or three-dimensional packing.

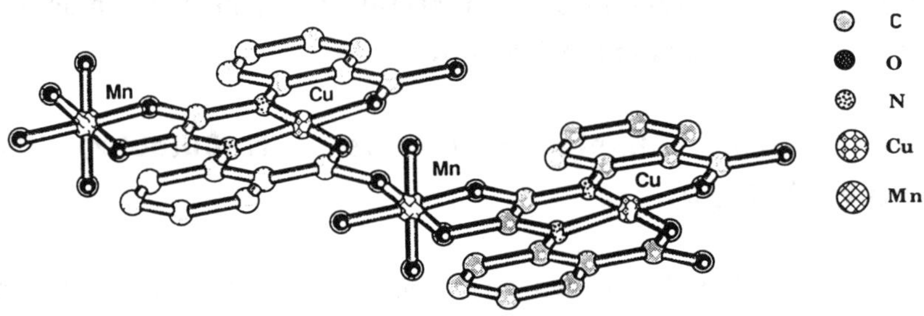

Figure 6. Basic structure proposed for **6** and **7**. The oxygen atoms occupying the apical positions around the Mn(II) ions belong probably to carboxylato groups of adjacent chains.

In spite of the structural analogies, **6** and **7** have completely different magnetic properties. While **6** orders antiferromagnetically at 2.3 K, the deshydrated form **7** presents a ferromagnetic transition at 14 K. The temperature dependence of the molar magnetization M in a field of 0.1 G is shown in Figure 7. The FCM shows the typical feature of a ferromagnetic transition, i.e. a rapid increase of M when T decreases below 15 K, then a break in a curve around T_c=14 K. We then measured the remnant magnetization by switching off the field at 5 K and warming up. The remnant magnetization, as expected, vanishes at T_c. Finally, we measured the ZFCM,; it exhibits a maximum around

T_c, as expected for a polycristalline ferromagnet [16].

We also studied the magnetic hysteresis for **7**. The hysteresis loop at 4.2 K is shown in Figure 8. The remnant magnetization is equal to 6.3×10^3 cm^3 G mol^{-1}, i.e. about 30% of the saturation magnetization (see below) and the coercitive field is about 60 G.

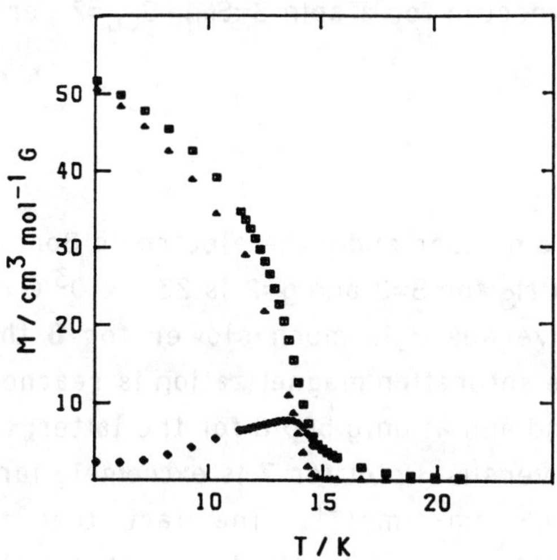

Figure 7. Temperature dependence of the magnetization M for **7** in the 5–20 K range and with a field of 0.1 G. ▫ , field–cooled magnetization; ▵, remnant magnetization; ◇, zero–field–cooled magnetization.

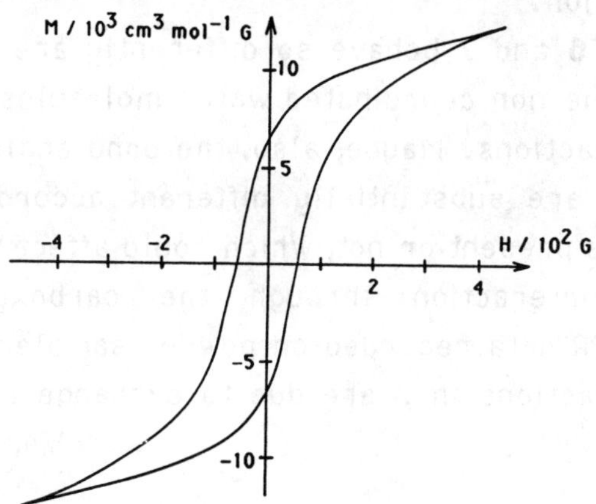

Figure 8. Hysteresis loop M=f(H) for a polycristalline sample of **7** at 4.2 K.

The last experiment concerning the magnetization compares the variations of M versus the applied magnetic field for **6** and **7**. The results at 4.2 K are shown in Figure 9. They reveal what follows : (i) the high-field limits are identical for both compounds and equal to about 23.0×10^3 cm^3 mol^{-1} G. This value agrees with the saturation magnetization M_S expected for a spin $S=S_{Mn}-S_{Cu}=2$ per MnCu unit. M_S is then given by :

$$M_S = Ng\beta S$$

where N is Avogadro number and β the electronic Bohr magneton. The theoretical value of M_S for $S=2$ and $g=2$ is 23.3×10^3 G cm^3 mol^{-1} ; (ii) the increase of M versus H is much slower for **6** than for **7**. For instance, half of the saturation magnetization is reached at 7500 G for the former compound and at only 500 G for the latter; (iii) the slope in zero-field of the M versus H plot for **7** is extremely large, actually of the order of 2×10^2 cm^3 mol^{-1}. The fact that the zero-field susceptibility $(dM/dH)_{H=0}$ is not infinite is essentially due to the demagnetization field created by the surface of the sample. Moreover, the experiment is carried out with a polycrystalline powder, so that the susceptibility is averaged over all directions, including the hard magnetization direction.

The reasons why **6** and **7** behave so differently are not perfectly clear yet. Maybe, the non coordinated water molecules in **6** prevent through-space interactions. Maybe, also, the bond angles around the carboxylato bridge are substantially different according as these water molecules are present or not, which could affect the magnitude of the Mn...Cu interaction through the carboxylato group. Furthermore, the EPR data recorded on powder samples suggest that the interchain interactions in **7** are due to exchange rather than to dipolar effects.

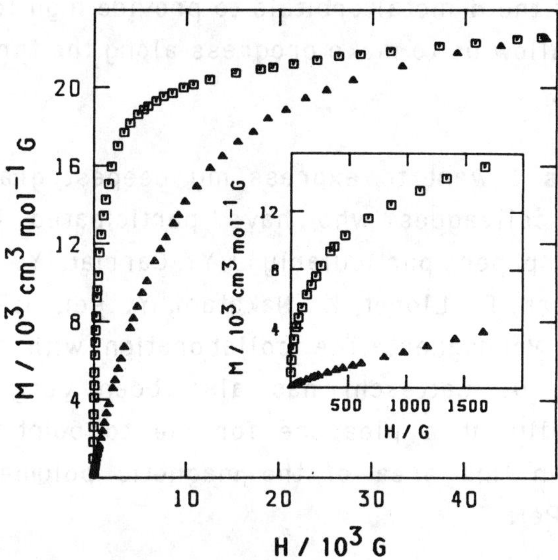

Figure 9. Field dependences of the magnetization M for polycristalline samples of **6** (▲) and **7** (◻).

5. Conclusion and outlook

The field of molecular-based magnetic materials is still in its first infancy. Several groups over the world work along this perspective, using various strategies. The emulation and the collaboration between them should favor a fast development. In our opinion, the main goals are the following : (i) to shift T_C toward higher temperatures. Reaching liquid nitrogen temperature would be obviously a major step in this area;
(ii) to design stable systems, easy to handle, giving well-defined hysteresis loop with large remnant magnetization and coercive fields;
(iii) to obtain transparent thin films retaining the ferromagnetic properties below T_C. Such films could be designed either from compounds soluble in organic solvents, or by using the Langmuir-Blodgett technique.

Our specific approach utilizing both the efficiency of organic-based ligands to transmit the electronic effects on long distances and the specific ability of the d-metal orbitals to provide high local magnetic moments should allow us to make progress along the three directions underlined above.

Acknoledgments I want to express my deepest gratitude to my coworkers and colleagues who have participated to the work described in this paper, particularly J.Y. Carriat, Y. Journaux, P. van Köningsbruggen, F. Lloret, K. Nakatani, Y. Pei, J.P. Renard, J. Sletten, and M. Verdaguer. The collaboration with the group of Florence led by D. Gatteschi has also been very fruitful and stimulating. Finally, it a pleasure for me to point out that the pionering work in this area of the magnetic polymers has been performed by Y. Pei.

[1]   Miller, J.S., Epstein, A.J., and Reiff, W.M., (1988) Chem. Rev., 88, 201. (1988) Acc. Chem. Res., 21, 114. (1988) Science, 240, 40.

[2]   Kahn, O., (1987) Struct. Bonding(Berlin), 68, 89. (1987) In P. Delhaes, and M. Drillon (eds.), Organic and Inorganic Low-Dimensional Crystalline Materials, NATO ASI Series, Vol. 168, Plenum, New York, , pp. 93–108.

[3]   Miller, J.S. and Epstein, A.J., (1987) J. Am. Chem. Soc., 109, 3850.

[4]   Miller, J.S., Calabrese, J.C., Rommelmann, H., Chittipeddi, S.R., Zhang, J.H., Reiff, W.M. and Epstein, A.J., (1987) J. Am. Chem. Soc., 109, 769.

[5]   Caneschi, A., Gatteschi, D., Renard, J.P., Rey, P. and Sessoli, R., (1988) Inorg. Chem., 27, 1756.

[6]   Kahn, O., Pei, Y., Verdaguer, M., Renard, J.P. and Sletten, J., (1988) J. Am. Chem. Soc., 110, 782.

[7]   Nakatani, K., Carriat, J.Y., Journaux, Y., Kahn, O., Lloret, F., Renard, J.P., Pei, Y., Sletten, J., Verdaguer, M., (1989) J. Am. Chem. Soc., 111, 5739.

[8]   Verdaguer, M., Julve, M., Michalowicz, A., and Kahn, O., (1983)

Inorg.Chem., 22, 2624.

[9]   Gleizes, A. and Verdaguer, M., (1984) J. Am. Chem. Soc., 106, 3727.

[10] Carlin, R.L., (1986) Magnetochemistry, Springer–Verlag, Berlin.

[11] Pei, Y., Verdaguer, M., Kahn, O., Sletten, J. and Renard, J.P. (1987) Inorg. Chem., 26, 138.

[12] Kahn, O., (1985) Angew. Chem. Int. Ed. Engl., 24, 834.

[13] Pei, Y., Kahn O., Sletten, J., Renard, J.P. Georges, R., Gianduzzo, R., Curely, J., and Xu, Q. (1988) Inorg. Chem., 27, 47.

[14] Gatteschi, D., Guillou, O., Zanchini, C., Sessoli, R., Kahn, O., Verdaguer, M., and Pei, Y. (1989) Inorg. Chem., 28, 287.

[15] Van Koningsbruggen, P.J., Nakatani, K., Pei, Y., Kahn, O., Drillon, M., unpublished results.

[16] See, for instance, Hitzfeld, M., Ziemann, P., Buckel, W., and Claus, H., (1984) Phys. Rev. B, 29, 5023.

ION IMPLANTED POLYPARAPHENYLENE : MODIFICATIONS OF LATERAL AND IN DEPTH CONCENTRATION PROFILES UPON ANNEALING

G. FROYER, Y. PELOUS, M. GAUNEAU, R. CHAPLAIN
C.N.E.T. LAB/OCM, B.P. 40
F-22301 LANNION Cedex
A. MOLITON, B. RATIER
LEPOFI, Faculté des Sciences
123, Rue A. Thomas
F-87060 LIMOGES

ABSTRACT. Films of poly(paraphenylene) were synthesized by electroreduction and doped by ion implantation. Evolution of concentration profiles upon annealing was followed by Secondary Ion Mass Spectroscopy on n-doped with Cs or Na or on p-doped with I samples. Experimental depth and lateral profiles indicated that diffusion was rather slow, if any, even at temperatures as high as 270°C. However in some cases plateaus were observed in the profiles which may be accounted for by some "grain boundaries" diffusion. This process was tentatively explained by the very slow diffusion of ions from the crystalline microdomains followed by a fast one at the interface between these crystallites.

1. Introduction

It has been known for several years that ion implantation may be useful in providing modifications of polymers [1] which then present new transport, magnetic or optical properties [2]. However these modifications are generally more pronounced when irridiations use high energy ions than with ions in the range 20-100 keV. Some recent works on electroactive polymers were carried out using low energy ion beams to avoid chemical modifications usually encountered [3] in organic materials. Though polyacetylene was first used to build junctions through appropriate ion implantation [4] the air stability of such devices was rather poor, rectifying characteristics remaining only for a few weeks. We have succeeded recently in building stable p-n junction by double implantation in the electroactive polymer poly(paraphenylene) PPP [5] and have correlated theoretical and experimental ion ranges with electrical conductivities of the materials [6]. We now present a study on annealing effects on depth and lateral profiles obtained by Secondary Ion Mass Spectroscopy (SIMS) as compared to theoretical profiles.

J. L. Brédas and R. R. Chance (eds.), Conjugated Polymeric Materials:
Opportunities in Electronics, Optoelectronics, and Molecular Electronics, 263–271.

2. Experimental

Experiments were performed on PPP films obtained by electrochemical reduction of dibromobiphenyl with the method already described [7]. The films were deposited on glass coated with ITO used as working electrode during the electrochemical process, their thicknesses ranged between 2000 Å and 5000 Å.

Ion implantation was carried out in a HVEE 400 ion implanter at 30 keV (exceptionally 200 keV) with Cs^+, Na^+ or I^+ at doses of $2\ 10^{15}$ ions/cm^2. The current density was intentionally kept well below 1 $\mu A/cm^2$ to avoid irreversible transformations which might arise under more drastic conditions [3b].

Metallic masks were tightly pressed onto the PPP films in a holder specially designed to avoid any problem during implantation : this set up allows irradiation through holes (diameter : 250 μm) which gives regularly arranged implanted dots on the sample surface, with a well defined geometry suitable to get informations on the lateral ion profiles.

Annealing was performed at 200 and 270°C under argon atmosphere in an oven allowing the sample to be installed in a cold region at room temperature without oxygen. The implanted samples were thus introduced under inert gas flow in the cold region of the oven, and, after a while, moved to the hot region at the setting temperature. After a given annealing time duration the sample was brought back to the cold region and allowed to cool down before removing it from inert atmosphere.

The depth profiling measurements were performed by SIMS using i.m.s. 3 f and 4 f Cameca ion microscope apparatus. The Cs and Na depth profiles were monitored under 0_2^+ primary ion bombardment by recording the $^{133}Cs^+$ or $^{23}Na^+$ secondary ion species. Typically a 1 μA oxygen primary beam was rastered over a 250 μm square area and secondary ions were extracted from a 60 μm-diam centered region. The iodine depth profiles were recorded under Cesium primary ion bombardment (50 nA rastered over a 125 μm square crater) by measuring the $^{127}I^-$ secondary ion signal.

The lateral distribution of the implanted ions (Na, Cs, or I) at the boundaries of the implanted dots was determined using low current small diameter primary beams of oxygen for cesium and sodium and cesium for iodine. Typically, we used 5 nA, 3-4-μm-diameter primary beams. The primary beam was line-scanned along a dot diameter in such a way to cross the boundary between the implanted and unimplanted zone. The scanning was performed step by step, each step equalling 3 μm. This means that the lateral resolution would be at best equal to about 3-4 μm.

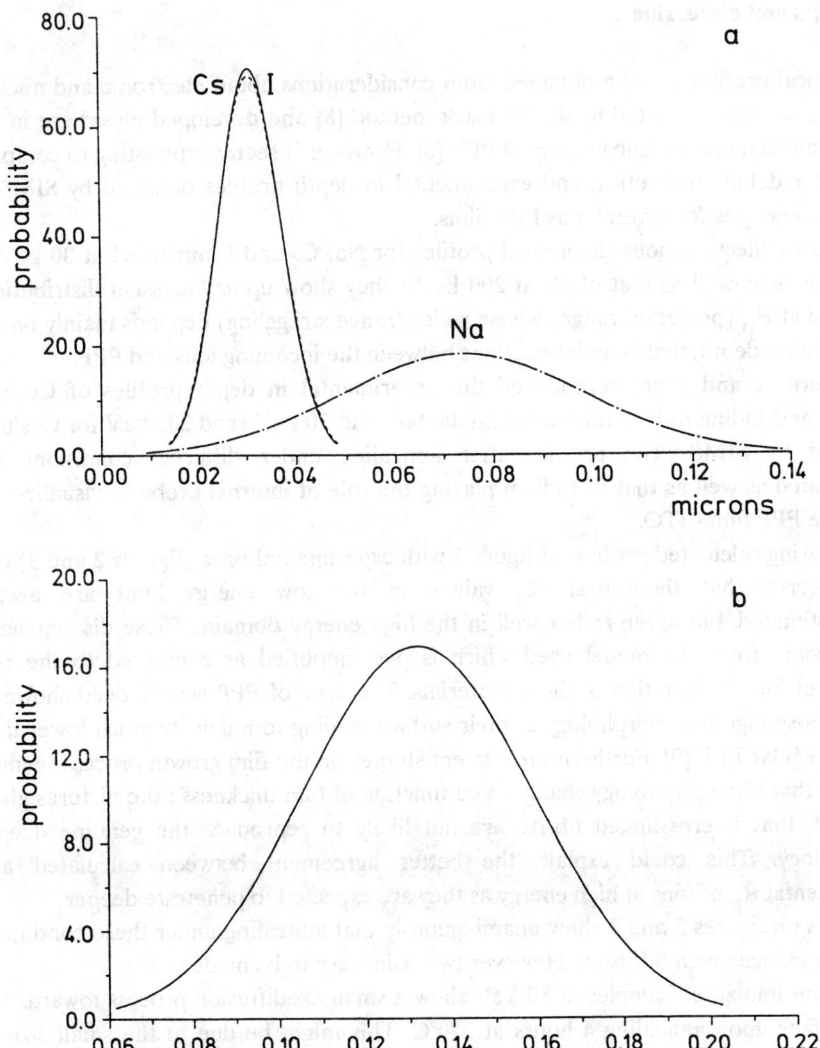

Figure 1. Calculated profiles
a) Energy 30 keV for Cesium and Iodine centered at
R_p = 30 nm and Sodium (R_p = 75 nm).
b) Energy 200 keV for Cesium centered at R_p = 135 nm.

3. Results and discussion

Theoretical profiles may be obtained from considerations about electronic and nuclear stopping powers calculated by the Biersack method [8] and developed elsewhere in the case of alkali metal or halogen doped PPP [6]. However it seems interesting to compare in greater details theoretical and experimental in depth profiles obtained by SIMS at different energies for several ions PPP films.

Figure 1 collects various theoretical profiles for Na, Cs and I implanted at 30 keV in PPP targets as well as that of Cs at 200 keV : they show up as Gaussian distributions centered at R_p (projected range) whose width (range straggling) depends mainly on the implantation depth, that is on interactions between the incoming ions and PPP.

In figures 2 and 3 are reproduced the experimental in depth profiles of Cesium, Sodium and Iodine in PPP films after implantation at 30 keV (and 200 keV for Cesium) obtained by SIMS. The profiles after annealing under different conditions are represented as well as that of Indium playing the role of internal probe to visualize the interface PPP film - ITO.

Comparing calculated profiles of figure 1 with experimental ones (figures 2 and 3) one can observe that theoretical R_p values in the low energy limit are always underestimated, but agree rather well in the high energy domain. These discrepancies might come from the model used which is oversimplified as compared to the real process of ion implantation in these materials. This type of PPP were indeed shown to present a sponge-like morphology at their surface leading to a density much lower than that of regular PPP [9]. Furthermore current studies on the film growth process seem to indicate that film morphology changes as a function of film thickness : the pictures given by SEM, that is crosslinked fibrils, are not likely to reproduce the genuine deeper morphology. This could explain the better agreement between calculated and experimental R_p of ions at high energy as they are expected to penetrate deeper.

Profiles on figures 2 and 3 show unambiguously that annealing under these conditions does not induce much diffusion. However two points are to be made :

- Sodium implanted samples at 30 keV show a small exodiffusion process towards the film surface upon annealing 4 hours at 270°C. This might be due to the small size of Sodium ions as compared to Cesium ions which do not behave this way.

- Iodine implanted samples at 30 keV are the only ones to present a net modification of the concentration profiles upon annealing for 4 hours at 270°C. The occurence of a flat concentration profile from 150 nm to the very interface film - ITO is the signature of some "grain boundaries" diffusion process throughout the film and even into ITO in some cases. However, the ion concentration maximum remains actually located roughly at the same depth R_p instead of shifting as one would expect and was observed on thicker films [6].

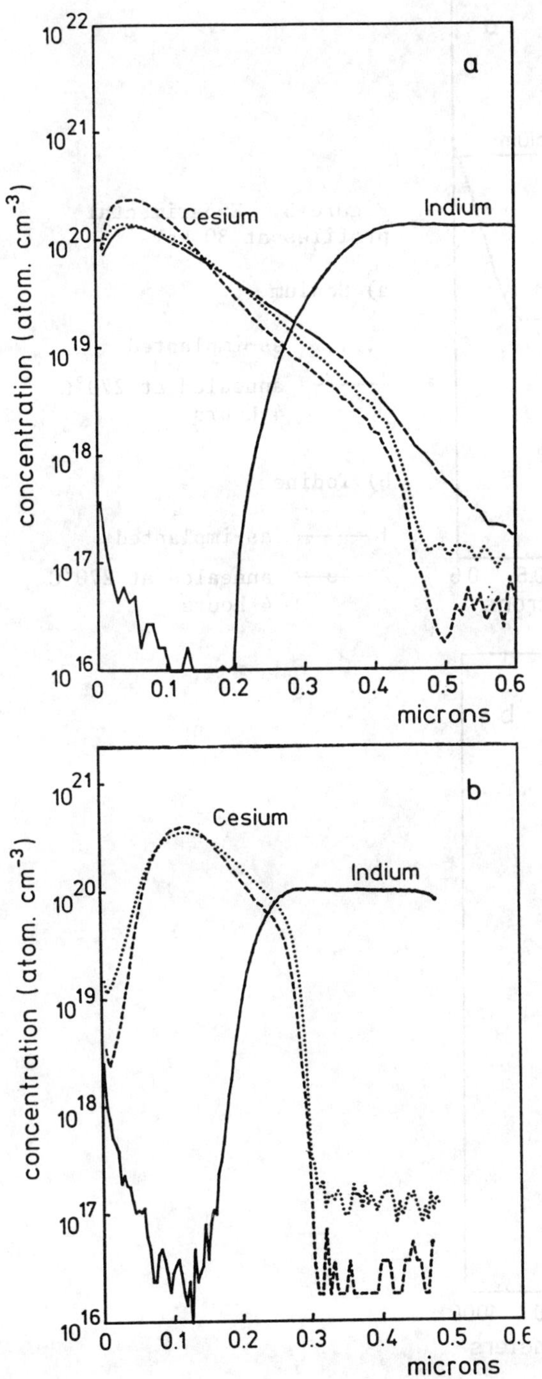

Figure 2. Experimental profiles for Cesium ions with or without annealing

.... as-implanted

---------- annealed 2 hours at 200°C

— . — . — annealed 4 hours at 270°C

a) Energy : 30 keV

b) Energy : 200 keV

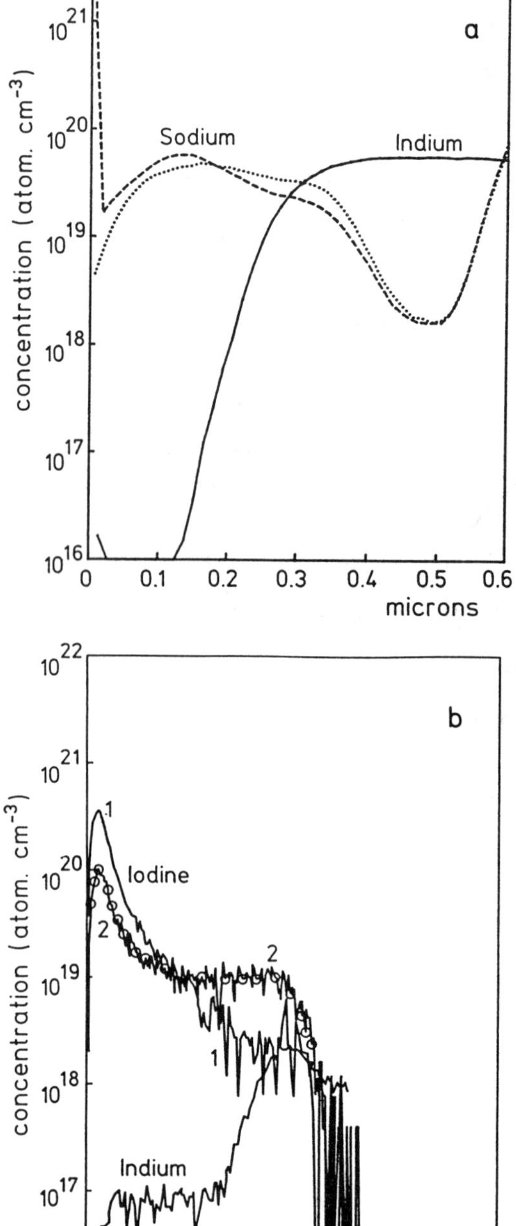

Figure 3. Experimental profiles at 30 keV

a) Sodium

...... as-implanted

------ annealed at 270°C 4 hours

b) Iodine

1 ——— as-implanted

2 —o— annealed at 270°C 4 hours

Lateral concentration profiles for Iodine, Sodium and Cesium implanted at 30 keV are represented in figure 4. The effect of annealing (4 hours at 270°C) on the ion concentration indicates clearly a slight lateral diffusion from the implanted domain through the virgin material : the difference between Iodine and Sodium may be accounted for by their size difference, the smaller the ion, the higher its mobility. But in both cases, the concentration level in the non-implanted PPP is higher after annealing and almost uniform on a long range far from the diffusion source. This observation is difficult to explain unless a fast diffusion process is postulated which might correspond to a sort of "grain boundaries" process showing up in the profile as a plateau. However the evolution of Cesium lateral profiles at 30 keV, upon annealing under different conditions, either 2 hours at 200°C or 4 hours at 270°C, show that walls between implanted and non-implanted domains become smoother and smoother as annealing proceeds featuring a true diffusion process in a bulk (figure 4 c).

In light of the experimental data presented above one sees that ion diffusion in PPP obey probably a double level process :

- a very slow diffusion process in the bulk of the material.

- a very fast one -with a diffusion coefficient several orders of magnitude larger- taking place at some place where the material is loosely arranged.

How could we rationalize these observations ?

We know from transmission electron microscopy work [10] on PPP powder obtained by the same technique on a mercury pool that the cristallinity of the material is quite high and comes from a large number of crystallites about 50 Å thick and between 200 to 300 Å large. They are more or less organized to a larger scale in sort of columnar assemblies which might reach 1000 Å in height. This organization leads to domains which possess a high crystallinity, that is a high density, through which diffusion should be hindered to some extent as compared to zones in between. As ions are implanted evenly through PPP films whatever their nature -amorphous or crystalline at the microscopic level- they are forced to enter as well in the crystallites where exodiffusion process is very slow. Yet as soon as they get out from these crystalline domains they may migrate much more easily at the "grain boundaries" corresponding to the zones in between crystallites described above, with a diffusion coefficient several orders of magnitude larger.

We think that these assumptions might explain in part the difference between stability of PPP doped by ion implantation as compared to doping by other methods in which ions penetrate into the material by a diffusion process and therefore reach only regions close to the diffusion source : thus the way back -exodiffusion- is somewhat easier in that case than when ions are implanted. Further studies are currently carried out aiming at the determination of ion diffusion coefficients in PPP films doped by ion implantation.

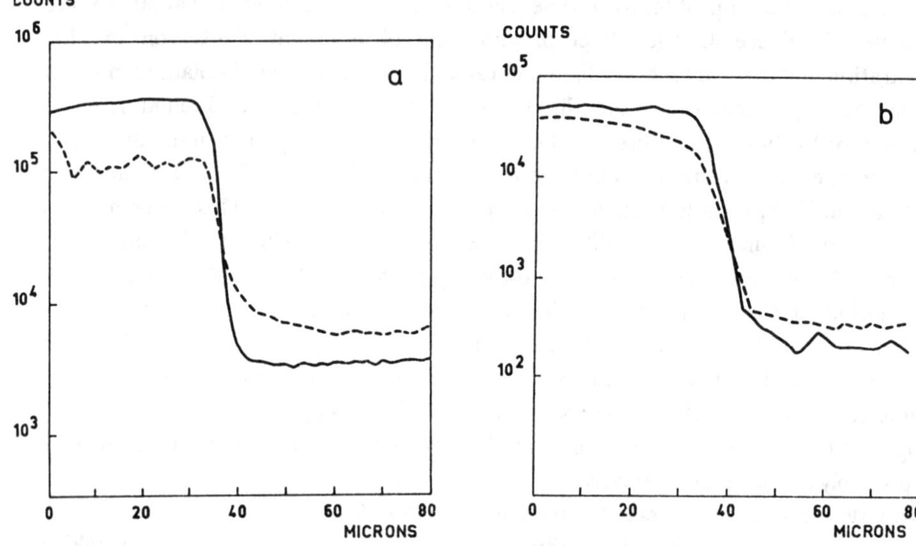

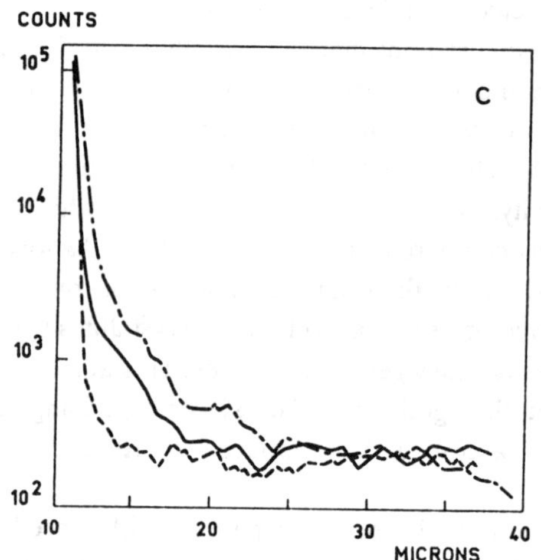

Figure 4. Experimental lateral profiles at 30 keV
a) Sodium : ———— as-implanted ; ————— annealed at 270°C 4 hours.
b) Iodine : ———— as implanted ; ————— annealed at 270°C 4 hours.
c) Cesium : 1 ————— non-annealed ;
            2 ———— annealed at 200°C 2 hours ;
            3 —.—.— annealed at 270°C 4 hours.

References

[1]   See for example :
Proceedings of the Sixth International Conference on Ion Beam Modification
of Materials, Tokyo, June 12-17 1988 in Nucl. Instr. and Meth. (1989), B 39.

[2]   MAZZOLDI P., ARNOLD G.W. Editors (1987)
Ion beam modification of insulators, Elsevier, Amsterdam.

[3]   a) ISOTALO H., STUBB H., KUIVALAINEN P. (1989)
"Electrical properties of ion implanted polythiophene"
Synth. Metals, 28, C 305.
b) DUROUX J.L., MOLITON A., FROYER G. (1988)
"Influence of implantation parameters on the poly(paraphenylene) electrical
conductivity", Nuclear Instr. and Meth., B 34, 450.

[4]   LIN S., SHENG K., BAO J., RONG T., ZHOU Z., ZHANG L., ZHU D.,
SHEN Z., YAN M. (1989)
"Ion beam modification of polyacetylene films"
Nucl. Instr. and Meth., B 39, 778 and references therein

[5]   MOLITON A., DUROUX J.L., RATIER B., FROYER G. (1988)
"pn$^+$ junction in an implanted electroactive polymer : poly(paraphenylene)"
Electronics Lett., 24, 383.

[6]   DUROUX J.L., HEJDUK A., MOLITON A., FROYER G., GAUNEAU M.
(1988), "Ion ranges in PPP and conductivity effects", Synth. Metals,
24,53.

[7]   FAUVARQUE J.F., PETIT M.A., DIGUA A., FROYER G. (1987)
"Electrochemical synthesis of poly(1,4-phenylene) films", Makromol. Chem.,
188, 1833.

[8]   BIERSACK J.P. (1982)
"New projected range algorithm as derived from transport equations"
Z. Phys. A. - Atoms and Nuclei 305, 95.

[9]   ABOULKASSIM A., PELOUS Y., CHEVROT C., FROYER G. (1988),
"Voltammetric studies on poly(paraphenylene) films obtained by electroreduction", Polymer Bull., 19, 595.

[10]   PRADERE P. (1985)
Thesis, Université Paul Sabatier, Toulouse.

THE ELECTROACTIVE NATURE OF POLYANILINE.
THE NATURE OF THE CHROMAPHORES

A P MONKMAN
Centre for Molecular Electronics
Applied Physics Group
University of Durham
Science Laboratories
South Road
Durham   DH1 3LE

ABSTRACT. Thin films of polyaniline (PANi) prepared electrochemically can readily be investigated insitu. In this way, both the oxidation state and level of protonation can be precisely controlled. Insitu UV/VIS/NIR and Resonance Raman spectra are presented here. These results show that the usual polaron/bipolaron models applied to other conducting polymers are not consistent in the case of PANi. The structures of the chromaphores associated with the observed 'dopant' induced optical absorptions in PANi are quite different, which is inconsistent with polaron theory. Resonance Raman spectra show that the structure of the chromophore giving rise to absorption at ca. 3eV is that of a simple p-di-substituted phenylene moiety, whereas that of the absorption at ca. 1.6 eV is consistent with a mixed benizonoid/quinoid radical (eg. a semiquinone moiety). In both cases the chromophores are localised. Therefore a new model for PANi is proposed.

1.   INTRODUCTION

The work reported in this paper comprises studies of the electronic properties of electrochemically prepared PANi, and the nature of the chromaphores on the chain. Further, the degree to which PANi protonation effects these properties was also investigated. The techniques used were insitu UV\VIS\NIR optical spectroscopy and insitu Resonance Raman spectroscopy (RRS). These techniques taken together give a coherent picture of the nature of the chromaphores along the PANi back bone and the ensuing electronic structure they give rise to.

2.   EXPERIMENTAL DETAILS

All PANi samples were synthesised electrochemically as thin films. The electrochemical conditions used were as previously reported [1], using a 2M HCl/aniline electrolyte. For optical studies, films were grown on ITO

J. L. Brédas and R. R. Chance (eds.), Conjugated Polymeric Materials:
Opportunities in Electronics, Optoelectronics, and Molecular Electronics, 273–283.

glass electrodes to facilitate transmission measurements. Films used for Raman studies were grown on Pt electrodes.

Cyclic voltammograms (CV's) of films deposited on Pt were recorded as a function of electrolyte acid concentration. All CV's were measured using an in house potentiostat. For continuity, the CV at each acid concentration was measured, and a series of CV's versus acid strength were thus built up for individual films. Before each measurement, the film was washed in distilled water and a batch of the new electrolyte. CV's were recorded starting with the weakest acid working up to the highest acid concentration. A scan rate of 50 mV/s was used throughout.

Insitu optical spectra were measured as a function of both applied oxidation potential and electrolyte acid concentration. The electrolytes used in these experiments were 2M, 1M and $^1/_4$ M HCl. All voltages quoted are with respect to SCE reference. A Pt/Rh coiled wire was used as an auxiliary electrode. Films deposited on ITO were washed in acetone and distilled water before being placed in the spectroelectrochemical cell. Samples were held at set oxidation potentials whilst spectra were recorded.

Resonance Raman spectra were measured using a 180° backscattering geometry. As the PANi samples were deposited on highly polished Pt, this method is particularly suited. RRS were recorded using several different laser excitation lines. The 2.71 eV and 2.41 eV lines from an Ar$^+$ ion laser were used together with the 1.96 eV (red) Helium Neon line. Typical laser beam powers incident on the sample were of the order 10 to 100 mW, in order to minimised polymer degradation. As before, samples were thoroughly washed before being placed in the Raman electrochemical cell, and held at the applied oxidation potential during collection of each spectrum. Only a 2M HCl electrolyte was used during RRS experiments.

3. DISCUSSION OF RESULTS

The effect of electrolyte acid concentration on the CV of PANi can be seen in Fig. 1. The first oxidation wave at ca. 0.2 V is seen to be little affected by acid concentration, and so must be only weakly dependent on polymer protonation. The small shift at the highest acid strengths reflects the amount of positive charge on the polymer chain, ie. it is more difficult to remove electrons from a system which already has a large positive charge. The increase in peak height at high acid strength is consistent with reduced IR drop caused by the increased conductivity of the PANi film.

The shift seen in the position of the second oxidation wave is far more dramatic, indicating a high degree of dependence on chain protonation. Huang et al [2] have shown that this shift is indicative of the loss of two protons per electron, consistent with the formation of neutral quinoid moieties during this second oxidation process.

On reduction, large changes are observed. At high pH, the reduction wave (ca. 0.6 - 0.8 V) is seen to be very sharp and symmetric, with a shoulder to the higher potential side. As the pH is lowered, the peak shifts to higher potential, indictive of greater chain protonation, and the shoulder is lost. At the highest pH (curve 1) the presence of a shoulder on the reduction peak suggests that two different moieties are

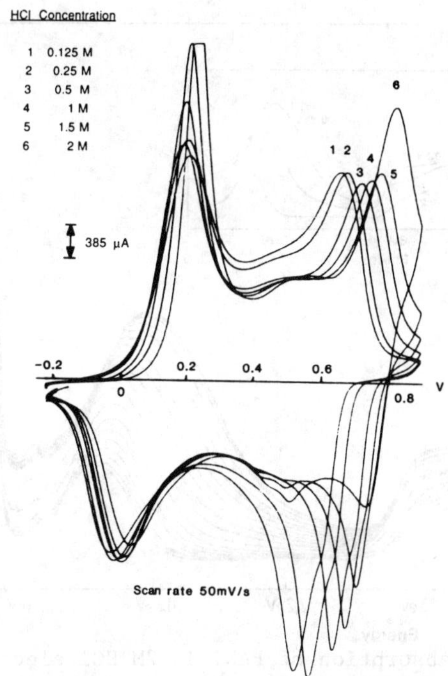

Figure 1.   Cyclic voltammograms of PANi as a function of electrolyte acid concentration.

being reduced, possible protonated and basic quinoid rings.

At low pH levels, 'new' redox peaks are seen to grow, at ca. 0.45 and 0.5 V, indicating that the formation mechanism of the moieties giving rise to these peaks must be pH dependent, as at all pH levels the films were oxidised to the same final potential. These new redox peaks are ascribed to the oxidation and reduction of degradation products, arising from hydrolysis of the polymer [3].

The insitu optical spectra of PANi in 2M, 1M and $^{1}/_{4}$ M HCl electrolytes respectively are shown in Figs 2-4. As can be seen from the spectra, the behaviour of PANi is extremely complex, and pH dependent. At the highest acid concentration used, Fig. 2, the peak observed at ca. 4.1 eV, monotomically decreased in intensity with increased oxidation, it also appears to blue shift slightly, by ca. 0.05 eV, between the range -0.2 to 0.3 V. At higher oxidation potentials, the peak disappears. This peak also seems to have a low energy shoulder at 3.9 eV, although this may be caused by the underlying absorption at ca. 3.65 eV. The peak at 3.65 eV is seen to grow in intensity and red shift with increased oxidation, shifting by some 0.2 eV from start to finish. In the oxidation range -0.2 to 0.3 V, a new peak, at ca. 3.0 eV grows, which then loses intensity on further oxidation. At these higher oxidation levels a further peak is seen, at ca. 2.8 eV. It is possible that the 3 eV peak red shifts to 2.8 eV, although from the shape of the spectra, this seems unlikely.

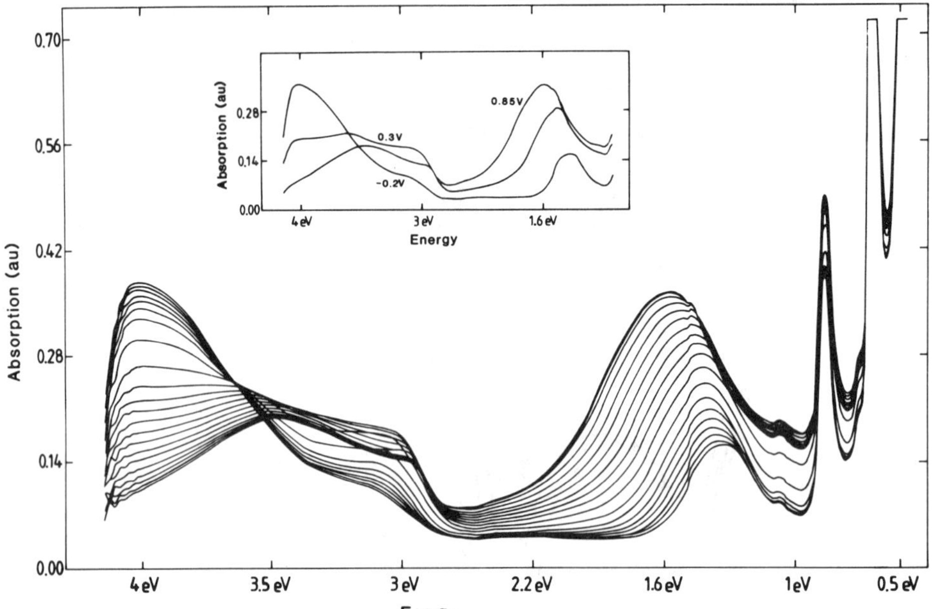

Figure 2. Insitu optical absorption of PANi in 2M HCl electrolyte.
Spectra were recorded every 0.05 V from -0.2 to 0.85 V. The insert
shows selected spectra.

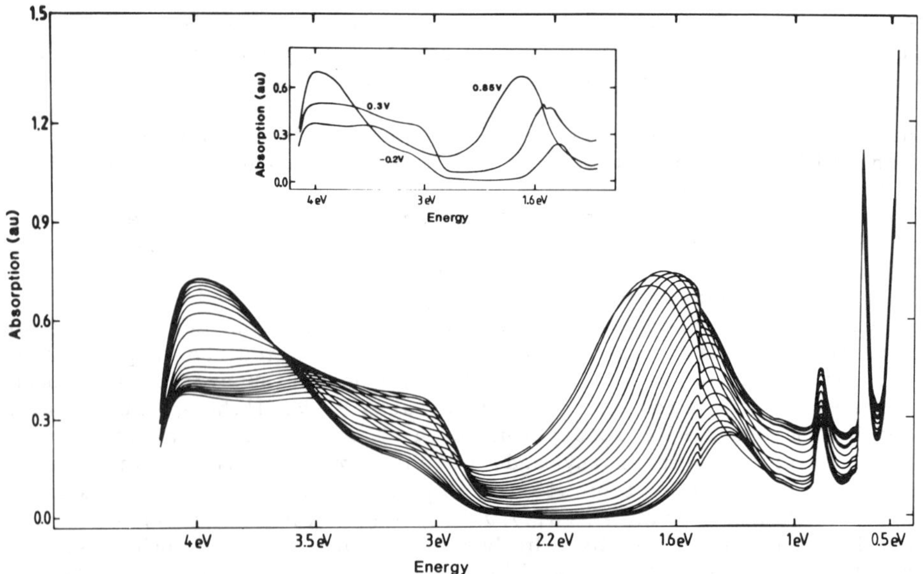

Figure 3. Insitu optical absorption of PANi in 1M HCl electrolyte.
Spectra were recorded every 0.05 V from -0.25 to 0.85 V. The insert
shows selected spectra.

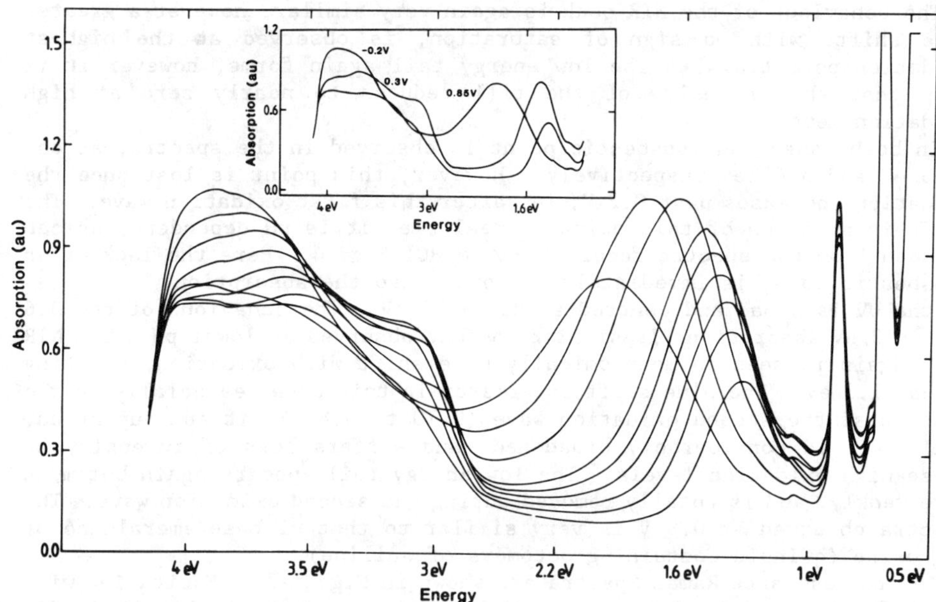

Figure 4. Insitu optical absorption of PANi in 1/4M HCl electrolyte.
Spectra were recorded every 0.1 V from -0.2 to 0.85 V, excluding -0.1
and 0.5 V. The insert shows selected spectra.

In the red/NIR region, an intense broad absorption is seen to grow with
oxidation.   The absorption forms at ca. 1.33 eV, as the film is
progressively oxidised the intensity of the absorption grows, and its peak
position monotomically blue shifts, ending up at 1.59 eV (0.85 V) in the
oxidised (0.85 v) state. Note, oxidation past 0.75 V does not cause any
loss of intensity, indicating that non of these moieties are degraded.
On the low energy side of this broad band a tail grows, the onset of its
formation is coincident with the first oxidation wave at ca. 0.0 V. This
absorption reaches a maximum at ca. 0.3 V, where it has approximately half
the oscillator strength of the NIR peak. Increasing the oxidation causes
a small loss in its intensity, but it is still quite strong at 0.85 V.
Although absorptions due to the electrolyte overlap with this region of
the spectrum, ie. the peaks at 0.86 eV, 0.63 eV and cut off at 0.52 eV,
it can be seen (in the windows between the electrolyte peaks) that the
tail intensity diminishes at low energy, ie. to about half of the maximum
value at 0.56 eV.
A very similar pattern is observed in molar HCl, Fig. 3. The 4.1 eV
peak again monotomically decreases with oxidation, although here the peak
is not entirely lost, the accompanying small blue shift and low energy
shoulder are also observed. The 3.65 eV peak behaves as before, but does
not become such a dominant feature. The 3 eV peak again forms during the
first oxidation wave, with accompanying loss of intensity at higher
oxidation more prominent.   The 2.8 eV peak is extremely weak.

278

The behaviour of the NIR peak is again very similar, however a greater blue shift, with no sign of saturation, is observed at the highest oxidation potentials. The low energy tail again forms, however it is seen that the intensity of the tail reduces to nearly zero at high oxidation levels.

In both cases, an isosbestic point is observed in the spectra, at ca. 3.63 eV and 3.67 eV respectively. However, this point is lost once the oxidation increases past 0.2 V, ie. after this first oxidation wave. The shift in position of this point is real, ie. it is pH dependent, as can be seen from the spectra record in $^1/_4$ M HCl Fig. 4. Here the lack of an isosbestic point is immediately obvious, also the absorption in the UV is broad and centred at ca. 4.05 eV. The behaviour of the 3.6 eV and 3 eV absorptions is similar to that observed at lower pH. The NIR peak again is seen to monotomically blue shift with oxidation, starting at ca. 1.3 eV. The blue shift is greater in this case, especially in the range past the second oxidation wave (0.70 to 0.8 V), it ends up at ca. 2 eV. It is considerably broadened, and suffers loss of intensity at these high oxidation levels. The low energy tail appears again but much more weakly, and is totally removed during the second oxidation wave. The spectra observed at 0.8 V is very similar to that of base emeraldine or nigrosine (quinoid containing octomers of aniline).

Insitu Resonance Raman Spectra are shown in Figs 5-7. Excitation with the 2.71 eV line, Fig. 5, show a RR spectrum characteristic of a p-di-substituted phenylene moiety. As can be seen in the series of spectra, the peaks ascribed to the presence of semiquinone radicals (ca. 1380 cm$^-$^{1}) are very weak, showing no enhancement. From the insitu optical spectra, it is seen that an optical absorption at ca. 3 eV, grows in and then dies away with increasing oxidation. This behaviour is also reflected in the RR spectra. The resonant enhancement rises and falls with increasing oxidation in a similar manner. Thus, this electronic absorption is due to a para-phenylene diamine (like) chromaphore, not a semiquinone radical.

It can be seen from the optical spectra that the 2.41 eV excitation line sits between two optical absorptions. Hence, at low oxidation potentials, up to 0.2 V, the line will be near to resonance with the absorption at ca. 3 eV, as in the previous case, and consequently a similar RR spectrum is observed. However, as the oxidation potential is increased further, the optical absorption of the film changes, and consequently new peaks appear, while others shift. These changes in the RR spectra have been proposed to be as a result of structural changes on the polymer backbone of a single moiety in resonance with the excitation line. However at higher oxidation potentials this excitation line comes into resonance with the NIR optical absorption, which has blue shifted with increased oxidation.

When in resonance with this lower energy absorption, a more complex spectrum is observed, no longer indicative of a simple p-di-subsituted phenyl ring. The characteristic bands of the di-substituted phenyl are present, eg. 1610, 1185 and 810 cm^{-1}, along with new peaks at 1565 and 1470 cm^{-1}, and a line at 1160 cm^{-1} which grows at the expense of the 1185 cm^{-1} line. These peaks are ascribed to the presence of quinoid imine

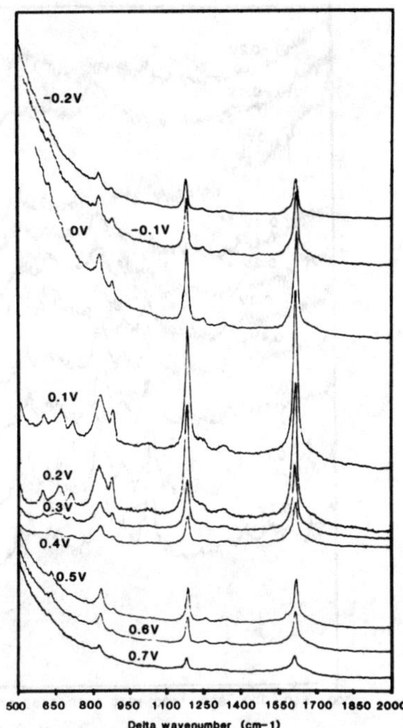

Figure 5. Insitu Resonance Raman spectra of PANi. 2.71 eV
excitation line.

moieties, as observed in the Raman spectra of model compounds [4]. In
addition, a broad band is seem centred at ca.
1315 cm^{-1}. This has been ascribed to the semiquinoid moiety [4]. The
line at ca. 1500 cm^{-1}, due to the presence of NH_2 stretch, indicates that
the chromaphore is fully protonated.

Unfortunately, it was not possible to oxidise the film any higher than
0.75 V, which is the onset of the second oxidation wave at this acid
concentration. The spectra become swamped by a large fluorescent
background. Such behaviour is observed for other conjugated polymers [5],
when the chains are degraded. Hence, the fluorescent background is
ascribed to film degradation.

On going to the 1.96 eV excitation line, the onset of resonance with
the red absorption of the polymer should occur earlier, ie. at lower
oxidation potentials. This is seen to be the case. However, as a HeNe
laser source had to be used, the available power was low and accordingly
the spectra at low oxidation potentials are extremely noisy. These low
oxidation state spectra show the evolution of broad band complex spectra
as expected. At higher oxidation potentials the spectra are very similar
to those observed in the previous experiment. This confirms that the
changes seen using the 2.41 eV excitation are resonance effects. Thus,
the optical absorption in the red, is ascribed to a chromaphore, proposed
to consist of a protonated amine-imine (two ring) radical.

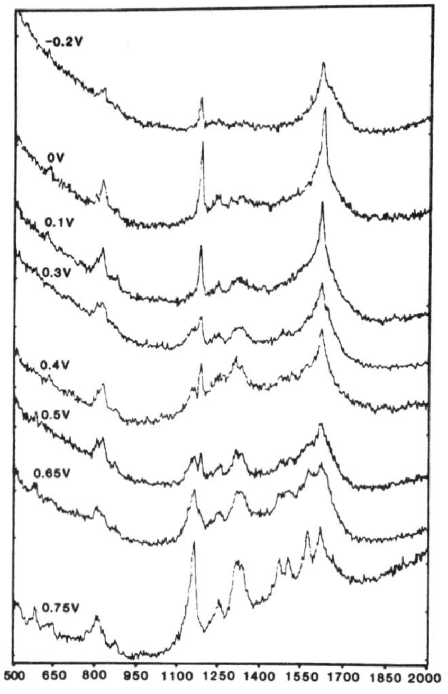

Figure 6. Insitu Resonance Raman spectra of PANi. 2.41 eV excitation line.

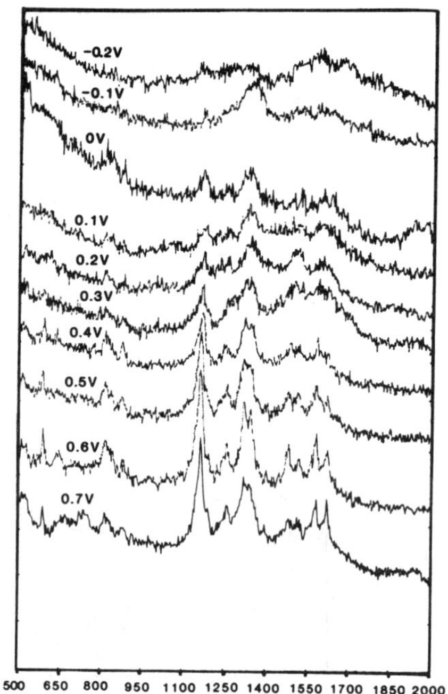

Figure 7. Insitu Resonance Raman spectra of PANi. 1.96 eV excitation line.

4. CONCLUSION

In all the optical spectra, of polymer and oligomers [6], both protonated and nonprotonated [7], an absorption at ca. 4.1 eV is observed. On comparison with model compounds, this absorption can be ascribed to the $\pi \to \pi^*$ transition, ie. formation of a phenyl ring exciton. During the first oxidation step it is seen that the 4.1 eV transition loses oscillator strength, consistent with the removal of HOMO electrons. Peaks at 3 eV and in the NIR form on oxidation. This behaviour is also observed in model semiquinones [6]. RRS confirms that the chromaphore giving rise to the NIR band has a semiquinoid like structure. ESR has also shown that spins are created during the first oxidation step [8]. However, the data form the investigation of the 3.0 eV peak is inconsistent with this interpretation. Raman spectroscopy shows the chromaphore associated with this transition to have a simple p-di-substituted phenyl ring structure. This cannot be reconciled if the 3eV and NIR absorptions are polaron bands, as can be seen from the definition of a polaron. It is known that polarons form on a local region of the polymer chain, ie. a single chromaphore, and so optical transitions between polaron energy levels should couple with the same chromaphore structure. The observed data would rather suggest that the 3 eV absorption is due to the formation of a localised self-trapped exciton on a phenyl unit, and the NIR band is due to the absorption of the semiquinoid moieties. Raman spectra suggests that these moieties are relatively localised and so are probably excitonic as well. The mechanism for the self trapping process would be the twisting out of plane of the ring on which the exciton was formed. If the ring twisted such that it was at 90° to its neighbours, the excitation would be well screened from the surrounding π system.

At higher oxidation levels, further changes are seen in the spectra. A new peak forms at 3.65 eV, and the 3 eV peak is either lost or shifts to 2.8 eV, where a new peak is observed. The NIR band is seen to blue shift further, into the visible, and its's low energy tail disappears in all cases except in the 2M HCl spectra. It is seen in spectra taken in $^1/_4$ M HCl at high oxidation levels, that a peak at 2 eV dominates. A gain in the intensity of the 4.1 eV absorption is observed together with the loss of all absorption in the region 3.6-2.6 eV. This spectrum is very similar to that of neutral quinoid/imine oligomer [9] (main absorption 2.2 eV, ascribed to the formation of a self trapped exciton). This indicates that deprotonation occurs during the second oxidation process, and that at low pH, protonation locks in some moieties preventing them from being oxidised further, leading to a mixed optical spectrum at low pH. Note also that decreasing pH causes the onset of the second oxidation wave to shift to higher potentials and thus it is harder to oxidise the polymer a second time.

An isosbestic point is observed during the first oxidation step at low pH, but is not seen at high pH ($^1/_4$ M HCl). This would indicate that at low pH, protonation causes only a single species to form during oxidation, ie. a simple A->B reaction. Once past this first wave, ie. ca. 0.2-0.3 V, the isosbestic point is lost, indicating that a mixed reaction is taking place. This could be due to the effects of deprotonation which leads to the formation of neutral quinoid moieties, as well as the formation of new charged species during the second oxidation step. The chemistry involved during the second oxidation process is complex and not well understood.   Further work is required to fully understand the role of the protons.

RRS has been used to probe the structures of the chromaphores associated with the observed optical absorptions of PANi. Excitation at 2.71 eV probes the optical absorption peak at 3 eV. The spectra show only those peaks ascribed to para-di-substituted phenylene moieties, hence the chromaphore must have a structure of a para-phenylene diamine unit; no semiquinone bands are observed. Those bands which are seen are sharp, consistent with a π electron system localised on one ring.

Using a lower energy excitation line, 2.41 eV, a mixed spectrum is seen. This arises because the excitation line lies between two oxidation dependent optical absorptions. At oxidation levels below 0.2 V, a p-di-substituted phenylene spectrum is seen. At higher oxidation levels, new Raman bands are seen to grow into the spectra. The observed mixed spectra can be seen to consist of a part ascribed to a para-di-substituted phenyl moiety, as before, bands consistent with the presence of a quinoid/imine moiety, and a band at 1315 cm^{-1}, ascribed to the presence of semiquinoid radicals. Spectra recorded with a 1.96 eV excitation line further support the hypothesis that the mixed spectra are caused by a changing resonance effect, not a structural change of a single chromaphore.   Thus, the chromaphore giving rise to the optical absorption in the red is proposed to have a structure of a radical imine-amine (protonated) two ring unit.

As before, the Raman bands are sharp indicative of a localised system.

These findings are in agreement with those of Wudl et al [10] who have shown that the physical properties of octomers of aniline are nearly identical to those of PANi, and thus involve only relatively localised species.

SUMMARY

The work presented here gives a model of PANi which is different to those proposed for other conducting polymers. Investigation of the oxidation induced chromaphores, and their dependency on protonation level show the polaronic model so far proposed for PANi to be inconsistent with the experimental data. RRS shows that the two main absorption bands formed in the visible during initial oxidation arise from different chromaphores on the back bone, one a p-di-substituted-phenyl ring, the other on imine/amine (two ring) radical semiquinone. If such absorptions were caused by transitions between polaron levels, they would have to show the same chromaphore structure. Similarly, they would have to behave in a similar fashion with respect to oxidation and protonation. This is not the case. An absorption band has also been observed at ca. 0.4 eV [11] which behaves in a similar fashion to the 3.6 eV band, with respect to oxidation, and is pH dependent. This band is very similar to those observed in charge transfer salts, which arise from the charge transfer process. It is therefore believed that the 0.4eV band in PANi is caused by the internal transfer of charge from a reduced ring to an adjacent oxidised ring. Such a moiety could be considered as a small, Holstein polaron. The absorptions in the visible region of the spectrum can be considered as excitonic in nature. Hence, conduction in PANi is mediated by the movement of these (Holstein like) charge transfer polarons.

In the fully oxidised form, PANi, exhibits an optical spectrum very similar to that of nigrosine base (fully oxidised octomer) and this is consistent with the hypothesis that during the second oxidation step PANi is deprotonated. In highly acidic electrolytes, some of the protonated form remains locked in during this second oxidation step. Note however, that in such highly acidic media, the second oxidation potential shifts to higher potentials, therefore it is harder to oxidise the polymer.

This very different nature of PANi arises from the heteroatomic coupling within the conjugation path, and the interplay of protonation and ring twist.

ACKNOWLEDGEMENTS

I would like to thank both the SERC and CEGB for funding this research. GEC Marconi for the use of their Lambda 9 Spectro-photometer. Further I would like to thank Dr D N Batchelder and R J E Smith for their help in obtaining the Raman spectra. Finally I wish to thank Professor D Bloor and Dr G C Stevens for all their useful comments and discussion.

REFERENCES

1. A P Monkman, D Bloor, G C Stevens J C H Stevens. J Phys. D: Appl. Phys. 20 (1987) 1337.

2. W S Huang, B D Humphrey and A G MacDiarmid, J Chem. Soc. Faraday Trans., 82 (1986) 2385. J C Chiang and A G MacDiarmid, Synth. Met., 13 (1986) 193.

3. A P Monkman, Thesis University of London. (1989)

4. Y Furakawa, F Ueda, Y Hyodo, I Harada, T Nakajima and T Kawayoe, Macromolecules, 21 (1988) 1297.

5. R K Day, Thesis University of London. (1988)

6. H Linschitz, J Rennet and T M Korn, J Am. Chem. Soc., 55 (1933) 1488.

7. F L Lu, F. Wudl, N Nowak and A J Heeglar, J Am. Chem. Soc. 108 (1986) 8311.

8. A P Monkman, D Bloor, G C Stevens, J C H Stevens and P Wilson, Synth Met. 29 (1989) E277. A P Monkman, D Bloor, and G C Stevens, J Phys. D: Appl. Phys. Lett. in press.

9. C B Duke, E M Conwell and A Paton, Chem. Phys. Lett., 131 (1986) 82.

10. F Wudl, R O Angus Jr., F L Lu, P M Allemand, D J Vachon, M Nowak, Z X Lui and A J Heegar, J Am. Chem. Soc., 109 (1987) 3677.

11. A P Monkman, Polymer, in press.

TRANSPORT IN ORIENTED POLYACETYLENE

M.GALTIER*, M.ROLLAND**, A.MONTANER*, A.ALIBENAMARA*
* G.D.P.C. (UA 233) Université des Sciences et Techniques
   du Languedoc – 34060 Montpellier Cedex

**Laboratoire   de Physique Moléculaire et Cristalline
   Université des Sciences et Techniques du Languedoc
   34060 Montpellier Cedex

ABSTRACT.   Highly oriented polyacetylene has been synthesized by Akagi's
method. After p and n doping they present a d.c. conductivity of about
20,000 Ω^{-1}cm^{-1} at room temperature. Between T = 300 K and T = 20 K σ
decreases slowly (one order of magnitude) for quasi metallic system and
drops between T = 20 K and T = 4K. For mean doping levels the temperature
dependence is well fitted by a Sheng formula. Optical conductivity has
been compared to d.c. conductivity for p doped $(CH)_x$. This leads to the
conclusion that d.c. conductivity is dominated by interchain or inter-
fibril hopping while optical conductivity is dominated by intrachain
mechanism.

INTRODUCTION

In the last years several routes have been proposed in order to synthesize highly ordered polyacetylene, leading to quasi metallic materials,
most of the papers dealing with stretched oriented polymer. In this paper
we report transport properties of polyacetylene prepared by Akagi's method and compare them to other materials ones.

1. THE MATERIAL

Polyacetylene studied in this paper was synthesized following Akagi's
procedure |1| using a nematic crystal as a solvent of the catalyst. The
combined effect of gravitational flow and magnetic field (47 KG) was used
to orientate the liquid crystal |2| .
     As grown self standing films have thicknesses between 0.5 and 2
micrometers. They are partially cis (60%) and partially trans (40%) due
to the temperature of the synthesis (10°C). They have been characterized
by different techniques as Infrared and Raman spectroscopies, X-Ray,
T.E.M and electronic diffraction |3| , leading to the following main
features.

285

J. L. Brédas and R. R. Chance (eds.), Conjugated Polymeric Materials:
Opportunities in Electronics, Optoelectronics, and Molecular Electronics, 285–291.
© 1990 *Kluwer Academic Publishers. Printed in the Netherlands.*

1.1. Undoped samples

They present well aligned fibrils with lengths varying between some micrometers to 200 micrometers. The misorientation of the fibrils spreads over 20° on each side of the average direction.

The infrared absorption anisotropy ($A_{//}/A_{\perp}$) is about 10.

The fibrils are built with several microfibrils with well crystallized domains of about 120 Å in the chains direction and 50 Å perpendicular to the chains.

1.2. Doped samples

For iodine doped samples, a good orientation remains with mean doping levels. On the other hand the crystalline order decreases with high doping levels (5 to 15 %).

2. EXPERIMENTAL

2.1. Doping

Iodine doping was achieved in iodine vapour. The samples are kept during 30 minutes in the vapour and then pumped so as to remove eventual molecular iodine crystallized on the surface. The doping level, expressed hereunder in iodine by CH unit was determined by wheight uptake, and monitored by doping temperature.

Doping with donors is more difficult. We performed potassium doping in the vapour phase as already described |4| , |5| . The sample is introduced in glass vessel together with freshly purified potassium, and then heated in a temperature gradient. The sample temperature is 10°C higher than potassium one which is fixed to 160°C. This avoids condensation of potassium vapour on the polymer.

The vessel is connected to secondary vacuum, and the experiment is controlled by measuring the resistivity of the sample. The doping process is rather fast, but it is followed by an other process, rather slow, which is a reorganization of the doped species. In both processes, the conductivity increases. We never observed a decrease in conductivity as reported by some authors |6| even after three days at that temperature.

It is not convenient to vary the doping level in that kind of experiment because at lower temperature the vapour pressure is too low to achieve doping and at higher temperature polyacetylene begins to be damaged. As a consequence the doping level is about 0,1 K/CH for all samples studied.

Handling n doped polyacetylene needs great caution because of the ultra fast oxydation by atmosphere. The resistivity measurements reported in this paper were done in the same vessel as doping, and due to this reason are limited to nitrogen temperature.

2.2. Resistivity measurements

Resistivity measurements were done by direct U versus I techniques.

Typical samples are 10 x 2 x 0.001mm, with the long edge along fiber direction. Four contacts are pasted with electrodag or silver paint using gold or copper wire. In all cases we have checked the linearity of the U versus I function, which is an indication of the doping homogeneity.

2.3. I.R. Measurements

Infrared absorption was measured on a Bruker Fourier transform spectrometer, with polarized light. The detector is a germanium bolometer operating at 2.7 K. This allows transmission measurements down to $10\,cm^{-1}$ from room temperature to about 30 K.

3. RESULTS AND DISCUSSION

3.1. d.c. Conductivity

Temperature dependence of the conductivity is reported in fig.1 for iodine doped species with various doping levels, and potassium doped species with 0,1 K by CH unit. In all cases the conductivity increases with temperature, with the following remarks.

Room temperature conductivity can reach values higher than $10^4\ \Omega^{-1}cm^{-1}$, but significantly lower than those reported by Naarman et al |6|. Our values are in the same range as those reported by Epstein or Mac Diarmid in this conference.

Temperature dependence does not show the same behaviour for quasi metallic materials ($y \geqslant 0.1$) and for other ones. For quasi metallic system, the conductivity decreases slowly by cooling from room temperature, then drops to rather low values in a rather narrow range of temperature. For $(CH, I_{0,15})_x$ for instance, the conductivity decreases by two orders of magnitude between 20K and 4K while the decrease is only one order of magnitude betweween 300K and 20K. This result is very similar to the one obtained by Schimmel et al |7| with Naarman's films.

For lower doping levels, the temperature dependence of the conductivity is well described by Sheng's Model |8| i.e.

$$\sigma(T) = \sigma_\infty \exp\left(-\frac{T_1}{T+T_o}\right)$$

The parameters deduced from the fit are listed in table 1

doping Level	$\sigma_\infty(\Omega^{-1}cm^{-1})$	$T_1(K)$	$T_o(K)$
0.02.7	13.7	1227	9,2
0.04	687	1190	82
0.07	1926	663	54

As can be seen in fig.1 where the fitted functions are reported together with the experimental values, the agreement is quite good.

288

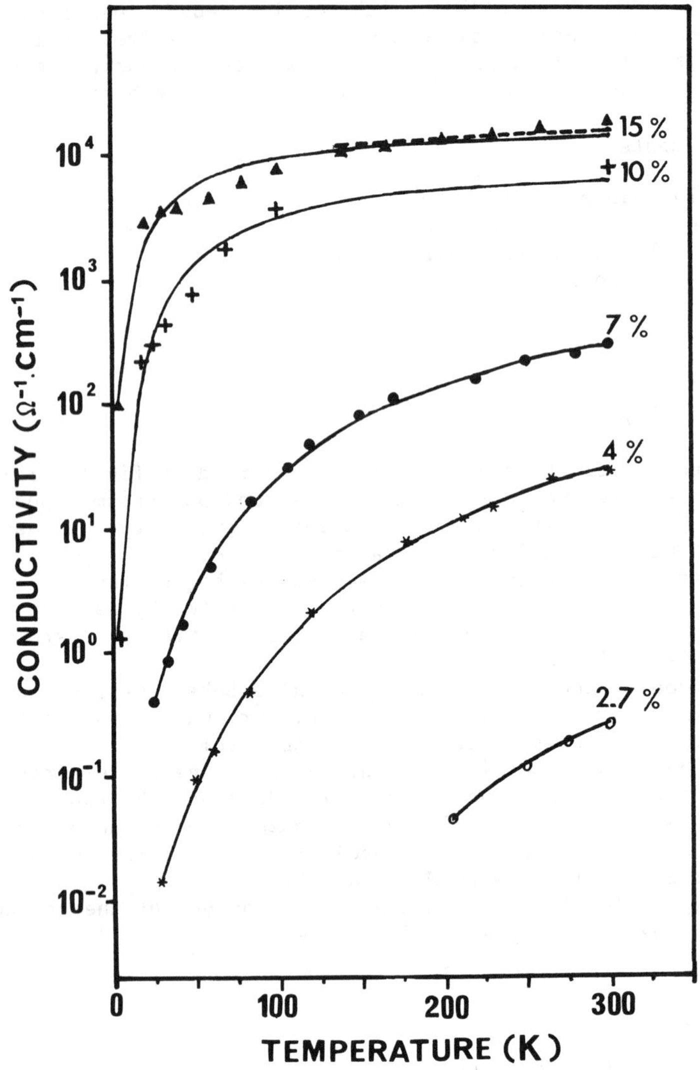

Figure 1. ▲,+,●,*, o Experimental values
Solid line : I_2 doped $(CH)_x$ d.c. conductivity fitted with Sheng formula
Dashed line: Experimental curve , K doped : $(CH,K_{0,1})_x$

It is difficult to insure that Sheng's model is a good microscopic description of our system, even if the fit is good. But it is reasonable to conclude that conductivity is limited by tunelling through barriers.

3.2. Optical conductivity

It is interesting to try to compare d.c. conductivity to optical conductivity in order to see if the two mechanisms are limited by the same processes. In fact two difficulties arise in this project. The first one is that optical conductivity contains other contributions than the motion of free carriers. The second one is that the determination of optical conductivity is not an easy experimental problem, because it supposes the determination of the complex optical index.

Fortunately the two difficulties can be overcome at the same time by inspection of the low frequency limit of the I.R. transmission of a thin film. Clearly, this limit is free from resonant contribution and involves only the motion of free carriers. On another hand, this limit can be calculated as

$$\lim_{\omega \to 0} T = \frac{1}{\left(1 + \frac{\sigma d}{2 \varepsilon_0 C}\right)^2} \tag{1}$$

where d is the thickness of the film , σ its conductivity
       C the velocity of the light
This limit is reached when the optical thickness of the film is much smaller than the wavelength λ. In the case of conducting systems, the optical index ñ is dominated by conductivity

$$\tilde{n} = \left(\frac{\sigma}{\varepsilon_0 \omega}\right)^{1/2} \exp\left(-i \frac{\Pi}{2}\right)$$

So that the limit is reached when

$$\left(\frac{\sigma}{\varepsilon_0 \omega}\right)^{1/2} d << \frac{2 \Pi C}{\omega}$$

for $d \sim 1$ micrometer and $\sigma \sim 10^4 \Omega^{-1} cm^{-1}$ this condition is equivalent to $\lambda >> 200$ μm.
The transmittance of an iodine doped sample is reported for different temperatures fig.2a in the range 15–50 cm^{-1}. To use equation (1), we have extrapolated the transmittance to zero frequency and obtained values of the conductivity which are probably underestimated. A second set of values of the transmittance was used replacing crudely the zero frequency limit by the value measured at 15 cm^{-1}. This is an over estimation of the conductivity.

The overestimated and underestimated values are plot fig.2b as a function of temperature, together with the d.c. conductivity. Clearly the two values exhibit the same behaviour which is quite different from d.c.

The optical conductivity is higher than d.c. one by a factor varying from 5 at room temperature to 10^3 at low temperature. Neverthless, the optical conductivity is an activated process, since it increases with temperature, but the barriers controlling this mechanism are not the same as those controlling d.c. motion.

It seems very likely that d.c. resistivity is dominated by interchain or interfibrils hopping, while optical conductivity is dominated
by intrachain mechanism.

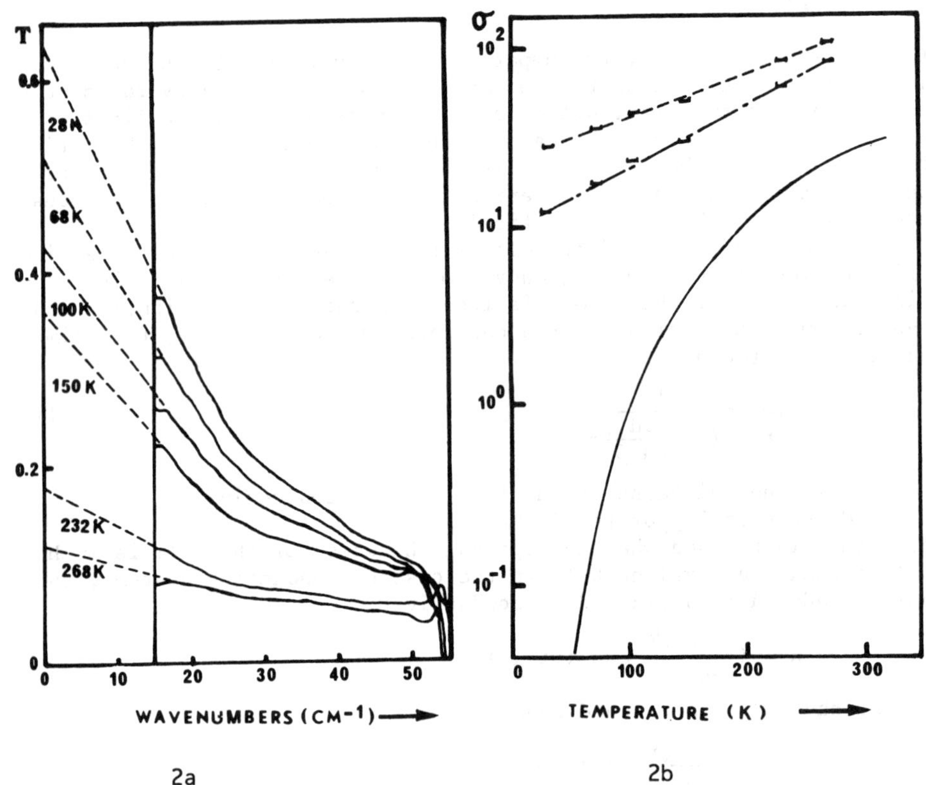

Figure 2a. Far infrared transmittance of I_2 doped sample $(CH\ I_{0.04})_x$
for different temperatures.
------ Extrapolation to zero frequency

Figure 2b. Evolution of d.c. and optical conductivities of $(CH,\ I_{0.04})_x$
with temperature
---- optical conductivity for $\omega = 15\ cm^{-1}$
—·— optical conductivity for $\omega = 0\ cm^{-1}$
——— d.c. conductivity

CONCLUSION

Akagi's method to synthesize polyacetylene provides a material with
transport properties very similar to Naarman's material ones, despite
the fact that the macroscopic orientation of the fibrils is much poorer.
This shows that local properties (cristallinity, Sp_3 content...) are the
relevant parameters for transport.
    In all situations under our investigations transport mechanisms are
dominated by thermally activated processes and never by electron phonon
interactions.

REFERENCES

1. Akagi K., Katayama S., Shirakawa H., Araya K., Mukoh A.,
    Narahara T., Synth.Met. 14-199 (1986)

2. Montaner A., Rolland M., Sauvajol J.L., Galtier M., Almairac R.,
    Ribet J.L., Polymers, 29, (1986) 1101

3. Montaner A., Rolland M., Sauvajol J.L., Meynadier L., Almairac R.,
    Ribet J.L., Galtier M., Gril C., Synth. Met.28, (1989) D 19-D 25

4. Moses D.,Colaneri N.,Heeger A.J., Sol. Stat. Comm. 58 (1985) 535

5. Ghanbaja J.,Maréché J.F.,Mac Rae E.,Billaud D., Solid Stat. Comm.
     58 (1986) 87

6. Naarman M., Theophilou N., Synt.Met. 22 (1987) 1.

7. Schimmel Th. et al, Synthetic Met. 28 (1989) D 11

8. Sheng P. , Phys. Rev. B 21 (1980) 2180

TRANSIENT PHOTOCONDUCTIVITY IN POLYACETYLENE AND MOLECULAR ELECTRONIC ASPECTS

Siegmar Roth, Jürgen Anders, and Jürgen Reichenbach
Max-Planck-Institut für Festkörperforschung,
Heisenbergstr. 1, D-7000 Stuttgart 80, FRG

ABSTRACT

Experimental data on the transient photoconductivity of trans-poly-acetylene are reviewed. Emphasis is layed on the schubweg and its dependence on temperature and on the applied electric field. The implications on "soliton switching" in molecular electronics are discussed.

INTRODUCTION

Polymers with conjugated double bonds (polyenes) have been intensively studied during the last decade, both because of the high metal-like electric conductivity which these substances acquire after doping - recently conductivity values nearly as high as that of copper have been obtained [1] - and because of the novel physics associated with electronic excitations strongly coupled to the lattice: Excess electrons in the conduction band (π^*-band) or holes in the valence band (π-band) distort the lattice (pattern of bond lengths) so strongly that actually new quasi-particles are formed, often described as "solitons" or as "polarons" [2].

As one of the approaches to molecular electronics it has been proposed to use conjugated polymer chains as "molecular wires" along which solitons run or to store information on polyene segments by modifying the pattern of bond alternations ("soliton switching") [3]. There has been a long discussion on whether excitations in polyenes are "really" solitons or not, but the molecular electronic switching process is only marginally touched by this controversy. On the other hand it is important to know about the mobility of these excitations: They must be mobile enough along the chains to ensure communication between functional groups and hopping between chains or over interruptions of the conjugation within a chain should be sufficiently rare to allow for a certain selectivity of switching.

293

J. L. Brédas and R. R. Chance (eds.), Conjugated Polymeric Materials:
Opportunities in Electronics, Optoelectronics, and Molecular Electronics, 293–297.
© 1990 *Kluwer Academic Publishers. Printed in the Netherlands.*

In this report we summarize our results on the motion of photoexcitations in polyacetylene and discuss the implications for molecular electronics.

Experimental Results

The experiment is that of transient photoconductivity. It is carried out on undoped trans-polyacetylene, this substance being the most simple of all polyenes. Gold electrodes are evaporated on the sample, across which an electric field of some 10^4 V/cm is applied. Short light flashes from a pulsed laser are shone onto the sample. If a photon is absorbed by a polyene chain an electron is lifted from the π to π^*-band, leaving back a hole in the former and creating an electron in the latter, similar to the photogeneration of an electron-hole pair in a classical semiconductor. In contrast to a classical semiconductor, however, the hole and the excess electron distort the underlying lattice severely, so that it is justified to speak of new quasi-particles (polarons, solitons). The photogenerated charged particles are seperated by the applied electric field and as long as they move a photocurrent flows in the external circuit.

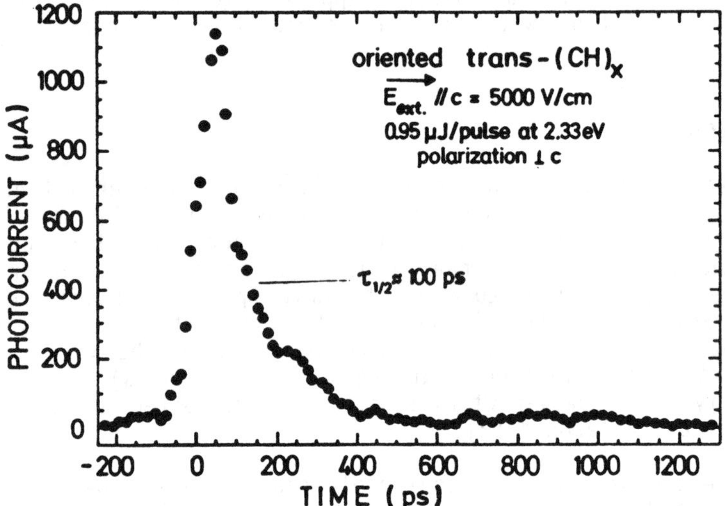

<u>Fig. 1:</u> Fast component of transient photoconductivity in undoped trans-polyacetylene [4]

Fig. 1 shows the photocurrent produced by laser pulses of 20 ps [4]. The time resolution of the electronic system in the external circuit is about 50 ps, in agreement with the signal rise-time in the figure. The photosignal decays within about 100 ps, but it is not clear whether we have really measured a "carrier life time" or whether the signal decay is distorted by the detecting circuit. In any case, the time integral over the photocurrent will be proportional to total charge displacement in the sample and if the quantum yield for photogeneration of carriers is known the schubweg can be figured out. (The schubweg is the average migration distance of a carrier in an electric field.)

Fig. 2 presents a more systematic study of the photocurrent in polyacetylene after the sample has been hit by light pulses. Here both axis scales are logarithmic and the current decay is followed from ns to ms. Now we see that the photosignal consists of two parts, a fast component decaying linearly in time and a slow component with a square-root behaviour. If the fast component is extrapolated to shorter times it matches the peak of Fig. 1. Fig. 2 also shows the photosignal at low temperatures (100 K). We notice that the slow component can be frozen out whereas the fast component is indipendent of temperature.

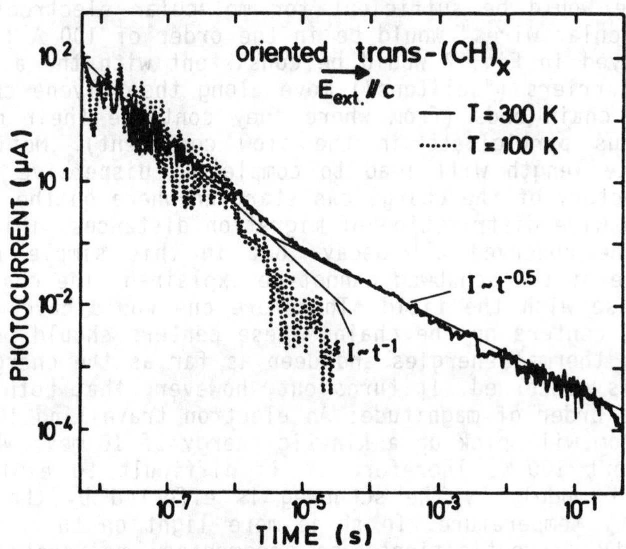

<u>Fig. 2:</u> Transient photoconductivity in trans-polyacetylene studied in an extended time regime [4]

The fast component of the photoconductivity has been carefully studied as a function of temperature, of the applied electric field, of the orientation of polyene chains with respect to the applied field (in stretch-oriented polyacetylene), of the polarisation of the laser light, and of the light intensity. Fig. 1 shows the photocurrent for the case where the applied field is parallel to the chain direction. For perpendicular orientation the signal almost vanishes in the noise and is at least by two orders of magnitude lower. The peak light (and the time integral over the peak) has been found to be proportional to the applied field, thus the schubweg increases linearly with the field.

DISCUSSION

It is reasonable to assume that the fast component of the photoconductivity is related to carrier motion along the polyene chains, whereas the slow component corresponds to hops from chain to chain (and perhaps over obstacles within a chain). Thus the fast component is what one would like to use in molecular electronics and the slow component will destroy the stored or processed information. (If this interpretation is correct, the fact that the slow component can be cooled away is good news.)

From the signal in Fig. 1 a schubweg of 400 Å has been calculated. Such a distance would be sufficient for molecular electronics. (The length of "molecular wires" would be in the order of 100 Å.) The t^{-1}-behaviour observed in Fig. 2 would be consistent with the assumption, that the photocarriers ("solitons") move along the polyene chains and are trapped at chain ends (from where they continue their motion by hopping and thus participate in the slow component). Motion along chains of finite length will lead to completely dispersive transport since the trajectory of the charge can start anywhere on the chain and there will be a wide distribution of migration distances. This is consistent with the observed t^{-1} decay. But in this simple model the field dependence of the schubweg cannot be explained. The chain length does not increase with the field. Therefore one would have to assume some scattering centers on the chain. These centers should be shallow with respect to thermal energies and deep as far as the energy of the applied field is concerned. It turns out, however, that both energies are of the same order of magnitude: An electron travelling 100 Å in a field of 10^4 V/cm will pick up a kinetic energy of 10 meV, which corresponds to about 100 K. Therefore it is difficult to explain in a simple transport model why the schubweg is effected by the electric field but not by temperature. To shine more light on this, transient photoconductivity investigations on segmented polyacetylene are planned, in which the chain length is an additional parameter that can be varied. Another approach is to study polyene oligomers of definite chain length and to investigate their prospects for molecular electronics [5].

REFERENCES

[1] H. Naarmann: in "Electronic Properties of Conjugated Polymers (Kirchberg II)", H. Kuzmany, M. Mehring, S. Roth (Eds.), Springer Series in Solid-State Sciences, Springer Verlag Heidelberg, 1987, p. 12

H. Naarmann, N. Theophilou: Synthetic Metals $\underline{22}$, 1 (1987)
Th. Schimmel, W. Rieß, J. Gmeiner, G. Denninger, M. Schwoerer,
H. Naarmann, N. Theophilou: Solid State Commun. $\underline{65}$, 1311 (1988)

[2] S. Roth, H. Bleier: Advances in Physics $\underline{36}$, 385 (1987)
A.J. Heeger, S. Kivelson, J.R. Schrieffer, W.-P. Su: Review of
Modern Physics $\underline{60}$, 781 (1988)

[3] F.L. Carter (Ed.): "Molecular Electronic Devices", Marcel Dekker,
New York and Basel, 1982, p. 51

[4] H. Bleier, H. Lobentanzer, G. Leising, S. Roth: Europhysics
Letters $\underline{4}$, 1397 (1987)
H. Bleier, S. Roth, Y.Q. Shen, D. Schäfer-Siebert, G. Leising:
Phys. Rev. $\underline{B38}$, 6031 (1988)
H. Bleier: "Photoleitung in Polyacetylen - zeitliches Verhalten und
Anisotropie", PhD Thesis, Stuttgart 1987
similar data have been obtained by M. Sinclair, D. Moses,
A.J. Heeger: Solid State Commun. $\underline{59}$, 343 (1986)

[5] S. Roth, G. Mahler, Y.Q. Shen, F. Coter: Synthetic Metals $\underline{28}$, C815
(1989)
S. Roth, H. Bleier: Proc. of the International Conference on
Molecular Electronics - Science and Technology, Hawaii, 1989, to be
published by IEEE

DEHYDROCHLORINATION OF PVC BY PHASE TRANSFER CATALYSIS

S. Dhainaut, A. Périchaud, P. Bernier*, S. Lefrant** and A. le Méhauté***

Laboratoire de Chimie Macromoléculaire, Université de Provence, 3, Place V. Hugo, 13331 Marseille Cedex 3.
* Groupe de Dynamique des Phases Condensées, USTL, Place E. Bataillon, 34060 Montpellier Cedex.
** Laboratoire de Physique Cristalline, IPCM, Université de Nantes, 44072 Nantes Cedex 03.
*** Laboratoires de Marcoussis, CGE, Route de Nozay, 91460 Marcoussis.

ABSTRACT. By dehydrochlorination of PVC by Phase Transfer Catalysis we obtained polyacetylene like materials. We discusse the experimental conditions and give spectroscopic results.

1. Introduction

Many kinetic investigations of chemical dehydrochlorination of polyvinyl chloride in solution have been performed by means of alcoholic potassium hydroxide [1-3] and potassium-t-butoxide [4,5]. In phase transfer catalysis (PTC) the kinetics of dehydrochlorination (DHC) have been studied by Kisé et al. [6,7]. These works concern small dehydrochlorination ratios, and the products obtained have not been characterized. In the realm of our present investigation on the low cost synthesis of polyacetylene we present here some results of optimization (to be published) concerning the effect of some experimental parameters on the dehydrochlorination rate in solid-liquid PTC and Raman spectroscopic characterization of the material obtained in order to assess the delocalization level in the $(CH)_x$ chain. It is shown that the delocalization kinetics may be analysed by NMR and that some fraction of Csp3 of the order of 20% remains. It is also shown that the number of delocalized bonds could be about 17.

2. Experimental Techniques

2.1 SAMPLE SYNTHESIS

In a Shlenk tube are added one gramme of PVC (PVC XII of the IUPAC Working) in powder together with a required amount of tetrabutyl ammonium bromide TBAB (Fluka). Both materials are out gassed simultaneously under high vacuum and then regassed with argon U. This operation is repeated three times. In the same way the

J. L. Brédas and R. R. Chance (eds.), Conjugated Polymeric Materials:
Opportunities in Electronics, Optoelectronics, and Molecular Electronics, 299–304.
© 1990 Kluwer Academic Publishers. Printed in the Netherlands.

concentrated NaOH solution required (Carlo-Erba) is out gassed in an ultrasonic vessel with argon bubbling for three hours. Both vessels are heated at the same temperature then the NaOH solution is added to the two powders through a system of teflon tubing under argon pressure. The reaction takes place at the chosen temperature then the dehydrochlorinated PVC (DPVC) is extracted, neutralized and conditioned under vacuum.

2.2 CHEMICAL AND PHYSICAL MEASUREMENTS

The progress of dehydrochlorination is followed by a coulometric titration of the eliminated chlorine ions. Conversion is also calculated by a weight decrease in the products or by elementary analysis.

^{13}C NMR spectra in the solid state are obtained using a Bruker CXP 200 spectrometer. Chemical shifts are measured versus TMS as reference. NMR spectra are performed in air at room temperature.

Raman spectra were recorded using a Jobin-Yvon HG2S Ramanor double monochromator, the signal was detected by a cooled GaAs photomultiplier coupled to a photon-couting system. Exciting lines were provided by a cw Ar^+ laser (457.9 nm) or a cw Kr^+ laser (676.4 nm). Experiments were always performed with a 90° scattering geometry, at room temperature.

3. Experimental Results and Discussion

In the optimization studies to be published we have attempted to improve the DHC fraction with respect to three parameters:

- the reaction temperature X_1
- the NaOH concentration X_2
- the amount of TBAB X_3

For this purpose we have used an optimization method of the Simplex type [8-10] yielding to a series of twenty experiments whose parameters are indicated in the following table (table 1).

The progression of the DHC reaction has been first measured by chemical analysis.

In its first phase the DHC reaction proceeds of a Zipper like mechanism, without a clear view of the partial order with respect to TBAB and NaOH announced by Kisé [7]; our conversion rates are much higher. For some experiments with small DHC rate it seems that we also deal with a constant speed regime. The propagation reaction takes place with widely variable rate according to the experimental conditions. Although for the three parameters considered here the reaction conditions are relatively close to those of Kisé [7], the conversion ratio measured here (20% vs 0.3 %) were much higher over five hours. The only relevant factors might be the nature of the PVC, the order in which the reactants are introduced, and especially the nature of the gas, (nitrogen or air). We had noticed previously for a DHC in solution that the Zipper mechanism was much less efficient under oxidizing atmosphere, because of the competition with the formation of the peroxide located in the last of the created double bonds [4].

TABLE 1. 5 hours dehydrochlorination: experimental conditions for 1g PVC and 30 ml NaOH.

| experiment | experimental conditions | | | dehydrochlorination (%) |
	$X1(°C)$	$X2(\%)$	$X3(g)$	after 5 hours
1	65	25	0,33	68
2	40	25	0,33	5
3	77,5	38	0,33	30
4	52,5	12	0,33	2
5	77,5	12	0,33	2
6	52,5	38	0,33	15
7	77,5	29,33	0,586	90
8	52,5	20,67	0,074	21
9	77,5	20,67	0,074	37
10	65	33,7	0,074	27
11	52,5	29,33	0,586	22
12	65	16,34	0,586	46
13	90	25	0,33	51
14	65	33,77	0,84	22
15	90	33,77	0,842	70
16	102,5	29,33	0,586	88
17	90	42,33	0,586	40
18	90	16,34	0,586	6
19	65	42,33	0,586	10
20	77,5	20,67	0,84	70

The extracted DPVC has been characterized by NMR of solid carbon 13, RPE, IRFT and Resonant Raman Spectroscopy. In the way it has been shown, by NMR for instance, that the relative intensity of the $C_{sp}2$ bands (136-140 ppm /TMS) increases with respect to those of the $C_{sp}3$ bands (60-80 ppm) with DHC production (Figure 1).

However even for 100% elimination - determined by potentiometric measures - an important number of $C_{sp}3$ remains. This can be explained by the collision of two propagating fronts CH* and or by interchain crosslinking reactions.

$$- CH = CH - CH = CH^* - CH_2 - \underset{\underset{Cl}{|}}{CH} - CH_2 - CH^* = CH -$$

$$\longrightarrow - CH = CH - CH = CH - CH = CH - CH_2 - CH = CH -$$

$$- CH = CH^* - CH_2 - \underset{\underset{\underset{\underset{-CH=CH^*-CH-}{|}}{H}}{Cl}}{CH} -$$

$$\longrightarrow$$

$$- CH = CH - CH_2 - \underset{\underset{- CH = CH - CH -}{|}}{CH} -$$

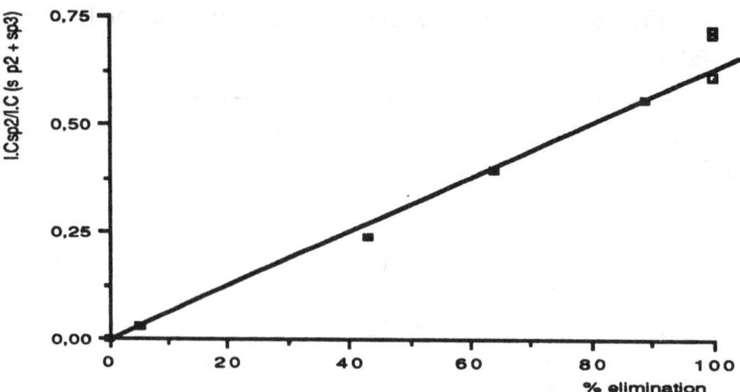

Figure 1. Relative intensity of $C_{sp}2$ bands vs elimination rate.

This property allows a clear distinction between the polyacetylene produced here and the one of the Shirakawa type.

The optimum experimental conditions of DHC are of the following:
Reaction temperature =75°C, NaOH concentration =27%; and amount of TBAB = O.73g.

After 13 hours the synthesized products are extracted, washed, dried, sealed under vacuum and then investigated by Raman Resonance Spectroscopy (table 2).

TABLE 2. Raman Resonance Spectroscopy measures on DPVC.

Experiment	% DHC	$\lambda_{ex} = 458$ nm		$\lambda_{ex} = 676$ nm	
		ν_1	ν_2	ν_1	ν_2
1	85	1141,0	1544,2	1112,7	1491,5
2	9	1136,0	1540,0	1106,3	1485,0
3	42	1142,5	1542,0	1109,1	1487,7
4	3	1137,6	1539,1	1101,4	1478,8
5	2	1142,4	1542,0	1099,4	1480,3
6	34	1138,8	1544,7	1111,0	1486,9
7	99	1140,0	1540,0	1122,0	1501,0
8	36	1140,5	1544,0	1104,8	1478,7
9	19	1140,9	1541,3	1104,0	1481,9
10	31	1142,7	1546,7	1114,2	1490,2
11	48	1141,0	1541,5	1106,9	1483,0
12	78	1139,1	1540,3	1112,8	1489,0
13	53	1144,5	1542,1	1098,6	1478,1
14	30	1140,5	1542,0	1111,3	1483,0
15	76	1140,3	1541,5	1103,0	1485,2
16	48	1139,2	1541,4	1104,7	1483,8
17	58	1141,8	1540,8	1103,0	1483,1
18	9	1138,8	1542,7	1105,7	1489,1
19	19	1145,1	1544,7	1106,8	1484,3
20	98	1144,3	1545,7	1125,3	1501,3

Raman resonance spectra of experiment 7 are given Figure 2.

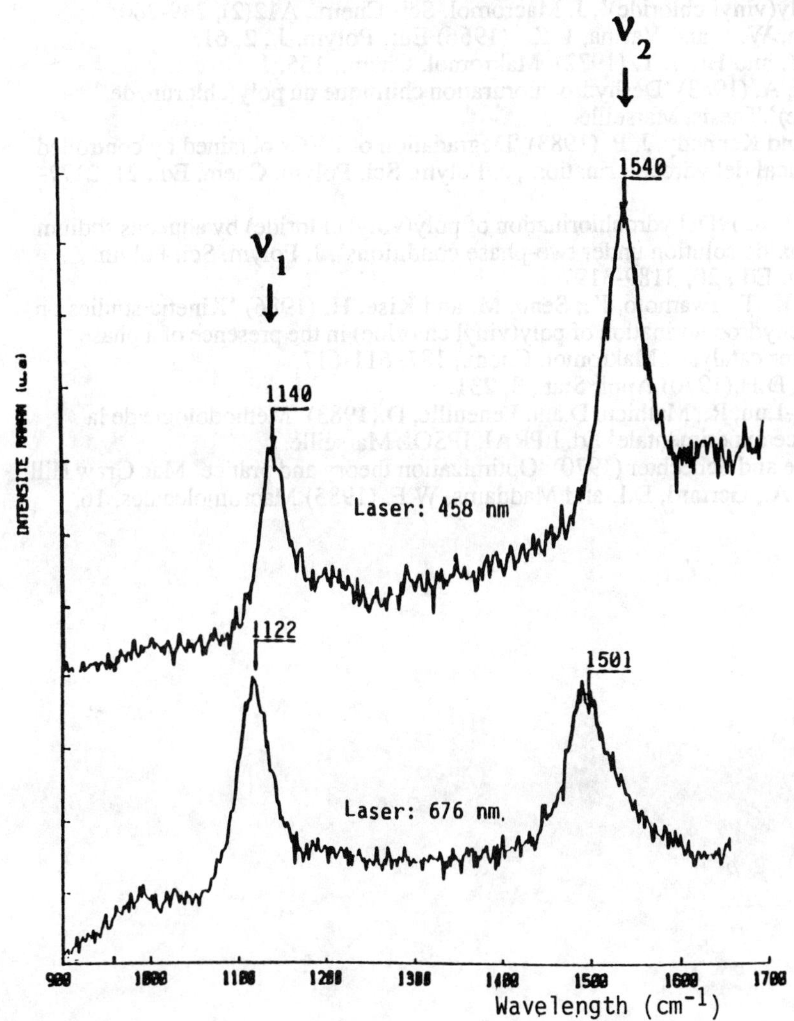

Figure 2. Raman resonance spectra of DPVC (DHC 99%).

Using a calculation method proposed by Baruya et al.[11] it appears that the average length of the conjugated polyenic chains measured from the resonance frequency ν_2, at $\lambda_{ex} = 676$ nm, about 1501 cm^{-1} would be of the order of 17.

A complete kinetic analysis is under way in the hope of determining the partial orders with respect to the reactants for high dehydrochlorination rate.

4. References

1 - Östensson, B. and Flodin, P. (1978) 'Kinetics of the alkaline dehydrochlorination of poly(vinyl chloride)', J. Macromol. Sci.-Chem., A12(2), 249-260.
2 - Bengough, W. I. and Varma, I. K. (1966) Eur. Polym. J., 2, 61.
3 - Shindo, Y. and Hirai, T. (1972) Makromol. Chem., 155, 1.
4 - Périchaud, A. (1983) 'Déshydrochloruration chimique du poly(chlorure de vinyle)',Thesis, Marseille.
5 - Ivan, B. and Kennedy, J. P. (1983) 'Degradation of PVCs obtained by controlled chemical dehydrochlorination', J. Polym. Sci. Polym. Chem. Ed., 21, 2177-2188.
6 - Kise, H. (1982) 'Dehydrochlorination of poly(vinyl chloride) by aqueous sodium hydroxide solution under two-phase conditions', J. Polym. Sci. Polym. Chem. Ed., 20, 3189-3197.
7 - Howang, K. T., Iwamoto, K., Seno, M. and Kise, H. (1986) 'Kinetic studies on the dehydrochlorination of poly(vinyl chloride) in the presence of a phase transfer catalyst', Makromol. Chem., 187, 611-617.
8 - Doehlert, D.H.(1970) Appl. Stat., 3, 231.
9 - Phan-Tan-Luu, R., Mathieu, D.and Feneuille, D.(1983) 'Méthodologie de la science expérimentale' Ed. LPRAI, IPSOI, Marseille.
10 - Beverdge and Schechter (1970) 'Optimization theory and pratice' Mac Graw Hill.
11 - Baruya, A., Gerrard, D.L and Maddams, W.F. (1985) Macromolecules, 16, 576.

VINYLENE-LINKED LOW-BAND-GAP CONDUCTING POLYMERS: ELECTRONIC STRUCTURE AND DEFECTS

H. ECKHARDT, K.Y. JEN,[*] AND L.W. SHACKLETTE
Allied-Signal Research and Technology
P.O. Box 1021R
Morristown, New Jersey 07962, U.S.A.

S. LEFRANT
Laboratoire de Physique Cristalline, I.P.C.M.
Universite de Nantes
44072 Nantes Cedex 03, France

ABSTRACT. The band structure and the electrochemical potentials of conductive polymers based on phenyl and thiophene backbone structures can be strongly modified by substituents on the benzene and thiophene rings in poly(p-phenylenes) and polythiophenes, respectively, and by interposing vinylene double bond linkages between rings. Substitution with a methoxy or ethoxy group on the ring further lowers the ionization potentials and band gaps in the vinylene polymers substantially. Excellent agreement is found between band gaps measured by electrochemical spectroscopy (ECPS) and optical absorption spectroscopy. The substituted thienylene vinylene polymers can be described as low-band-gap semiconductors with their absorption edge largely in the near IR. In the heavily doped state, these polymers become "transparent conductors". A detailed vibrational analysis of PPV yields valuable information about the structural distortion which occur upon doping. The results favor the formation of bipolarons over polarons in PPV and are in good agreement with semiempirical self-consistent (MNDO) defect calculations on large PPV model compounds.

1. Introduction

Processible, high molecular weight conducting polymers such as poly(p-phenylene vinylene) and poly(thienylene vinylene) have attracted considerable interest recently. This interest is rooted in the good stability and relative ease of processing compared to the early polymers, i.e. polyacetylene, polyphenylene etc. The vinylene linked polymers are synthesized via a high molecular weight, soluble precursor yielding high quality material which can be easily formed into films and oriented to a high degree. The common backbone of these polymers is made up of an aromatic ring bridged in the para position by a vinylene linkage. Unlike in polyacetylene, these polymers have a non-degenerate groundstate which can support polarons and bipolarons as charged excitations. We have recently shown[1] that the band gap and ionization potentials can be adjusted over a wide range by a choice of appropriate electron donating substituents. In particular the substituted thienylene vinylenes can be described as low-band-gap semiconductors which, upon photoexcitation or chemical doping, support mainly bipolarons as basic excitations. Like all conducting polymers, the oscillator strength of the interband transition, which is dominant in the spectra of the neutral form, is reduced substantially by the charge transfer process and shifted to the transitions in the near IR. As a result many of these low band gap materials transmit light in the visible and can be called transparent conducting polymers. Because of these rather unusual properties, the use of these materials as conducting coatings for window applications and as

305

J. L. Brédas and R. R. Chance (eds.), Conjugated Polymeric Materials:
Opportunities in Electronics, Optoelectronics, and Molecular Electronics, 305–320.
© 1990 *Kluwer Academic Publishers. Printed in the Netherlands.*

306

coatings to selectively block solar energy and/or IR radiation is being explored.

In this paper we will consider the effect of substitution on the electronic structure of these materials. The attachment of electron donating groups on the ring part of the polymer backbone will change the electron density and modify parameters like ionization potential (IP) and band gap (E_g). We will also discuss how modifications in the backbone itself affect these parameters. The polymers of interest are poly(phenylene vinylene), poly(thienylene vinylene), and poly(furylene vinylene), and derivatives with electron donating moieties such as methyl, methoxy and ethoxy groups. The polymer repeat units are shown in figure 1. In poly(furylene vinylene) the sulfur atom is replaced by oxygen.

The structures and the substitution effects were modeled using the semiempirical quantum chemical modified neglect of differential overlap (MNDO) method[2], which yields good approximations to the ground state geometry of these polymers. The MNDO-determined geometry was used as input for the band structure calculations employing the valence effective Hamiltonian (VEH) technique. This method has been used in the past to predict band structures, ionization potentials and band gaps of conjugated polymers giving excellent agreement with experimental values[3].

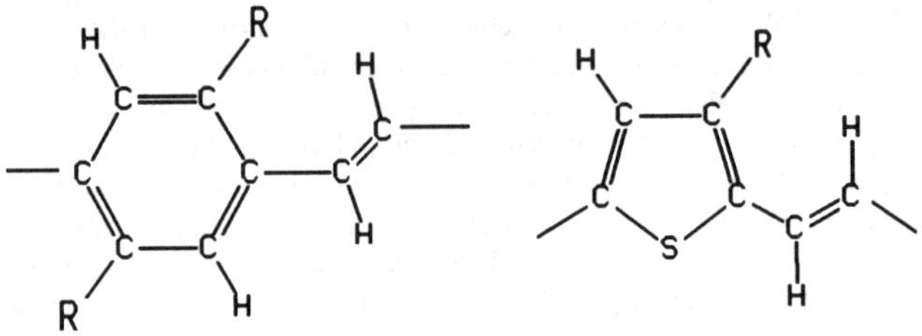

Figure 1. Polymer repeat units for phenylene- (I) and thienylene vinylene (II) polymers. The substituent groups R are methyl, methoxy and ethoxy.

In the second part we will discuss some of the vibrational properties of the unsubstituted undoped and doped vinylene polymers and compare results obtained from IR and Raman spectroscopy to MNDO based defect calculations.

2. Electrochemical and Optical Properties

2.1 ELECTROCHEMICAL POTENTIAL SPECTROSCOPY (ECPS)

Absorption spectroscopy and electrochemical potential spectroscopy (ECPS) are two complementary techniques to probe the electronic structure of conducting polymers and determine band gaps and ionization potentials. A detailed description of ECPS as well as method for the synthesis of these materials can be found in reference 1. Valuable information regarding the band structure of conducting polymers can be obtained from electrochemical curves. The measured onset potentials for the oxidative and reductive cycles can be related to the ionization potential, band gap and electron affinity. The relationship between these parameters and the electrochemical potentials is drawn in a simplified schematic fashion in figure 2. The left side shows the density of states for the two relevant bands, the valence and conduction band. The energy difference between the ionization potential, IP, and the band gap, E_g, can be taken , in a first approximation, as the electron affinity. The right side of the figure shows the potentials for a reduction or oxidation cycle as a function of accumulated charge. The onset potential for the oxidation process, $(E_0)_{ox}$, corresponds to the highest occupied energy level in the valence band. At this point charges are being removed from the polymer electrode and more and more states are emptied as the voltage is further increased. In the reduction process unoccupied states in the conduction band are filled with electrons when the applied voltage is lowered sufficiently to reach the onset

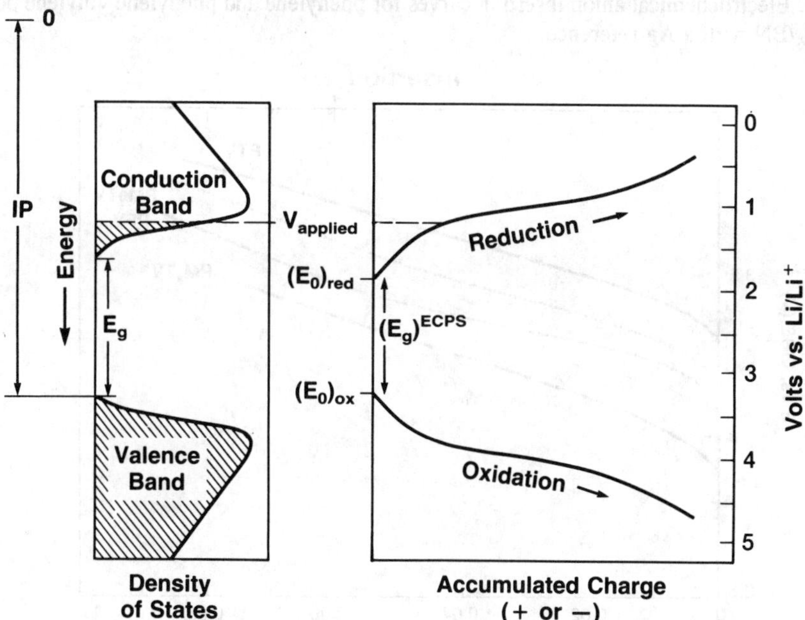

Figure 2. Schematic drawing of band structure for polymers and the relationship to electrochemically measured onset potentials for oxidation and reduction cycles. The electrochemical band gap, $(E_g)^{ECPS}$ is the difference between the two onset potentials. The ionization potential, IP, is also indicated. IP - $(E_g)^{ECPS}$ gives the electron affinity.

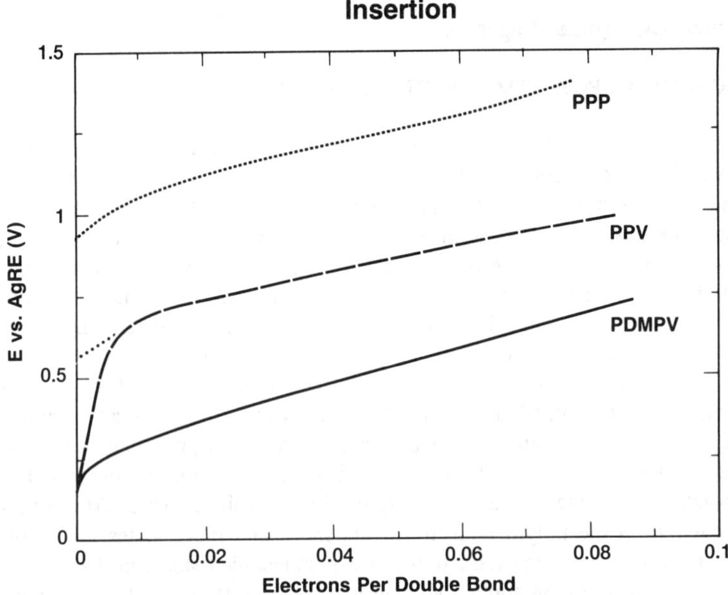

Figure 3. Electrochemical anion insertion curves for phenylene and phenylene vinylene polymers in $NaPF_6/BN$ with a Ag reference.

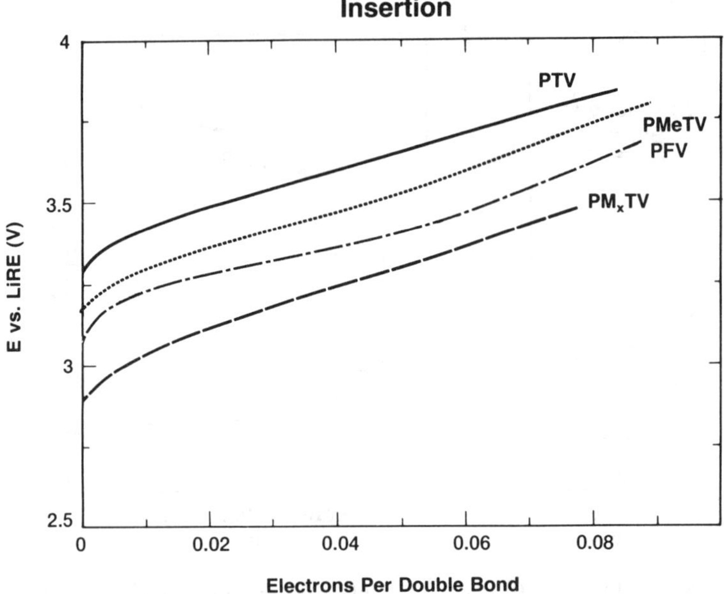

Figure 4. Electrochemical anion insertion curves for thienylene vinylene polymers in $LiClO_4/PC$ with a Li reference.

potential $(E_0)_{red}$. The difference between the two onset potentials gives a good value for the band gap. This is particularly true when the oxidation and reduction cycles are performed in the same electrolyte. When electrolyte stability is a problem and different electrolytes and reference electrodes have to be used for the two cycles, the dissimilar electrochemical media have to be calibrated by comparing the $E_{1/2}$ potentials of a standard redox couple which is relatively solvent independent. We have used ferrocene (Fcn/Fcn$^+$) for this purpose. A larger error for the band

TABLE I. ECPS data for the oxidation (p-doping) of conductive polymers.[a]

Polymer	Ref.	Electrolyte	E_o(V)	E_{pa}(V)	E_{pc}(V)	$E_{1/2}$(V)	E_g(eV)[b]
PPP	Ag	NaPF$_6$/BN	3.94 ± 0.01	4.23			2.84 ± 0.11
PPV	Ag	NaPF$_6$/BN	3.63 ± 0.02	3.82	3.72	3.77	2.40 ± 0.10
PDMPV	Ag	NaPF$_6$/BN	3.24 ± 0.01	3.54	3.53	3.54	2.05 ± 0.08
	Na	NaPF$_6$/DME	3.31 ± 0.02				2.11 ± 0.03
PTV	Li	LiClO$_4$/PC	3.28 ± 0.01	3.56	3.51	3.54	
PMeTV	Li	LiClO$_4$/PC	3.16 ± 0.01	3.43	3.39	3.41	
PMxTV	Li	LiClO$_4$/PC	2.88 ± 0.01	3.26	3.19	3.23	
PExTV	Na	NaPF$_6$/DME	2.85 ± 0.03	3.12	3.09	3.11	1.31 ± 0.05
PFV	Li	LiClO$_4$/PC	3.04 ± 0.05	3.30	3.26	3.28	1.71 ± 0.11
	Na	NaPF$_6$/DME	3.05 ± 0.02	3.26	3.26	3.34	1.71 ± 0.04
PA	Ag	NaPF$_6$/BN	3.25 ± 0.10	3.66	3.37	3.52	1.36 ± 0.18

[a]Potentials referenced to Li assuming E_{Ag}-E_{Li}=3.06V and E_{Na}-E_{Li}=0.065V.
[b]E_g determined by E_0(ox) - E_0(red)
The following abbreviations for the polymers are used:
PPP   poly(p-phenylene)
PPV   poly(p-phenylene vinylene)
PDMPV   poly(2,5-dimethoxyphenylene vinylene)
PTV   poly(thienylene vinylene)
PMeTV   poly(3-methylthienylene vinylene)
PMxTV   poly(3-methoxythienylene vinylene)
PExTV   poly(3-ethoxythienylene vinylene)
PFV   poly(furylene vinylene)
PA   polyacetylene

gap values is then introduced. However, these band gap values still agree well with optically determined band gap numbers. The errors associated with the electrolyte/reference electrode calibration apply also to the determination of reliable values for IP.

    Our main focus of discussion is on substitution effects. Within a given class of polymers, i.e. phenylene or thienylene vinylenes, we were able to use the same electrolyte and reference electrodes with the exception of poly(3-ethoxythienylene vinylene) (PExTV), for which we used a different electrolyte/reference electrode system. For the phenylene series, $NaPF_6$ in benzonitrile (BN) with a Ag wire as reference electrode provided a stable electrolyte over a wide voltage range. $LiClO_4$ in propylene carbonate (PC) with Li as a reference electrode was the electrolyte of choice for the thienylene vinylenes.

Anion insertion curves (p-type doping) for polymers with a phenylene-type backbone are shown in figure 3. The equivalent curves for the thienylene vinylene series are drawn in figure 4. The effect of substituent groups is evident from these plots. Addition of the strong electron donating groups (methoxy and ethoxy) to the ring part of the backbone reduces the electrochemical potentials by about 0.4 V. Modifications in the backbone itself also strongly influence the potentials. The increased delocalization of π-electrons induced by the vinylene linkage leads to a reduction of 0.3 to 0.4 volts for both types of polymers.

Table I summarizes the measured potentials. Also included are the electrochemical potentials for poly(furylene vinylene) (PFV), a polymer with a similar structure to PTV in which the heteroatom in the ring is oxygen instead of sulfur. For comparison, the potentials for polyacetylene are included as well. The potential values in Table I were not directly deduced from the above figures, but taken from an inverse derivative plot dy/dE versus E.[1] This derivative represents the change in dopant level per unit voltage as a function of applied potential for ion insertion and extraction and gives more precise onset potential values. For PDMPV, PExTV and PFV the oxidation and reduction cycles were performed in the same electrolyte system, and yield therefore, as pointed out before, the electrochemical band gap directly. The ECPS determined band gaps are more reliable in these cases.

In general, the potentials as well as the band gaps of the phenylene polymers are higher than the corresponding values for the thienylene polymers. The potentials for the thienylene vinylenes are close to the potentials for polyacetylene. This correspondence is not surprising since geometry optimization yields bond distances and bond angles for PTV close to those of polyacetylene and polythiophene.[1] The table also shows that the combined effect of derivatization with vinylene linkages and dimethoxy substituents is sufficient to bring the potential values for the phenylenes down to the level of polyacetylene. The onset potentials and the ECPS band gaps for the substituted thienylene vinylene polymers are below the values of PA. They even begin to approach the E_o value of the most easily oxidized polymer, polypyrrole. The onset potential of polythiophene (PT)[4] is about 0.4 V above E_0 of PTV. It is well known that heteroatom substitution strongly affects band gaps and electrochemical potentials. Replacement of sulfur by oxygen lowers E_o by about 0.2 V. This is still substantially above the onset potential for polypyrrole $(E_0 = 2.60$ V), a polymer with nitrogen as heteroatom and no vinylene linkage. We do not have an ECPS determined band gap for PTV.

2.2 OPTICAL SPECTROSCOPY

Thin film optical spectroscopy corroborates the substitution effects on band gaps observed with ECPS. The optical properties of PPP and PPV have been studied extensively and have been described in the literature.[1,5] The band gaps determined from absorption onsets are 2.8 eV and 2.4 eV for PPP and PPV respectively. Attaching two methoxy groups as in PDMPV reduces the band gap by another 0.4 eV.[1] Similar substitution effects are observed in the spectra of the thienylene polymers. Figure 5 shows the absorption spectra of undoped PT, PTV, PExTV and PFV. The molecular weight of PExTV (and PMxTV), synthesized by a Grignard coupling reaction, is lower than the molecular weight of PTV and PFV, which were prepared by the soluble precursor route which is known to yield very high molecular weights. The absorption edge of truly high molecular weight PExTV could conceivably be shifted to even lower energy. Again incorporating a vinylene double bond into the polymer backbone lowers the band gap by almost 0.4 eV and ring substitution lower band gaps even further. The polymer with the lowest onset absorption and lowest peak maximum is poly(3-ethoxythienylene vinylene) in spite of the lower molecular weight. The onset and the absorption maximum are similar to what has been observed in trans-$(CH)_x$. The absorption spectra of all methoxy and ethoxy substituted polymers are characterized by a large degree of disorder. No vibronic fine structure as seen in PPP and PPV is observed. Not shown is the absorption spectrum of PMxTV. The measured absorption edge for this polymer is above the edge of PTV, reflecting the high degree of short chain oligomers present in the thin films. In contrast to what has been observed for the electrochemical potentials, heteroatom substitution in the five member ring dos not lower the band gap. The absorption edge of PFV appears to be slightly above the edge of PTV. The strong effect on band gaps by methoxy and ethoxy substitution is also present in the monomer of the thienylene polymers. The solution spectrum of the methoxy substituted 1,2-bis-(2-thienyl)-ethylene, a molecule in which two thiophene rings are linked by a vinyl double bond, is shifted by 0.25 eV to lower energy compared to the spectrum of the unsubstituted monomer.

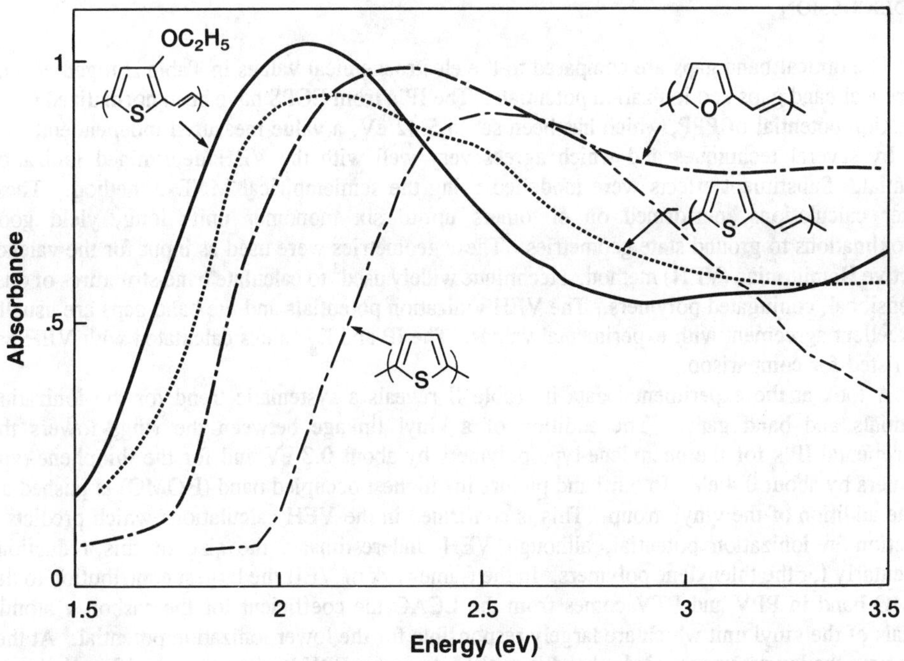

Figure 5. Absorption spectra of thienylene- and furylene vinylenes. Also shown is the absorption spectrum of polythiophene.

Chemical and electrochemical doping changes the absorption spectra of the polymers drastically. As with all conducting polymers, the interband transition is reduced and new transitions in the IR, resulting from the creation of polarons and bipolarons, are observed. At the highest doping levels, an absorption gap is opened up which permits a high degree of transparency in the visible part of the spectrum. The changes in the absorption spectra are accompanied by dramatic color changes which depend on the location of the absorption edge of the neutral polymer. The fine tuning of the absorption properties through appropriate substitutions enable one to obtain films which appear almost transparent in their highest doped state but which possess an increased absorption and reflection in the IR, hence the term "transparent conducting" polymers. These properties make the substituted phenylene vinylenes and thienylene vinylenes interesting candidates for coatings on solar screening windows, where high reflection and/or absorption of solar near IR radiation is required while maintaining transparency in the visible to achieve a low level of tint. An additional advantage is the blocking of UV radiation which occurs in these polymers even at very high doping level, provided the band gap of the neutral polymer is shifted far enough into the red.

To summarize the results so far, control of molecular architecture allows the engineering of conducting polymers with desirable electrochemical and optical properties. Selected polymers discussed here offer special advantages for the design of active or passive solar screening windows. These advantages range from their unique optical and electrochemical properties to the ease of processing and their environmental stability.

2.3 DISCUSSION

The optical band gaps are compared to the electrochemical values in Table II together with theoretical band gaps and ionization potentials. The IP's from ECPS have been normalized to the ionization potential of PPP, which has been set to 5.42 eV, a value measured independently for PPP by several techniques and which agrees very well with the VEH determined ionization potential. Substituent effects were modelled using the semiempirical MNDO method. These cluster calculations, performed on oligomers up to six monomer units long, yield good approximations to ground state geometries. These geometries were used as input for the valence effective Hamiltonian (VEH) method, a technique widely used[3] to calculate band structures of one-dimensional, conjugated polymers. The VEH ionization potentials and the band gaps are usually in excellent agreement with experimental values. The IP and E_g values calculated with VEH are also listed for comparison.

A look at the experimental data in Table II reveals a systematic trend for the ionization potentials and band gaps. The addition of a vinyl linkage between the rings lowers the experimental IP's for the phenylene-type polymers by about 0.3 eV and for the thiophene-type polymers by about 0.4 eV. In our band picture the highest occupied band (HOMO) is pushed up by the addition of the vinyl group. This is confirmed in the VEH calculations which predicts a reduction in ionization potential, although VEH underestimates the size of this reduction, particularly for the thienylene polymers. In the framework of VEH the largest contribution to the HOMO band in PPV and PTV comes from the LCAO the coefficient for the carbon π atomic orbitals of the vinyl unit which are largely responsible for the lower ionization potential. At the same time the lowest unoccupied π band is pushed down for PPV leading to a significantly lower band gap in PPV compared to PPP. Experimentally, this band is lowered by 0.13 eV in going from PPP to PPV. As a consequence, the band gaps are lowered progressively from 2.84 eV to

2.40 eV in the ECPS data and from 2.80 eV to 2.40 eV from absorption data. There is excellent agreement between the optical band gaps and the ECPS band gaps for the phenylene series. The theoretical ionization potentials and band gaps are also in very good agreement with the measured values. Similar shifts in IP and E_g with vinylene substitution are observed for the thienylene polymers. However, the agreement between electrochemical energies, optical data and theoretical values is not as good.

Substitution with electron donating groups such as methoxy or ethoxy groups on the ring part of the polymer backbone leads to additional lowering of ionization potentials and band gaps for both, the substituted phenylene vinylenes and thienylene vinylenes. The combined effect of vinyl bridging and methoxy or ethoxy substitution lowers the ionization potentials by 0.7 eV and 0.87 eV and the band gap by 0.73 eV and 0.93 eV for the phenylene and thienylene series, respectively. It is important to point out, that the ECPS determined band gaps for the substituted thienylene vinylene are more representative of conjugated, infinite chain polymers than the optical band gaps, which reflect the absorption of the relatively short chains characteristic of the soluble fraction used to prepare the thin film samples for optical measurements. This point is particularly

TABLE II. Experimental and theoretical ionization potentials and band gaps[a]

Polymer	IP^{ECPS}	IP^{VEH}	E_g^{opt}	E_g^{ECPS}	E_g^{VEH}
PPP	5.42[b]	5.42[c]	2.80	2.84±0.11	2.90
PPV	5.11±0.03	5.21	2.40	2.40±0.10	2.21
PDMPV	4.72±0.03	5.03	2.00	2.11±0.03	2.27
PT	5.20±0.1[d]	5.07	2.00	2.24±0.08	1.69
PTV	4.76±0.02	4.95	1.64	----	1.61
PMeTV	4.64±0.02	4.92	----	----	1.62
PMxTV	4.36±0.02	4.92	1.67	----	1.69
PExTV	4.33±0.04	----	1.48	1.31±0.05	----
PFV	4.52±0.01	4.46	1.76	1.71±0.11	1.47
PA	4.73±0.1	4.77	1.42[e]	1.36±0.18	1.46

[a]All values in eV. Optical gaps determined from absorption onsets
[b]IP^{ECPS} normalized to IP for PPP
[c]Theoretical IP's adjusted by 1.9 eV to account for polarization energy
[d]Reference 4
[e]Reference 6

true for PExTV, where the ECPS gap is substantially below the optical gap, and even below both, the electrochemical and optical gap of polyacetylene.

The effect of the methyl and methoxy groups on the electronic structure of the vinylenes at the VEH level is very small. While the shift of IP with substitution is correctly predicted, the calculated shifts are too small compared to the measured shifts. The HOMO band is pushed upwards by about 0.2 eV for PDMPV, which has two methoxy groups attached to the benzene ring, and by 0.1 eV for PMxTV, but at the same time the LUMO band is also pushed up, yielding a slight increase in band gap. Methoxy substitution appears to lower the IP's by about 0.1 eV per methoxy group within our theoretical model, much smaller than the measured values for PDMPV (0.2 eV per methoxy) and PMxTV (0.4 eV per methoxy). The small effects on ionization potential and band gaps within VEH can be traced to the small contribution of the LCAO coefficients for the π atomic orbitals on the methyl carbons and methoxy oxygens to the HOMO. It appears that VEH underestimates the relative weight of the substituent groups in the HOMO and LUMO. However, not too much emphasis should be placed on the values of the band gaps, since the theoretical foundation for the band gaps within VEH are not well established.

The large effect of the electron donating groups on the electronic structure of the phenylene vinylenes and thienylene vinylenes is partially electronic and partially steric. It is noteworthy that methyl and methoxy substitution in polymers, in which the aromatic or weakly aromatic rings are not linked by a vinyl double bond such as polyisothianaphthene (PITN)[7] and polypyrrole, does not show the same strong electronic effect. The separation of phenyl or thiophene rings by the vinyl group in substituted PPV and PTV provides the necessary flexibility for the chain to assume a coplanar chain conformation. This contention is strongly supported by our MNDO geometry optimization on phenylene and thiophene oligomers. For example, in methyl and methoxy thiophene model compounds the equilibrium geometry is a conformation in which the rings are nonplanar and twisted by about 37 degrees. In contrast thienylene vinylene oligomers assume a lowest energy conformation in which the rings are coplanar. A similar effect has been observed in methyl pyrroles where the rings are strongly twisted out of the plane.[8] To fully utilize the electron donating function of methyl and methoxy groups for modification of band gaps and ionization potentials in polymers with aromatic ring structures a mechanism is required to sufficiently reduce the steric hindrance of the substituents on neighboring rings which drives the rings into a non-planar conformation. The vinylene linkage in phenylene vinylene and thienylene vinylene provides this mechanism.

The effect of heteroatom substitution in the five-member ring is also apparent from Table II. Replacement of sulfur (PTV) by oxygen (PFV) lowers the IP by 0.25 eV, not nearly as much as nitrogen substitution. A large shift in the same direction (0.45 eV) is also predicted by VEH. The calculated and measured IP values are in good agreement for PFV, and so are the two experimental band gaps, while the VEH gap appears to be too low. The lower ionization potential for PFV is a true oxygen substitution effect and does not arise from nuclear relaxation and a different bond length alternation pattern of the carbon skeleton. We have simulated the heteroatom effects within VEH by maintaining the same carbon backbone for PTV and PFV. Oxygen substitution leads to a stronger perturbation of band gap and ionization potential. Lee and Kertesz[8] see similar heteroatom effects on band gaps in their extended Hückel calculations on polythiophene and polypyrrole.

3. Vibrational Analysis and Defect Structure

Poly(p-phenylene vinylene) and poly(thienylene vinylene) have a non-degenerate groundstate which supports mainly bipolarons as charged excitations.[9] In this section we will relate the defect structure of PPV and, to a lesser degree, PTV as obtained from a detailed vibrational analysis to the polaron and bipolaron structure calculated by the MNDO method. Our approach is to deduce structural information for neutral and doped chains from a force constant analysis and compare it to bond length alternation patterns calculated for cations and dications of large model oligomers.

3.1 VIBRATIONAL ANALYSIS OF PPV AND PTV

We have performed a complete experimental and theoretical analysis of undoped PPV and PTV and of doped PPV.[10,11] A complete valence force field was developed for PPV and PTV utilizing the experimental Raman and IR frequencies. The theoretical model uses a dynamic Fourier matrix to describe the vibrations of a one-dimensional crystal. Details of this approach are published elsewhere.[12] To arrive at a reliable force field for the polymers and to refine the force constants a detailed normal mode analysis of model compounds for PPV (and PTV) such as trans-stilbene was performed first using the experimental frequencies of the oligomers. The resultant model compound force field served as starting point for the polymer calculations. This approach allowed us to make a reliable assignment of observed frequencies for undoped and doped PPV. At this time no complete vibrational data for doped PTV are available, although the complete force field for undoped PTV has been determined. We will focus on PPV and the changes in force constants upon doping.

Our valence force field model is restricted to the in-plane vibrations. The rings are assumed to be coplanar. In this model the force constants can be divided into four different groups: the first group is associated with the unperturbed ring, the second one with the ring perturbed by the para substitution of the vinylene bridge, the third describes the coupling between the vinyl group and the ring and the fourth relates to the vinylene group itself. The four major C-C stretch force constants for PPV are shown in figure 6.

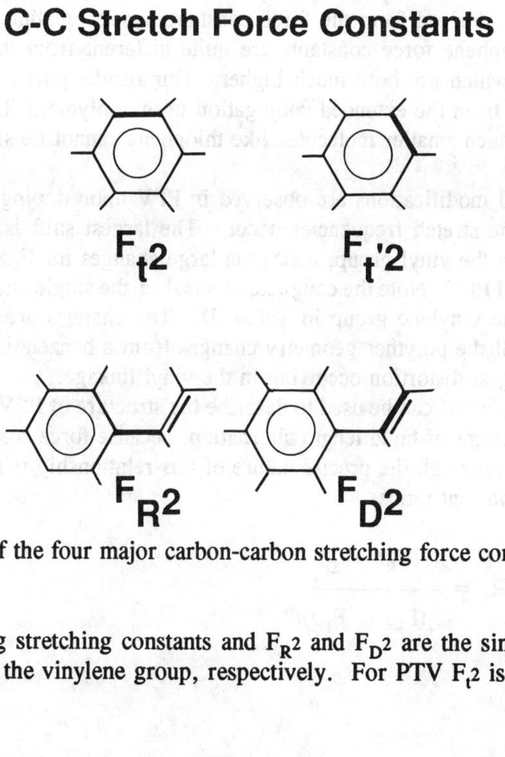

C-C Stretch Force Constants

Figure 6. Description of the four major carbon-carbon stretching force constants.

F_t2 and $F_t'2$ are the ring stretching constants and F_R2 and F_D2 are the single and double bond stretching constants for the vinylene group, respectively. For PTV F_t2 is the C-C single bond stretching force constant and $F_{t'}2$ the $C=C$ double bond stretching force constant in the thiophene ring. The values for these four force constants are tabulated in Table III for neutral and doped PPV and undoped PTV.

Table III. Carbon-carbon stretching force constants for PPV and PTV

Polymer	F_R2	F_D2	F_t2	$F_{t'}2$
Undoped PPV	4.80	7.06	6.32	6.32
Doped PPV	6.28	4.59	6.45	5.64
Undoped PTV	4.80	7.06	5.77	7.55

[a]Force constants in mdyn Å^{-1}

The two force constants F_t2 and $F_{t'}2$ in PPV are the same, reflecting the aromatic character of the benzene ring with all bond lengths equal. The same force constants in PTV characterize a weakly aromatic ring with strong bond length alternation in the ring, although not as strong as the one for the vinylene group. The two force constants for the C-C stretching of the thiophene ring are stronger than the corresponding vinylene force constants. The electronic structure of neutral PPV and PTV derived from the force constants agree qualitatively with the proposed structures.[1] The thiophene force constants are quite different from the values of Scott[13] for molecular thiophene which are both much higher. Our results portray a thiophene ring with weaker bonds arising from the extended conjugation in the polymer. The results make it clear that force fields for much smaller molecules like thiophene cannot be simply transferred to the polymers.

Strong structural modifications are observed in PPV upon doping. Experimentally large shifts in carbon-carbon stretch frequencies occur. The largest shift is recorded for the $C=C$ double bond stretch in the vinyl group, hence the large changes for F_D2 and F_R2 in going from undoped PPV to doped PPV. Note the complete reversal in the single and double bond stretching force constants for the vinylene group in Table III. The changes are less pronounced in the benzene ring. Overall the polymer geometry changes from a benzenoid-type to a quinoid-type structure with the largest distortion occurring in the vinyl linkage.

The results in Table III can be used to describe the structure of PPV and the dopant induced structural changes in terms of bond length alternation, because force constants and bond lengths are inversely related, although the precise nature of this relationship is not known.   We define the following force constant ratios:

$$R_v = \frac{F_R2 - F_D2}{(F_R2 + F_D2)/2}$$

$$R_B = \frac{F_t^2 - F_{t'}^2}{(F_t^2 + F_{t'}^2)/2}$$

$$R = \frac{F_R^2 + F_t^2 - [F_D^2 + F_{t'}^2]}{(F_R^2 + F_t^2 + F_D^2 + F_{t'}^2)/4}$$

R_v is a measure of the relative bond length alternation in the vinylene group normalized to an average force constant (or bond length), R_B characterizes the bond length alternation in the ring and R gives the average bond length alternation along the polymer backbone. With these definitions a value of $R < 0$ describes an aromatic structure and $R > 0$ a quinoid structure.

Using the results of Table III we calculate $R_v = -0.095$ and 0.0775, $R_B = 0$ and 0.0335 and $R = -0.023$ and 0.0275 for the neutral and doped structure, respectively. For the neutral polymer the average bond length pattern can be described as aromatic with zero bond length alternation in the ring and strong alternation in the vinylene bridge. Doping induces a quinoidal character in the ring and completely reverses the bond length alternation in the vinylene linkage between the rings. The overall structure of doped PPV is quinoidal. The structural distortion between the rings is much stronger than in the ring itself. Most likely the charge transfer occurs at the site of the vinylene double bond leading to the observed large structural changes which then extend into the ring. This observation is consistent with our discussion of the electronic structure of PPV. The largest contribution to the highest occupied band in PPV, which determines the ionization potential, comes from the two carbon atoms which form the vinylene double bond. They may be more susceptible to electron removal than is the benzene ring.

3.2 POLARON AND BIPOLARON DEFECT STRUCTURE IN PPV

We have carried out semiempirical self-consistent-field calculations (MNDO) on large PPV oligomers to study the defect geometry and charge distribution. Our work follows the approach taken by Boudreaux et al.[13] in their investigation of solitons and polarons in trans-$(CH)_x$ and, more recently, by Stafstrom and Bredas[14] in their study of polythiophene. The methodology and a detailed discussion of this approach is given in these two references. It has been shown recently,[15] that the MNDO calculated charge distribution in diphenylpolyenes, which are good model oligomers for polyacetylene, is in good agreement with the measured charge distribution, lending validity to this methodology.

Geometry optimizations in the presence of polaron and bipolaron defects were performed on the following molecules: (i) $[C_8H_7\text{-}(C_8H_6)_n\text{-}C_6H_5]^{+\bullet}$ for the polaron and (ii) $[C_8H_7\text{-}(C_8H_6)_n\text{-}C_6H_5]^{++}$ for the bipolaron with n=0 (trans-stilbene) to 4. (C_8H_6) denotes the PPV repeat unit. The longest oligomer is therefore a molecule with 6 phenyl rings connected by 5 vinylene linkages. The size of this oligomer assures that end effects do not have a major influence on the geometry of the defect. We will therefore concentrate our discussion here on the longest PPV oligomer. To determine the geometry of the cation, which is an open shell system, we have used the half-electron method (HE) in MNDO as well as unrestricted Hartree-Fock (UHF). Since the HE method, a restricted Hartree-Fock treatment, modified to deal with an open shell system, is believed to be superior to the UHF scheme, we will restrict our discussion of the cation structure to the former method. The dication calculation for the bipolaron does not present any special

318

problems since it is a closed shell system.

In figure 7 we show the excess atomic charge, Δq, relative to the neutral oligomer along the carbon backbone. The carbon sites are plotted along the x-axis. The center of the molecule, which is also the center of the defect, are carbon sites 16 and 17. These two carbons form the double bond in the central vinylene group. Removal of one or two electrons sets up a charge density wave (CDW) along the carbon skeleton, which is delocalized over the whole length of the molecule and, most likely, limited by end effects. The amplitude of the CDW is much larger for the bipolaron than for the polaron.

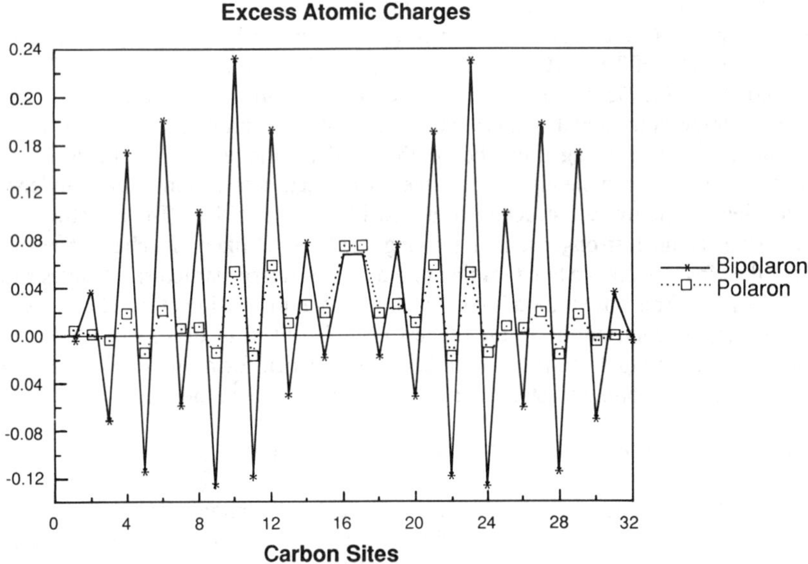

Figure 7. Polaron and bipolaron charge distribution in PPV oligomers. The excess atomic charge from MNDO relative to the neutral system is plotted versus the carbon sites.

The largest charge oscillations are found away from the center of the defect. A similar CDW is predicted for polythiophene[14] and trans-$(CH)_x$[13] and has been observed experimentally in solutions of long cation polyenes.[15] It would be of great interest to see if thiophene or PPV oligomers show the same charge oscillations.

We are interested in the geometry of the defects at the center of the molecule and the bond length alternation patterns of the neutral, singly and doubly charged molecule to relate the calculated structures to the results discussed in the previous section. The bond lengths at the center of the oligomer are given in the top part of Table IV for the neutral, polaron and bipolaron case. Analogous to the procedure in the previous section we define bond length alternations and relative bond length alternations normalized to average lengths for the two subunits, the vinylene group and the ring, and for the average bond length alternation along the backbone.

$$\Delta r_v = R - D \qquad \Delta r_v / r_v = \frac{R - D}{(R + D)/2}$$

$$\Delta r_B = t - t' \qquad \Delta r_B / r_B = \frac{t - t'}{(t + t')/2}$$

$$\Delta r = R + t - [D + t'] \qquad \Delta r / r = \frac{R + t - [D + t']}{(R + t + D + t')/4}$$

The bond lengths R, D, t and t' are defined through the force constants in figure 6. Note that, because of the inverse relationship between force constants and bond lengths, $\Delta r > 0$ describes an aromatic structure and $\Delta r < 0$ a quinoidal structure. The MNDO-determined bond length alternations are given in Table IV.

TABLE IV. MNDO geometries and bond length alternations at center of defect

Bond[a]	neutral	Polaron	Bipolaron	
D	1.344	1.389	1.428	
R	1.451	1.408	1.369	
t'	1.405	1.430	1.454	
t	1.390	1.377	1.357	
Δr_v	0.107	0.019	-0.059	$\Delta r > 0$: aromatic
Δr_B	-0.015	-0.053	-0.097	
Δr	0.092	-0.034	-0.156	$\Delta r < 0$: quinoid
$\Delta r_v / r_v$	0.0766	0.0136	-0.042	
$\Delta r_B / r_B$	-0.0107	-0.0380	-0.069	
$\Delta r / r$	0.0658	-0.0240	-0.111	

[a] Bond lengths and bond length alternations in Å.

In the neutral molecule the benzene rings possess a slight quinoidal character while a strong positive bond length alternation (aromatic) exists between the rings arising from a strong C=C double bond and a long single bond. This pattern is uniform across the whole length of the model oligomer. Within the above definitions of aromatic and quinoidal structure the average bond length pattern in the backbone is strongly aromatic. The largest changes occur in the bipolaron model. At the center of the molecule a complete reversal in bond length pattern is observed. The vinylene single and double bonds are interchanged and the benzene rings become strongly quinoidal. The distortion in the polaron case is less dramatic. Even at the center of the defect

320

a slight aromatic character is preserved with the two vinylene bonds almost equal. Both, the bipolaron and the polaron defect exhibit a smooth transition to the neutral geometry when moving away from the center of the molecule. The geometry for the two terminal PPV units is almost identical for the three cases. The bipolaronic distortion extends over approximately 4 PPV units while the width of the polaron is slightly less. Over the length of the central PPV repeat unit MNDO predicts a strong quinoid distortion for the bipolaron and a weakly quinoidal geometry for the polaron. In our vibrational analysis the bond order for the vinylene group shows a complete reversal, i.e. R_v changes from -0.095 (aromatic) to 0.0775 (quinoid). At the same time the geometry of the benzene rings is driven to a weak quinoid structure leading to a quinoidal distortion for doped PPV. These vibrational results strongly favor the bipolaron picture over the polaron structure in doped PPV.

In summary, valuable structural information can be obtained through a detailed vibrational analysis of neutral and doped PPV. The deduced vibrational defect structure supports the formation of bipolarons in the p-type doped polymer and agrees well with the bipolaron structure and the bond length alternation pattern calculated in the framework of MNDO. In particular, the strong bond order reversal of the central vinylene group observed in the vibrational spectra upon doping favors the bipolaron structure over the polaron structure.

4. References

1. H. Eckhardt, L.W. Shacklette, K.Y. Jen, and R.L.Elsenbaumer, J.Chem.Phys. <u>91</u>, 1303 (1989).
2. M.J.S. Dewar and W. Thiel, J.Am.Chem.Soc. <u>99</u>, 4899 (1977).
3. J.L. Bredas, in *Handbook on Conducting Polymers*, edited by T. Skotheim (Dekker, New York, 1986), Vol.2 pp.859-913.
4. S.J. Porter, Mater.Sci.Forum <u>21</u>, 43 (1987)
5. B. Tieke, C. Bubeck, and G. Lieser, Makromol.Chem.Rap.Commun. <u>3</u>, 261 (1982); J. Obrzut and F. Karasz, J.Chem.Phys. <u>87</u>, 2349 (1987).
6. H. Eckhardt, J.Chem.Phys. <u>79</u>, 2085 (1983).
7. J.L. Bredas, A.J. Heeger, and F. Wudl, J.Chem.Phys. <u>85</u>, 4673 (1986).
8. Y.S. Lee and M. Kertesz, J.Chem.Phys. <u>88</u>, 2609 (1988).
9. C.M. Foster, Y..H. Kim, N. Uotani and A.J. Heeger, Synthetic Metals <u>29</u>, E135 (1989).
10. S. Lefrant, E. Perrin, J.P. Buisson, H. Eckhardt and C.C. Han, Synthetic Metals <u>29</u>, E91 (1989).
11. S. Lefrant, J.Y. Mevellec, J.P. Buisson, E. Perrin, H. Eckhardt, C.C. Han and K.Y. Jen, Springer Series in Solid State Science Vol.<u>91</u>, pp. 123-126, (1989), H. Kuzmany, M. Mehring, S. Roth (Eds.),; J.Y. Mevellec, J.P. Buisson, S. Lefrant, H. Eckhardt and K.Y Jen, Synthetic Metals, in press.
12. S. Lefrant, E. Perrin, J.P. Buisson, H. Eckhardt and C.C. Han, to be published.
13. D.S. Boudreaux, R.R. Chance, J.L. Bredas and R. Silbey, Phys.Rev.B, <u>28</u>, 6927 (1983).
14. S. Stafstrom and J.L. Bredas, Phys.Rev.B, <u>38</u>, 4180 (1988).
15. L.M. Tolbert and M.E. Ogle, J.Am.Chem.Soc. <u>111</u>, 5958 (1989).

*Present address: Enichem Americas Inc., Monmouth Junction, N.J. 08852 (USA)

ORGANIC CONJUGATED MATERIALS FOR OPTOELECTRONIC AND APPLICATIONS

J.C. DUBOIS
Thomson-CSF
Central Research Laboratory
Domaine de Corbeville
91404 ORSAY Cédex, France

ABSTRACT. This conference deals with organic conjugated materials for non linear optics and their applications. The first part in devoted to the presentation of molecular structures and some of the parameters allowing to optimize the hyper-polarisability ß. The second part describes the possibility to obtain optical quadratic coefficient in polymers by using guest dyes or copolymers. The last part is the presentation of an electrooptic modulator made with this polymer. First results on a directional coupler are also presented.

INTRODUCTION. The use of the light as a vector for information in place of the electron allows to increase the speed of the signal processing. Optoelectronic system using optical films and laser diodes are now widely developped [1]. Now components for integrated optics such as modulators or directional couplers use inorganic crystals such as $LiNbO_3$ or KDP. In these materials Pockel effect or non linear optical response effects are used.

The Propagation of light waves in a transparent media is governed by the dielectric properties of this media.

- At low intensity the polarisation induced by the light is linear and expressed by :

$$P = P_o + \chi E$$

where χ = susceptibility
           E = electric field
           P_o = static polarisation with an isotropic medium
           $\chi = n^2 - 1$ (n = refractive index)

- At high intensity level (lasers) the induced polarisation is non linear and we have :

$$P = P_o + \chi^{(1)} E + \chi^{(2)} EE + \chi^{(3)} EEE + ...$$

where $\chi^{(2)}$, $\chi^{(3)}$ = susceptibility of second and third order.

321

J. L. Brédas and R. R. Chance (eds.), Conjugated Polymeric Materials:
Opportunities in Electronics, Optoelectronics, and Molecular Electronics, 321–340.
© 1990 *Kluwer Academic Publishers. Printed in the Netherlands.*

$\chi^{(2)}$ is important for second harmonic generation (SHG) (frequency doubling) or other applications concerning Pockel effect or parametric oscillator.
$\chi^{(3)}$ is important for third harmonic generation (THG) or Kerr effect.

The second harmonic properties can be described by a figure of merit and compared between inorganics and organic materials (see Table 1). This is obvious that the organic materials such as POM or NPP are the best in the frequency range visible and near infrared and specially, the doped polymers.

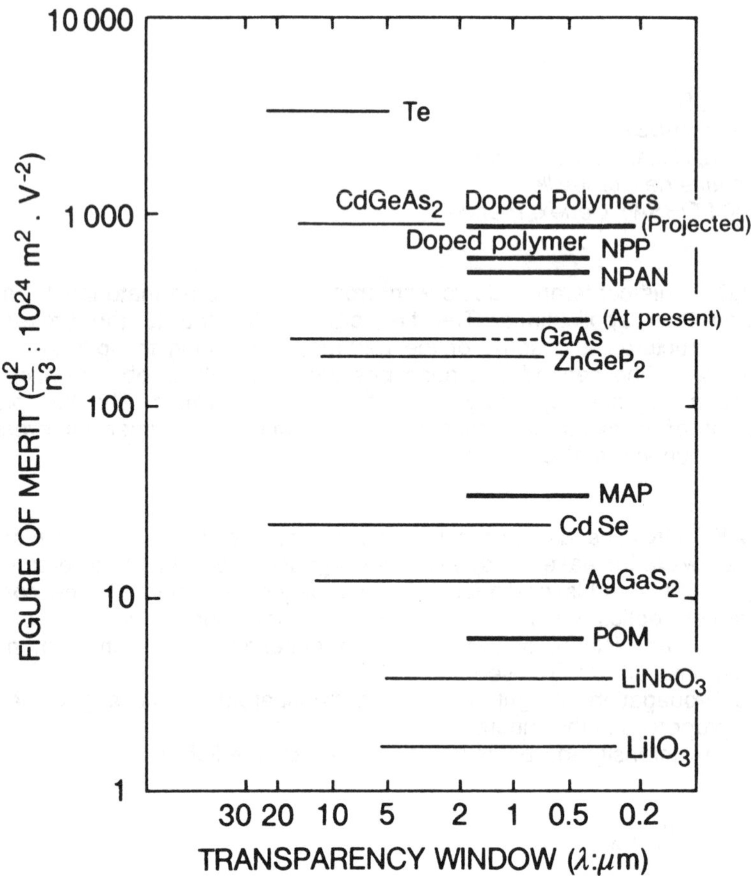

Table 1 : Figure of Merit for nonlinear material versus wavelength λ.

It would be too simple to describe the second harmonic properties by the figure of merit. One have also to include the processability, the optical damage threshold, the transparency and the ageing. It is well known that organics can resist as high power beam (> GW.cm^{-2}). This property is important for the second harmonic generation efficiency that is proportional to the square of the incident power and for the high power beam modulation specially in integrated optical applications.

So the advantages of organic materials are :

- At the optical frequencies the NL effects are purely electronic, it is why the response is very rapid.

- There is an enormous potential of molecular engineering to fit the NL properties and transparency.

- The low dielectric constant (ε = 3 to 4) with $\sqrt{\varepsilon}$ = n), reduces the mismatch between optical wave and electrical wave at high frequency.

- Processability is also an important quality.

The disadvantages are :

- a limited optical window (due to visible absorption) and the life time has to be tested.

Both electronic and chemical companies develop research in this field of organic materials for NLO. Moreover, in Europe CEE support in ESPRIT[1] or RACE such research programs. In Japan it is to mention for example, the "Frontier program" with the Riken Institute.

This paper will give some aspects of organic materials for integrated optics : molecular structures, polymers, example of application with a modulator.

1. Molecular Structures for Quadratic Non Linear Effects

1.1. GENERAL STRUCTURE.

Moleculars having a strong second order non linearity are conjugated molecules of general formula :

$$ \text{A} - \boxed{\pi \text{ links}} - \text{D} $$

where A = electron acceptor group

       D = electron donating group

such as :

$$ \underset{O}{\overset{O}{\diagdown}} N - \bigcirc - N \underset{H}{\overset{H}{\diagup}} $$

The dielectric response of such a molecule submitted to an electric field E(w) is :

- at low intensity level the induced polarization is :

$$ p = p_0 + \alpha\, E(\omega) $$

---

[1]ESPRIT research program with Thomson (F), ICI (G.B.), CNET (F), Université Notre-Dame Namur (B), CEA (F)

324

- at high intensity level the polarization becomes :

$$p = p_0 + \alpha \, E(\omega) + \beta \, E(\omega) \, E(\omega) + \gamma \, E(\omega) \, E(\omega) \, E(\omega)$$

with β, γ : hyperpolarisability of second and third order.

The non linear coefficients will depend on the values of the hyperpolarizability of the molecule, a strong charge transfer interaction between the D substituent and A substituent takes places giving rise to a large β enhancement. In that case, the quadratic polarizability can be decomposed into two parts :

$$\beta = \beta_{add} + \beta_{C.T}$$

where β_{add} is an additive part due to substituent induced asymmetry in the charge distribution

$\beta_{C.T}$ is the charge transfer contribution.

A useful expression of $\beta_{C.T}$ has been derived from the perturbation approach employing a two-level model :

$$\beta_{C.T} = \frac{3 \, e^2 \hbar^2 \, F \, \Delta\mu}{2 \, m \, [\, W^2 - (2\hbar\omega)^2 \,] \, [\, W^2 - (\hbar\omega)^2 \,]}$$

where W is the energy of the optical transition (charge transfer band)
$\hbar\omega$ is the fundamental photon energy
F   is the oscillator strength of the optical transition
$\Delta\mu$ is the difference between ground and exciteds tate dipole moment
$(\mu_g - \mu_e)$.

Since the early observations, considerable efforts (both at the theoretical and the experimental levels) have been made to establish some relationships between the chemical structure of the organic molecules and their efficiency as materials for quadratic effects. Theoretical calculations are now a reliable tool to predict the magnitude of the molecular hyperpolarizabilities and provide useful guidelines for the synthetic chemist.

Quadratic non linear materials have to exhibit a large effective non linear coefficient d. This can be achieved by maximizing the second order hyperpolarizability β of the molecules and by optimizing the material structure : bulk monocrystal, Langmuir-Blodgett films, monocrystalled thin films and poled polymers. One will focus on the design of organic compounds for quadratic effects and their related properties.

1.2. NATURE OF THE SUBSTITUENTS.

Use of more effective electron attractor donor substituents increase the second order non linearity of the molecules [3] as it is exemplified in Table 2.

The classification of functional groups as electron withdrawing or electron donating groups, relative to hydrogen and their efficiency are wellknown from the synthetic chemist (Table 3).

A \ D	CH$_3$	OCH$_3$	NH$_2$	N(CH$_3$)$_2$
CN	12	20	56	60
CHO		35		96
COCH$_3$				
NO$_2$	30 to 38	59 to 73	192 to 199	215

Chemical structure: D—⟨benzene ring⟩—A **3**

Table 2 : β values (10^{-40} m^4 V^{-1}) for a series of p-disubstitued benzene derivatives obtained by the EFISH technique at $\lambda = 1,89$ μm in DMSO (10).

Acceptor groups (A)	N$_2^{\oplus}$ > NO$_2$ > COCH$_3$ > CHO > CN
Donor groups (D)	O$^{\ominus}$ > N(CH$_3$)$_2$ > NH$_2$ > OCH$_3$ > CH$_3$

Table 3 : Classification of electron donors and electron attractors.

The fully charged substituents N$_2^+$ and O$^-$ are among the most interesting A and D groups but have found very limited use in NLO molecules today. More sophisticated acceptor groups [3] such as the dicyano-vinyl and tricyano-vinyl moieties have recently proved their high efficiencies (Table 4) [4, 5].

1.3. EFFECT OF THE LENGTH OF THE CONJUGATED π SYSTEM.

The effect of increasing the conjugated π-system length on ß is clearly demonstrated in table 5.

An order of magnitude increase in ß can be obtained in going from one benzene ring to three.

But it seems that it exists an optimum value for the length of the molecule when the effective hyperpolarizability per unit volume $\rho = ß/V$ is considered [6]. For example, the following disubstituted polyphenyls [6] reach the optimum value of ρ when $n = 3$. Figure 1 shows that the ß value increases with n (number of ring) but the ρ value decreases drastically for $n = 3$.

$(CH_3)_2N$ —⟨◯⟩— A			$\underset{\sim}{4}$
A	NO_2	$-\underset{H}{\overset{}{C}}=C\underset{CN}{\overset{CN}{}}$	$-\underset{CN}{\overset{}{C}}=C\underset{CN}{\overset{CN}{}}$
$ß\,(10^{-40}\ m^4/V)$	88	130	326

Table 4 : ß values (10^{-40} m⁴/V) for a series of p substitued nitro benzene derivatives obtained by the EFISH technique at $\lambda = 1.35\ \mu$m in DMSO.

H_2N —(⟨◯⟩)ₙ— NO_2	$\underset{\sim}{5}$
n	$ß\ (10^{-40}\ m^4/V)$
1	24
2	84
3	210

Table 5 : ß values (10^{-40} m⁴/V) for a series of p-disubstitued polyphenyl derivatives obtained by the solvatochromism method (12).

1.4. INFLUENCE OF THE PLANARITY OF THE MOLECULES.

In order to achieve conjugation with the best efficiency, molecules have to be plane. Table 6 shows some examples of the influence of the planarity on ß.

The dimethylamino group in compound 7 is not quite coplanar with the rest of the molecule. Forcing the nitrogen into coplanarity through a covalent bonding to the benzene ring (compound 8) results in an effective doubling of ß. The azomethine linkage is not planar and will disrupt conjugation in molecule 9. Incorporation of an ortho hydroxyl group (molecule 10) creates an intra molecular hydrogen bond which forces the molecule into planarity and results in a threefold increase in ß.

MOLECULES	$\beta\ (10^{-40}\text{m}^4\text{V}^{-1})$
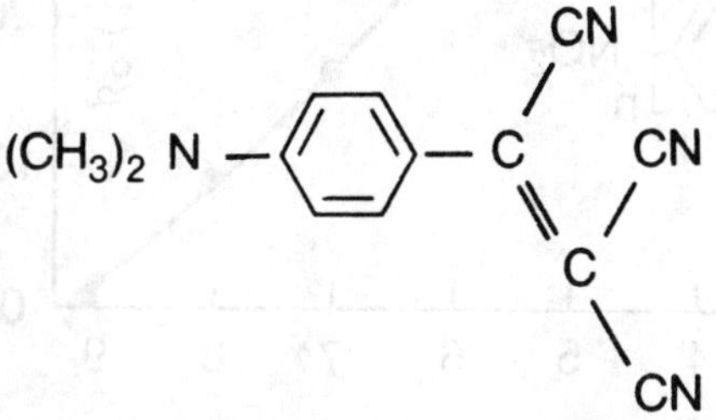 $(CH_3)_2N$—⟨benzene⟩—C(CN)=C(CN)—C(CN)(CN) **7**	90
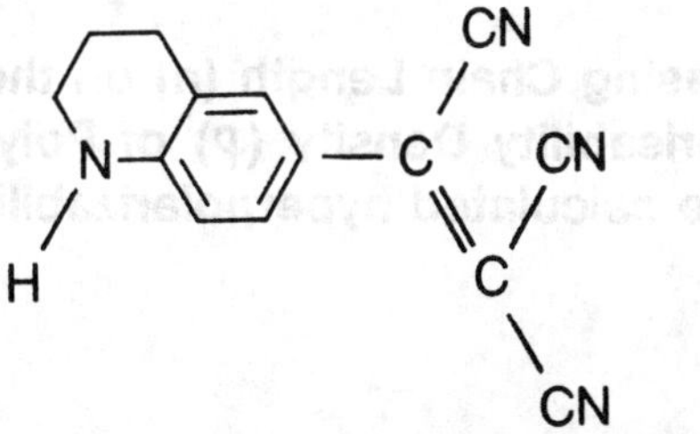 (fused N–H ring)—C(CN)=C(CN)—C(CN)(CN) **8**	174
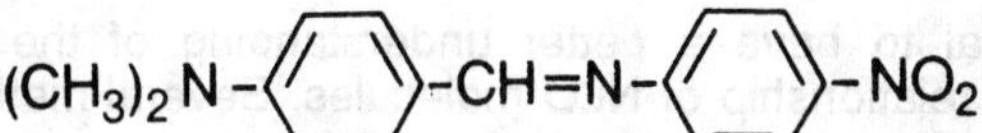 $(CH_3)_2N$—⟨benzene⟩—CH=N—⟨benzene⟩—NO_2 **9**	98
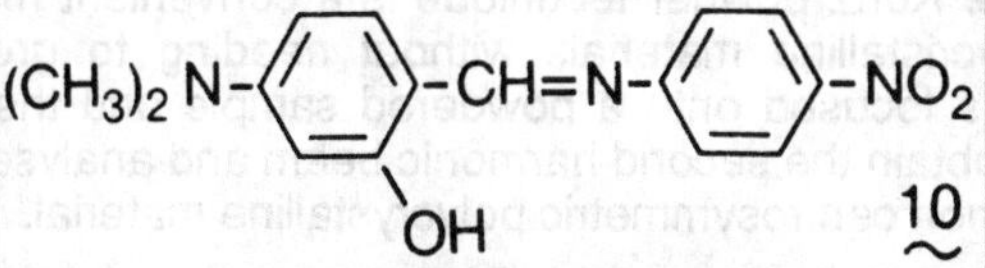 $(CH_3)_2N$—⟨benzene⟩—CH=N—⟨benzene(OH)⟩—NO_2 **10**	258

Table 6 : Influence of the planarity of the molecules on ß values (10^{-40} m^4/V) obtained by solvatochromism method (12).

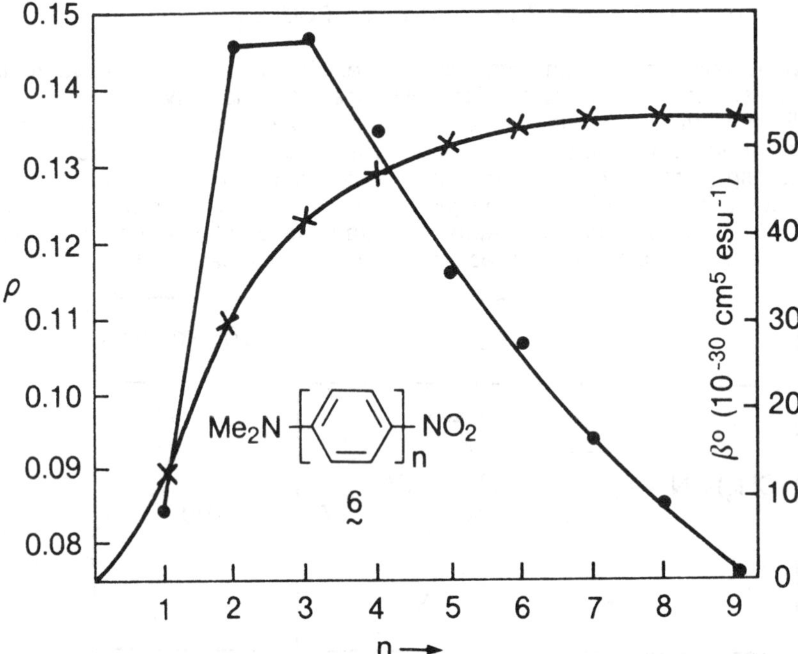

Fig.1 : Effect of Increasing Chain Length (n) on the Calculated Hyperpolarisability Density (ρ) of Polyphenyls (•) and on the calculated hyperpolarizabilities (β°) (X)

1.5. DETERMINATION OF THE QUADRATIC HYPERPOLARIZABILITY ß

Determination of ß is essential to have a better understanding of the chemical structure - quadratic properties relationship of NLO molecules. Several methods are now available for measuring ß.

1.5.1. *Kurtz Powder Technique.* Kurtz powder technique is a convenient method for testing large numbers of polycristalline materials without needing to grow large single crystals. A laser beam is focused onto a powdered sample and the emitted light is collected and filtered to obtain the second harmonic beam and analysed.
This technique requires a non centrosymmetric polycrystalline material.

I.5.2. *Solvatochromism Method.* Solvatochromism is a simple but approximate method for evaluating the ß of organic molecules by using a UV - visible spectrophotometer and a Abbe refractometer [7]. The dependence of the optical absorption band on solvent polarity, allows the determination of Δμ recording the UV - visible spectra of the studied molecule in two different solvents.

The absorption band shift Δv is expressed by :

$$\Delta v \sim \Delta f \, (\Delta \mu)^2$$

with $\quad \Delta f = \dfrac{\varepsilon - 1}{2\varepsilon + 1} - \dfrac{n^2 - 1}{2n^2 + 1}$

where ε is the dielectric constant
$\quad\quad n$ is the refractive index
This method requires obviously a non centrosymmetric material.

I.5.3. *Langmuir-Blodgett Film*. ß can be determined from a SHG (Second Harmonic Generation) experiment performed on a monolayer L.B. film.

This method which has recently been used [8, 9] for that purpose. The amphiphilic molecule with a high ß value has been designed : the DPNA (4-[4-N-n-dodecyl-N-methylamino) phenylazo]-3 nitrobenzoic acid.

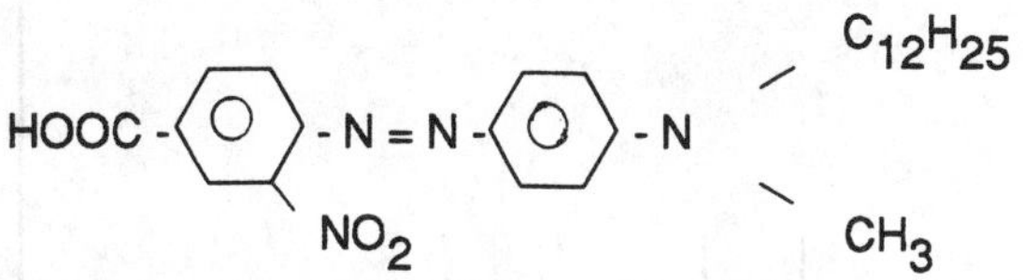

HOOC -⟨O⟩- N = N -⟨O⟩- N $\diagup$ $C_{12}H_{25}$ $\diagdown$ CH_3, with NO_2 group

DPNA with $\beta = 1200 \times 10^{-30}$ e.s.u.[2]
to be compared to :

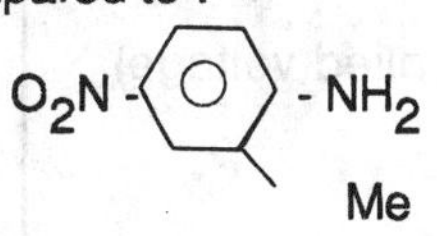

O_2N -⟨O⟩- NH_2 , Me

MNA $\quad\quad \beta = 17 \times 10^{-30}$ e.s.u.

I.5.4. *EFISH Method*. The electric field induced second harmonic (EFISH) method is an accurate technique to determine ß. In that case, the material has to be noncentrosymmetric. The compound under investigation is put into solution and a strong DC electric field E induces an orientation. The induced quadratic non linearity can then produce the second harmonic of a laser beam, from which ß can be extracted, by taking into account the second term γE due to the third hyper-polarizability γ. ß depends on the solvent.

---

[2]Foot note :

Three different units can be used to express ß which sometimes gives rise to some difficulties. The conversion of unit is the following :

$$\beta \; (m^4 \, V^{-1}) = 4\pi/3 \; 10^{-10} \quad \beta \; (cm^5 \, esu^{-1})$$

$$\beta \; (C^3 \, J^{-2} \, m^3) = 3.7 \; 10^{-21} \quad \beta \; (cm^5 \, esu^{-1})$$

330

2. Quadratic Non-Linear Effects in Polymers

2.1. POSSIBLE STRUCTURES.

One of the application of material having a high $\chi^{(2)}$ is frequency doubling. This $\chi^{(2)}$ can be obtained from molecules with high ß value. This requires a non centrosymmetric organization of the molecules. This organization can be obtained in a crystal, but 80 % of the compounds crystallise in a non centrosymmetric system. This organization can be obtained in a Langmuir-Blodgett film as we have seen. Another way is to use the orientation of this molecule with an electric field in a polymer matrix.

Such a system is formed by dissolving N polarizable molecules in a polymer, the polymer solution is spin-coated and a film is formed. The non centrosymetric system is obtained by application of a transverse electric field (Fig. 2).

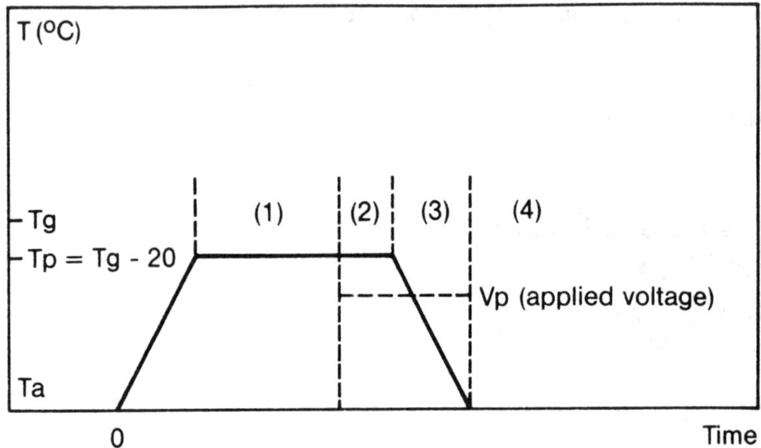

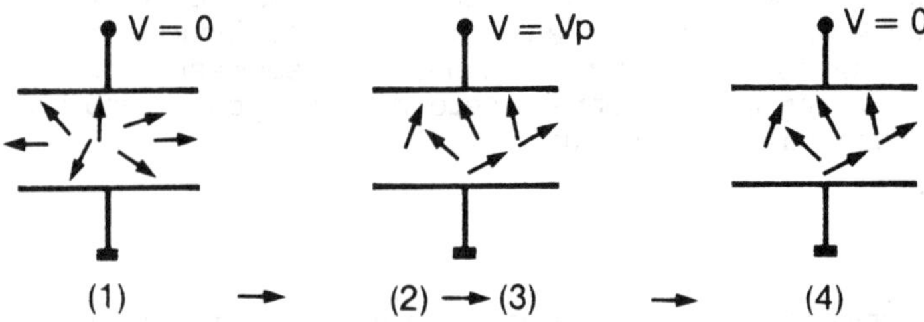

Fig. 2 : Schematic representation of molecular orientation under an electric field.

The advantages of such a system is that it can be easily manufactured into different shapes such as cylinders, fibers, thin films. They can be applied over large areas and over various substrates. They can be produced much faster than crystals at a reduced cost.

There are different major types of polymer structures :
- crystalline and semi-crystalline polymers
- amorphous polymers
- liquid crystalline polymers.

Some elements concerning these polymers are described now.

. Concerning liquid crystalline polymers [10] :
- they can exhibit a variety of mesophases (N, N*, S, D)
- the degre of orientation may be preserved by cooling it down below the Tg

. Concerning crystalline polymers : they must be fabricated into a single crystal format in order to avoid light scattering, but crystal growth is not consistent with optimal molecular packing for getting NLO properties. This is particularly true for the molecular structures designed for high ß. These structures involve the existence of large ground state dipoles moments in the molecules. The molecular packing which achieves the lowest crystal energy state, usually involves packing of dipoles in opposing directions which give centrosymmetric structures.

. Semi-crystalline polymers :
- consist of crystalline domains separated by amorphous regions
- they are not interesting for NLO applications, due to intense light scattering from the crystalline polymer regions.

. Amorphous polymers :
- they adopt a nearly random chain conformation
- the polymer chain does not interfere with the alignment of the NLO molecules in the poling field.

The magnitude of non linear optic coefficients in amorphous polymers can be estimated from thermodynamic considerations that include the effect of local fields [11].

$$\chi^{(2)}_{zzz} \simeq NF^{2\omega} F^{\omega} \beta \left[\frac{\varepsilon (n^2 + 2)}{n^2 + 2\varepsilon} \right] \frac{\mu E_p}{5 kT}$$

(we will use also d_{33} coefficient in place of $\chi^{(2)}_{33}$)

N : number density of NLO molecules
ß : hyperpolarizability
μ : static dipole moment
E_p : electric field
f : local field factor (due to the contribution of neighbour molecules).

The dye can be introduced in solution in the polymer used as a matrix. The system is called "guest host system" or the dye can be introduced via a comonomer in a copolymer. We will describe now examples of these systems.

332

2.2. GUEST HOST AND COPOLYMER SYSTEMS

The guest host system is prepared by mixing active molecules in polymethyl metacrylate Elvacite 2041 (Dupont) M_W = 950 000. A 8 % (W/W solution of PMMA in methylisobutylketone (MIBK) is first prepared. The dopant, an azo dye provided by ICI (SC 113 172) is then dissolved in the previous solution (figures 3, 4).

To obtain the copolymer system, the monomeric dye is prepared by esterification of Diperse Red 1 with methacryloyl chloride in pyridine at 100°C for 30 mn. After cooling, the reaction mixture is poured into water. The crude product is purified by chromatography on silicon gel (using toluene as eluant and crystallization from cyclohexane. Yield 24 % melting point 97.5°C, UV (chloroform) : λ_{max} = 474 nm, ε = 32 000 lmole^{-1} cm^{-1} (figure 4) instead of 488 nm obtained with the guest host system.

Copolymerizations are carried out at 60°C in DMF for 24 hours. AIBN is used as an initiator. The solutions of monomers (containing AIBN) are thoroughly outgassed by the conventional freeze - thaw technique, before sealing and removing from the vacuum apparatus. Concentration in monomers is equal to 10 % in weight, the [monomers] / [AIBN] molar ratio was equal to 400. Purification is accomplished by two reprecipitations in ether, after which the copolymers are dried under vacuum. The yield is about 70 % in all cases.

The compositions of the copolymers are determined by UV - visible measurements performed in chloroform as solvent. The molar dye content of the copolymers is slighly higher than of the feed.

The glass transition temperatures of the copolymers are in the order of 130°C.

Thin films (1 to 2 μm) of doped polymers or copolymers are obtained by spin coating onto indium tin oxide (ITO) coated on glass and a transparent ITO counter-electrode is deposited under vacuum. Then these thin films are poled under high electric field (> 0.8 MV/cm) at Tg - 20°C temperature. The absorption spectra gives the absorption coefficients $\alpha^{2\omega}$ and α^{ω} (figure 4) and show a down-shift of λ_{max} in the case of copolymer. This is due the change of electronic structure and environment between guest host and copolymer systems.

At 1.06 μm wavelength, different dye weight percentages have been studied (27,30 and 10 %) ; whatever this percentage is, the same typical curve shape has been obtained. The second harmonic intensity increases according to the square of the poling field Ep (fig. 5). These results are in good agreement with the model introduced by Singer and al [12] featuring a linear dependence of d_{eff} with E_p.

The effect of poling temperature on d_{eff} has also been studied for PMMA doped with 1.5 or 10 % dye in weight. Poling field of 90 V per micrometer has been applied which is a little bit less than the breakdown electric field (100 V per micrometer).

The plots of $I^{2\omega}$ versus poling temperature exhibit a maximum around 80 degrees (fig. 6). This temperature is slightly lower than the glass transition temperature of the guest host system which is around 100 degrees because of the dye plasticizing effect. On one hand, at low temperature a high viscosity of the medium prevents the NLO molecules from moving easily, on the other hand at higher temperature the thermal randoming effect disturbs the molecules orientation. 80 degrees seems to be a good compromise between these two effects.

The second harmonic generation coefficients obtained with optimal poling conditions have been quantitatively studied for several contents of NLO molecules.

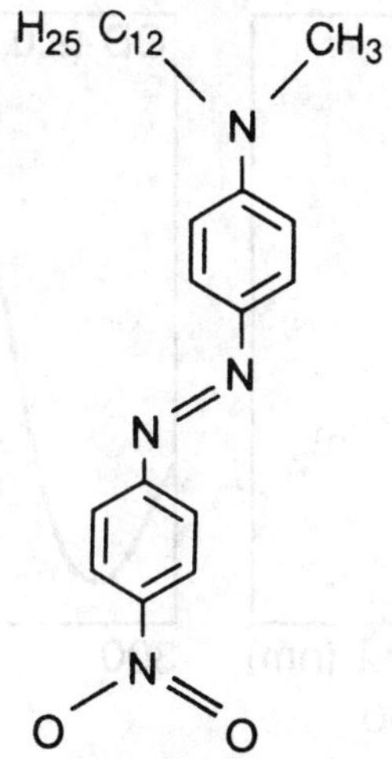

Fig. 3 : Molecular structures of dopant (11) and copolymers (12) used as experimental samples.

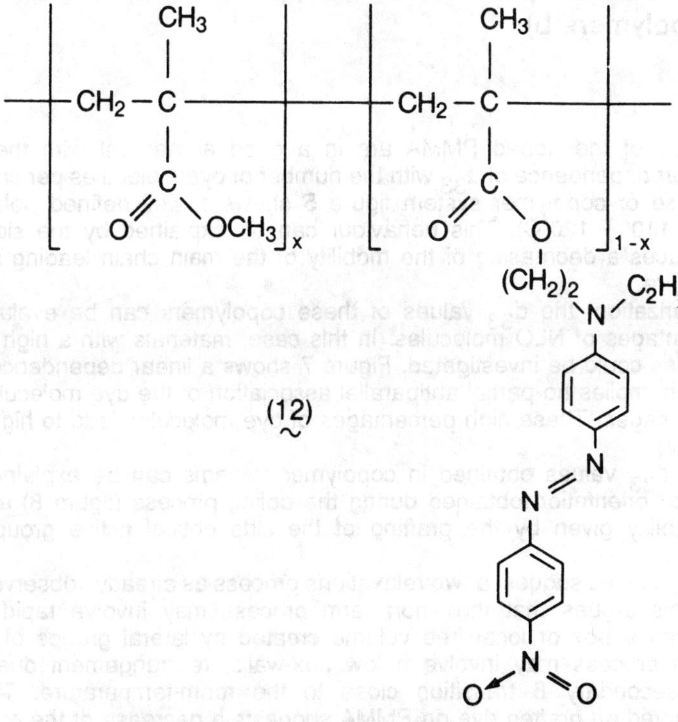

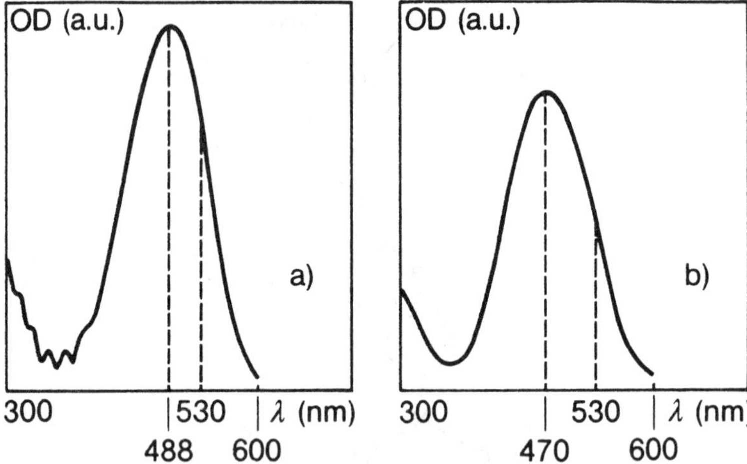

Fig. 4 : Absorption band for doped PMMA a) and for copolymers b).

The d_{33} values of the doped PMMA are in a good agreement with the model featuring a linear dependence of d_{33} with the number of dye molecules per cm^3.

In the case of copolymer system figure 5 shows a well defined polarization temperature ($\sim110°$ - $120°C$). This behaviour can be explained by the side optic group that induces a decreasing of the mobility of the main chain leading a higher glass temperature.

After polarization, the d_{33} values of these copolymers can be evaluated for different percentages of NLO molecules. In this case, materials with a high number of dye molecules could be investigated. Figure 7 shows a linear dependence of d_{33} versus N, which implies no partial antiparallel association of the dye molecules even for large percentages. These high percentages of dye molecules lead to high values of d_{33} (57 pm/V).

The best d_{33} values obtained in copolymer systems can be explained by a better molecular orientation obtained during the poling process (figure 8) and by a better time-stability given by the grafting of the side optical active group (figure 18).

The decay curves suggest a two relaxations process as already observed by Ye et al. [13]. This argues that the short term process may involve rapid dopant reorientation into a box or local free volume created by lateral groups of PMMA. The long term process may involve a low box-walls, rearrangement due to the existence of secondary ß transition close to the room-temperature. The lower decrease observed on grafted dye on PMMA suggests a decrease of the correlation length between two consecutive lateral PMMA groups weakening the ß transition effect. New polymers actually show a good time stability (figure 9).

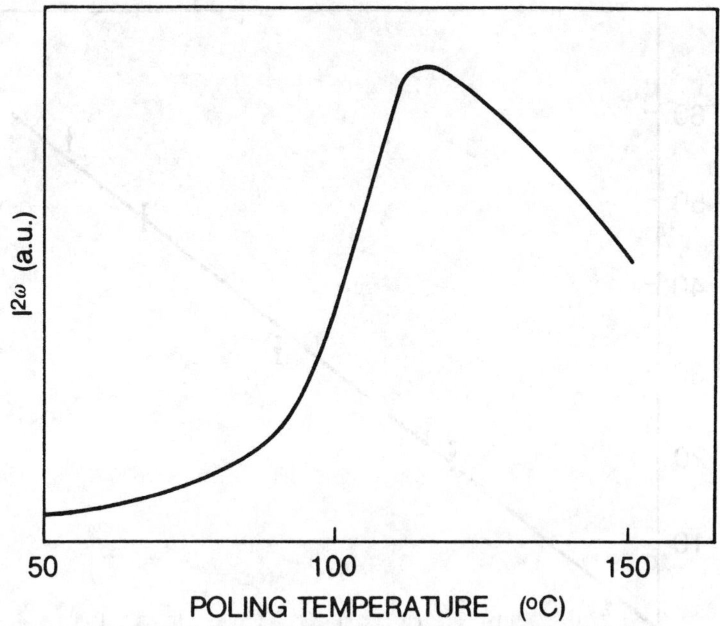

Fig. 5 : $I^{2\omega}$ versus poling temperature for a copolymer with a Tg = 130°C.

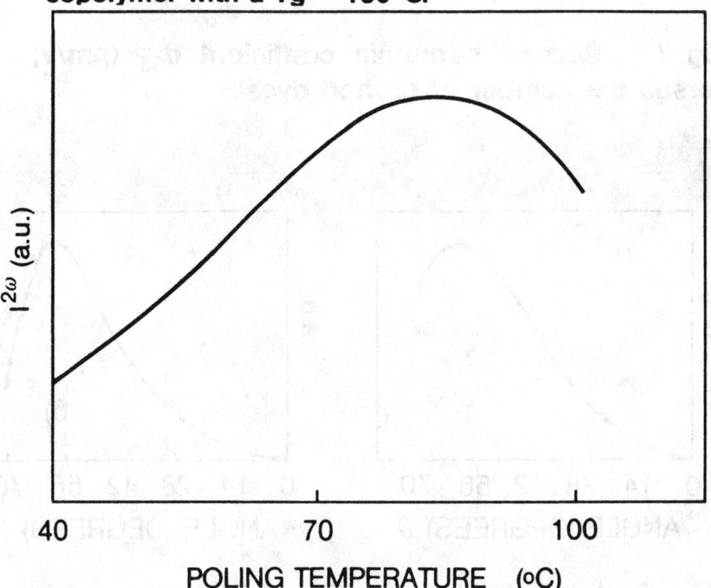

Fig. 6 : $I^{2\omega}$ versus poling temperature for a doped polymer with a Tg = 110°C.

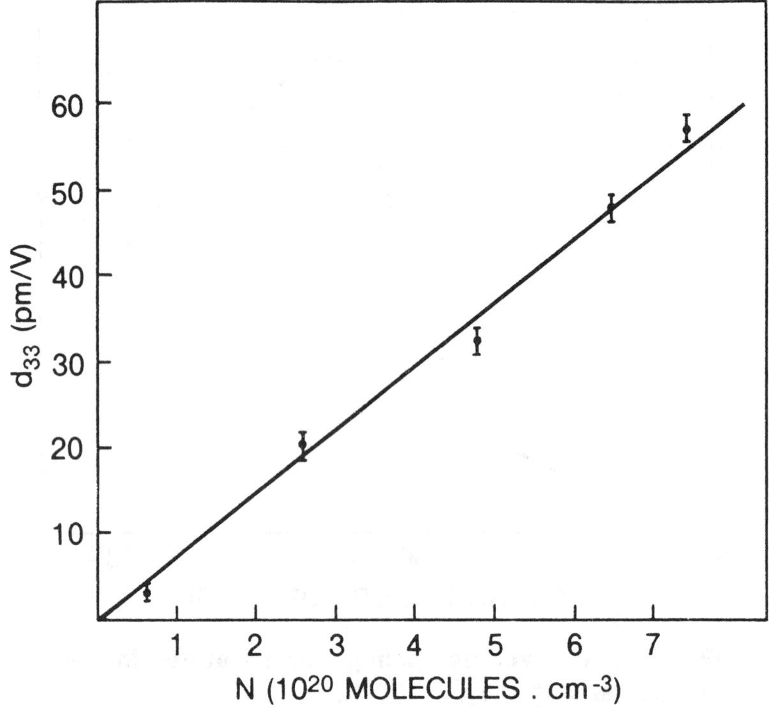

Fig. 7 : Second harmonic cœfficient d_{33} (pm/v) versus the number of grafted dyes.

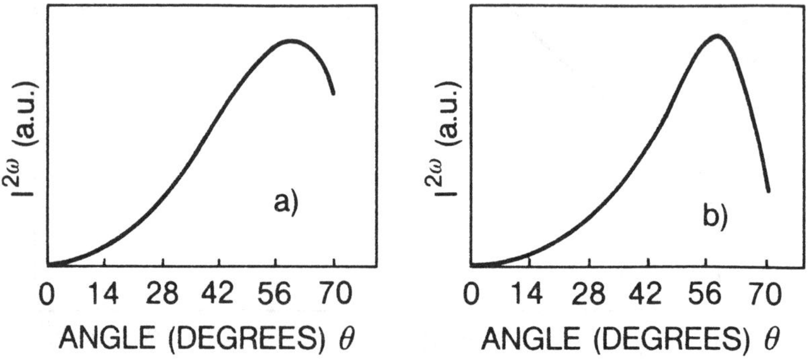

Fig. 8 : Intensity at 2ω for doped PMMA a) and for copolymers b).

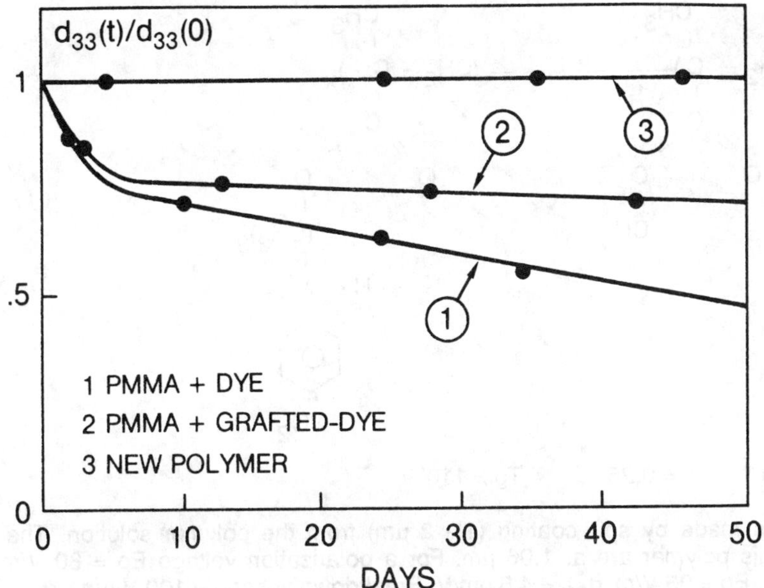

Fig 9 : Temporal behaviour of the second harmonic cœfficient at room-temperature of a doped PMMA with dye ① a grafted dye on PMMA ② and a new polymer ③.

3. Electrooptic Modulator (14)

We had the objective to realize an electrooptic modulator working in visible wavelength (670 nm) with low losses and with a modulating voltage of some ten volts. We describe here one of the component.

3.1. POLYMER STRUCTURE.

We have made this component using the following polymer :

$\lambda_{max} = 370$ nm $\qquad x = 0.25 \qquad T_g = 110°$

The film is made by spin-coating (1 - 2 µm) from the polymer solution. The properties of this polymer are at 1.06 µm. For a polarization voltage $E_p = 80$ V/m $d_{33} = 3.2$ pm/V, $E_p = 95$ V/m $d_{33} = 4.5$ pm/V. Breakdown voltage = 100 V/µm.

3.2. MODULATOR STRUCTURE (Fig. 10)

The active layer is sandwiched between two buffer layers of refractive index inferior to the refractive index of the active layer (n = 1.567). The glass structure substrate sustain the first electrode (chromium gold layers). The first buffer is a photocrosslinked polymer of refractive index 1.51. The active polymer is spin coated on this layer. The second buffer is a fluorinated polymer of refraction index n = 1.42. The electrode 2 is evaporated aluminium.

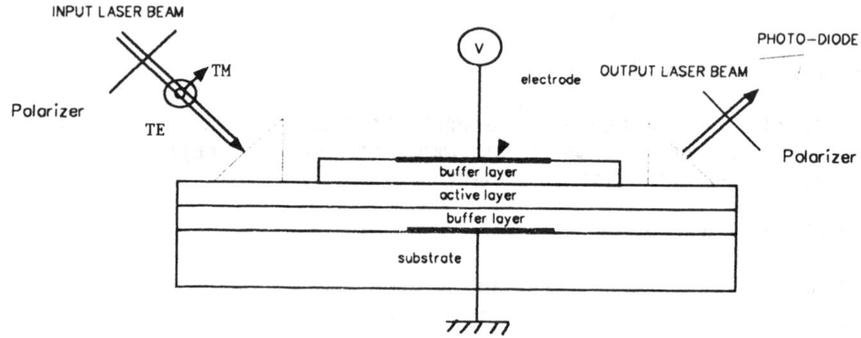

Fig. 10 : Polymer electrooptic modulator

3.3. MODULATOR CHARACTERISTICS

The coupling of the laser beam is achieved by a prism into the active layer. The light intensity (measured by the photodiode) is (fig. 11) : $I = \sin^2 \varphi/2$ where φ = birefringence, with $\varphi = \varphi_o + \varphi_{created}$ when φ_o = static birefringence. By variation of the modulating electric field we obtain curves of different shapes. v_π correspond to the voltage difference between 2 curve of constant shape.

The characteristics of the electrooptic modulator are typicaly :
- total thickness : 3 μm
- driving voltage : v_π = 40 V (λ = 0.6328 μm)
- length : 1 cm
- losses = 3 - 4 dB/cm

(with coefficient : d_{33} = 3.2 pm/V).

Other results have been obtained by using a silicon - silica substrate. A directional coupler have been made with direct laser beam insertion.

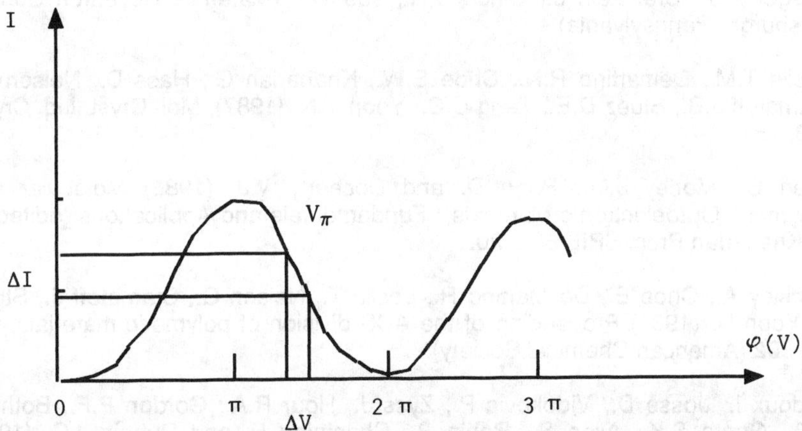

Fig. 11 : Electrooptic modulator : interferences as a function of V. I = light intensity voltage ; vπ : modulating voltage

4. Conclusion

Doped amorphous polymers and copolymers are good materials for integrated optical devices. Coefficient d_{33} as high as 60 pm/V can be obtained. This is very interesting specially for electrooptic modulation because of the potentially low driving power and large bandwith. A first electrooptic modulator has been made with a driving voltage of some ten volts this performance can be improved. Other potential applications involve frequency doubling. No doubt that organic materials due to their large possibilities will find applications in integrated optics.

340

References

[1] Dubois J.C. (1985) "The Future of Chemistry in Electronics", Info. Chimie n°
     259 Proceedings Forum France - USA.

[2] Broussoux D.and al. (1989) "Organic Materials for Non Linear Optics", Revue
     Technique Thomson, vol. 21-22, 3, p. 151.
     Esselin S., Le Barny P., Robin P., Broussoux D., Dubois J.C. and Raffy J.
     (1988), SPIE Proc. vol. 971 p. 120.

[3] Zyss J. and Chemla D.S. (1987) "Non linear optical properties of organic
     molecules and crystals"
     Chemla D.S., Zyss J. (1987), eds. 23 (Academic Press New York).

[4] Katz J.E., Dirk C.W., Schilling M.L., Singer K.D. and Sohn J.E. (1988) "Non
     linear optical properties of polymers" Mat. Res. Soc. Symp· Proc. vol. 109,
     Heeger A.J., Drenstein J., Ulrich D.R., eds 127 (Materials Research Society
     Pittsburgh, Pennsylvania).

[5] Leslie T.M., Demartino R.N., Choe E.W., Khanarian G., Hass D., Nelson G.,
     Stamatoff J.B., Stuez D.E., Teng C.C., Yoon Y.N. (1987), Mol. Cryst. Liq. Cryst.,
     153, 451.

[6] Allen S., Morley J.O., Pugh D. and Docherty V.J. (1986) Molecular and
     Polymeric Optoelectronic Materials : Fundamentals and Applications, edited by
     G. Khanarian Proc. SPIE 682, 20.

[7] Buckley A., Choe E., De Martino R., Leslie T., Nelson G., Stamatoff J., Stuetz
     D. Yoon H. (1986) Proceeding of the ACS division of polymeric materials, vol.
     54, 502 (American Chemical Society).

[8] Ledoux I., Josse D., Vidokovic P., Zyss J., Hour R.A., Gordon P.F., Bothwell
     B.D., Gapta S.K., Allen S., Robin P., Chastaing E. and Dubois J.C. (1987)
     Europhysic Letters 3 (7) p. 803-809 (1987).

[9] Murray R.T. (1988) Phys. Sc. vol. T23.

[10] Dubois J.C. (1988) Phys. Scr. vol. 23, 299-305.

[11] Singer D., Kuzyk M.G. and Sohn J.E. (1987) J. Opt. Soc. Am., B4, 968.

[12] Singer K.D., Lalama S.J. and Sohn J.E. (1985), SPIE Proc. 578 "Integrated
     optical circuit engineering II".
     Singer K.D., Sohn J.E. and Lalama S.L. (1986), Appl. Phys. Lett. 49, 5, 248.

[13] Ye C., Minami N., Marks T.J., Yang J. and Wong G.K. (1988) to be published in
     NATO Symposium Proceedings, Nice.

[14] Bourbin Y., Papuchon M. and coll. (1987), Revue Technique Thomson-CSF, n°
     3-4, vol. 19, p. 519-613.

NLO of Conjugated Polymers:
Progress in Science and Prospects for Technology

S. Etemad, W-S. Fann, P.D.Townsend, G.L.Baker and J. Jackel
Bell Communications Research,
Red Bank, NJ 07701, USA

ABSTRACT. In this chapter we highlight two studies carried out at Bellcore in an effort to better understand the nonlinear optical (NLO) properties of conjugated polymers, and to establish an existence proof of an organic-based all-optical switching device. The first is the spectrum of $\chi^{(3)}$ in polyactylene. Using an infrared free electron laser (FEL) we have measured the spectrum of $\chi^{(3)}$ to energies below $\sim E_g/4$. We have identified both two- and the three-photon resonance enhancements of $\chi^{(3)}$, and have discussed their implications on the role of Coulomb correlations in this simplest of all conjugated polymers. The second is characterization of a polydiacetylene-based directional coupler that can be used as a all-optical switching device. Nonlinear optical transmission and switching phenomena are observed in this prototype device. At $\lambda = 1.06$ μm, we find the transmission losses are dominated by an adverse two-photon absorption process. Such adverse effects can be eliminated by decreasing the photon energies to $h\nu < E_g/2$.

INTRODUCTION

The nonlinear optics (NLO) of lower dimensional systems continue to be an exciting field of research. There are still many issues that challenge scientific curiosity at the same time as being of interest for potential applications. Most recent studies have focused on two distinct class of materials, the two dimensional quantum wells based on artificially structured layered compounds such as GaAs:GaInAs and on conjugated polymers which are analogues for one dimensional semiconductors. While the optical nonlinearities associated with excitons in 2-D quantum wells have been extensively investigated, experimental studies of optical nonlinearities in 1-D systems have only recently become widespread. The overall NLO properties of the two classes of materials appear to be similar and follow the trend observed in going from 3-D (bulk) to 2-D (layered) materials. If it were possible to construct 1-D systems based on the GaAs:GaInAs superlattices, it is conceivable that the trend would have continued. Comparison of the optical nonlinearities of the 1-D and 2-D excitons has been instructive, since important differences have been found in the NLO properties of the two classes of materials. For example, in certain energy range near resonance, because of the strong electron-phonon interaction the phonon-mediated optical nonlinearities are

J. L. Brédas and R. R. Chance (eds.), Conjugated Polymeric Materials:
Opportunities in Electronics, Optoelectronics, and Molecular Electronics, 341–352.

the dominant processes in conjugated polymers.[1] [2]

In this chapter we highlight two of our recent studies on the $\chi^{(3)}$-based NLO of conjugated polymers.[1][3] [4] The first is the spectrum of $\chi^{(3)}$ in polyacetylene, the infinite polyene.[3] This study is important because it provides a measure of the relative size of gap for charged and neutral excitations (also discussed as the ionic and covalent gaps[5], respectively) in this simplest of all conjugated polymers. The second is a progress report on efforts to construct and demonstrate an organic-based all-optical switching device that uses the large optical nonlinearity of a conjugated polymer.[4] We have now an existence proof of such a device, a polydiacetylene-based directional coupler. Optically induced switching in *relative* intensities of two waveguides is observed. However, the total transmission losses are dominated by a two-photon absorption process at the operating wavelength of 1.06 μm. We discuss how to avoid the adverse effects of such a nonlinear absorption process in construction of a "useful" polydiacetylene-based all-optical switching device.

PROGRESS IN SCIENCE

The extensive recent interest in the nonlinear optical (NLO) properties of conjugated polymers has focused on studies of their third order susceptibility, $\chi^{(3)}(\omega)$, since almost all of these materials possess a center of symmetry. Most studies of the NLO properties of the polymeric semiconductors have been concerned with third harmonic generation (THG) experiments. Although many of these recent studies have been fueled by the technological implications of the large NLO coefficients in conjugated polymers,[6] an understanding of the large values of $\chi^{(3)}(\omega)$ in polyenes has historically been important because they are the simplest conjugated systems.[7] Equally important is the extenive body of works on one- and two-photon allowed transitions in polyenes that has promoted the role of Coulomb correlations in the development of quantum chemistry.[8] Polyacetylene, the infinite polyene, deserves special attention since it is the prototype conjugated polymer, and is useful in modeling the NLO properties of conjugated polymers in general. Unfortunately most of these studies have been based on single frequency measurements. In contrast, the spectrum of $\chi^{(3)}(\omega)$ not only provides insight into the origin of such large nonlinearities, it also determines the operating wavelength-window for potential applications of conjugated polymers as NLO materials. This is clearly seen in the strong energy dependence of the calculated $\chi^{(3)}(\omega)$ and $n_2(\omega)$, the intensity dependent part of the index of refraction, for polyacetylene.[9] [10]

Our early measurement of the $\chi^{(3)}$ spectrum of polyacetylene used a Nd:YAG laser in conjunction with three stages of nonlinear optical processes in order to achieve "tunability".[11] The spectrum had gaps in its energy coverage and did not extend to energies lower than $h\nu \sim 0.8$ eV.[11] However, these results hinted at a further rise in the magnitude of $\chi^{(3)}$ for pump photon energies below ~ 0.8 eV, and identified the low energy region of the $\chi^{(3)}$ spectrum as an important region for further studies.

We have now completed our measurement of the spectrum of $\chi^{(3)}(\omega)$ in *trans*-polyacetylene down to 0.38 eV (3.3 μm) using an infrared free electron laser (FEL).[3] FEL's have been demonstrated as sources of *intense and tunable* radiation in the infrared and far infrared parts of the electromagnetic spectrum. In the far infrared range, FEL's do not have competition from conventional sources, since by and large the conventional sources are fixed in wavelength. In the IR range, the important energy range for most of the NLO activities associated with conjugated polymers, tunable light is available through a succession of nonlinear optical processes following a conventional source, but at the expense of the loss of much power and stability. The combination of wide tunability in a single source, short pulse duration, good stability, high peak power and an extremely large average power has made infrared FEL's a unique source of radiation for studies of nonlinear optics in conjugated polymers.

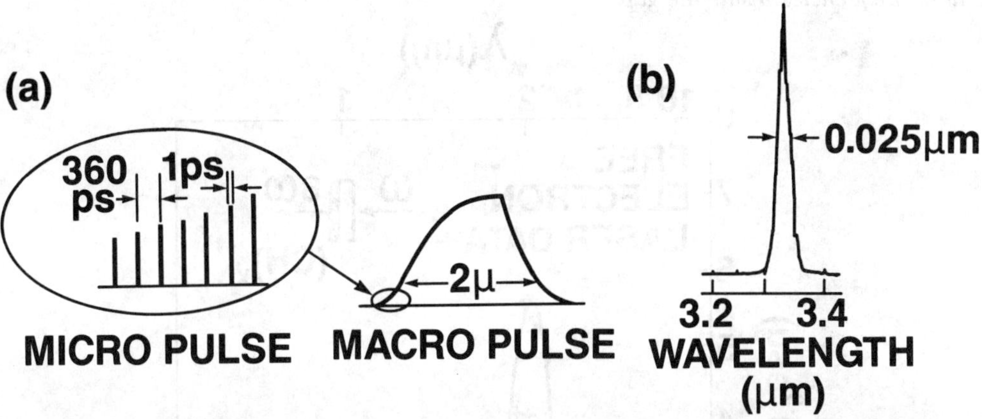

Figure 1: (a) The structure of a 2 μs duration *macropulse* and ~2 ps duration *micropulses* in the Mark III infrared Free Electron Laser. (b) The spectral content of a typical *macropulse* that contains 10^4 Fourier transform limited *micropulses*.

The Mark III infrared FEL used in our studies is continuously tunable from 2 to 8 μm.[12] Its output is in the form of a series of 1 to 8 μs macropulses separated by 67 ms. Each macropulse contains a train of micropulses each approximately 2 ps in duration and separated by 350 ps from each other. The micropulses typically have Fourier-transform limited spectra, however, a uniform chirp can be introduced in their wavelength within the macropulse envelope. The chirp can be used for pulse compression, and has been used in high-resolution time-resolved vibrational spectroscopy using an infrared FEL.[13] Nontheless, the monochromaticity of the macropulses is adequate for most studies, $\delta\lambda/\lambda \sim 10^{-2}$. Typical micropulse powers range from 10^5 to 10^6 watts.

The significance of the energy range covered by the infrared FEL is that it extends to an energy well below $E_g/4$ for polyacetylene. *Therefore, all of the multiphoton resonances associated with a THG process are expected to lie in this energy range.* This point is readily seen by examining the following expression for THG[14]

$$\chi^{(3)}(-3\omega;\omega,\omega,\omega) \sim \sum_{nmn'} \Omega_{gn}\,\Omega_{nm}\,\Omega_{mn'}\,\Omega_{n'g} \qquad (1)$$

$$\left[\frac{1}{(E_{ng}-3\omega)(E_{mg}-2\omega)(E_{n'g}-\omega)} + \cdots\cdots \right]$$

where g is the ground state, n and n' are excited states of opposite symmetry and m is an excited state of the same symmetry as g. The Ω's are the dipole matrix elements between any two states of opposite symmetry. As seen in Eqn.(1), resonance-enhancements in $\chi^{(3)}$ occur whenever the energy of one, two, or three photons couples two states with a large transition probability. *Clearly, a full under-standing of the nonlinear optical response of a semiconductor requires measurements at all frequencies inside the gap.*

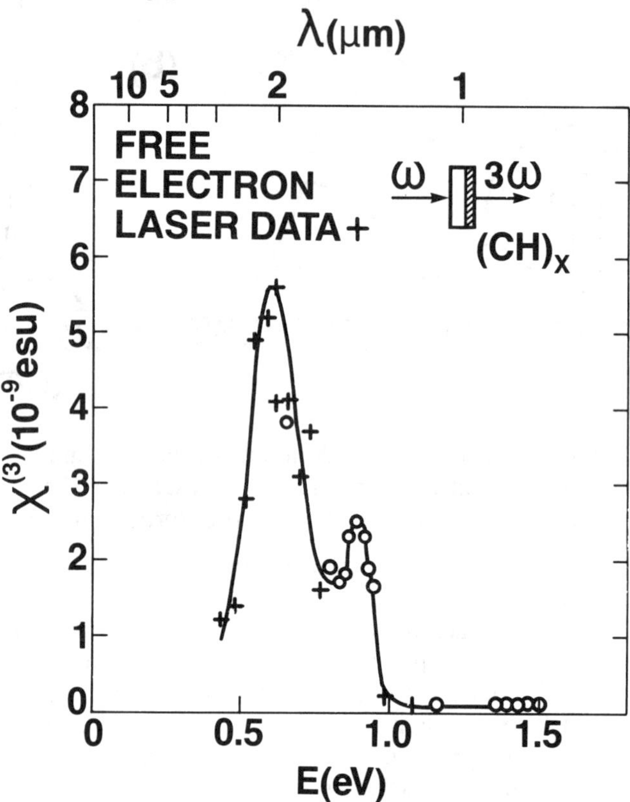

Figure 2. The $\chi^{(3)}(-3\omega;\omega,\omega,\omega)$ spectrum in *trans*-(CH)ₓ; (+) free electron laser data, (O) from Ref. 10.

Figure 2 shows the spectrum of $\chi^{(3)}(\omega)$ from 0.4 eV to 1.5 eV. The pluses are the results obtained using the infrared FEL, and the open circles are from our earlier work.[11] We identify two distinct regions in the spectrum of $\chi^{(3)}(\omega)$; the off resonance regime that extends from ~ 1.0 eV to >1.5 eV, and the multi-photon resonance regime that extends from ~0.4 eV to ~1.0 eV. In addition to the previously found peak in $\chi^{(3)}(\omega)$ at 0.89 eV, the FEL measurements have uncovered a broader and much stronger peak at 2/3 of that value, 0.6 eV. Both peaks correspond to multiphoton resonance enhancements of $\chi^{(3)}$ since there is not a similar structure in the spectrum of $\chi^{(1)}$, the absorption coefficient.[15] We note that no other structure of comparable intensity and width appears in the energy range extending down to 1/4 of E_g, the optical gap.

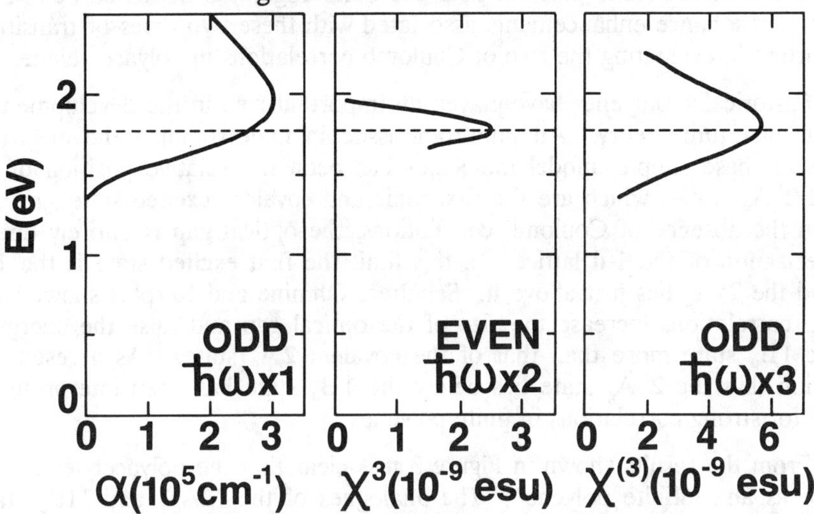

Figure 3: (a) The spectrum of absorption coefficient $\chi^{(1)}(\omega)$ in TR$-$(CH)$_x$ as a function of ω.[15] (b) The spectrum $\chi^{(3)}(\omega)$ in the vicinity of the two-photon resonance as a function of $2\times\omega$. (c) The spectrum $\chi^{(3)}(\omega)$ in the vicinity of the three-photon resonance as a function of $3\times\omega$. The dashed line is the position of the 1-d energy gap.

Since our measurements (see Figure 2) extend below $E_g/4$, the intense three-photon resonance enhancement of $\chi^{(3)}(\omega)$ that couples the even-symmetry ground state to the odd-symmetry band edge must lie in this energy range. This statement is true for a free electron model,[9] as well as a highly correlated-electron model for polyacetylene.[16] Since *both* $\chi^{(3)}$ and $\chi^{(1)}$ can correspond to transitions between states of opposite symmetry and $\chi^{(1)}(\omega)$ peaks at ~2 eV,[15] (see Figure 3a) we assign the newly found peak at 0.6 eV to the three-photon resonance enhancement of $\chi^{(3)}(\omega)$. The dashed line shows the position of 1-d energy gap in *trans*-(CH)$_x$ at 1.8 eV.[17] In Figure 3c we replot the spectrum of this peak in $\chi^{(3)}(\omega)$ as a function of $3\times\omega$. This is to be compared with the spectrum of $\chi^{(1)}$ in Figure 2a. We note that consistent with the assignment of the 0.6 eV feature as a 3-photon resonance, the two spectra shown in Figures 3a and 3c are quite similar. The faster drop in the intensity of $\chi^{(3)}(3\times\omega)$ compared to $\chi^{(1)}(\omega)$ at higher energies results from extra energy-dependent terms in the expression for $\chi^{(3)}$ that decrease with increasing energy above the band edge.[14]

The peak in $\chi^{(3)}(\omega)$ at 0.89 eV is assigned to a two-photon resonance enhancement since the spectrum of $\chi^{(1)}$ does not show a similar structure at the same or three times that energy.[15] The significance of the 2/3 ratio in the positions of the two peaks in $\chi^{(3)}(\omega)$ becomes clear by plotting the spectrum of the two-photon resonance peak as a function of $2 \times \omega$ in Figure 3b and comparing the result with the other two spectra in Figures 3a and 3c. It is clear that the excited state energy levels responsible for the parity conserving and parity nonconserving transitions are at the same position near the band edge. As discussed below, the positions of resonance enhancements associated with these two types of transitions are important in evaluating the role of Coulomb correlations in polyacetyelene.

Historically, polyenes have played an important role in the development of molecular quantum theory. An important issue in understanding the electronic structure of these simple, model molecules has been the relative position of the $1B_u$ and 2^1A_g states, which are the first ionic and covalent excited states, respectively. In the absence of Coulomb correlations, the optical gap is entirely due to the dimerization of the 1-d lattice. In this limit the first excited state is the $1B_u$ state, and the 2^1A_g lies just above it. Schulten, Ohmine and Karplus showed that Coulomb correlations increase the size of the optical gap and raise the energy of the ionic $1B_u$ state more than that of the covalent $2A_g$ state.[8] As a result, the observation that the 2^1A_g state lies below the $1B_u$ state has been interpreted as evidence for strong correlations in finite polyenes.

From the results shown in Figure 3 it is clear that the polyacetylene is not the same as an "infinite polyene": The analogues of the "$2A_g$" and "$1B_u$" states are at the same energy. In other words the gap to charged (ionic) and neutral (covalent) excitations have comparable values. Clearly the Coulomb correlations are screened in the "actual" polyacetylene film used in our study. We believe that this deviation from the expected behavior based on extrapolation from finite polyenes to "an infinite polyene" is because of the interchain screening of the long range Coulomb interaction. The studies of the relative position of the $2A_g$ and $1B_u$ in polyenes have primarily been done on *isolated* polyene molecules in solution, where the only screening is due to weak dielectric screening by the solvent. In polyacetylene films, the $(CH)_x$ chains are not isolated, and the large dielectric constant of neighboring chains provides a strong screening mechanism for transitions that involve separation of charge. In fact, it is known that the screening from polar (large dielectric constant) solvents decreases the energy of $1B_u$ relative to that of $2A_g$.[18] Since neighbroring chains have very large effective dielectric constant, it is reasonable that the ionic and covalent gaps have comparable values in polyacetylene.[19] [20]

Based on simple extrapolations from finite polyenes[21] and ignoring the importance of screening effects,[22] [23] [20] Kohler and coworker have favored an alternative interpretation of the results shown in Figure 2.[24] Within this interpretation the 0.6 eV resonance enhancement peak in $\chi^{(3)}$ (see Figure 2) is due to the $2A_g$ state in polyacetylene that lies 1.2 eV above the ground state.[24] Since there is three-photon resonance enhancement at $E_g/3 \sim 0.6$ eV irrespective of the strength of Coulomb interactions,[9][16] the peak in $\chi^{(3)}$ must contain both the two- and the three-photon resonances. Unfortunately, this interpretation is inconsistent with the predictions of the exact calculation of $\chi^{(3)}$ for the strongly correlated-electron model that it is based on.[16] The $\chi^{(3)}$ for this model has *two strong* resonance enhancements, a three-photon resonance with the $1B_u$ and a two-photon resonance with the $2A_g$ states.[16] Besides the fact that this alternative interpretation does not allow for any screening, it completely ignores the presence of the experimentally observed peak at 0.9 eV.

In summary, we have extended the spectrum of $\chi^{(3)}$ in polyacetylene to an energy less than $E_g/4$ using an infrared free electron laser, the first use of such a laser for nonlinear optical spectroscopy. The spectrum of $\chi^{(3)}$ shows an off-resonance regime from 1.0 eV to > 1.5 eV, and two resolvable multiphoton resonances at 0.6 eV and 0.89 eV. The magnitude of $\chi^{(3)}$ reaches $\sim 10^{-8} esu$ at the peak of a strong three-photon resonance at 0.6 eV. The position of the three-photon resonance is 2/3 of the position of the two-photon resonance, implying that states of opposite parity lie at the band edge. Our finding that the onset of the two-photon accessible excitations starts at the same energy as the optical gap suggests that the effective Coulomb correlations in polyacetylene are screened.

PROSPECTS FOR TECHNOLOGY

In the off-resonance regime, a large $\chi^{(3)} \equiv \chi^{(3)}(-3\omega;\omega,\omega,\omega)$, measured through THG efficiency experiments in a material implies a large $n_2 \equiv \chi^{(3)}(-\omega;\omega,-\omega,\omega)$, the coefficient of intensity dependent change in the index of refraction,

$$n = n_o + n_2 \times I \qquad (2)$$

Whereas $\chi^{(3)}$'s are in the scientific *e.s.u.* units, the n_2's are usually in more practical units. For a material with an index refraction n_o, $\chi^{(3)}$ is related to n_2 through

$$n_2(cm^2/W) = (16\pi^2/n_o^2 c) Re\ \chi^{(3)}(esu) \qquad (3)$$

where c is the velocity of light. The large values of $\chi^{(3)}$ in polyacetylene at all frequencies (see Figure 2), therefore, might be utilized in all-optical switching devices. The building blocks of such devices are waveguides of a few microns in size suitable for confining, *over large distances,* the intense light. The basic requirement for operation of an all-optical device is a light induced change in phase of the optical field of a magnitude $\sim \pi$. Such a change can be induced by a small change in the index of refraction through n_2. Therefore, for the operation of such a device we must have,

$$n_2 \times I \times L/\lambda \sim \pi \qquad (4)$$

where L is the length of the waveguide that confines the intense radiation. Note

348

that the small change in the index of refraction, $\delta n = n_2 \times I$, is amplified through the interaction length by L/λ.

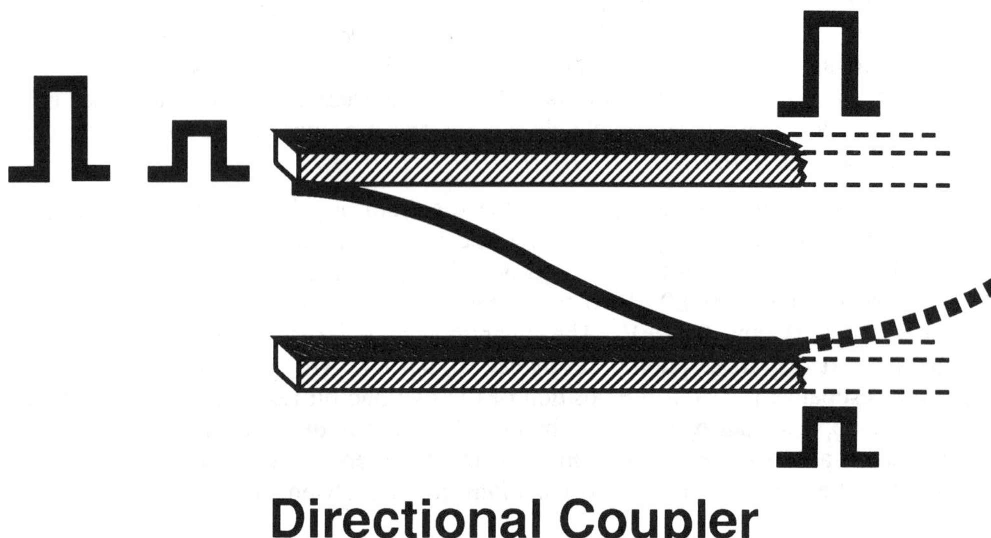

Directional Coupler

Figure 4. Operation of a directional coupler: Pulses injected at the upper guide at low (high) intensities appear at the out put of the lower (upper) guide. The switching time is dictated by the time scale needed to change the index of refraction; i.e. inverse of the optical frequencies.

To illustrate the operation of an all-optical switching device let us consider, as an example, a directional coupler shown schematically in Figure 4.[25] The device consists of a pair of waveguides coupled together through the overlap of the evanescent tails of their respective guided modes. When light is coupled into the input of the upper guide, power is transferred *periodically* between the two guides as the wave propagates along the device. The period of oscillation of the power is twice the coupling length, L_c. In the absence of transmission losses, the optical field oscillates indefinitely between the two guides as long as they stretch. L_c is determined by geometry, mode structure and separation of the two guides that is fixed, and by an effective index of refraction that can be changed though n_2. For a "tuned" device, the length of the device is adjusted to L_c. At low light intensities full power transfer occurs after one L_c. However, at high intensities phase mismatch prevents coupling and the power transfer is frustrated.[25] As a result, depending on the input pulse intensity, the output pulse switches between the two guides. Since what determines the switching speed is the *electronic* n_2, it can be as fast as the inverse optical transition rate; i.e. faster than 10^{-14} seconds.

Operation of an all-optical switching device has recently been demonstrated using a dual-core optical fiber made of glass.[26] Since the optical nonlinearity in conjugated polymers is orders of magnitude larger than in glass, a conjugated polymer-based device should operate at substantially lower power levels that may be obtained using a diode laser, for example. (see Equation 4) However, no such devices have been developed up to now, and important questions concerning their linear and nonlinear optical properties remain unanswered. This has primarily been due to the fact that many conjugated polymers are highly intractable materials which are difficult to process and patterned to construct *low loss* waveguide structure required for this type of application. Polyacetylene which seems to have the largest optical nonlinearity, for example, is practically useless because of the high transmission losses due to scattering.[27] In fact the first demonstration of all-optical switching device using glass fiber owes its success to the small transmission losses associated with fiber optics.[26]

We have approached this problem by developing a composite channel fabrication technique to construct low-loss, single-mode channel waveguide structures in thin films of the solution-processable polydiacetylene, 4BCMU.[28] These channel waveguides form the basic building block of an all-optical device, (see Figure 4) and by evaluating the linear and nonlinear optical properties we have attempted to address some outstanding questions regarding the possible application of this class of materials. The composite technique for defining channel waveguides in 4BCMU-films consists of first patterning the substrate with high-index ion-exchanged channels prior to spin coating.[29] [30] In this technique, lateral confinement of the optical fields in the polymer film is provided by the underlying channel and no patterning of the polymer itself is required. Importantly, since the refractive index of the polymer is significantly higher that of the ion-exchanged channel, almost all of the optical field intensity is confined to the polymer film.

Figure 5 shows the intensity dependence of the fractional output from a directional coupler made of a pair of composite channel waveguides(see Figure 4). These reasults have been obtained by reducing the repetition rate of the input pulses from 82 MHz to 1 KHz. We observed a stronger nonlinearity at full repetition rate of 82 MHz that has been attributed to differential heating effects.[4] The result shown in Figure 5 is the first demonstration of intensity dependent transmission and switching phenomena in a device based on a conjugated polymer.[4] However, at this stage this is not a "useful" device since the total transmission is adversely dominated by a two-photon absorption process.[4] The two-photon absorption process is due to parity conserving transitions between the ground state and states above the band edge. At the operating wavelength of $\lambda = 1.06$ μm, its magnitude is $\alpha_2 \sim 10^{-9}$ cm/W.[4] Note that at the intensities used here, the magnitude of nonlinear transmission loss is comparable to linear transmission loss of ≤ 5 db/cm.[31]

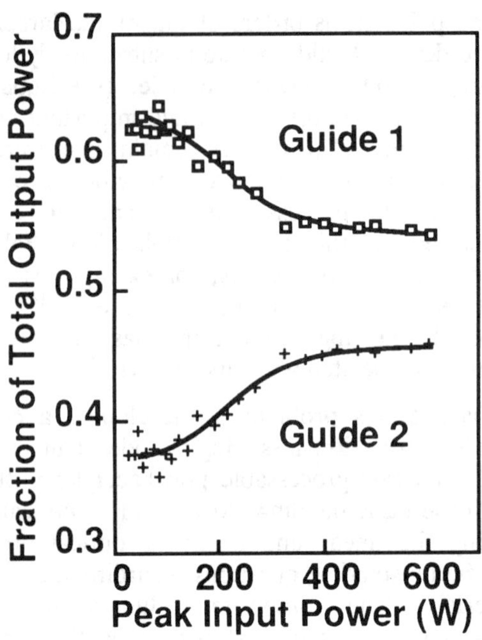

Figure 5. The fractional output in the two guides shown in Figure 4 as a function of peak power input to guide 1.

Can a "useful" all-optical switching device be constructed? We believe that it is possible. Whether or not it will have a widespread usage depends on development of a system architecture that takes advantage of such a fast $\chi^{(3)}$-based switch. We have already shown how to avoid the adverse two photon absorption effects. The theoretical calculation of n_2 for any two-level system, shows that Imn_2 is zero for photon energies less than $E_g/2$. For example, for transitions between the valence and conduction bands in polyacetylene this has recently been demonstrated.[9][10] However, the same calculations show that the "useful" Ren_2 that controls the phase rather than the amplitude of the optical fields at high intensities is still large.[9][10] By decreasing the photon energies to less that $E_g/2$ in 4BCMU, we have shown that the two-photon absorption process is undetectable at the operating wavelength of 1.3 μm.[32]

In summary, nonlinear optical transmission and switching phenomena have been observed in directional coupler devices fabricated from soluable polydiacetylene 4BCMU. Effects due to both slow thermal nonlinearities and ultrafast electronic nonlinearities have been identified.[4] At the operating wavelength of 1.06 μm, the ultrafast electronic nonlinear phenomena originates from intensity-dependent changes in the imaginary part of the refractive index due to two-photon absorption process. At an operating wavelength of 1.3 μm, the two-photon absorption process is not operative and switching due to the real part of the change in refractive index may be possible in 4BCMU-based directional couplers.

REFERENCES

1. G.J.Blanchard, J.P.Heritage, A.C.Von Lehmen, M.K.Kelly, G.L.Baker and S.Etemad, Phys. Rev. Lett. **63**, 890(1989).

2. B.I.Greene, J.F.Mueller, J.Orenstein, D.H.Rapkine, S.Schmitt-Rink, and M.Thakur, Phys. Rev. Lett. **61**, 325(1988).

3. W-S. Fann, S.Benson, J.M.J.Madey, S.Etemad, G.L.Baker and F.Kajzar, Phys. Rev. Lett. **62**, 1492(1989).

4. P.D.Townsend, J.L.Jackel, G.L.Baker, J.A.Shelburne, III, and S.Etemad, Appl. Phys. Lett. **55**, 1829(1989).

5. Z. Soos and L.R.Ducass, J. Chem. Phys. **78**, 4092(1983).

6. For recent developments see MRS Symposium Proceedings, **109** "Optical Properties of Polymers", Eds: A.J.Heeger, J.Orenstein and D.R.Ulrich, 1988, MRS Publications, Pittsburgh, Pa.

7. J.-P. Hermann, D.Richard and J.Ducuing, Appl. Phys. Lett. **23**, 178(1973).

8. K. Schulten, I. Ohmine, and M.Karplus, J. Chem. Phys. **64**, 4422(1976).

9. Weikang Wu, Phys. Rev. Lett. **61**, 1119(1988).

10. G.P.Agrawal, C.Cojan and C.Flytzanis, Phys. Rev. **B17**, 776(1985).

11. F. Kajzar, S. Etemad, G.L. Baker and J. Messier, Synth. Met. **17**, 563 (1987); Solid State Commun. **63**, 1113(1987).

12. S.Benson, J.Schults, B.A. Hooper, R. Crane, and J.M.J. Madey, Nucl. Instum. Methods Phys. Res. Sect. A **272**, 22 (1988).

13. W-S Fann, L. Rothberg, M. Roberson, S.Benson, J. Madey, S. Etemad and R.Austin, Phys. Rev. Lett. **63**, (1989).

14. D. C. Hanna, M. A. Yuratich and D. Cotter, in Nonlinear Optics of Free Atoms and Molecules, Springer-Verlag, Berlin (1979).

15. B. R. Weinberger, C. B. Roxlo, S. Etemad, G. L. Baker, and J. Orenstein, Phys. Rev. Lett. **53**, 86 (1984), and N.Suzuki, M.Ozaki, S.Etemad, A.J.Heeger, A.G.MacDiarmid, Phys. Rev. Lett. **45**, 1209(1980).

16. Z.G. Soos and S. Ramasesha, J. Chem. Phys. **90**, 1067 (1989).

17. D. Moses, A. Feldblum, E. Ehrenfreund, A.J. Heeger, T_C. Chung, A.G.MacDiarmid, Phys. Rev. **B26**, 3361(1982).

18. K.L.D'Amico, C.Manos, R.L.Christeinsen, J. Am. Chem. Soc. **102**, 1777(1980).

19. W. Wu and S. Kivelson, Phys. Rev. **B33**, (1986).

20. Z.Soos, G.W.Hayden and S.Etemad, Bull. Am. Phys. Soc. **34**, 770(1989), and to be published.

21. B.S.Hudson, B.E.Kohler and K.Schulten, Excited States **6**, 1(1982).

22. S. Basu, Adv. Quantum Chemistry, **1**, 145(1964).

23. J.R.Andrews and B.S.Hudson, J. Chem. Phys. **68**, 4587(1978).

24. B.E.Kohler et. al. these proceedings.

25. S.M.Jensen, IEEE J. Quantum Electron. **QE-18**, 1580(1982).

26. S.R.Friberg, Y.Siberberg, M.K.Oliver, M.J.Andrejco, M.A.Saifi and P.W.Smith, Appl. Phys. Lett. **51**, 15(1987).

27. S.Etemad, G.L.Baker, D.Jaye, F.Kajzar and J.Messier, SPIE **682**, 44(1986).

28. G.N.Patel, Poly. Prepr. Am. Chem. Div. Polym. Chem. **19**, 155(1978).

29. J.L.Jackel, N.E.Schlotter, P.D.Townsend, G.L.Baker and S.Etemad, SPIE **971**, 239(1988).

30. N.E.Schlotter, J.L.Jackel, P.D.Townsend, G.L.Baker, Appl. Phys. Lett. (in press).

31. P.D.Townsend, G.L.Baker, N.E.Schlotter, C.F.Klausner, and S.Etemad, Appl. Phys. Lett. **53**, 1782(1988).

32. P.D.Townsend, W-S. Fann, S.Etemad, G.L.Baker, J.L.Jackel, J. Shelburne III, and Z. Soos (to be published).

TRANSITION DIPOLES OF POLYACETYLENE OLIGOMERS

BRYAN E. KOHLER AND JOHN A. PESCATORE, JR.
Department of Chemistry
University of California, Riverside
Riverside, CA 92521 U.S.A.

ABSTRACT

Quantum mechanical calculations of nonlinear optical susceptibilities often involve perturbation sums over transition dipoles. Unfortunately, there is little experimental data that can be used to evaluate the reliability of the calculated dipoles. To provide some reference data and answer questions about the dependence of these dipoles on polyene chain length we have measured 1^1A_g to 1^1B_u transition dipoles for the α,ω-diphenyl polyenes with 2 through 8 double bonds in the polyene chain. The α,ω-diphenyl polyenes were dissolved in N,N-dimethylformamide, methylene chloride and toulene. The <u>in vacuo</u> electronic transition dipole moment from the ground state (1^1A_g) to the second excited singlet state (1^1B_u) was determined to converge to an asymptotic value of 11.1 ± 0.4 Debye.

1.   Introduction

Our previous spectroscopic investigations of linear polyene electronic structure have largely focused on determining the excitation energies and potential energy surfaces (the dependence of electronic energy on conformation) of the low lying singlet states [1]. While questions about what happens to these basic electronic properties when the polyenes are strongly perturbed remain open, there is also considerable interest in other observables.

The proceedings of this workshop testify to the increasing attention that is being given to understanding the linear and non linear optical responses of these species. In many of the theoretical treatments, the non linear susceptibilities are developed as a perturbation sum over terms involving products of transition dipoles divided by products of electronic energy differences [2]. Because any non linear optical measurement in these systems will include resonance contributions, it is reasonable to be skeptical of theoretical methods that cannot predict the observed ordering

J. L. Brédas and R. R. Chance (eds.), Conjugated Polymeric Materials:
Opportunities in Electronics, Optoelectronics, and Molecular Electronics, 353–364.
© 1990 *Kluwer Academic Publishers. Printed in the Netherlands.*

of the low lying singlet states. Then, since energy is only
one measure of the quality of a calculated wavefunction, even
calculations that give reasonably accurate energies
(denominators) may still give relatively poor descriptions of
such subtle observables as the off-diagonal dipole moment
matrix elements that are the numerators in these expressions.
In order to provide some reference data on transition moments
for well defined oligomers of polyacetylene, we have made
quantitative measurements of the intensity of the 1^1A_g to 1^1B_u
transition in a series of diphenylpolyenes with 4 through 16
carbon atoms in the polyene chain. These measurements were
made for three different solvents. When the transition
dipole magnitudes were corrected to vacuum by a classical
Lorentz cavity correction [3], the resulting values were
independent of solvent within the precision of our
measurements. We found that the magnitude of the 1^1A_g to 1^1B_u
transition moment per repeat unit in the polyene chain
decreases with increasing chain length so that the magnitude
of the transition dipole per polyene repeat unit is maximum
at an effective chain length of 5 double bonds.

2.   Experimental

2.1. DATA COLLECTION

The α,ω-diphenylpolyenes with two through four double bonds
in the polyene chain were purchased from Aldrich Chem. Co.;
those with five through eight double bonds in the polyene
chain were synthesized at Northern Illinois University by C.
Spangler [4]. Except for the almost insoluble
diphenyltetradecaheptaene and diphenylhexadecaoctaene, all of
the diphenylpolyenes were recrystallized from methylene
chloride. Solvents were toluene, methylene chloride and
N,N-dimethylformamide (Fisher Scientific Co., 99% purity) and
used without further purification. Solutions were prepared
by sonicating carefully weighed amounts (0.3 to 3 mg,
accurate to 50 µg) in solvent at room temperature and then
diluting to reach a final concentration (1 to 100 µM) for
which the maximum absorbance for the 1^1A_g to 1^1B_u transition
was approximately 1 for cell pathlength.
  For the shorter polyenes we were able to rule out
aggregation by making further dilutions and verifying that
absorbance was strictly proportional to concentration. In
the case of the sparingly soluble heptaene and octaene, we
determined the optical density of a saturated solution and
then took care to prepare only solutions with significantly
smaller absorbances. When sonication is used to speed the
rate of solution, there is always the danger that colloids
will be produced. In fact, centrifugation of some of the

most concentrated polyene solutions at 30,000 times gravity
and 300K did reduce the absorbance, presumedly through the
removal of very small colloidal particles from the clear
solutions. We only collected data for solutions in a
concentration range that strictly followed Beer's law and/or
proved to not be affected by this kind of centrifugation.

At least three solutions of each of the diphenylpolyenes
were made in each of the three solvents. Absorbance spectra
were measured with a diode array spectrophotometer (Hewlett-
Packard model 8451A) and transferred to a microcomputer
(Hewlett-Packard model 310) for computing $\int \varepsilon d\ln\nu$. Integration
limits were easily set since the absorption band for the 1^1A_g
to 1^1B_u transition is well defined (that is, goes from
baseline on the long wavelength edge to baseline at the short
wavelength edge) for all the compounds except
diphenylbutadiene. In the case of diphenylbutadiene, high
solvent absorbance on the short wavelength side of the 1^1A_g to
1^1B_u absorption band made the short wavelength limit somewhat
arbitrary, but a systematic change of the lower limit showed
that the errors in the integrated intensity for
diphenylbutadiene cannot exceed 10%.

2.2. DATA ANALYSIS

In general, molecular electronic transitions exhibit complex
band profiles. In room temperature solutions this reflects
both inhomogeneous environmental effects (different spectra
for different molecules because of a variation in local
structure) and unresolved structure (bands reflecting the
coupling of electronic excitation to intermolecular and
intramolecular vibrations). Intense absorptions like the
polyene 1^1A_g to 1^1B_u transition are adequately described
within the crude adiabatic Born Oppenheimer framework with
the radiation field treated semiclassically. At this level
the interaction between molecules and photons is limited to a
purely electronic matrix element (the transition dipole) and
integration over the absorption band is equivalent to summing
over the complete set of states that describe nuclear motion
[5]. Even though the derivation of the connection between
the fundamental quantum mechanical quantity, the transition
dipole, and the absorption observed for a macroscopic sample
by time dependent perturbation theory is a standard exercise
in basic courses on quantum mechanics, there are a plethora
of formulae in the literature. The one that we have derived
and used expresses the electric field at the molecule in
terms of the classical Lorentz field

$$E_{eff} \approx E_{in\ vacuo}(2n^2+1)/3n^2 \tag{1}$$

where n is the refractive index of the solvent. This formula specifically refers to absorption intensity in terms of the decadic molar extinction coefficient ε. Our expression for the relation of transition dipole to the absorbance seen for a randomly oriented ensemble of molecules in a solvent of refractive index n, derived by the standard time dependent perturbation approach, is

$$|\underset{\sim}{\mu}|^2 = 9.186 \times 10^{-3} \text{ Debye}^2 \text{mol } 1^{-1} \text{ cm } [(2n^2+1)/3n^2]^2 \int \varepsilon \, d\ln\bar{\nu} \qquad (2)$$

Refractive indexes used for toulene, methylene chloride and N,N-dimethylformamide were 1.4961, 1.4242 and 1.4305, respectively.

3.   Results and Discussion

Figure 1 shows the absorption spectra, plotted as molar extinction coefficient versus wavelength, for the diphenylpolyene series in N,N-dimethylformamide. A typical absorption spectrum showing the actual area integrated is shown in Figure 2. Table 1 summarizes the average values and uncertainties of the maximum extinction coefficients, integrated absorption intensities, and transition moment magnitudes for the diphenylpolyenes with 2 through 8 double bonds in the polyene chain.

While the value for diphenylhexadecaoctaene was obtained with reasonable precision in repeated trials, the extremely low solubility of this molecule and the fact that this compound could not be further purified by recrystallization leaves open the possibility that actual concentrations were significantly lower than those calculated from the solution composition. Accordingly, we take the measured transition dipole for the octaene to be a lower bound to the actual moment, and have excluded this data point from the analyses that follow.

The most striking feature of the 1^1A_g to 1^1B_u transition diple magnitudes is their relatively weak dependence on polyene chain length (Figure 3). It is clear that an asymptotic value is rapidly attained and that the transition dipole per polyene repeat unit actually decreases with increasing chain length. Given that even rather sophisticated molecular orbital theories cannot even properly order the low lying excited singlet states of linear polyenes [1], it is not surprising that the proper functional form for a plot of transition dipole versus polyene chain length is an open question. An empirical approach often used to fit the

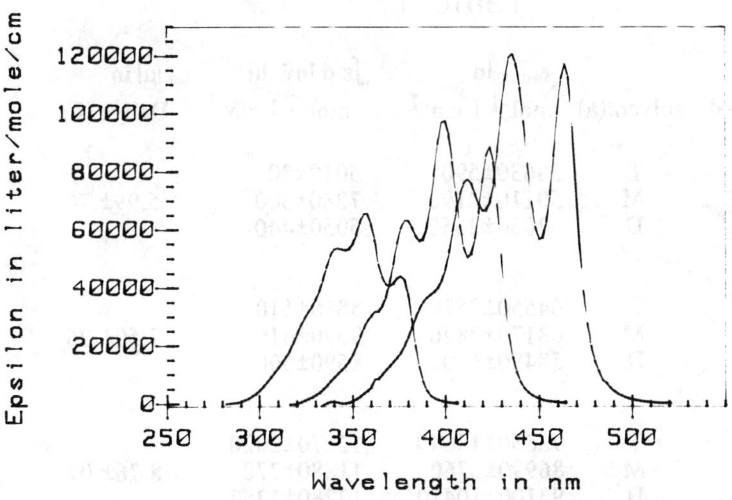

Figure 1. Absorption spectra of diphenylhexatriene (left curve), diphenyldecapentaene (middle curve) and diphenyltetradecaheptaene (right curve) in room temperature N,N-dimethylformamide.

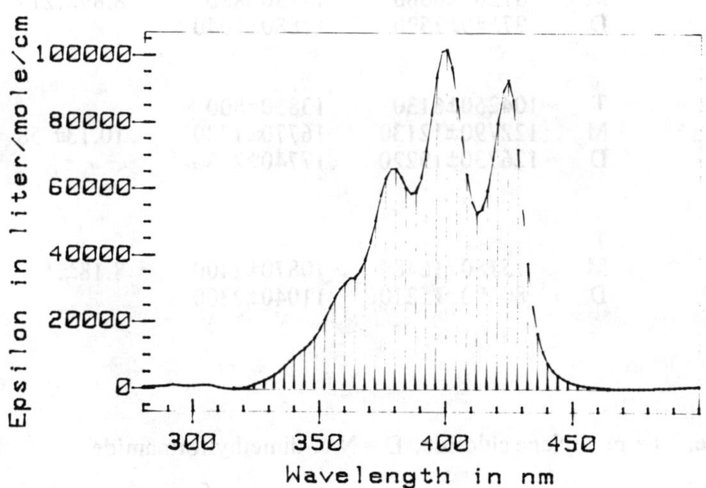

Figure 2. Absorption spectrum of diphenyldecapentaene in room temperature N,N-dimethylformamide. To obtain the 1^1A_g to 1^1B_u transition dipole $\int \varepsilon d\ln\tilde{\nu}$ was computed for the shaded area.

Table 1

N for $\phi-(CH=CH)_N-\phi$	solvent(a)	ε_{max} in mol^{-1} l cm^{-1}	$\int \varepsilon \, d\ln\tilde{\nu}$ in mol^{-1} l cm^{-1}	$|\mu|$ in Debye (b)
2	T	38630±590	5010±30	
	M	50110±2190	7280±300	5.99±.70
	D	37030±2650	5050±440	
3	T	64550±3520	8810±510	
	M	68170±2820	9590±410	7.50±.26
	D	58480±4000	8590±500	
4	T	93830±14640	12670±2020	
	M	86990±5760	11880±770	8.76±.07
	D	93100±10410	12280±1350	
5	T	95630±3130	12770±430	
	M	10130±14760	13540±1850	8.99±.24
	D	93400±8470	12480±1100	
6	T	99620±2670	13250±390	
	M	87260±6060	11780±880	8.89±.21
	D	97540±7590	12880±1040	
7	T	104250±6130	13850±800	
	M	122790±12130	16770±1780	10.13±.58
	D	126730±19220	17740±2530	
8	T	-	-	
	M	75180±16480	10870±2400	8.18±.18
	D	80690±15210	11040±2300	

a) T = toluene, M = methylene chloride, D = N,N-dimethylformamide

b) $|\mu|^2 = 9.186 \times 10^{-3}$ Debye^2 mol l^{-1} cm $[(2n^2+1)/3n^2]^2 \int \varepsilon \, d\ln\tilde{\nu}$; n values used were 1.4961 for toluene, 1.4242 for methylene chloride and 1.4305 for N,N-dimethylformamide.

chain length dependence of excitation energies, is to fit the data with a power series in 1/N, stopping at first order (N is the number of repeat units in the polyene chain: the general substituted polyene is written as R-(CH=CH)$_N$-R'. An empirical fit of the transition dipole magnitude for the diphenylpolyene series by

$$|\underset{\sim}{\mu}|=A+B/N \tag{3}$$

gives A≅-B=11 Debye. For these parameters, the maximum transition dipole magnitude per polyene repeat unit is 3.0 Debye which is reached at N=2.

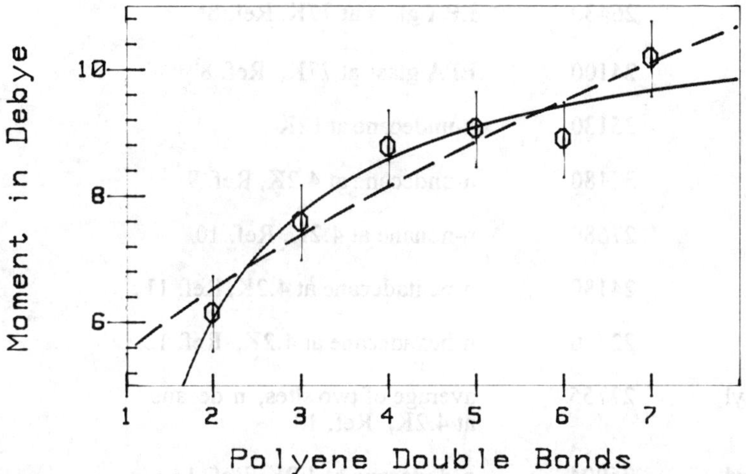

Figure 3.   Transition dipole magnitudes versus polyene double bonds for the diphenylpolyene series. The points are the best values of the vacuum 1^1A_g to 1^1B_u transition dipole magnitudes. The solid line is the best fit to these data by $|\underset{\sim}{\mu}|=A+B/N$; the dashed line is the best fit to these data by $|\underset{\sim}{\mu}|=(A+BN)^{1/2}$.

The degree to which the phenyl rings extend the effective conjugation length of the substituted polyene chain can be empirically determined by simultaneously fitting 1^1A_g to 1^1B_u excitation energies of both series by A+B/N for the unsubstituted polyenes and A+B/(N+2C) for the diphenylpolyenes. The data for this exercise are summarized in Table 2, and the fit is displayed in Figure 4. Fitting the transition dipole magnitude by

Table 2

Dependence of 1^1B_u 0-0 Excitation Energies on Chain Length for Molecules in the Series R-(CH=CH)$_N$-R

N	R	$\tilde{\nu}_{0-0}$ (cm^{-1})	Conditions
3	H	36850	3-methylpentane at 77K, Ref.6
4	H	32100	n-octane at 4.2K, Ref. 7
6	H	26430	EPA glass at 77K, Ref. 8[a]
7	H	24100	EPA glass at 77K, Ref. 8[b]
3	CH$_3$	35130	n-undecane at 12K
4	CH$_3$	32180	n-undecane at 4.2K, Ref. 9
5	CH$_3$	27680	n-nonane at 4.2K, Ref. 10
7	CH$_3$	24180	n-pentadecane at 4.2K, Ref. 11
8	CH$_3$	22770	n-hexadecane at 4.2K, Ref. 12
2	phenyl	27755	average of two sites, n-decane at 4.2K, Ref. 13
3	phenyl	25904	n-dodecane at 4.2K, Ref. 13
3	phenyl	26046	n-tridecane at 4.2K, Ref. 13
4	phenyl	24381	n-dodecane at 4.2K, Ref. 14
5	phenyl	23068	n-decane at 4.2K, Ref. 15
6	phenyl	21930	n-dodecane at 4.2K, Ref. 15

[a]) taken from Figure 2 by digitizing a xerox copy of the original figure and then least squares fitting a Gaussian to the 0-0 band.

[b]) taken from Figure 3 by digitizing a xerox copy of the original figure and then least squares fitting a Gaussian to the 0-0 band.

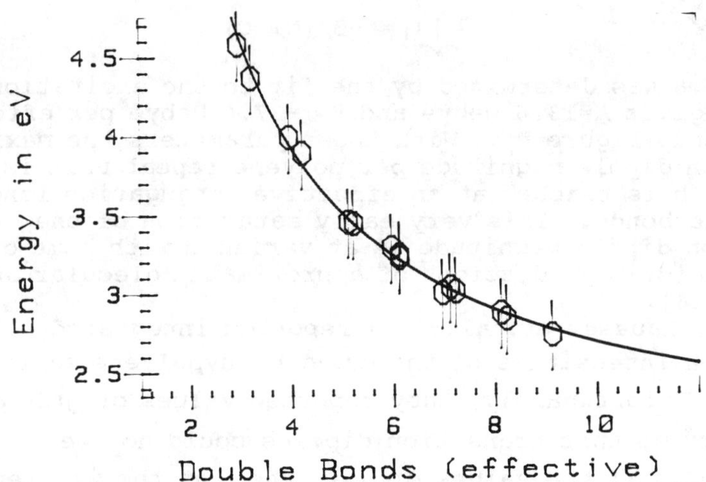

Figure 4. Polyene 1^1A_g to 1^1B_u excitation energies versus N_{eff}, the effective number of double bonds. N_{eff} for unsubstituted polyenes is just the number of double bonds in the polyene chain; for diphenyl polyenes it is the number of double bonds in the polyene chain plus 3.2 for the two phenyl rings. With these choices for N_{eff}, all points are well fit by E = 1.9 eV + $8.3/N_{eff}$.

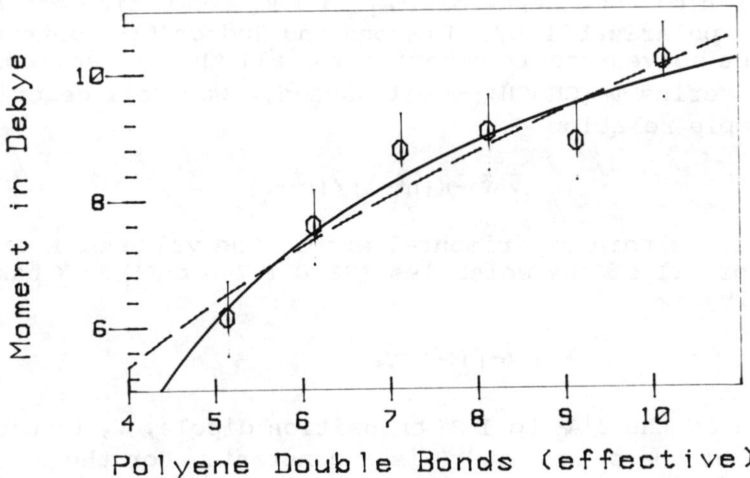

Figure 5. Transition dipole magnitudes versus N_{eff}, the effective number of double bonds for the diphenylpolyene series. The points are the best values of the 1^1A_g to 1^1B_u transition dipole magnitudes. The solid line is the best fit to these data by $|\mu|=A+B/(N+3.2)$; the dashed line is the best fit of these data by $|\mu|=(A+B(N+3.2))^{1/2}$.

$$|\underset{\sim}{\mu}|=A+B/(N+2C) \tag{4}$$

where C=1.6 was determined by the fit to the excitation energies gives A=13.6 Debye and B= -37.6 Debye per effective double bond (Figure 5). With these parameters the maximum transition dipole magnitude per polyene repeat unit is 1.2 Debye which is reached at an effective conjugation length of 5.5 double bonds. This very early saturation of the transition dipole magnitude is at variance with some of the more optimistic predictions of approximate molecular orbital methods [16].

In 1935, Hausser et. al. [17] reported integrated absorption intensities of the α,ω-diphenypolyene series in benzene. Unfortunately, they reported values of $\int \varepsilon dv$ rather than $\int \varepsilon d\ln\tilde{v}$ so that transition dipoles could not be calculated. If the values of ε_{max} (the ε at the wavelength for maximum absorption) that we measured for toluene solutions of the diphenylpolyenes are compared with the values that they report, our values are on average 10% lower. Furthermore, while they argued for a linear dependence of both ε and $\int \varepsilon dv$ on the number of polyene double bonds, we find a sub linear dependence.

Finally, we note that the values for the transition dipole magnitudes that we have determined are fully consistent with the observed dependence of 1^1A_g to 1^1B_u excitation energy on solvent polarizability. Diamond and Hudson [18] determined that the solvent shift behavior for all the diphenylpolyenes in the series $\phi-(CH=CH)_N-\phi$ with N=1-6,8 was well described by the simple relation

$$\tilde{v}=\tilde{v}_0-K(n^2-1)/(n^2+2) \tag{5}$$

and that, within experimental error, the value of K was the same for all of the molecules (9870 ± 740 cm^{-1}). K has the form [19]

$$K=|\underset{\sim}{\mu}|^2P/V_s \tag{6}$$

where μ is the 1^1A_g to 1^1B_u transition dipole, V_s is the volume of the solute and P is a constant. For the diphenylpolyenes

$$V_s=V_{phenyls}+V_{chain}=V_{phenyls}+NV_{chain\ unit} \tag{7}$$

so the constancy of K implies that the transition dipole magnitude should be fit by

$$|\underset{\sim}{\mu}| = (A+BN)^{1/2} \qquad (8)$$

Figures 3 and 5 show that the transition dipoles reported here are consistent with this expectation, although the standard deviation of fit for this form is approximately 20% higher that that obtained when the data are fit by equation 3. When equation 8 is used the maximum transition dipole magnitude is 4.4 Debye at a chain length of 3 double bonds for both Figures 3 and 5. The important point is that, when either form is fit to the data, the transition dipole magnitude per polyene repeat unit decreases with increasing chain length.

4.   Conclusions

The 1^1A_g to 1^1B_u transition dipoles clearly show a sub linear dependence on the polyene chain length. Furthermore, the deviation from linearity at high chain length suggests that the largest off-diagonal dipole moment matrix elements are obtained for dienes. Thus, for the linear response there is no advantage to increasing the chain length of polyacetylene oligomers. The question of how the non linear susceptibilities depend on chain length remains an open and very interesting one.

5.   References

1.   Hudson, B.; Kohler, B.E. and Schulten, K. (1982), in Excited States, edited by E.C. Lim, Academic, New York, 6, 1-52.

2.   Orr, B.J. and Ward, J.F. (1971), Molec. Phys. 20, 513.

3.   Böttcher, C.J.F. in Theory of Electric Polarization, (1973), American Elsevier, New York, 1, Chap. II.

4.   Spangler, C.W.; Nickel, E.G. and Hall, T.J. (1987), Polym. Prepr. (Am. Chem. Soc., Div. Poly. Chem.) 28, 219.

5.   Strickler, S.J. and Berg, R.A. (1962), J. Chem. Phys. 37, 814.

6.   Granville, M.F.; Kohler, B.E. and Snow, J.B. (1981), J. Chem. Phys. 75, 3765.

7.   Granville, M.F.; Holtom, G.R. and Kohler, B.E. (1980), J. Chem. Phys. 72, 4671.

8.   Snyder, R.; Arridson, E.; Foote, C.; Harrigan, L. and Christensen, R.L. (1985), J. Am. Chem. Soc. 107, 4117.

9.   Andrews, J.R. and Hudson, B.S. (1980), J. Chem. Phys. 72, 4671.

10. Christensen, R.L. and Kohler, B.E. (1975), J. Chem. Phys. 63, 1837.

11. Simpson, J.H.; McLaughlin, L.; Smith, D.S. and Christensen, R.L. (1987), J. Chem. Phys. 87, 3360.

12. Kohler, B.E.; Spangler, C. and Westerfield, C. (1988) J. Chem. Phys. 89, 5422.

13. Heatherington III, W.M. (1977) Thesis, Stanford University, Stanford, CA.

14. As measured in this laboratory by C. Westerfield.

15. Horwitz, J.S.; Itoh, T.; Kohler, B.E. and Spangler, C.W. (1987) J. Chem. Phys. 87, 2433.

16. Pierce, B.M. (1989), J. Chem. Phys. 91, 791.

17. Hausser, K.W.; Kuhn, R. and Smakula, A. (1935), Z. Physikal. Chem. B 29, 384.

18a. Diamond, J. (1978) Thesis, Stanford University, Stanford, CA.

18b. Sklar, L.A.; Hudson, B.S.; Petersen, M. and Diamond, J. (1977), Biochem. 16, 813.

19. Amos, A.J. and Burrows, B.L. (1973), Advan. Quantum. Chem. 7, 303.

LINEAR OPTICAL PROPERTIES OF A SERIES OF POLYACETYLENE OLIGOMERS

H. E. Schaffer, R. R. Chance
Exxon Research and Engineering Company
Corporate Research Laboratories
Annandale, NJ   08801

and

K. Knoll, R. R. Schrock, and R. Silbey
Massachusetts Institute of Technology
Department of Chemistry and Center for Materials
Science and Engineering
Cambridge, MA   02139

ABSTRACT. A homologous series of polyacetylene oligomers having up to 13 double bonds has been prepared, thus allowing, for the first time, an examination of optical properties of polyenes having enough conjugation length to suggest extrapolation to the polymer.   We have measured two linear optical properties, the uv-visible absorption spectra in solid and solution form and the Raman scattering spectra in solid form.   Both the lowest energy electronic absorption peak and the frequency of the Raman band associated with the carbon-carbon double bond stretch become linear in 1/n for n≥7. Assuming the validity of the 1/n extrapolation, the results suggest that the familiar forms of polyacetylene have effective conjugation lengths of no more than approximately 30 double bonds.

I. Introduction

Polyacetylene has attracted considerable attention over the past decade as the prototypical conjugated polymer.[1] It is now well established that an understanding of the electronic properties of oligomers can contribute significantly to the understanding of conjugated polymers such as polyacetylene. This is particularly true given the highly disordered nature of polyacetylene and other conjugated polymers, since disorder leads to a dispersion in conjugation length, defined as the length over which the π electron structure remains ordered (i.e. planar). Both a chemical defect (chain end, crosslink, or impurity) or a structural defect (rotation out of planarity) can serve to interrupt conjugation. A segment of conjugation length n, where n is the number of double bonds, in a long polymer chain is proposed to have electronic behavior similar to that of an oligomer of length n.   There is now much support for and general acceptance of this idea in the literature.[2-4]

The electronic properties of oligomers generally extrapolate close to those of the corresponding polymer with a 1/n dependence.   This has been shown, at least semi-quantitatively, for numerous conjugated polymers (polyacetylene, polydiacetylenes, polythiophene, polyphenylene, etc.) for a variety of

J. L. Brédas and R. R. Chance (eds.), Conjugated Polymeric Materials:
Opportunities in Electronics, Optoelectronics, and Molecular Electronics, 365–376.

electronic properties, such as optical properties (linear and nonlinear), oxidation/reduction potentials, and ionization potentials.[3] Detailed examination of the n dependence of the electronic properties requires preparation of oligomers over a large range in n approaching as close as possible (in a 1/n sense) to the infinite chain limit. They must also be chemically and isomerically pure.

In this paper we present such a detailed examination of the linear optical properties of a homologous polyene series prepared as described elsewhere by Knoll and Schrock:[5]

$$H_3C-\underset{\underset{CH_3}{|}}{\overset{\overset{CH_3}{|}}{C}}-\left(C=\underset{n}{C}\right)-\underset{\underset{CH_3}{|}}{\overset{\overset{CH_3}{|}}{C}}-CH_3$$

These molecules have well-defined n values that range from 1 to 13 and are pure all trans. Specifically, we present and discuss (Section III) the UV-visible adsorption spectra in various solvents and in the solid state. We also present and discuss (Section IV) the n dependence of Raman frequencies and relate these results to the dispersion in these vibrational frequencies that are observed in polyacetylene. Experimental details are given in Section II.

II. Experimental Section

The t-butyl-capped polyenes were prepared via ring-opening of 7,8-Bis(trifluoromethyl)tricyclo[4.2.2.0^{2,5}]deca-3,7,9-triene by W(CH-t-Bu)(N-2,6-C$_6$H$_3$-i-Pr$_2$)(O-t-Bu)$_2$ followed by reaction with pivaldehyde or 4,4-dimethyl-trans-2-pentanal. After heating the resulting mixture, the t-butyl capped polyenes, with n from 1 to 13, were isolated by column chromatography and characterized by ^{1}H and ^{13}C NMR studies.[5]

The samples, in the form of powder or small crystallites, used for Raman scattering and optical absorption were stored under argon in glass vials having Teflon-lined screw-on tops. When not in use in the experiments, the vials were stored at dry ice temperature (-78°C) in the dark. The quality of the seal was checked by noting the absence of any ice on the inside of the cooled vials, as well as the reproducibility of spectra taken before and after several weeks of storage. For the Raman scattering measurements, the vials were allowed to warm to room temperature and then placed directly into the spectrometer such that the sample within the vial was at the focal point of the collecting optics, which were set in near-backscattering geometry. The n=1 polyene was, unlike all the other molecules, a liquid at ambient temperature, and condensing the sample on a wall of the vial with dry ice enabled collection of scattered radiation with the same backscattering geometry. For the optical absorption studies, the vials were opened in an argon dry box to allow removal of small amounts (<1 mg) of sample to several vials, one for each solvent. These vials were capped in the above described manner and transported into the spectroscopy laboratory. There, they were opened to allow introduction of a several mls of spectroscopy grade solvent (one of pentane, cyclohexane, benzene, or carbon disulfide), or of a solution of sodium dodecylsulfate in water, and the resulting solution (or, in the case of SDS/H_2O, the suspension obtained after brief sonication of the mixture) was immediately transferred to a quartz cell for absorption measurement with a Perkin-Elmer Lambda 9 spectrophotometer. In some cases, measurements were made on thin films deposited on glass slides by evaporation of dilute solution in a volatile solvent.

Raman spectra were recorded with either of two instruments. The first was a Spex Triplemate spectrometer terminated with a EG&G Princeton Applied Research model 1420 silicon photodiode array detector. The diode array provided a range of approximately 350 cm-1 centered at the set wavelength of the Triplemate; the range was calibrated for wavenumber Raman shift from a particular exciting laser wavelength using known emission lines[6] from low pressure inert gas bulbs. The data collection was managed by a PAR model 1460 optical multi-channel analyzer. The resolution of the diode array was approximately 0.7 cm-1. However, the relatively large slit settings needed to illuminate enough of the diode array to provide the above spectral range contributed a signifi-cant instrumental line width to the spectra.

The second instrument was a Spex scanning triple monochromator terminated with a C31034 photomultiplier tube operating in photon-counting mode. In this case the collection of data, at 2 cm-1 resolution, was managed by a Spex Datamate computer. The entrance slit was set at 1 mm, providing a band pass of approx-imately 10 cm^{-1}.

In both cases, the resulting spectra were transferred to a PC for analysis. No correction for absorption of scattered light by the sample was attempted. For some of the samples that showed significant fluorescence below the Raman bands, a polynomial-fit baseline was subtracted before peak center determina-tion. Peak positions were determined using a center of mass algorithm with the half-maximum intensity used as the cut-off, coded for SpectraCalc soft-ware.

III. UV-Visible Spectra

UV-visible spectra of the our model polyenes have been measured in a variety of solvents and in the solid state. As noted in the experimental section, solid-state spectra were recorded for thin films on glass slides or for water dispersions. In both cases, scattering distorts the spectra considerably, though we have been able to ascertain the peak positions for the lowest energy absorption peaks.

Some examples of solution spectra are shown in Figure 1. All solution spectra show a well-defined 0-0 transition followed by a vibrational progression with a spacing on the order of 0.2 eV. As expected, the position of the 0-0 band (E_0) decreases smoothly with increasing chain length. The vibrational spacing also decreases somewhat as chain length increases.

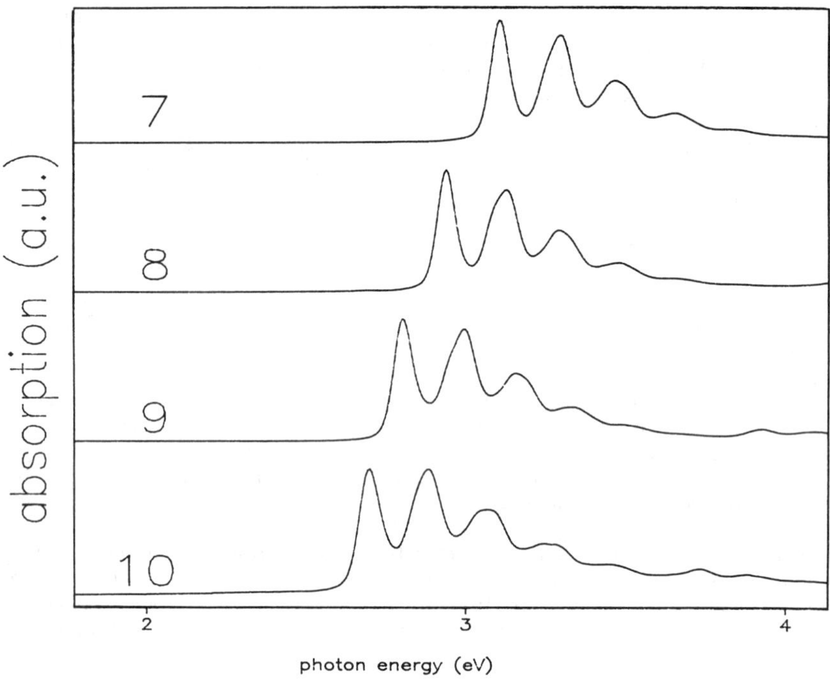

Figure 1. Solution phase uv-vis absorption spectra of four of the t-butyl capped polyenes, n=7-10, as labelled.

Figure 2 summarizes some of the E_0 data for our model polyenes. The data for pentane solution are consistent with literature results for other polyene series,[3,8] though the Figure 2 data are more complete and extend to larger n. For example, the data for dimethylpolyenes for n=1 to 10 would essentially superimpose on our pentane data.[8] We have also studied the solvent dependence of electronic absorption. The solvent series chosen was pentane, cyclohexane, benzene, and carbon disulfide; this series covers a large range in refractive index (1.36 to 1.62) so that solvent polarizability effects on electronic absorption can be determined. We find results which are quite consistent with those of Sklar et al.,[9] who studied diphenyl polyenes over the chain length range 1 to 8, and of D'Amico et al.[10] for regular polyenes with n=4-6.

As can be seen in Figure 2, E_0 is linear in 1/n at large n, becoming sublinear for n less than roughly 7. This is true for all solvents studied. Solvent polarizability has a substantial effect on E_0 but little or no effect on the

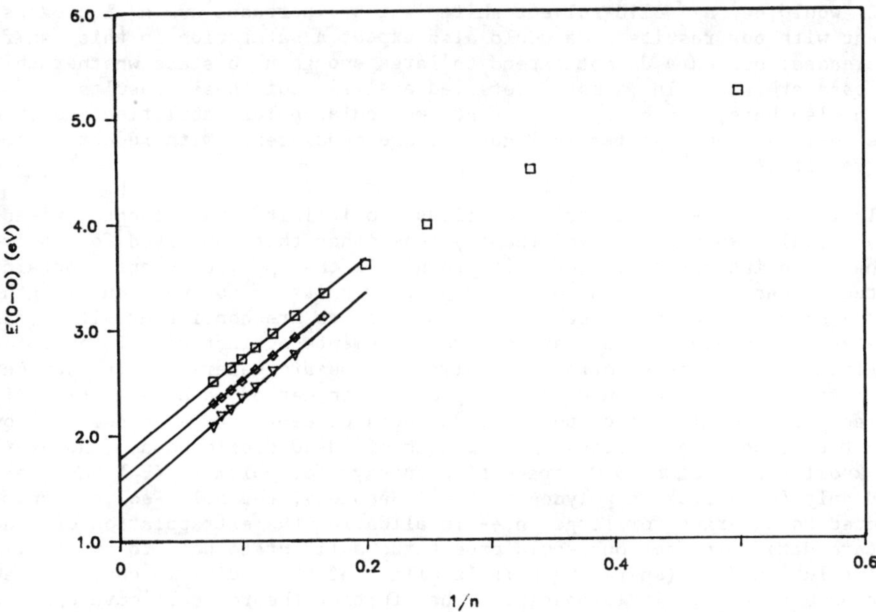

Figure 2. Energy of 0-0 transition (E_0) vs. the inverse of the number of double bonds in the molecule. Squares are in pentane solution, diamonds are in CS_2 solution, and inverted triangles are in solid form.

slope of E_0 versus $1/n$ in the linear regime. Extrapolation of the curves to the infinite chain limit ($1/n=0$) yields E_0 values which are greater than that of polyacetylene ($E_p=1.55$ eV).[11] This is to be expected since the polymer data are only available for the solid state where the surrounding molecules are polymers with polarizability substantially greater than that of the solvents we have used.

The E_0 data for thin films of the model polyenes are significantly lower than those for solutions again because of the higher polarizability of the medium. Analysis of the solid-solvent E_0 shifts is complicated by the fact that polyene polarizabilities are expected to vary as a function of n. According to recent theoretical studies,[12] the molecular polarizability of polyenes increases roughly as n^2 for small n and begins to approach a linear dependence at roughly n=10. The latter is the trivial increase due to the increase in molecular size; therefore in the range the polarizability of the medium would be approaching a constant which would also be expected to be characteristic of the polymer. The consequences for solid-state E_0 data such as those shown in

Figure 2 would be a solid-solvent shift which increases as n increases, consistent with our results. We would also expect a saturation in this shift as n increases; our data do not extend to large enough n to state whether this is the case or not. In a more detailed analysis of these results to be published elsewhere, we have shown that molecular polarizabilities derived from the solid-solvent shifts in Figure 2 are consistent with recent theoretical results.[12]

Extrapolation of the E_0 data for the solids to infinite chain length yields 1.33 eV, a value which is significantly less than that observed for polyacetylene. One interpretation of this result is that polyacetylene contains conjugated segments that are no more than 20-30 units in conjugation length. As will be seen in the next section, Raman results are consistent with this interpretation. However, there are several arguments against this interpretation. First, absorption spectra for polyacetylene are generally independent of preparation method; we would expect preparation method to have a significant effect on chemical and physical defects present at a concentration sufficient to produce a limiting chain length of 20-30 double bonds. However, to our knowledge a clear 0-0 transition energy for polymer (E_p) has been obtained only for Shirakawa polyacetylene.[11] Secondly, the solid-solvent shift is expected to saturate for large n,[12] invalidating the extrapolation of the solid-state data. In fact one would expect the solid-state data to eventually parallel solution data (where the polarizability of the medium is constant) as n increases. Finally, it is possible that further theoretical developments will reveal higher order terms in the n dependence of electronic adsorption which will cause a sublinear dependence at small values of n. Therefore the theoretical justification for the 1/n extrapolation is not rigorous.

IV. Raman Spectra

Raman scattering has been a prime probe of the structure of conducting polymers, in particular of both cis and trans isomers of polyacetylene.[14-18] The most salient feature of the Raman spectra obtained from undoped trans-polyacetylene, as by now observed by numerous workers, is the dependence of Raman shift upon excitation wavelength, or dispersion. The two strongest Raman bands in the range corresponding to mid-infrared strength vibration are nominally the carbon-carbon single bond stretch and double bond stretch, although both modes involve a mixture of single and double bond stretches and CCH angle bend. For red excitation, the first appears as an asymmetric band at about 1060 cm^{-1} and the second similarly at about 1460 cm^{-1}; these frequencies are referred to as the primary peaks. As the excitation wavelength is decreased toward the blue, high frequency shoulders develop and increase both in intensity and frequency with decreasing excitation wavelength. For violet excitation they are resolved as satellites on the high energy side of the primary peaks. It has been found that the shapes of the Raman bands, and in particular the relative strengths of the primary and satellite peaks, are dependent upon sample preparation and history, sometimes varying for different sections of a single sample. However, the frequencies of both primary and satellite peaks depend only upon excitation wavelength.

A number of models[18-23] have been proposed to account for this dispersion, and two are currently in the greatest favor.[23,24] In the first,[20,21] the Raman frequency of either of the two strongest modes is assumed to decrease with increasing conjugation length as is found for the double bond vibration in finite molecular polyenes, according to the form

$$\nu_{1,2} = (A_{1,2} + B_{1,2}/n) \tag{1}$$

where 1,2 denote the "single bond" and "double bond" stretch, respectively. The most recently quoted values of the parameters are $A_1 = 1060$, $B_1 = 600$, $A_2 = 1450$, $B_2 = 500$, all in cm^{-1}.[21] As noted above, the lowest electronic excitation energy is also assumed to decrease with increasing conjugation length, consistent with experimental and various theoretical models. According to this model, red excitation probes only those segments with relatively long conjugation lengths, providing their characteristic Raman frequency. As the excitation wavelength is decreased, excitation of the shorter, segments is also allowed, thus showing the the characteristic Raman shifts of both. When applied to the Raman spectra of polyacetylene, this model suggests a bimodal distribution of conjugation lengths.[24] The details of the determined distribution of conjugation lengths depends upon the details of the model connecting the shape of the measured Raman spectrum to the excitation wavelength via the conjugation length, and in particular upon the assumed dependencies of excitation energy and Raman shift with respect to that length. It may also be noted that another recent model has proposed a direct relationship between the double bond stretch frequency and the excitation energy,[25] as was previously suggested.[26]

In the second model[23] the cross-section for Raman scattering is expressed as the product of two functions. The independent variable is not the conjugation length, but rather a parameter, $\tilde{\lambda}$, which is indirectly related to the electron-phonon coupling constant λ; the form of the relationship of these two parameters is determined by the specific model for electron-phonon interaction. The Raman frequencies (or, more precisely, the product of the frequencies of three coupled modes) are found to increase with increasing $\tilde{\lambda}$, while the lowest electronic excitation energy is found to depend upon the related parameter λ, e.g., as in a Peierls model.[23] Thus, a correspondence between Raman shift and electronic excitation is determined, as in the previous model. A (unimodal) distribution of values of $\tilde{\lambda}$ within a particular sample is then allowed. The first component function is peaked at the Raman frequency characteristic of chain segments described by the mode of the population distribution in $\tilde{\lambda}$. The second component function is peaked at the Raman shift characteristic of chain segments having $\tilde{\lambda}$ corresponding to the value of λ corresponding to an electronic excitation equal to the photon energy. Thus, red excitation provides only one line (at least for samples prepared via the usual Shirakawa route) because the most probable chain segments have an electronic excitation energy in the red, so that both component functions are peaked at the same Raman shift. As the excitation wavelength is decreased, the sparser population of chain segments having higher excitation energy are allowed to come into resonance, exhibiting their characteristic Raman frequencies as well; multiplication of the two component functions with different peaks provides a double-peaked function consistent with experiment. In this scenario, the interpretation of the Raman dispersion

372

is critically dependent upon the model used for electron-phonon coupling and
the model used for the electronic excitation as a function of that coupling.[23]
One could, if desired, decorate this model with the assumption that the
parameters λ and $\tilde{\lambda}$ are primarily dependent upon some form of effective conjugation length.   This approach allows a perhaps more easily rationalizable
assumption of unimodal population distribution to explain the data.

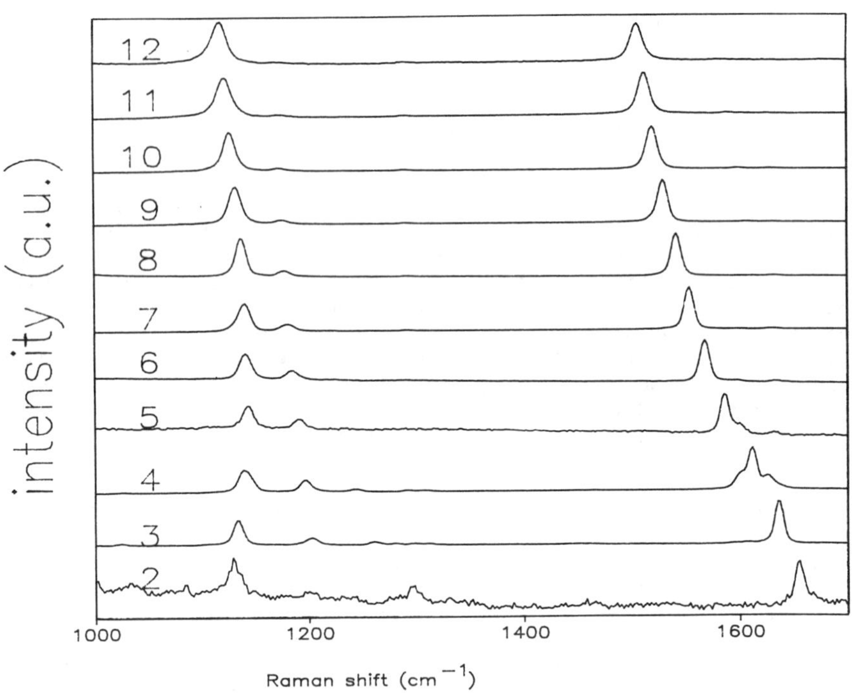

Figure 3. Raman spectra between 1000-1700 cm-1 of t-butyl capped polyenes,
for n=2-12, as labelled. These spectra were recorded for materials as solids
at room temperature with excitation wavelength 6470.9 Å.

At the present time, it is difficult to choose between these models; however,
it is clear that both of them would benefit from calibration information:
dependence of measured Raman spectra upon polyene segments with known conjugation length, or, through measurement of uv-visible absorption, of known
electronic excitation energy.   For this reason, we have measured the Raman
spectra of these isolated compounds having known conjugation length and
absorption energies. In figure 3, we show the the Raman spectra between 1000
and 1700 cm^{-1}, for excitation at 6470.9 Å, of the t-butyl capped polyenes for
n from 2 to 12.   As noted above for polyacetylene, two bands are most prominent for the 12-ene, at the top of the figure. The "single bond stretch" is
at 1120 cm^{-1}, while the "double bond stretch" is at 1501 cm^{-1}. As the conjugation length is decreased, moving down the figure, the shift of both bands are

apparent. The simpler behavior is that of the double bond stretch, which moves monotonically to higher energy as n decreases and is seen centered at 1654 cm^{-1} for the 2-ene. The single-bond stretch similarly begins to increase in frequency as n decreases. It is also seen, however, that another band, which for the 12-ene is not visible on the scale of this figure but is found at 1167 cm^{-1}, grows in magnitude as conjugation length is decreased. This band is, in the 12-ene, attributed to the carbon-carbon single bond stretch of the capping t-butyl groups, and as the central portion of the molecule becomes shorter, containing fewer single bonds, the outer segments become relatively more significant. This band is also seen to monotonically increase in frequency. For n<5, however, the "chain" single bond band begins to decrease in frequency.

In fact, higher sensitivity observation of these spectra show that the situation is more complicated. The single bond region has between two and five bands, depending on the molecule, while the double band has between one and three. The number of bands, their position, and their relative intensities for each conjugation length, the correlation of these data with optical absorption energies, and the dependence of oscillator strengths upon excitation wavelength, will be reported in a following publication. Here we focus on the relevance of the conjugation behavior of the three most prominent bands.

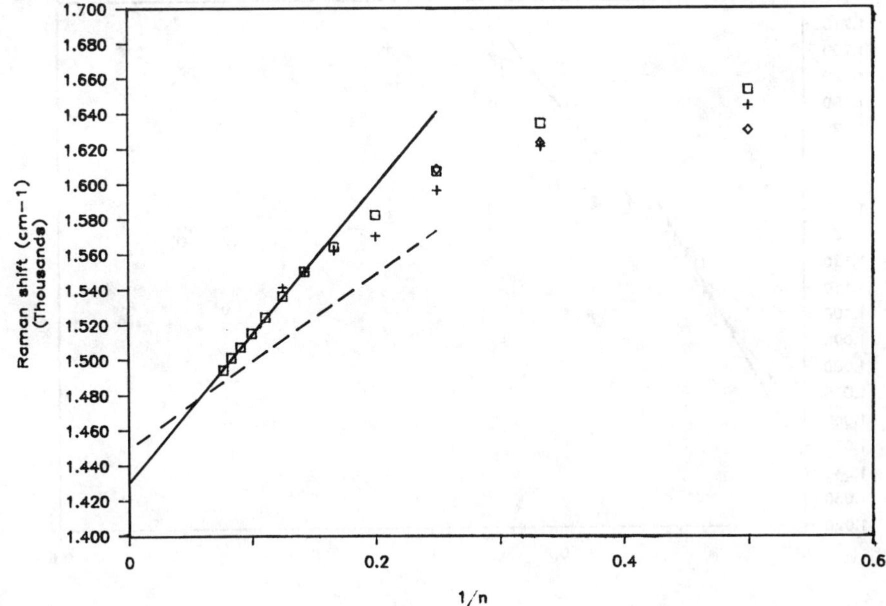

Figure 4. Raman frequencies of the strongest band in the carbon-carbon double bond stretch region. The squares are the data for the t-butyl capped poly-enes, the +'s are data reported for ethyl ester capped polyenes (Ref. 27) and the diamonds are data reported for unsubstituted polyenes (Ref. 28). The full line is the linear fit to our data for n=7-12, and the dashed line is given by the parameters reported in Ref. 21.

Figure 4 shows our measured values of the double bond stretch, along with similar values for polyenes either totally unsubstituted[27] or terminated with ethyl ester groups,[28] as a function of inverse conjugation length 1/n. Good agreement between data sets for shorter conjugation is found; however, our data provide for the first time, an extension of molecular polyenes to a conjugation length that can support extrapolation to polyacetylene. The full line is a linear (in inverse conjugation) fit to our data for n≥7. The dashed line is determined from the parameters as given above for ν_2 as given in Ref. 21. In comparison to these values, we find $A_2 = 1430$ cm^{-1} and $B_2 = 840$ cm^{-1}. The ultimate origin of the parameters quoted in Ref. 21 is as follows:[19,20] the slope was determined by fitting of the values for n between 4 and 7; as can be seen in the figure, this would provide a shallower slope than we have found. The intercept was set to mimic the value for trans-polyacetylene under red excitation, upon the assumption that this corresponded to infinite conjugation. One can deduce from our determined fit, and the good linearity of the measured data for n≥7, that a measured Raman shift of 1460 cm^{-1} for polyacetylene corresponds to a conjugation length of about 30 double bonds.

Figure 5 shows the dependence of the two most prominent single bond bands in the single bond region as a function of inverse conjugation. Also included

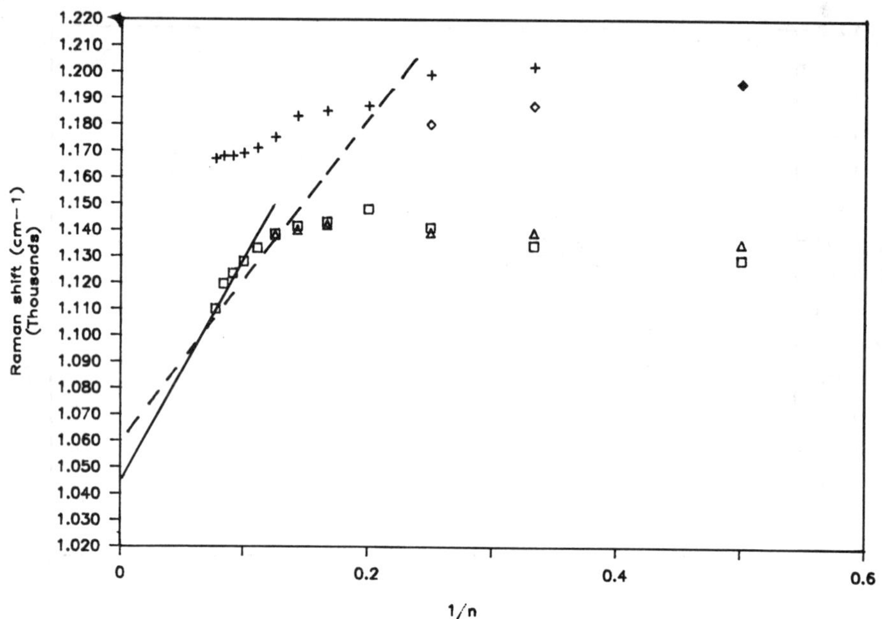

Figure 5. Raman frequencies of the two strongest bands in the carbon-carbon single bond region. The squares and +'s are data measured for t-butyl capped polyenes. The triangles are data reported for ethyl ester capped polyenes (Ref. 27) and the diamonds are data reported for unsubstituted polyenes (Ref. 28). The full line is the best fit to our data assuming the same slope as found for the double bond frequencies; the dashed line for comparison is given by the parameters reported in Ref. 21.

for comparison are values reported by other workers for unsubstituted polyenes[27] and for di-ethyl ester capped polyenes.[28] Attempts to fit the repulsion of the two single bond band frequencies as a two-level system with an
n-dependent coupling parameter were unsuccessful. For ν at large n, our data
are not precise enough and the linear regime not extended enough for determination of the linear fit parameters A_1, B_1. To estimate them, we have assumed
$B_2 \simeq B_1$, since both ν_1 and ν_2 contain strong admixture of single and double
bonds. With this assumption we find for the intercept $A_1 = 1045$ cm^{-1}. We
have shown this as a full line, and the line determined from the parameters of
Ref. 21 as a dashed line. The implications of these data for the second model
will be discussed with the full Raman data in a forthcoming publication.

V. Conclusions

In this paper we have described the optical and resonance Raman spectra of a
series of polyenes with the number of double bonds varying from 1 to 13.
These data serve to calibrate the conjugation length dependence of the electronic properties of polyacetylene oligomers beyond the previously existing
data base. Extrapolation of these data to infinite conjugation length allows
comparison to the corresponding values in polyacetylene. This comparison
suggests that the longest conjugation lengths in polyacetylene are only about
30 double bonds. Alternative models, which would allow longer conjugation
lengths, are possible, but they require more elaborate relationships among
conjugation length, optical gap and resonance Raman spectrum, based on assuming a form for the electron phonon coupling.

Acknowledgements: We thank Professor Bryan Kohler for many helpful discussions. R. R. S. thanks the Director, Office of Basic Energy Research, Office
of Basic Energy Sciences, Chemical Sciences Division of the U.S. Department of
Energy (Contract DE-FG02-86ER13564) for support. K. K. thanks the Deutscher
Akademischer Austauschdienst for a NATO fellowship. Part of this research was
funded by a grant from the NSF (to Silbey and Schrock), #DMR87-19217.

References:

1. A. J. Heeger, S. Kivelson, J. R. Schrieffer, and W.-P. Su, Rev. Mod.
   Phys. **60**, 781 (1988), and references therein.
2. See, e.g., B. E. Kohler, C. Spangler, and C. Westerfield, J. Chem. Phys.,
   89, 5422 (1988).
3. J. L. Bredas, R. Silbey, D. Boudreaux, and R. R. Chance, J. Amer. Chem.
   Soc., **105**, 6555 (1983).
4. B. E. Kohler, J. Chem. Phys., **88**, 2788 (1988).
5. K. Knoll and R. R. Schrock, J. Amer. Chem. Soc., **111**, 7989 (1989).
6. A. R. Striganov and N. S. Sventitskii, Tables of Spectral Lines of
   Neutral and Ionized Atoms. (IFI, Plenum, NY, 1968).
7. D. G. Cameron, J.K. Kauppinen, D. J. Moffatt and H. H. Mantsch. Applied
   Spectroscopy, **36**, p. 245 (1982).
8. L. Salem, The Molecular Orbital Theory of Conjugated Systems. (Benjamin,
   Reading, 1966) p. 366, and references therein.
9. L. A. Sklar, B. Hudson, M. Petersen, and J. Diamond, Biochemistry **16**, 813
   (1977).

10. K. L. D'Amico, C. Manos, and R. L. Christensen, J. Amer. Chem. Soc. 102, 1777 (1980).

11. H. Eckardt, J. Chem. Phys., 79, 2085 (1983); S. D. Phillips, R. Worland, G. Yu, T. Hagler, R. Freedman, Y. Cao, V. Yoon, J. Chiang, W. C. Walker, and A. J. Heeger, Phys. Rev. B, 40, 9751, (1989).

12. C. P. de Melo and R. Silbey, J. Chem. Phys., 88, 2558 (1988).

13. J. M. André, C. Barbier, V. Bodart, and J. Delhalle in <u>Nonlinear Optical Properties of Organic Molecules and Crystal</u>, 2, Ed. D. Chenla and J. Zyss, (Academic Press, NY, 1987) p. 137; Z. Soos and G. W. Hayden in <u>Electroresponsive Molecular and Polymeric Systems</u>, Ed. T. Skotheim (Marcel Dekker, NY, 1988) p. 197; Z. Soos and G. W. Hayden, preprint, 1989.

14. I. Harada, Y. Furukawa, M. Tasumi, H. Shirakawa, and S. Ikeda, J. Chem. Phys., 73, 4746 (1980).

15. L. S. Lichtmann, A. Sarhangi, and D. B. Fitchen, Solid State Comm., 36, 869, (1980).

16. H. Kuzmany, Phys. Status Solidi B, 97, 521 (1980).

17. S. Lefrant, J. Phys. (Paris) Colloq., 44, C3-C247 (1983).

18. Z. Vardeny, E. Ehrenfreund, O. Brafman, and B. Horovitz, Phys. Rev. Lett., 51, 2326 (1983).

19. L. S. Lichtmann, Ph.D. thesis, Cornell University (1981).

20. G. P. Brivio and E. Mulazzi, Chem. Phys. Lett., 95, 555 (1983).

21. G. P. Brivio and E. Mulazzi, Phys. Rev. B, 30, 876 (1984).

22. B. Horovitz, Solid State Commun., 41, 729, (1982).

23. E. Ehrenfreund, Z. Vardeny, O. Brafman, and B. Horovitz, Phys. Rev. B, 36, 1535 (1987); B. Horovitz, Z. Vardeny, E. Ehrenfreund, and O. Brafman, J. Phys. C.: Solid State Phys. 19, 7291 (1986).

24. M. A. Schen, J. C. W. Chien, E. Perrin, S. Lefrant, and E. Mulazzi, J. Chem. Phys., 89, 7615 (1988).

25. F. Zerbetto, M. Z. Zgierski, and G. Orlandi, Chem. Phys. Lett., 141, 138 (1987).

26. L. L. Rimai, M. E. Heyde, and D. Gill, J. Amer. Chem. Soc., 95, 4493 (1973).

27. B. S. Hudson, B. E. Kohler, and K. Schulten, \it Excited States, 6, 1 (1982).

28. T. M. Ivanova, L. A. Yanovskaya, and P. P. Shorygin, Opt. Spectr., 18, 115 (1965).

29. G. Rossi, R. R. Chance, and R. Silbey, J. Chem. Phys., 90, 7594 (1989).

NOVEL LINEAR AND NONLINEAR OPTICAL EFFECTS IN POLYDIACETYLENES

D. BLOOR[1], D. J. ANDO, P. A. NORMAN
Department of Physics,
Queen Mary College,
London, E1 4NS, U.K.

A. F. DRAKE
Department of Chemistry,
Birkbeck College,
London, WC1H OAJ, U.K.

S. MANN, A. R. OLDROYD,
GEC-Marconi Research Centre,
Chelmsford,
Essex, CM2 8HN, U.K.

B. S. WHERRETT, A. K. KAR, W. JI and T. HARVEY
Department of Physics,
Heriot-Watt University,
Edinburgh, EH14, 4AS, U.K.

ABSTRACT Organic materials with useful nonlinear optical properties are
being developed. Practical utilisation requires special combinations of
properties that facilitate the fabrication of device structures as well
as providing large nonlinearities. Two classes of polydiacetylenes
suitable for different end uses are discussed. First, soluble
polydiacetylenes with helical backbones have been synthesised and their
linear optical properties determined. Low loss slab and mono-mode optical
waveguides have been produced and their optical nonlinearities studied.
Defocussing of laser beams by polydiacetylene single crystals has been
observed to result in new beam break-up effects. The crystals have been
shown to possess a large thermo-optic coefficient and their use in optical
bistable devices is considered.

1. INTRODUCTION

Although it has been known for more than a decade that polydiacetylenes
(PDAs), Figure 1, have large, fast third order nonlinear optical response

---

[1] Present Address: Applied Physics Group, University of Durham,
Durham

J. L. Brédas and R. R. Chance (eds.), Conjugated Polymeric Materials:
Opportunities in Electronics, Optoelectronics, and Molecular Electronics, 377–386.
© 1990 Kluwer Academic Publishers. Printed in the Netherlands.

[1], it is only over the last few years that detailed experimental studies have been made of the origins of the nonlinearity [2-7]. During the same period there has also been considerable activity in the preparation of PDAs tailored for particular applications, e.g. as active integrated optical wave guides [8-12]. In this context it is necessary to produce thin films that can guide light with a minimum of loss. Since the wavelength of the guided radiation can be chosen to lie outside the absorption band of the polymer, i.e. in the infra-red, the principle cause of loss is scattering from inhomogeneities in the film. Thus, for optimum performance a glassy PDA film is required, despite the natural tendency for the rigid, conjugated polymer chains to aggregate into micro-crystals. Careful control of film precipitation can lead to the freezing in of the disorder present in the chains of PDAs, such as the n BCMU's (Figure 1) dissolved in chloroform [8]. We describe an alternative approach through the production of non-planar PDA molecules for which the growth of crystallites is inhibited by the polymer conformation [12,13]. In addition the ability to obtain helical conformations with a definite chirality enables us to produce chiral optical waveguides which should possess distinctive properties.

The emphasis of the studies of the non-linear optical properties of PDAs has been on electronic processes and the extremely fast response that ensues. Experimental response times are femtoseconds for excitation out of resonance with the principal absorptions and are still picoseconds under resonant conditions. The magnitude of the non-linearity is such, i.e. in the range 10^{-10} to 10^{-8} esu, that high optical power levels are required to produce usable effects. Because of the concentration on electronic nonlinearities great lengths have been taken to eliminate any thermal effects which might affect the results. We have, however, discovered that PDA crystals possess extremely large thermo-optic

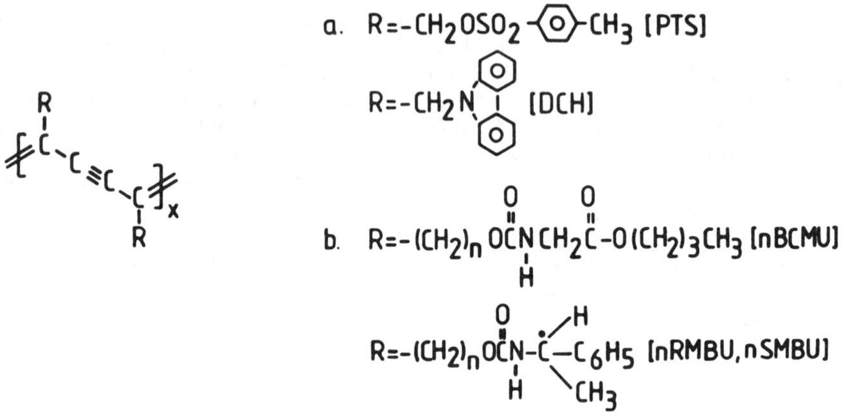

Figure 1. Schematic structure of the polydiacetylene chain with examples of side groups found in (a) crystalline and (b) soluble polymers.

coefficients. Furthermore, the intrinsic anisotropy of PDA single crystals leads to a new, novel self defocussing phenomenon for light beams propagating through thin single crystals normal to the polymer chains [14].

In the following we give a brief account of these novel linear and nonlinear optical effects.

2. CHIRAL POLYDIACETYLENES

2.1. Introduction

Previously reported syntheses of PDAs with chiral substituents have produced insoluble crystals with little evidence that the substituents affect the polymer backbone [15]. The soluble PDAs with optically pure, chiral methylbenzylurethane substituents which we have synthesised possess phases in which the PDAs adopt a helical backbone conformation with the chirality absolutely determined by that of the substituents. These polymers have been used to produce monomode slab and channel waveguides and their optical nonlinearities have been determined.

2.2 Materials and Methods

Chiral diacetylene monomers were prepared by the condensation reaction of either R-(+)-methylbenzyl isocyanate or the S-(−)-enantiomer with diacetylene diols which have -(CH$_2$)- sequences containing between 2 and 9 repeat units between the diacetylenic moiety and the alcohol end groups. The monomers were polymerized by ^{60}Co γ-ray irradiation and unheated monomer removed by solvent extraction. Circular dichroism (CD) spectra were recorded on a JASCO J40CS CD spectrometer. Thin films suitable for optical guiding were formed by dip coating from chloroform solutions onto glass and silicon oxynitride coated Si substrates. Further details of the nonlinear optical and spectroscopic characterisation are given in reference refs. [9,13].

2.3 Results

Solutions of nRMBU and nSMBU exhibit similar solvato- and thermo-chromism to that reported for the achiral nBCMU polymers [16]. When filtered, the yellow chloroform solutions show no circular dichroism in the vicinity of the polymer absorption. However, red solutions obtained by either cooling the solution or adding a non-solvent and red films obtained by precipitation, exhibit intense CD spectra in the same spectral region as the linear absorption band. The CD spectra show sharp monosignate features and broad bisignate bands, which correlate with structures of similar width in the linear absorption, see Figure 2. The implications of these results for the molecular conformation of the polymer chains have been discussed in reference [13].

Dip coated films of these polymers appear visually to be free of light scattering defects. At 1.3 μm guiding was observed using prism coupling in films of 9-SMBU between 1 and 3 μm thick. Optical loss is below 2 dB/cm with refractive indices of 1.567 in the plane of the film

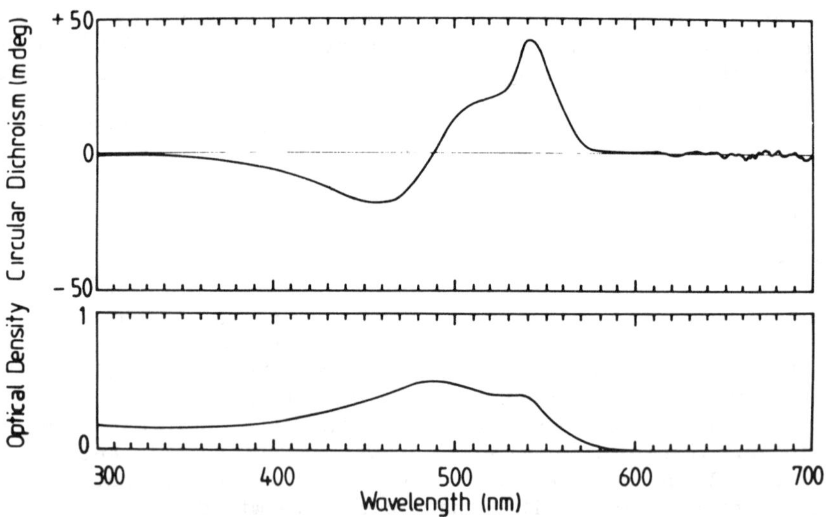

Figure 2. Linear and CD absorption spectra of 9 RMBU in tetrahydrofuran-water (1:2) solution

and 1.555 normal to the substrate. Similar birefringence has been observed for nBCMU polymer films [8] though whether this is due to either stress induced or intrinsic alignment of the polymer chains in the plane of the films has not been established. The nonlinear optical properties of these high quality films have been determined by through plane third harmonic generation (THG) [9] and electro-optic measurements on slab guides. At 1.064 μm the value of $\chi^{(3)}$ determined by THG is 7 x 10^{-12} esu and at 1.3 μm the effective $\chi^{(3)}$ determined from the quadratic electrooptic effect is 2 x 10^{-13} esu [17].

Mono-mode waveguides have been fabricated using the same polymer. Channels 2 μm deep and from 2 to 10 μm wide were etched into a 4 μm thick silicon oxynitride layer on a silicon substrate. These were filled with polymer by dip coating giving a partially planarising layer. Cleaving the substrate gave polymer end faces of sufficient optical quality that no end polishing was required. Observation of end-fire coupled 1.3 μm radiation gave mode profiles, recorded with an IR vidicon, extremely close to those predicted by a finite element analysis. Optimum coupling from optical fibres was obtained for a 5 μm wide channel and the coupling loss was below 4 dB. The channel loss deduced from these measurement was less than 6 dB, which is reasonable in view of the roughness of the reactive ion etched channels [9].

2.4 Discussion

The excellent optical quality of films of the chiral PDAs suggest that these are predominantly amorphous. This is almost certainly a consequence of the helical conformation which must severely hinder the aggregation of the polymer chains into crystalline arrays. Such behaviour is different from that observed for the planar nBCMU polymers where great care is needed to inhibit the formation of strongly scattering microcrystallites and retain the disorder present in dissolved polymer on precipitation [9,11,12]. Comparison of the ordinary and CD spectra of the chiral PDAs reveals common features that can be interpreted as arising from a major amorphous component, the broad absorption at circa 22,000 cm^{-1} and the associated bisignate CD feature, and a minor crystalline component, the sharper, lower energy absorption band and the monosignate CD feature. This interpretation is based on the fact that the former spectral features are characteristic of isolated molecular species and the latter of chiral aggregates. From estimation of the integrated absorption intensities the crystalline content is of the order of a few percent and, from the low optical loss, probably consists of particles much less than 1 μm in size.

The low $\chi^{(3)}$ value determined by THG is commensurate with the low filling factor of active conjugated chains resulting from the use of large pendent groups, the randomisation of chain orientation in the largely amorphous films and the blue shift of the exciton absorption relative to single crystal PDAs. While a similar reduction in quadratic electro-optic coefficient could be expected the available data suggests comparable values at 1.3 μm, i.e. $\Delta n \sim 10^{-5}$ at 10^7 V/m applied field, for both 9SMBU films and PDA crystals [4,17]. These values, though too small for practical device applications, are sufficient to enable a number of active integrated optic devices to be demonstrated.

3. THERMO-OPTIC EFFECTS

3.1 Introduction

The emphasis on studies of very fast optical nonlinearities has resulted in strenuous efforts to eliminate unwanted thermal effects. This is necessary in studies of refractive index changes produced by third order nonlinearity since heating of most materials results in a change of refractive index. The resulting effective, third order nonlinearity is often larger than the electronic effects being studied. Because of this these troublesome effects can often be useful in their own right. In particular they have been exploited in optical bistable etalons fabricated from compound semiconductors and liquid crystals [18]. Thus we have investigated thermal nonlinearities in thin single crystals of the PDA PTS, Figure 1(a). This study has revealed novel nonlinear effects, which result from the highly anisotropic nature of these crystals, and the occurrence of large thermo-optic coefficents.

3.2 Materials and Methods

Thin single crystals of the PDA PTS were obtained by cleavage of larger single crystals of either the monomer or the polymer. In either case the polymer was obtained by thermal polymerisation at 60°C in a vacuum oven. Typical samples had dimensions of circa 1 sq.cm on the (100) facet by 50 to 150 μm thickness. Samples were illuminated normal to the (100) facet with light of wavelength near 700 nm from a dye laser, which was polarized along the polymer chain axis. The beam was focused to a spot size of 2 μm diameter with maximum irradiance of 1 kW/cm^2. The transmitted irradiance profiles in the far field were recorded with a CCD camera and a frame store unit, as indicated in Figure 3(c).

3.3 Results

At low input irradiance the Gaussian profile of the input beam is preserved in the transmitted beam, Figure 3(a), with slight distortion due to imperfections in the crystal, e.g. surface cleavage steps, etc. This situation changes dramatically at a critical threshold above which the transmitted beam splits into two areas of high irradiance symmetrically placed about the original beam direction, Figure 3(b). In this condition there is a minimum at the centre of the intensity distribution in sharp contrast with the central maximum observed in the far field intensity distribution for self-defocussing of Gaussian laser beams by isotropic media.

3.4 Discussion

Irradiation of the PTS crystals at wavelengths near 700 nm leads to local heating of the sample since the absorption coefficient for light polarized parallel to the polymer chains is greater than 100 cm^{-1} [4,14]. Because the thermal conductivity of the crystals is anisotropic, being greater parallel to the polymer chains than for perpendicular orientations, the resulting temperature and nonlinear refractive index distributions are also anisotropic. Existing models for pure phase defocussing, i.e. pure

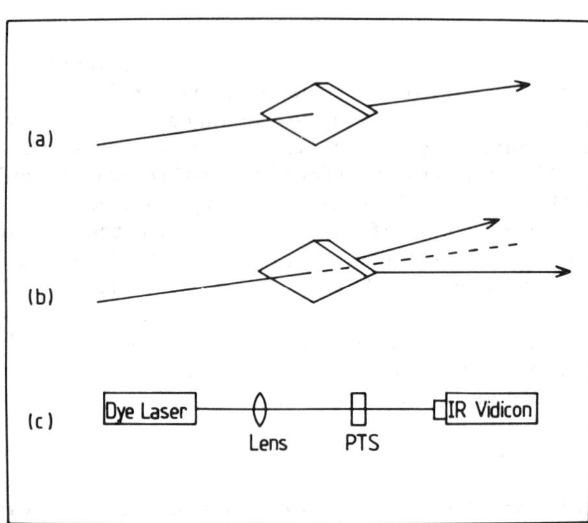

Figure 3   Transmission of near infra-red radiation at wavelengths near 700 nm through a thin PDA crystal: (a) at low input intensity and (b) at high input intensity; (c) experimental arrangement.

refractive index nonlinearity, for isotropic materials which predict cylindrically symmetric far-field distributions, with a maximum on the optical axis, are clearly inappropriate. These can, however, be extended for an elliptical refractive index nonlinearity by constructing the far-field pattern from a sum of elliptical Gaussian functions rather than from simple Gaussian functions [19].

In the geometric optics approximation, if the phase change in the medium is small compared to the ratio between the diffraction length within the medium and the sample thickness then the electrical field at the rear face of the sample can be expressed as a sum of Gaussian beams of different spot size. Each component beam can then be propagated to some distance z, behind the sample and the resultant diffraction pattern calculated numerically. However, in the present case the nonlinear refractive index profile within the medium has an elliptical Gaussian form (i.e. $\exp[-(x^2/r_x^2 + y^2/r_y^2)]$). The electric field at the exit plane of the sample is then given by:

$$E(x,y,0) = E(0,0,0) \; \exp\left[-\left(\frac{x^2 + y^2}{r_o^2}\right) + iX \; \exp\left[-\left(\frac{2x^2}{r_x^2} + \frac{2y^2}{r_y^2}\right)\right]\right] \qquad (1)$$

where

$$X = \frac{\omega}{c} \int_0^D n_2 I_o(z) \; dz,$$

r_o is the $1/e^2$ radius of the incident Gaussian beam, r_x and r_y are the axial extents of the induced elliptical Gaussian refractive index profile, I_o is the irradiance at the beam centre in the medium and n_2 the refractive index nonlinear coefficient. This electric field can be expressed as a sum of elliptical Gaussian beams of decreasing radii $W_{mj}^2(0)$:

$$E(x,y,0) = E(0,0,0) \sum_{m=0}^{\infty} (iX)^m \; \exp\left[-\left(\frac{x^2}{W_{mx}^2(0)} + \frac{y^2}{W_{my}^2(0)}\right)\right] \qquad (2)$$

where now, for j=x or y:

$$W_{mj}^2(0) = \frac{r_j^2 r_o^2}{(2mr_o^2 + r_j^2)}$$

Propagation of these elliptical Gaussian modes gives, at a distance z behind the sample, an electric field:

$$E(x,y,z) = E(0,0,0) \sum_{m=0}^{\infty} \frac{(iX)^m}{m!} (1 + (z^2/d^2_{mx}))^{-1/4} (1 + (z^2/d^2_{my}))^{-1/4}$$

$$x \exp\left[- iP_m(z) - x^2\left(\frac{1}{W^2_{mx}(z)} + \frac{ik}{2R_{mx}(z)}\right) - y^2\left(\frac{1}{W^2_{my}(z)} + \frac{ik}{2R_{my}(z)}\right)\right] \quad (3)$$

Where the following definitions are used to simplify equation (3):

$$k = \frac{\omega}{c} \;, \quad W^2_{mj}(z) = W^2_{mj}(0)\,(1 + (z/d_{mj})^2) \;, \quad d_{mj} = \frac{kW^2_{mj}(0)}{2} \;,$$

$$R_{mj}(z) = j\,(1 + (d_{mj}/z)^2) \;, \quad P_m(Z) = -1/2\,\tan^{-1}(z/d_{mx}) - 1/2\,\tan^{-1}(z/d_{my})$$

The integrated phase change through the sample is related to the average temperature along the light beam. On the optical axis for a wavelength of 696 nm ($\alpha = 264$ cm^{-1}) this is calculated to be of the order 10^{-2} K at 1 W/cm^2 incident irradiance allowing for nonlinear absorption, surface loss and longitudinal thermal diffusion. Using selected values for the anisotropy of the thermal conductivities parallel and normal to the polymer chains, the elliptical temperature and the nonlinear refractive index (Δn) distributions are calculated. The irradiance level at which the switch over from one to two beams occurs, i.e. beam break-up, depends on both dn/dT and the width of the Δn distribution. A best fit to the experimental results is obtained with a thermal conductivity anisotropy in accord with reported experimental values and with a dn/dT of $-(2\pm1) \times 10^{-3}$ K. This value is about 10 times larger than that observed for compound semiconductors. Finally it should be noted that the separation of the centres of the two beams increases continuously as the beams propagate beyond the sample and the intensity on the optical axis remains a minimum.

The theoretical analysis has been carried out for incident beams that are themselves anisotropic ($r_{oy} \neq r_{ox}$). In this case, if the medium is anisotropic then there is no two-beam break-up. This follows since on scaling the y-coordinate by r_{oy}/r_{ox}, the problem is identical to that of the Gaussian beam with an isotropic phase distribution. This propagates in the near-field as an annular structure and in the far-field as a single peak; on transforming the coordinates back equivalent elliptical annular and peaked structures are obtained. If both the incident irradiance and the nonlinearity are anisotropic than the largest beam-splitting occurs when the major axes of the radiation and conductivity distributions are perpendicular to one another. Thus the material anisotropy plays a central role in the break-up in the spatial profile of a cw laser beam transmitted through media with a nonlinear, power-dependent, refractive index. In marked contrast to the annular structure usually observed in the far-field for cylindrically symmetric (isotropic) media, nonlinear beam-splitting can occur with separate beams propagating to long distances for anisotropic media.

The large thermo-optic non-linearity suggests that PTS crystals should be suitable for inclusion in nonlinear Fabry-Perot etalons. Such a device has been constructed and sharply defined optical bistability hysteresis loops observed. Further details of these experiments will be published elsewhere.

Acknowledgements

This work was funded by the Science and Engineering Research Council and the Department of Trade and Industry under the Joint Opto-electronics Research Scheme.

References

1. Sauteret, C., Hermann, J-P., Frey, R., Pradere, F., Ducuing, J., Baughman, R. H. and Chance, R. R. (1976), 'Optical nonlinearities in one-dimensional-conjugated polymer crystals', Phys. Rev. Lett. 36, 956-959.

2. Green, B. I., Orenstein, J., Millard, R. R. and Williams, L. R. (1987), 'Nonlinear optical response of excitons confined to one dimension', Phys. Rev. Lett. 58, 2750-2753.

3. Greene, B. I., Mueller, J. F., Orenstein, J., Rapkine, D. H., Schmitt-Rink, S. and Thakur, M. (1988), 'Phonon-mediated nonlinearity in polydiacetylene', Phys. Rev. Lett. 61, 325-328.

4. Greene, B. I., Thakur, M. and Orenstein, J. (1989), 'Quadratic electro-optic effect in poly-diacetylene single crystals', Appl. Phys. Lett. 54, 2065-2067.

5. Blanchard, G. J., Heritage, J. P., Von Lehmen, A. C., Kelley, M. K., Baker, G. L. and Etemad, S. (1989), Excitonic and phonon-mediated optical Stark effect in a conjugated polymer Phys. Rev. Lett. 63, 887-890.

6. Blanchard, G. J., Heritage, J. P., Baker, G. L. and Etemad, S. (1989), 'The picosecond spectroscopy of a polydiacetylene in the small signal limit', Chem. Phys. Lett. 158, 329-333.

7. Nowak, M. J., Blanchard, G. L., Baker, G. L., Etemad, S. and Soos, Z. G. (1989), 'Interchain dynamics and side-group modulation of excitons in polydiacetylene PTS', Phys. Rev. Lett. in the press.

8. Townsend, P. D., Baker, G. L., Schlotter, N. E., Klausner, G. F. and Etemad, S. (1988) 'Waveguiding in spun films of soluble polydiacetylenes', Appl. Phys. Lett. 53, 1782-1784.

9.   Mann, S., Oldroyd, A. R., Bloor, D., Ando, D. J. and Wells, P. J. (1988). 'Fabrication and characterization of processable polydiacetylene waveguides', Proc. SPIE 971, 245-251.

10.   Kanetake, T., Ishikawa, K., Hasegawa, T., Koda, T., Takeda, K., Hasegawa, M., Kubodera, K. and Kobayashi, H., (1989), 'Nonlinear optical properties of highly oriented polydiacetylene evaporated films', Appl. Phys. Lett. 54, 2287-2289.

11.   Krug, W., Miao, E., Derstine, M. and Valera, J. (1989) 'Optical absorption and scattering losses of PTS and poly(4-BCMU) thin film waveguides in the near infra-red', J. Opt. Soc. Amer. B. 6, 726-732.

12.   Bloor, D. (1989) 'Progress towards nonlinear optical devices employing polydiacetylenes' in A. Aviram (ed) 'Molecular Electronics : Science and Technology', Eng. Foundation, N.Y. in the press.

13.   Drake, A. F., Udvarhelyi, P., Ando, D. J., Bloor, D., Obhi, J. S. and Mann, S. (1989), 'Chiroptical spectroscopic studies of polydiacetylenes', Polymer 30, 1063-1067.

14.   Harvey, T. G., Ji, W., Kar, A. K., Wherrett, B. S., Bloor, D. and Norman, P. A. (1989) ' Experimental observation of elliptical defocusing in the organic material pTS', Opt. Lett in the press.

15.   Kronkhe, K. (1987), 'Single crystals from optically active diacetylenes and polydiacetylenes', Ber. Bunsenges, Phys. Chem. 91, 982-984.

16.   Patel, G. N., Chance, R. R. and Witt, J. D. (1979), 'A planar-nonplanar conformational transition in conjugated polymer solutions', J. Chem. Phys. 70, 4387-4392.

17.   Oldroyd, A. R., Mann, S. and McCallion, K. J. (1989), 'Measurement of the quadratic electro-optic effect in a polydiacetylene optical waveguide', preprint.

18.   Wherrett, B. S., Chow, Y. T., Rhoomy-Darzi, A. K. and Lloyd, A. D. (1989), 'Optical bistable devices for memory and logic', Physica Scripta T25, 247-253.

19.   Harvey, T. G., Ji, W., Kar, A. K. and Wherrett, B. S. (1989), 'Theory of elliptical defocussing in anisotropic nonlinear optical media, Opt. Lett. in the press.

NONLINEAR OPTICAL PROPERTIES OF ULTRATHIN POLYMER FILMS

D. NEHER, A. KALTBEITZEL, A. WOLF, C. BUBECK[*],
G. WEGNER
Max-Planck-Institut für Polymerforschung,
Postfach 3148, D-6500 Mainz, Fed. Rep. of Germany

ABSTRACT. Third harmonic generation (THG) and degenerate
four wave mixing (DFWM) experiments are used to study
magnitude and response time of the third order nonlinear
susceptibility of spin cast films of poly (p-phenylene
vinylene) and poly (phenyl acetylene) derivatives with
various substituents at the phenyl ring. The contribution of
two and three photon resonances will be discussed. In the
DFWM experiments we see picosecond response times for these
polymers even under resonance conditions. DFWM experiments
have been performed with phthalocyanine (PC) thin films. The
relative distances of the PC rings were varied by dissolving
or copolymerizing PC in polystyrene and by building up
Langmuir-Blodgett films of PC monomers or polymers. The
linear and nonlinear optical properties of the PC films
depend strongly on the distance of the PC rings and
therefore on the electronic coupling between them. Increased
electronic coupling leads to a spectral broadening of the
optical absorption bands and to a reduction of the response
time to some picoseconds. Influences of energy migration and
transfer to trap states are discussed.

1. Introduction

Basic research in the field of linear and nonlinear optical
properties of thin polymer films has increased considerably
in view of the possibilities for applications in
optoelectronics or integrated optics. A search for new
materials with high nonlinear optical susceptibilities and
approaches for a basic understanding of their structure
property relationship are necessary. For quantitative
studies and for possible combinations with planar waveguides
the processing of these materials to thin films is crucial
(Zyss 1985). Furthermore, good optical material properties
especially low losses by absorption or stray light are
required.

J. L. Brédas and R. R. Chance (eds.), Conjugated Polymeric Materials:
Opportunities in Electronics, Optoelectronics, and Molecular Electronics, 387–398.

Third harmonic generation (THG) experiments are used in our group to obtain quantitative data of the nonlinear optical susceptibility $\chi^{(3)}$. Degenerate four wave mixing (DFWM) of picosecond laser pulses is further used to study the time dependence of the nonlinear optical phenomena. The high values of $\chi^{(3)}$ which are required for applications may only be achieved by means of resonance enhancements, because the nonresonant $\chi^{(3)}$ values reported recently for conjugated polymers are in the typical order of 10^{-12} to 10^{-10} esu (Williams 1983, Chemla and Zyss 1987, Heeger et al 1988, Prasad and Ulrich 1988, Messier et al 1989). Therefore the measurement and basic understanding of multiphoton resonances in nonlinear optics is an important task. These resonances are schematically shown in Fig.1. In the case of resonance the excitation and relaxation dynamics in condensed molecular systems become crucial clues for ultrafast optical switching processes.

2. Optical Experiments

The experimental setup for THG was described recently (Bubeck et al 1989b,c). Infrared light pulses (1064nm, 0.4mJ, pulsewidth 30ps) are generated by an active/passive modelocked Nd:YAG laser. The pulses were focused on the sample mounted on a rotation stage in an evacuated chamber. The DFWM experiments were performed using a folded BOXCARS configuration (Bubeck et al 1989b). The output of a synchronously pumped and cavity dumped dye laser system was split into three beams with variable delay and focused on the sample. The tuning range was given by R6G, DCM and Py1 laser dyes. The intensity at the sample position was in the order of 1 GW/cm^2.

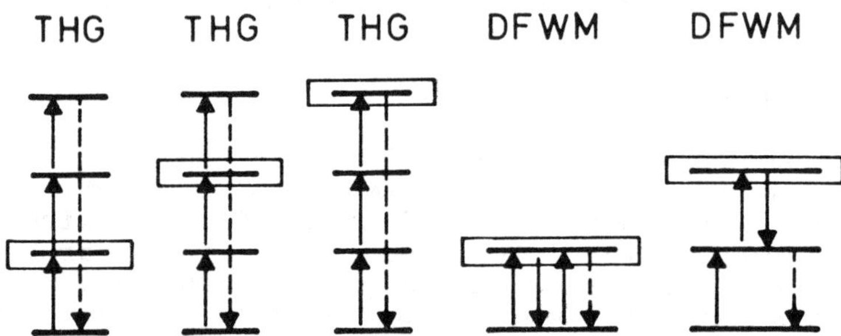

Figure 1. Energy level diagrams showing one, two and three photon resonances with real electronic excitation states (in frames). The nonlinear optical processes are third harmonic generation (THG) and degenerate four wave mixing (DFWM).

3. Polymers With Conjugated π-Electron Systems

3.1 SUBSTITUTED POLY (PHENYL ACETYLENES)

The poly (phenyl acetylene) (PPA) derivatives shown in Tab.1 were synthesized following the method of Masuda (1974) with different catalysts (Wolf 1989). Thin films of PPA were prepared by spin casting. The film thicknesses were measured with a step profiler. The absorption spectra of some films are shown in Fig.2. The shape of the absorption bands in the solid state is almost the same as in solution (Wolf 1989). The positions of the absorption maxima are plotted in Fig.3.

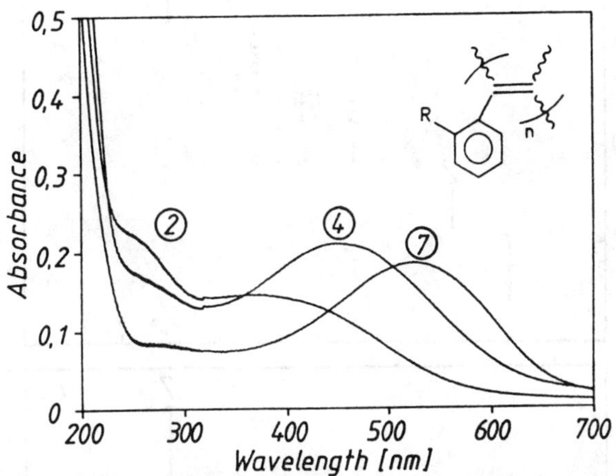

Figure 2.   Absorption spectra of ultrathin poly (phenyl acetylene) films (thicknesses approximately 40nm) on fused silica substrates. The numbers correspond to the PPA systems of Tab.1.

Table 1. List of substituted poly (phenyl acetylenes)

1	R=H	PPA	catalyst: $MoCl_5$
2	R=H	PPA	catalyst: WCl_6
3	R=CH_3	ortho-methyl-PPA	catalyst: $MoCl_5$
4	R=CH_3	ortho-methyl-PPA	catalyst: WCl_6
5	R=C_2H_5	ortho-ethyl-PPA	catalyst: WCl_6
6	R=C_8H_{17}	ortho-octyl-PPA	catalyst: WCl_6
7	R=$Si(CH_3)_3$	ortho-trimethylsilyl-PPA	catalyst: WCl_6

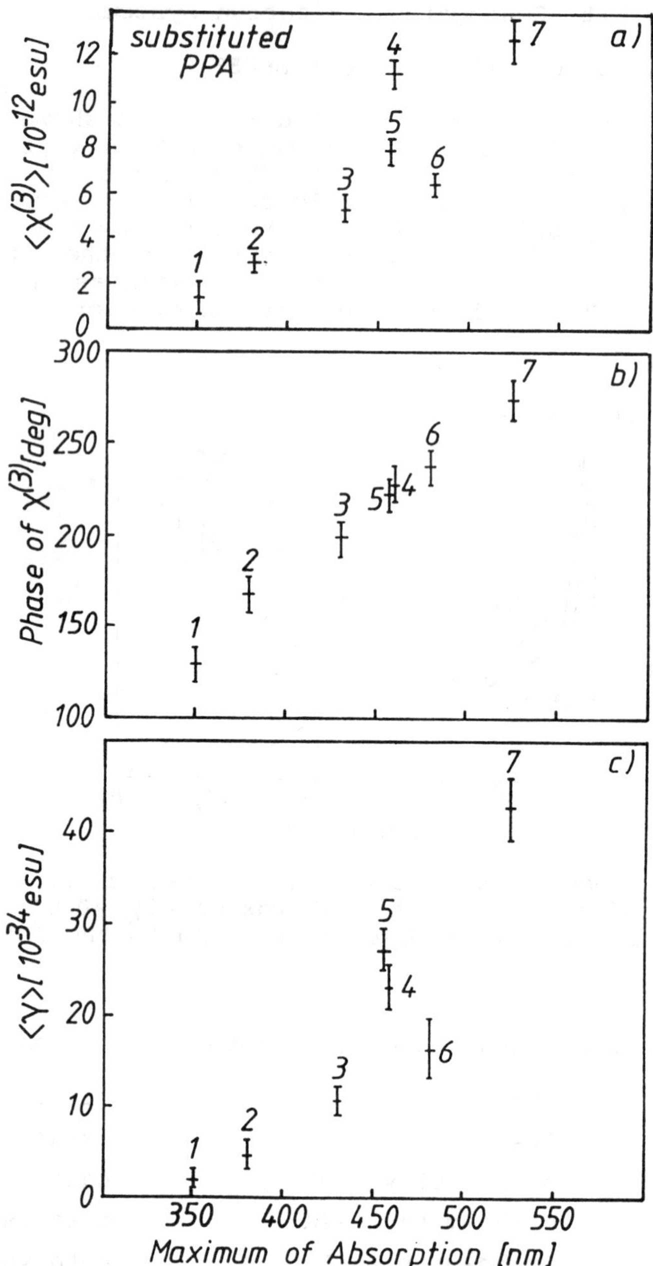

Figure 3. Results of THG investigations of ultrathin films of some substituted poly (phenyl acetylene) thin films. The numbers refer to Tab.1 (from Neher 1989b).

3.2 THIRD HARMONIC GENERATION

The harmonic intensity was analyzed as a function of the incidence angle via a formalism including all bound waves (Neher 1989a). The film thickness and refractive indices at ω and 3ω were taken into account. We used $\chi^{(3)}(-3\omega;\omega,\omega,\omega) = 3.11 \ 10^{-14}$esu at 1064nm for fused silica glass as a reference value (Kajzar 1985a). For the determination of both the absolute value and the phase of $\chi^{(3)}$ we first investigated a sample with the film on the front side of the substrate. Finally the Maker fringe pattern of the substrate alone was measured after removal of the film without displacement of the sample.

The averaged absolute values and the phases of $\chi^{(3)}$ as a function of the absorption maximum λ_{max} of the polymer absorption band are shown in Fig.3a,b. By a comparison of the absorption in solution and in the solid state, the molecular second order hyperpolarizability $\gamma(-3\omega;\omega,\omega,\omega)$ per monomer unit is calculated for every PPA system and shown in Fig.3c. If we assume that only the component of γ parallel to the polymer backbone is dominant, the local field factors at the fundamental and the harmonic frequency are identical to one (Cojan 1977). The increasing size of the substituent reduces the averaged nonlinear susceptibility by deminishing the density of polymer chains in the film. This influence can be seen by comparing systems PPA-4 and PPA-5 in Fig.3a,c.

The phase of $\chi^{(3)}$ increases with λ_{max} and approaches 270° for system PPA-7, which has an absorption maximum near to the second harmonic frequency of the Nd:YAG laser. We conclude that in system PPA-7 a two photon resonance occurs with an electronic excitation state close to 532nm. The accurate spectral position and the symmetry properties of this state need further consideration.

A three photon resonance should result in a phase of 90°. The main absorption bands of PPA-1 and PPA-2 coincide with 3ω of the laser fundamental frequency. Neither the phase nor $\chi^{(3)}$ show this resonance very distinct. This fact is related to the rather broad absorption maximum of PPA-1 and PPA-2 with a less pronounced absorption maximum as compared to the other systems. Therefore contributions of a two photon resonance with states at the long wavelength tail of the absorption spectrum give an additional contribution to the phase of $\chi^{(3)}$. Investigations of the dispersion of $\chi^{(3)}$ and the phase are necessary to unravel this question.

3.3 COMPARISON OF $\chi^{(3)}$ OF SOME CONJUGATED POLYMERS

To gain an overview of some presently available data of $\chi^{(3)}$ of conjugated polymers, Fig.4 shows these data as a function of the absorption maxima λ_{max}. In addition to the PPA systems described above, three forms of poly (3-decylthiophene) (PT) were prepared and investigated (Neher 1989b). PT-1 was electropolymerized in acetonitrile/methylene chloride and has $\lambda_{max} \approx$ 455nm (neutral form). PT-2 was electropolymerized in nitrobenzene and has $\lambda_{max} \approx$ 485nm. The neutral form of a chemically prepared PT-3 gives $\lambda_{max} \approx$ 505nm (Leclerc 1989). In contrast to the electrochemically prepared forms the latter is completely soluble in common organic solvents and has a low amount of irregular coupling.

Thin films of poly (p-phenylene vinylene) (PPV) were obtained by spin coating the tetramethylene sulfonium chloride precursor polymer (Lenz 1988), followed by thermal treatment in vacuum (24h at 220°C). The THG experiments were performed as described recently (Bubeck 1989a,b). With an improved evaluation method that takes into account all bound waves in the layer system (Neher 1989a), the $\chi^{(3)}$ value of PPV at 1064nm is 8∓3 10^{-11}esu. Within the experimental error this value is in accordance with other recently reported results (Bradley 1989). As compared to the PPA and PT systems, the larger $\chi^{(3)}$ of PPV can be interpreted with a lack of bulky substituents, resulting in a higher density of conjugated π-electrons per unit volume.

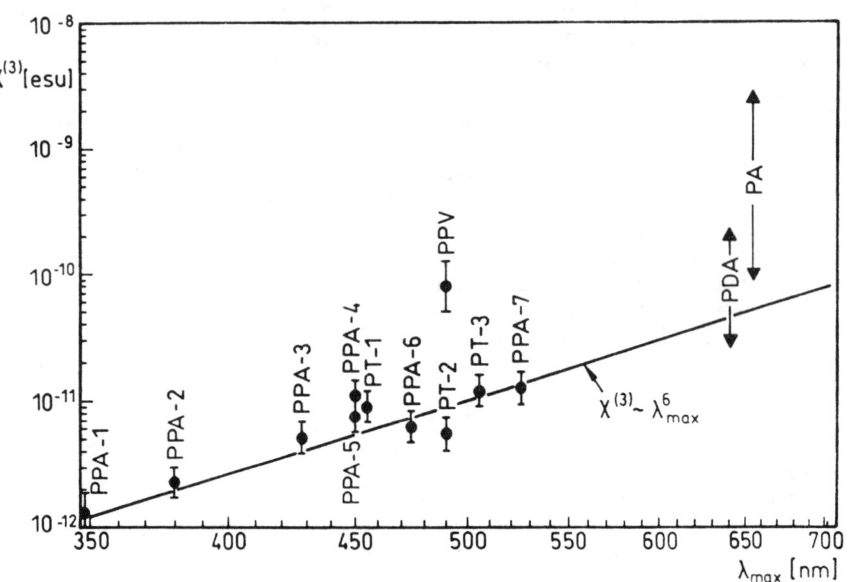

Figure 4. Survey of THG experiments with some conjugated polymers described in the text (double logarithmic plot). Circles are measurements at a laser wavelength of 1064nm, the triangles represent an estimate of the nonresonant and the two photon resonantly enhanced values of PDA and PA respectively.

Some wavelength dependent THG investigations of poly diacetylenes (PDA) and trans poly acetylene (PA) have been reported recently (Sauteret 1976, Kajzar 1985b, Messier 1989b, Kajzar 1987a,b, Fann 1989). The nonresonant and two photon resonant values of $\chi^{(3)}$ are also shown in Fig.4.

Treating the conjugated polymers as one-dimensional systems with a π-electron delocalization length L_d, Agrawal (1978) and Flytzanis (1987) have derived a scaling law for the nonresonant $\chi^{(3)}$ being proportional to $(L_d)^6$. If L_d is inversely proportional to the optical gap, $\chi^{(3)}$ should scale with $(E_0)^{-6}$ or $(\lambda_{max})^6$. Although Fig.4 shows a qualitative good agreement between theory and experiment, for a more accurate comparison and a real proof of the scaling law, a plot of the nonresonant γ values against the optical gap would be required.

3.4 DEGENERATE FOUR WAVE MIXING

The intensity of the DFWM signal can be used to evaluate $\chi^{(3)}(-\omega;\omega,\omega,-\omega)$ by a comparison with a CS_2 standard. Strong enhancements are observed if the laser wavelength is tuned to the long wavelength shoulder of the absorption bands of PPV, PT-3 and PPA-7 in the spectral range of 560 to 600nm. The spectral dependence of $\chi^{(3)}$ follows the absorption spectra. At a laser wavelength of 570nm we obtain $\chi^{(3)}$ values of $8*10^{-11}$esu, $1*10^{-9}$esu and $2*10^{-9}$esu for PPV, PPA-7 and PT-3, respectively. The response time of the DFWM signal is shorter than the laser pulsewidth of 1.5ps at 570nm for these polymers.

PPV thin films were studied with even shorter pulses of approximately 0.5ps at 650nm (Bubeck 1989b), but the response time was still limited by the laser pulsewidth. In the case of PPV this short response time is observed under resonance conditions at this wavelength, because we have shown that within an excitation range of 580nm to 760nm a two photon induced luminescence can be observed (Bubeck 1989c). This fluorescence has a quadratic dependence on the laser intensity in the range of $1GW/cm^2$. Its spectrum is similar to the PPV fluorescence that can be excited at 440nm with cw light of low intensity. Luminescence studies of PPV were reported recently (Friend 1987, Bradley 1989). A luminescence lifetime of 40-60ps was measured at fully converted samples (Wong 1987), which is two orders of magnitude longer than the response time measured in DFWM. Therefore a luminescent decay cannot be the dominant channel that leads to the subpicosecond response time in our experiment. We assume that nonradiative decay processes and ultrafast energy migration may account for this as will be discussed in more detail below.

4. Phthalocyanine Thin Films

For a study of one photon resonances, thin films of phthalocyanines (PC) are interesting candidates because of their good light and temperature stability. Their nonlinear optical properties have raised considerable interest recently (Ho 1987, 1988). In the case of resonance, the time response is determined by the photophysical relaxation processes of the electronic excited states. In an attempt to study the influence of the molecular environment and especially the intermolecular electronic coupling between the PC rings, the systems PC-1 to PC-4 shown in Tab.2 have been investigated (Kaltbeitzel 1989).

System PC-1 consisted of 50 layers of a polymer with a phthalocyanine moiety as depicted in Tab.2 connected by a flexible spacer with the chemical structure (Sauer 1989):

$$R^3 = -O-(CH_2)_4-Si(CH_3)_2-\text{\textcircled{}}-Si(CH_3)_2-(CH_2)_4-O-$$

In system PC-4 the PC moiety was attached as sidegroup in a copolymer with a styrene. The viscous solution of the copolymer was cast between glass slides. Evaporation of the residual monomer gave a solid film of several micron thickness. System PC-3 was prepared in a similar way by polymerizing styrene containing the phthalocyanine PC-2.

The absorption spectra and decay curves of the transient grating experiments with DFWM at approximately 650nm are shown in Fig.5. The transient grating results from a saturable absorption of the PC molecules.

Chemical Structure		System	R^3
(phthalocyanine structure with R^1, R^2, R^3 substituents; R^1= O-CH$_3$, R^2= O-C$_8$H$_{17}$ in all permutations)	1	LB-Film of the polymer molecules 50 layers	see text
	2	LB-Film of the monomer, 80 layers	Cl
	3	Polystyrene doped by monomer	Cl
	4	Styrene copolymer in bulk	O-C$_4$H$_7$

Table 2. Chemical structure and preparation method of the investigated phthalocyanine films.

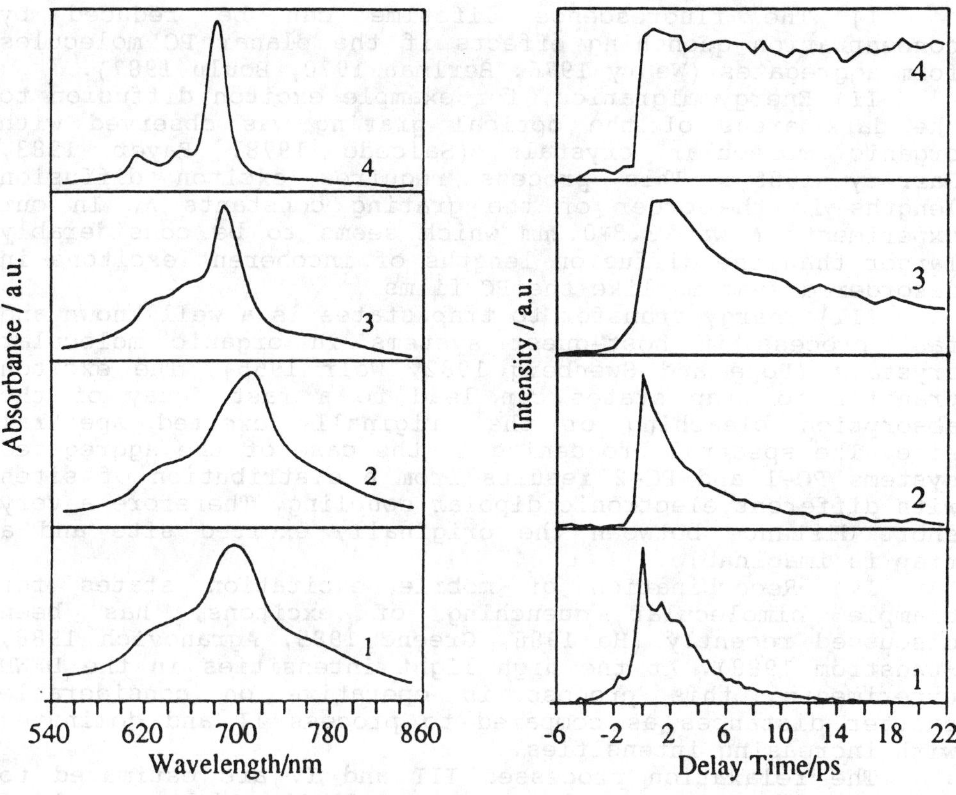

Fig.5. Absorption spectra (a) and decay of transient gratings (b) of some thin films of phthalocyanine systems PC-1 to PC-4 as shown in Tab.2.

In the case of isolated PC molecules for example in system PC-4, a sharp absorption band is observed. This corresponds to a slow decay time of the transient grating beyond the time resolution of our experiment. With increasing aggregation and finally a dense packing of PC molecules in the case of LB films an inhomogenous broadening and a considerable reduction of the response time to a few ps occurs. Obviously the electronic coupling of the transition dipole moments (Kasha 1976) has severe consequences not only on the absorption spectra, but also on the dynamical processes in PC aggregates. The following photophysical processes may account for the very fast relaxation processes of transient gratings in such aggregates.

I) The fluorescence lifetime can be reduced by concentration quenching effects if the planar PC molecules form aggregates (Wehry 1976, Berlman 1970, Boulu 1987).

II) Energy migration, for example exciton diffusion to the dark areas of the optical grating was observed with organic molecular crystals (Salcedo 1978, Fayer 1983, Garrity 1985). This process requires exciton diffusion lengths in the order of the grating constants Δ. In our experiments Δ was $3.3 \mp 0.2 \mu m$ which seems to be considerably larger than the diffusion lengths of incoherent excitons in disordered systems like the PC films.

III) Energy transfer to trap states is a well known and fast process in host-guest systems in organic molecular crystals (Pope and Swenberg 1982, Wolf 1965). The exciton transfer to trap states can lead to a fast decay of the absorption bleaching of the originally excited spectral site. The spectral broadening in the case of the aggregated systems PC-1 and PC-2 results from a distribution of sites with different electronic dipolar coupling. Therefore a very short distance between the originally excited site and a trap is imaginable.

IV) Recombination of mobile excitation states for example bimolecular quenching of excitons, has been discussed recently (Ho 1988, Greene 1985, Agranovich 1988, Sundström 1988). At the high light intensities in the DFWM experiments, this process is operative on considerable shorter distances as compared to process II and dominates with increasing intensities.

The relaxation processes III and IV are estimated to dominate for the systems PC-1 to PC-3. The tighter packing of the PC rings in systems PC-1 and PC-2 is responsible for the increasing electronic coupling between the rings and consequently for a reduced relaxation time. The discrimination between the processes is currently under investigation.

The ultrafast response times of the conjugated polymers described above are probably also related to the relaxation processes III and IV. The extended $\pi-$ electron conjugation, the tight coupling via intra- or interchain energy migration processes and the manifold of radiationless decay channels lead to the ultrafast response times of conjugated polymers even under resonance conditions.

Acknowledgment

The cooperation and helpful discussions with J.D. Stenger-Smith, T. Sauer and W. Caseri are gratefully acknowledged. Financial support was given by the german ministry for research and technology (BMFT) under project No. 03M4008E9

References

Agranovich, V.M., A.M. Ratner, M.Kh. Salieva (1988) Chem. Phys. 128, 23.

Agrawal, G.P., C. Cojan, C. Flytzanis (1978) Phys. Rev. B 17, 776.

Berlman, I.B. (1970) J. Phys. Chem. 74, 3085.

Boulu, L.G., L.K. Patterson, J.P. Chauvet, J.J. Kozak (1987) J. Chem. Phys. 86, 503.

Bradley, D.D.C., R.H. Friend (1989) J. Mol. Electronics 5, 19.

Bradley, D.D.C., Y. Mori (1989) Jap. J. Appl. Phys. 28, 174.

Bubeck, C., D. Neher, A. Kaltbeitzel, G. Duda, T. Arndt, T. Sauer, G. Wegner (1989a) in J. Messier, F. Kajzar, P. Prasad, D. Ulrich (ed.): Nonlinear Optical Effects in Organic Polymers, Kluver Acad. Publ., Dordrecht, p.185.

Bubeck, C., A. Kaltbeitzel, R.W. Lenz, J.D. Stenger-Smith, G. Wegner (1989b) in J. Messier, F. Kajzar, P. Prasad, D. Ulrich (ed.): Nonlinear Optical Effects in Organic Polymers, Kluwer Acad. Publ., Dordrecht, p.143.

Bubeck, C., A. Kaltbeitzel, D. Neher, J.D. Stenger-Smith, G. Wegner, A. Wolf (1989c) in H. Kuzmany, M. Mehring, S. Roth (ed.): Electronic Properties of Conjugated Polymers, Springer Solid State Sci., in press.

Chemla, D.S., J. Zyss (1987) Nonlinear Optical Properties of Organic Molecules and Crystals, Academic Press.

Cojan, C., G.P. Agrawal, C. Flytzanis (1977) Phys. Rev. B 15, 909.

Fann, W.S., S. Benson, J.M.J. Madey, S. Etemad, G.L. Baker, F. Kajzar (1989) Phys. Rev. Lett. 62, 1492.

Fayer, M.D. (1983) in V.M. Agranovich, R.M. Hochstrasser (ed.): Spectroscopy and Excitation Dynamics of Condensed Molecular Systems, North Holland, 233.

Flytzanis, C. (1987) in D.S. Chemla, J. Zyss (ed.): Nonlinear Optical Properties of Organic Molecules and Crystals, Vol. 2, Academic Press, 121.

Friend, R.H., D.D.C. Bradley, P.D. Townsend (1987) J. Phys. D: Appl. Phys. 20, 1367.

Garrity, D.K., J.L. Skinner (1985) J. Chem. Phys. 82, 260.

Greene, B.I., R.R. Millard (1985) Phys. Rev. Lett. 55, 1331.

Heeger, A.J., J. Orenstein, D.R. Ulrich (1988) Nonlinear Optical Properties of Polymers, Materials Research Soc..

Ho, Z.Z., N. Peyghambarian (1988) Chem. Phys. Lett. 148, 107.

Ho, Z.Z., C.Y. Ju, W.M. Hetherington III (1987) J. Appl.
        Phys. 62, 716.
Kajzar, F., J. Messier (1985a) Phys. Rev. A 32, 2352.
Kajzar, F., J. Messier (1985b) Thin Solid Films 132, 11.
Kajzar, F., J. Messier (1987) Polymer Journ. 19, 275.
Kajzar, F., S. Etemad, G.L. Baker, J. Messier (1987) Synth.
        Met. 17, 563.
Kaltbeitzel, A., D. Neher, C. Bubeck, T. Sauer, G. Wegner,
        W. Caseri (1989) in H. Kuzmany, M. Mehring, S. Roth
        (eds.): Electronic Properties of Conjugated Polymers,
        Springer Ser. Solid State Sci., in press.
Kasha, M. (1976) in B. DiBartolo (ed.): Spectroscopy of the
        Excited State, Plenum Press, 337.
Leclerc, M., F.M. Diaz, G. Wegner (1989) Makromol. Chem. in
        press,
Lenz, R.W., C.C. Han, J.D. Stenger-Smith, F.E. Karasz
        (1988) J. Polymer Sci. A 26, 3241.
Masuda, T., K. Hasegawa, T. Higashimura (1974)
        Macromolecules 7, 728.
Messier, J., F. Kajzar, P. Prasad, D. Ulrich (1989a)
        Nonlinear Optical Effects in Organic Polymers, NATO
        ASI Series E 162, Kluwer Acad. Publ., Dordrecht.
Messier, J. (1989b) in Messier et al (1989a) p. 47.
Neher, D., A. Wolf, C. Bubeck, G. Wegner (1989a) Chem.
        Phys. Lett. in press,
Neher, D., A. Wolf, M. Leclerc, A. Kaltbeitzel, C. Bubeck,
        G. Wegner (1989b) Synth. Met. in press,
Pope, M., C.E. Swenberg (1982) Electronic Processes in
        Organic Crystals, Clarendon Press, Oxford.
Prasad, P.N., D.R. Ulrich (1988) Nonlinear Optical and
        Electroactive Polymers, Plenum Press.
Salcedo, J.R., A.E. Siegman, D.D. Dlott, M.D. Fayer (1978)
        Phys. Rev. Lett. 41, 131.
Sauteret, C., J.-P. Hermann, R. Frey, F. Pradere, J.
        Ducuing, R.H. Baughman, R.R. Chance (1976) Phys.
        Rev. Lett. 36, 956.
Sundstöm, V., T. Gillbro, R.A. Gadonas, A. Piskarskas
        (1988) J. Chem. Phys. 89, 2754.
Wehry, E.L. (1976) Modern Fluorescence Spectroscopy,
        Heyden, London.
Williams, D.J. (1983) Nonlinear Optical Properties of
        Organic and Polymeric Materials, ACS Symp. Ser..
Wolf, A. (1989) PhD thesis, Mainz.
Wolf, H.C. (1965) in F. Sauter (ed.): Festkörperprobleme 4,
        Vieweg, Braunschweig.
Wong, K.S., D.D.C. Bradley, W. Hayes, J.F. Ryan, R.H.
        Friend, H. Lindenberger, S. Roth (1987) J. Phys. C:
        Solid State Phys. 20, L187.
Zyss, J. (1985) J. Mol. Electr. 1, 25.

THIRD HARMONIC GENERATION OF POLYTHIOPHENE DERIVATIVES

H. SASABE, T. WADA, T. SUGIYAMA, H. OHKAWA,
A. YAMADA and A. F. GARITO
Frontier Research Program, RIKEN
(The Institute of Physical and Chemical Research)
2-1 Hirosawa, Wako, Saitama 351-01, JAPAN

ABSTRACT. THG of electrochemically and/or chemically prepared polythiophene and poly(3-alkyloxymethylthiophene) were investigated by means of Maker fringe technique. The application of these polymers for waveguide is also discussed.

1. Introduction

Photoactive materials have been developed extensively in the fields of electrophotography, optical communication, display, and so forth. Among them the nonlinear optically (NLO) active compounds are key materials for "photonics" application [1], that is, optical image processing and/or switching elements in the next generation optical computing systems.

From these viewpoints it is very important to investigate new organic materials which show nonlinear optical behaviors, especially third-order optical nonlinearity. Recently it has been established experimentally and theoretically that organic intramolecular charge transfer compounds have anomalously large optical nonlinearity and show ultra-fast response. From the chemistry approach, systematic ideas of molecular design have succeeded in the enhancement of molecular susceptibilities to some extent, *i.e.*, extending the conjugation length, introduction of electron donative and acceptive groups, and reducing the dimensionality of π-electron system. Polydiacetylene (PDA) obtained by solid state polymerization was firstly reported to have large $\chi^{(3)}$ comparable to that of semiconductors, especailly parallel to the conjugated main chain [2]. On the other hand, from the physics approach, several techniques have been developed to obtain a thin film of PDA such as epitaxial growth [3], liquid crystal polymerization [4], the solution-shear technique [5] and the Langmuir-Blodgett (LB) method [6].

Electrochemical polymerization is known as the method to obtain conjugated polymeric films, and the electronic state of conjugated polymers can be controlled by an electrochemical procedure. Polythiophene (PTh) is stable in the reduced state and so expected to exhibit large third-order optical responses. In our previous study [7], it was found that $\chi^{(3)}$ of PTh is as large as that of polyacetylene. The value of $\chi^{(3)}$ was determined by optical third harmonic generation (THG) as

J. L. Brédas and R. R. Chance (eds.), Conjugated Polymeric Materials:
Opportunities in Electronics, Optoelectronics, and Molecular Electronics, 399–408.

3.52×10^{-10} esu at a fundamental wavelength of 1907 nm. In this study, soluble polythiophene derivatives, poly(3-alkyloxymethylthiophene) prepared electrochemically and chemically, were investigated on third-order nonlinear responses and applied to a slab-type optical waveguide. The nonlinear optically active polymer shows the large Kerr effect due to $\chi^{(3)}$, and hence has the intensity dependent refractive index change. Since the efficiency of nonlinear optical interaction depends strongly on the local intensity of the optical beam, the polymer waveguide is a desirable configuration for the enclosure of light.

2. Experimental

2.1. Samples

2.1.1. Polythiophene

Electrochemical polymerization of the monomer was carried out as follows: The solution of thiophene monomer of 0.2 mol/l with nitrobenzene was prepared. Tetramethylammonium perchlorate (TMAP) of 0.05 mol/l was added to the solvent as a supporting electrolyte. The working electrode was an ITO glass (area of 3 cm^2). The counter electrode was a Pt plate. On polymerization, the current density of 6.7 mA/cm^2 was applied between the electrodes at the duration of 20 sec. The synthesized polymer film was washed in nitrobenzene solvent, and then perchlorate ions were extracted from the film in the solvent of nitrobenzene/TMAP. In this case the voltage of -0.8 V between reference and working electrodes was applied for 2 days. The reference was a standard calomel electrode (SCE). After extraction, the undoped film was washed in nitrobenzene solvent and dried up in a vacuum chamber.

2.1.2. Poly(3-alkyloxymethylthiophene)

Poly(3-alkyloxymethylthiophene) was prepared as follows: A solution of $NaBH_4$ (5g, 0.125 mol) in 50 ml of 50% NaOH aq. was added to 3-thiophenealdehyde (28 g, 0.25 mol) in 100 ml of MeOH. The mixture was refluxed for 3 hrs and evaporated. The residue was extracted with methylenedichloride (300 ml x 3) and distilled under the reduced pressure. Yielded 3-thiophenemethanol (5.6 g, 0.05 mol) was dissolved in a mixture of 1-bromo-hexane and triethylbenzylammonium chloride in 100 ml of 50% NaOH aq., and refluxed for 5 hrs. Organic layer was washed with water, dried with Na_2SO_4 and distilled under reduced pressure. The monomer, 3-hexyloxymethythiophene (bp. 94 °C at 3 mmHg) was obtained (8.2 g, yield 84%). 2 g of the monomer was dissolved in 200 ml of MeOH which was saturated with $FeCl_3$ and refluxed for 24 hrs. Precipitated was collected by filtration and washed by Soxhlet extraction for 2 days. 1.2 g of red polymer was obtained (conversion 60%). Poly(3-dodecyloxymethylthiophene)(PDTh) has also synthesized via almost same procedure. Thickness-controlled thin films were made by spin coating of polymer solution.

In the case of alkyloxymethylthiophene, the polymers were cross-linked in some degree. Figure 1 shows the optical absorbance of poly-

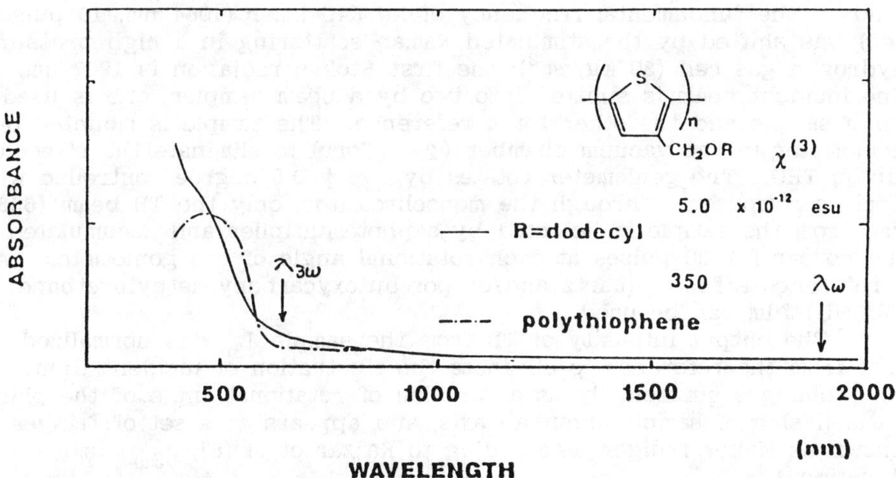

Figure 1. Absorption spectra of polythiophene and poly(3-dodecyloxy-methylthiophene).

thiophene derivatives. As the alkyloxymethyl pendants become longer, the absorption maximum shifts towards shorter wavelength (hyposochromic shift). This indicates the shortening of conjugation length due to the rotational motion of thiophene rings caused by longer alkyl chain.

2.2 THG Measurement

Transmitted third harmonic generation (THG) from the thin film deposited on a quartz substrate was observed at a wavelength of 1907 nm. Figure 2 shows schematically the layout of THG measuring system used in this

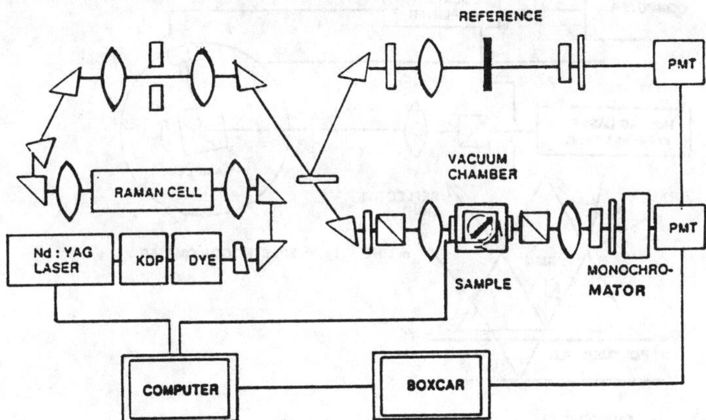

Figure 2. Schematic representation of the third harmonic generation (THG) measuring system.

402

study. The fundamental frequency of Nd:YAG laser (1064 nm, 10 pulse/
sec) was shifted by the stimulated Raman scattering in a high pressure
hydrogen gas cell (30 kg/cm^2): the first Stokes radiation is 1907 nm.
The incident beam is splitted into two by a beam sampler, one is used
for a sample and the other for a reference. The sample is mounted on a
goniometer in the vacuum chamber (2-4 Torr) to eliminate the effect of
air on THG. The goniometer rotates by every 0.5 degree controlled with
a micro-computer. Through the monochromator, only the TH beam (636
nm) from the sample is detected by a photomultiplier and accumulated in
the boxcar for 30 pulses at each rotational angle of the goniometer. As
a reference a fused quartz and/or polybutoxycarbonylmethylurethane
(3BCMU) film can be used.

The output intensity of TH from the sample $J_{3\omega}$ is normalized with
that from the reference to eliminate the fluctuation of incident light.
$J_{3\omega}$ changes sinusoidally as a function of rotational angle of the plane
paralell slab of sample about an axis, and appears as a set of fringes
known as Maker fringes. According to Kajzar et al [8], $J_{3\omega}$ is given as
Equation 1,

$$J_{3\omega} = \frac{256\pi^4 J_\omega^3}{C^2} \left| \frac{A\chi^{(3)}}{\Delta\varepsilon} \right|^2 \sin^2 \frac{\Delta\Phi}{2} \tag{1}$$

where J and n are respectively the light intensity and refractive index
for fundamental (subscript ω) and harmonic (3ω) frequencies, $\Delta\varepsilon$
$=n_\omega^2 - n_{3\omega}^2$, A a factor arising from transmission and boundary conditions, $\Delta\Phi$ a phase mismatch ($=\Delta kl$), l the sample thickness and c is the
speed of light in the free space. Under the condition of $l \ll l_c$ (coherence length), Equation 1 can be simplified as

$$J_{3\omega} = \frac{2304\pi^6 J_\omega^3}{C^2} \left| \frac{A\chi^{(3)}}{n_\omega + n_{3\omega}} \right|^2 \left(\frac{l}{\lambda_\omega} \right)^2 \tag{2}$$

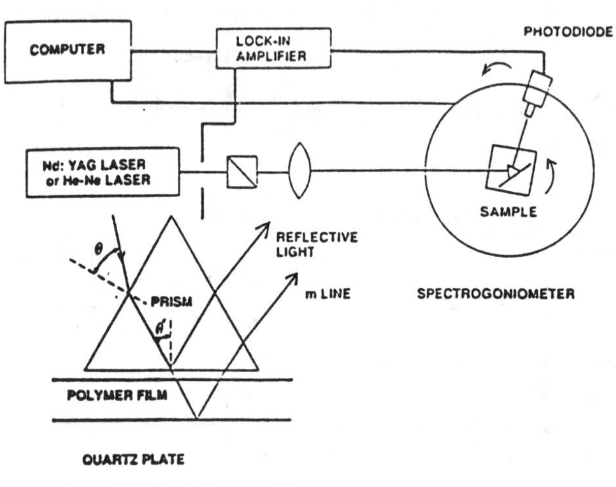

Figure 3. Schematic representation of the prism-film coupler and the
mode-line measuring system.

where λ_{ω} is the wavelength of fundamental light.

2.3 Mode-lines Measurement

In order to check the thin film light guide, the prism-film coupling is an essential technique. The coupling between the prism (refractive index of n_p) and the film waveguide (n_f) formed on the substrate (n_s) takes place along the bottom plane of the prism through a thin gap layer (air) of index n_c, as shown in Figure 3. The propagation constant β of the incident light beam along the waveguide film is given by

$$\beta = n_p k \sin\theta' \quad ; k = 2\pi / \lambda \tag{3}$$

where θ' is the incident angle onto the bottom of the prism, which is related to the incident angle θ outside the prism through Snell's law as

$$n_c \sin(\theta - \alpha) = n_p \sin(\theta' - \alpha) \tag{4}$$

α denotes the toe angle of the prism. If the propagation constant is in a range $n_c < \beta/k < n_f$ (guided mode), the guided wave is excited through the distributied coupling and penetrates as an evanescent wave into the film. In the waveguide film, β is expressed as

$$\beta = k n_{eff} \quad ; n_{eff} = n_f \sin\theta'' \tag{5}$$

where θ'' is the total reflection angle.

From the field equations for TE mode, the wave number in the x direction k_x can be derived in the form of *eigen value equation*,

$$k_x T = m\pi + \tan^{-1}(r_s/k_x) + \tan^{-1}(r_c/k_x) \tag{6}$$

where $k_x = k(n_f^2 - n_{eff}^2)^{1/2}$, $r_s = k(n_{eff}^2 - n_s^2)^{1/2}$, $r_c = k(n_{eff}^2 - n_c^2)^{1/2}$, T is the thickness of the waveguide film (PDTh) and m=0, 1, 2,... denotes the mode number. The incident beam focused on the prism base in a synchronous direction can feed the optical energy into one of the waveguide modes of the film. Since the film scatters the optical energy in the excited mode into other modes, then the beam is coupled back to the outside medium by the same prism. This phenomenon gives a series of bright lines on the screen with a bright spot on one of these lines (m-lines). From the positions and the widths of m-lines, we can determine the mode spectra, the refractive index and the film thickness. Therefore, if we can observe m-lines by means of the prism-film coupler method, this is a good evidence of guided waves. Figure 3 also indicates the schematic diagram of m-line measuring system used in this study.

2.4 Waveguide Measurement

We used a slab-type waveguide consisting of quartz substrate as cladding, Corning-7059 glass thin layer as a guiding layer, and PDTh thin film as a top layer whose thickness changes linearly along the y direction. Air on the top layer behaves as a forth layer, so this is called a four layer type waveguide. Laser light of Nd:YAG (1064 nm) was coupled and decoupled to a glass layer by a prism coupling technique and the output intensities were normalized by the intensity of reflection

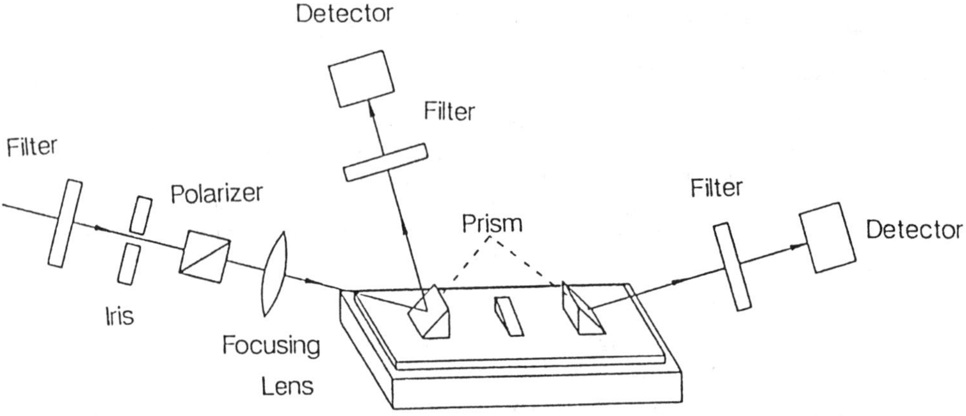

Figure 4. Schematic representation of the waveguide measuring system.

from the surface of the coupling prism or that of splitted light by a beam sampler. Figure 4 shows the layout of waveguide measurement system schematically.

3. Results and Discussion

3.1. THG: Maker Fringes

Poly(3-dodecyloxymethylthiophene)(PDTh) was spin-coated on the quartz substrate. The film was quite homogeneous in thickness of ca. 1500 nm. Figure 5 shows the incident angle dependences of third harmonic (TH) intensities of quartz substrate (a) and of PDTh film and quartz substrate (b). In the case of quartz, the TH intensity changes sinusoidally as a function of incident angle, and coincides well with the calculated curve of Equation 1. This oscillating pattern indicates the consecutive variation of optical pathlength equal to one, two, or more coherence length l_c, respectively. The minimum points are zero except the normal incident (angle=0), which indicates the quality of substrate being quite good. From the periodicity the coherence length l_c can be estimated as 18.3 μm. This value is in good agreement with the reported value by Meredith et al [9].

On the contrary, the PDTh film with quartz substrate shows the TH intensity being superposed of two components; one is a sinusoidally oscillating component and the other is an envelope of minimum points of the first component. The former is exactly the same as the quartz substrate. The latter is due to the film itself, and its intensity follows Equation 2. In the calculation it is necessary to know the refractive indicies of the film at the incident (1907 nm) and TH (635 nm) wavelengths, respectively. We assumed that these are almost the same as the index at 632.8 nm, then the value of $\chi^{(3)}$ for PDTh is estimated as 5×10^{-12} esu by using that of quartz (2.8×10^{-14} esu)[9].

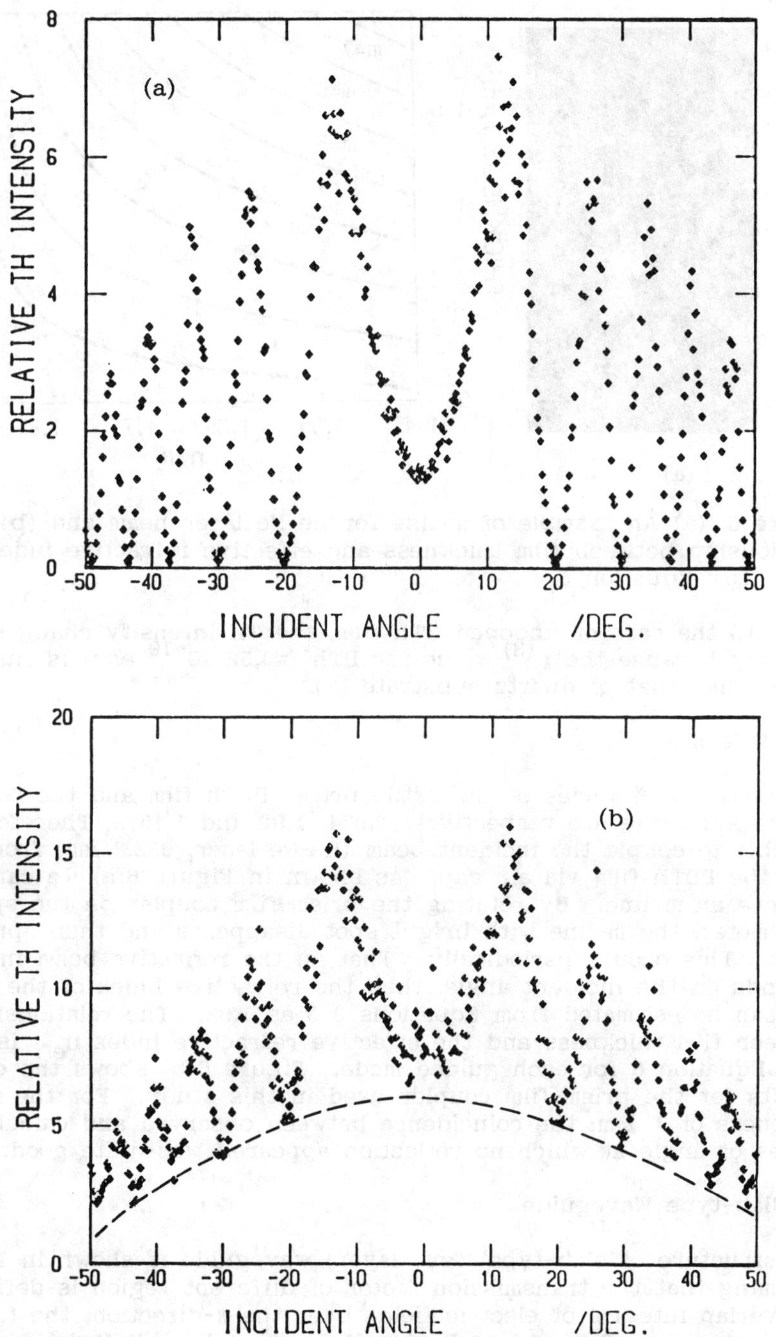

Figure 5. Incident angle dependence of TH intensity for (a) quartz substrate and (b) PDTh film and quartz substrate.

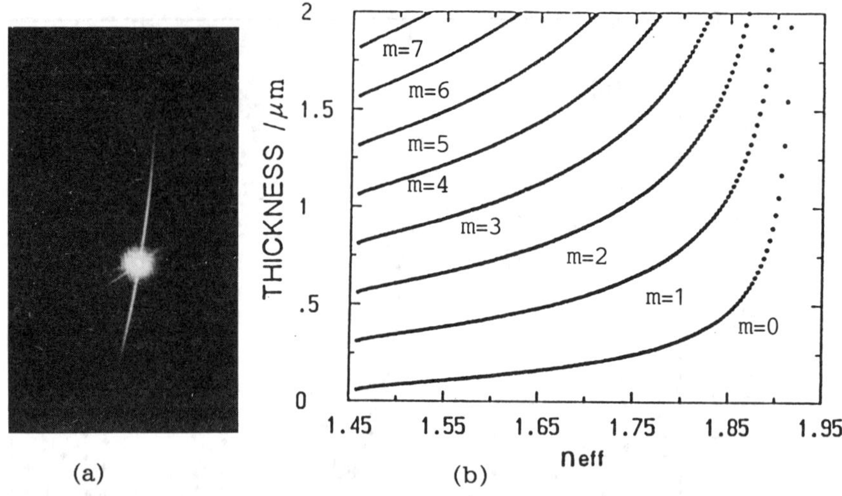

(a)   (b)

Figure 6. (a) An example of m-line for He-Ne laser beam and (b) the relationship between film thickness and effective refractive index calculated from Equation 6.

In the case of undoped PTh film, the TH intensity changes monotonically because the $\chi^{(3)}$ value for PTh (=3.52×10^{-10} esu) is much larger than that of quartz substrate [7].

3.2. Mode-line

The refractive indicies of the rutile prism, PDTh film and the fused quartz substrate are respectively 2.874, 1.92 and 1.457. Therefore it is possible to couple the incident beam (He-Ne laser; 632.8 nm) from prism with the PDTh film via air gap. As shown in Figure 6(a) we can clearly observe an m-line. By rotating the prism-film coupler on the spectrogoniometer, the m-line with bright spot disappears and then appears again. This occurs periodically. That is, the reflective beam intensity depends on the incident angle, then the refractive index of the PDTh film can be estimated from Equations 3-5 as 1.92. The relationship between film thickness and the effective refractive index n_{eff} is given from Equation 6 for each guided mode. Figure 6(b) shows the calculated results for the prism-film coupler used in this study. For the film thickness of 1 μm, the coincidence between observed and calculated values of angle at which no reflection appeared was quite good.

3.3 Slab-type Waveguide

The structure of slab-type (four layer) waveguide is shown in Figure 7. Assuming that the transmission factor of different region is defined as an overlap integral of electric fields over the x-direction, the transmission factor from Region I to Region II is given by $\int E_1 E_2 dx$, where E_i is the distribution of an electric field in the i'th Region. An observed output intensity is in proportional to $(\int E_1 E_2 dx)^4$. The output intensity

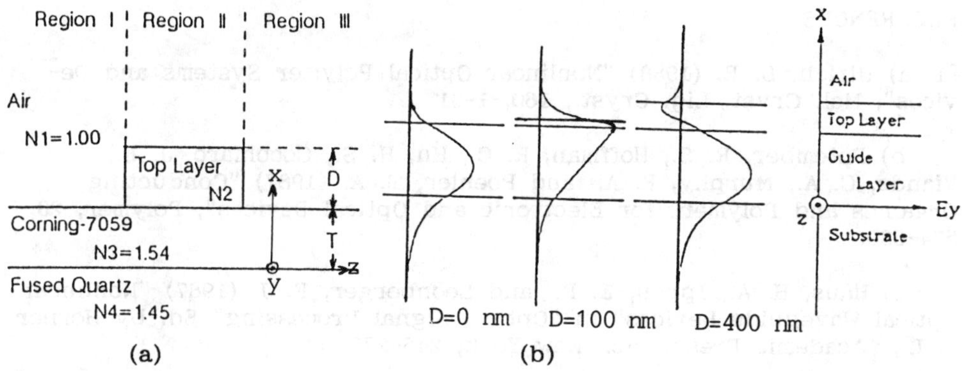

Figure 7. Structure of slab-type waveguide (a) and electric field distribution in the x direction as a function of top layer thickness D.

changes periodically as a function of the top layer thickness. Since the
top layer (PDTh) has a gradient of thickness along the y direction
(lateral) of 3 mm (maximum thickness is 1 μm), the field distribution of
propagation mode in the waveguide layer (Corning 7059) changes periodically in the Region II with an increase of the top layer thickness. That
is, when the field distribution in Region I is similar to that in Region II,
the output intensity becomes maximum, otherwise the light cannot propagate well from Region I to Region II and/or Region II to Region III as
shown in Figure 7(b). By moving the sample stage in the y direction
we can observe the periodical change of output beam intensity. Figure
8 is an example of the guided beam in the waveguide layer. In this
case the decoupling prism was not used to observe the emission from
the guide layer edge.

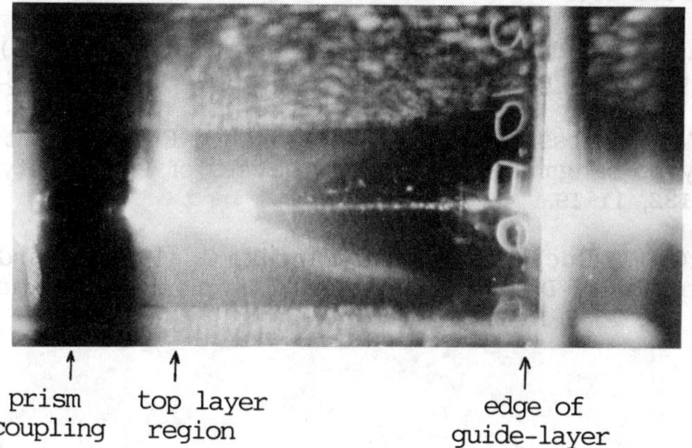

Figure 8. An example of guided beam (He-Ne laser) in the waveguide
layer (Corning 7059).

REFERENCES

[1] a) Ulrich, D. R. (1988) "Nonlinear Optical Polymer Systems and Devices", Mol. Cryst. Liq. Cryst., 160, 1-31

   b) Potember, R. S., Hoffman, R. C., Hu, H. S., Cocchiaro, J. E.,
Viands, C. A., Murphy, R. A. and Poehler, T. A. (1987) "Conducting
Organics and Polymers for Electronic and Optical Devices", Polymer, 28,
574-580,

   c) Haus, H. A., Ippen, E. P. and Leonberger, F. J. (1987) "Nonlinear
Optical Waveguide Devices" in "Optical Signal Processing" Ed. by Horner,
J. L., Academic Press, Inc., New York, 245-277.

[2] Sauteret, C. Hermann, J.-P., Frey, R., Pradere, F., Ducuing, J.,
Baughman, R. H. and Chance, R. R. (1976) "Optical Nonlinearities in One-
Dimensional-Conjugated Polymer Crystals", Phys. Rev. Lett., 36, 956-959

[3] Rickert, S. E., Lando, J. B. and Ching, S. (1983) "Epitaxial Growth of
Polydiacetylenes" in "Nonlinear Optical Properties of Organic and Polymeric Materials" Ed. by Williams, D. J., ACS Symp. 233, 229-233.

[4] Garito, A. F., Teng, C. C., Wong, K. Y. and Zammani'Khamiri, O., (1984)
"Molecular Optics: Nonlinear Optical Processes in Organic and Polymer
Crystals", Mol. Cryst. Liq. Cryst., 106, 219-258.

[5] Thakur, M. and Meyler, S. (1985) "Growth of Large-Area Thin-Film
Single Crystals of Poly(diacetylenes)", Macromolecules, 18, 2341-2344.

[6] Kajzar, F., Messier, J., Zyss, J. and Ledoux, I. (1983) "Nonlinear
Interferometry in Langmuir-Blodgett Multilayers of Polydiacetylene", Opt.
Commun., 45, 133-137.

[7] Sugiyama, T., Wada, T., Yamada, A and Sasabe, H. (1989) "Optical
Nonlinearity of Conjugated Polymers", Synthetic Metals, 28, C323-C328.

[8] Kajzar, F. and Messier, J. (1985)" Resonance Enhancement in Cubic
Susceptibility of Langmuir-Blodgett Multilayers of Polydiacetylene", Thin
Solid Films, 132, 11-19.

[9] Meredith,G. M., Buchalter, B. and Hanzlik, C. (1978) "Third-order
Optical Susceptibility Determined by Third Harmonic Generation I", J.
Chem. Phys., 78, 1533-1542.

THIRD ORDER HYPERPOLARIZABILITY OF POLYTHIOPHENE AND THIOPHENE OLIGOMERS

F. CHARRA and J. MESSIER
CEA-CEN SACLAY - D.LETI/DEIN
Laboratoire de Physique Electronique des Matériaux
91191 GIF-SUR-YVETTE CEDEX
FRANCE

ABSTRACT

We have performed a CNDO calculation on polythiophene. A new representation of the virtual excited states in terms of electron-hole distance is proposed. A careful size consistent calculation of γ is presented and leads to a finite value per unit volume for an infinite chain ($\chi^3 = 1.0\ 10^{-11}$ esu at $\omega = o$). The same procedure is used to check the influence of the polymer configuration.

1. INTRODUCTION

The thiophene derivatives possess a lot of universal features common to all one dimensional conjugated polymers : the linear and nonlinear optical properties depend strongly upon the π-electron delocalization length. This length obviously decreases with the molecular size but can also be significantly affected by molecular deformations which in the case of polythiophene may be important in many experimental situations (polythiophene in solution or thin films). To address this problem it is important to develop computational procedures which can be confronted with experimental situations.

In this paper we are dealing with electronic third order hyperpolarizabilities especially in large molecules. Numerous authors have studied theoretically the second order polarizabilities in short noncentrosymmetric molecules [1], where near exact methods can be used ; but these methods nead to consider multiple excitations states, the number of which becomes rapidly prohibitive for

409

J. L. Brédas and R. R. Chance (eds.), Conjugated Polymeric Materials:
Opportunities in Electronics, Optoelectronics, and Molecular Electronics, 409–420.
© 1990 *Kluwer Academic Publishers. Printed in the Netherlands.*

increasing sizes. So these methods were applied to third order polarizability (γ) only for very small molecules [2], up to about ten carbon atoms. In this size range the γ-coefficient per unit volume is still rapidly increasing as a power law of the molecular length and thus can't be used for extrapolating the limiting value of an infinitely long polymer chain.

In this paper we discuss the applicability of a CNDO/s method which only account for monoexcited states to the γ determination of very large molecules and we present the results for thiophene oligomers up to 16 repeat units, including possible deformations.

2. EXPERIMENTAL MOTIVATIONS

Thiophene α- oligomers are available with different number of repeat units N (N = 5, 6, 8) and soluble polybutylthiophene in solution or in thin films. The thin films exhibit broad absorption spectra ranging from 0,3 µm to an upper limit λ_{lim} increasing with N. For N$\geqslant$6 λ_{lim} reach a constant value of $\simeq 0,6$ µm. The maximum absorption wavelength λ_{max} is much lower than λ_{lim} and range from 0,38 µm (N=5) to 0,46 µm (N=∞). In solution we have a narrower line shape with maximum still much lower than λ_{lim} .

These observations have been interpreted [3] as being mainly to a broad distribution of geometrical conformations able to reduce the delocalization lengths. Under this assumption, λ_{lim} corresponds to the absorption of the more delocalized flat molecule.

Such a reduction of the π-electron delocalization length is expected to alter more drastically also the nonlinear susceptibilities. This effect has been measured experimentally [4]. To interpret these experimental results we have to relate to a given deformation a variation in the linear and nonlinear spectra.

3. ELECTRONIC STRUCTURE OF EXCITED STATES

We have performed CNDO/s calculations [5] on α-oligomers of thiophene assuming an interring distance of 1,43 Å (fig. 1).

The coulomb interaction between electrons was given by Mataga formula [6]. Only monoexcited states has been considered in the configuration

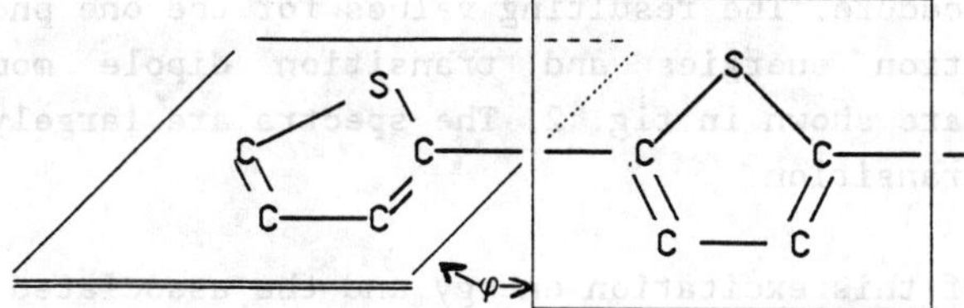

Figure 1 : Model of thiophene oligomers
The interring distance is 1,43 Å.
φ is the angle between two successive rings

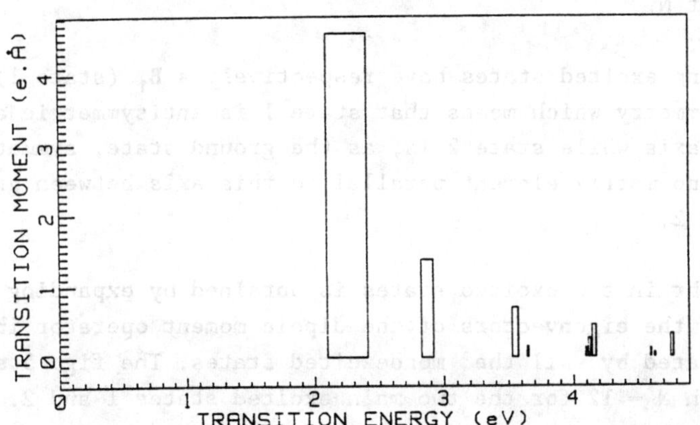

Figure 2 : Excitation energies (E_1) of the one-photon allowed
transition dipole moment for N = 12. The rectangle surface is
proportional to the square of the transition moment.

interaction procedure. The resulting values for the one photon allowed excitation energies and transition dipole moment in the example N = 12 are shown in fig. 2. The spectra are largely dominated by the lowest transition.

The evolution of this excitation energy and the associated transition dipole moment μ_{01} are shown in fig. 3. One can notice that μ_{01}^2 becomes proportional to N for N $\geqslant$ 6 which ensures constant linear susceptibility per unit volume, in the limit N $\rightarrow$ ∞. Since we are interested by the computation of nonlinear susceptibilities by a time dependant perturbation method [7] the program has been extended to the calculation of transition dipole moment between excited states.

From the first excited state (1) only one transition dipole moment has a noticeable value μ_{12} the variation of which is shown in fig. 4 as a function of N.

The corresponding excited states have respectively a B_1 (state 1) and A_1 (state 2) symmetry which means that state 1 is antisymmetric along the molecular axis while state 2 is, as the ground state, symmetric. So, there is no matrix element parallel to this axis between ground state and state 2.

A better insight in the excited states is obtained by expanding them in the basis of the eigenvectors of the dipole moment operator in the subspace generated by all the monoexcited states. The fig. 5 shows the results with N = 12 for the two main excited states 1 and 2.

Each eigenvector of the dipole moment operator X with eigenvalue $\varkappa$ can be visualized as electron-hole pair separated in space by a distance $\varkappa$ and delocalized on the molecule. The coefficients on these eigenstates reported on the figure 5 can be seen as the excitonic wavefunction. As a consequence of the configuration interaction, the electron-hole distance is confined to 5.5 Å for N = 12. The mean square distance is plotted in fig. 6 as a function of N. A limiting value of about 6 Å can be extrapoled for an infinite chain for the state 1.

The state 2 has a larger extension and slower convergence with increasing N. As will be shown in the following this is the main reason for a positive hyperpolarizability.

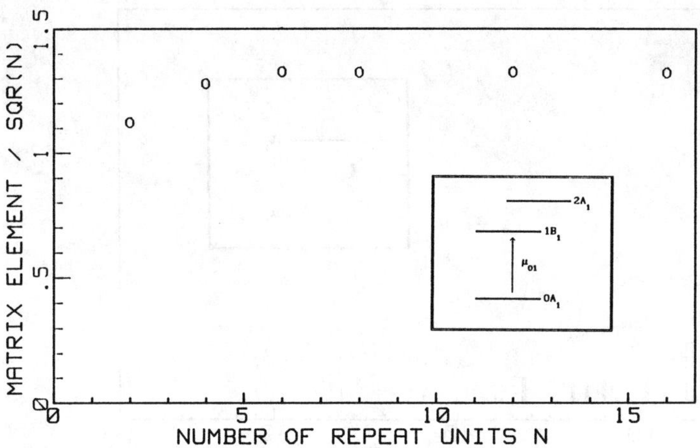

<u>Figure 3</u> : Variation with the number N of repeat units of excitation energies E_1 and associated transition dipole moment divided by $\sqrt{N}$

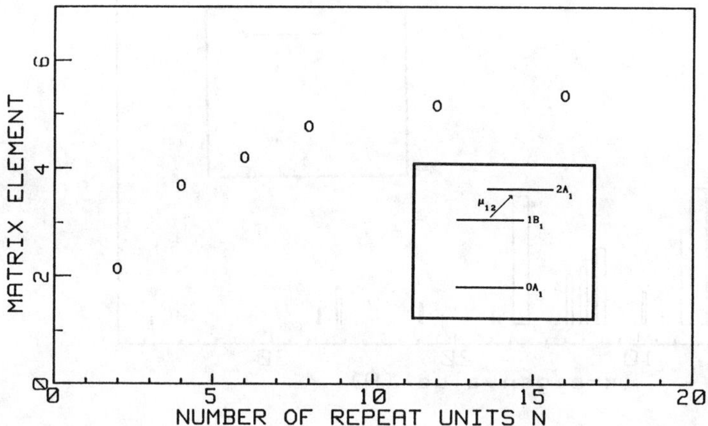

<u>Figure 4</u> : Variation of the transition dipole moment, μ_{12} , from the one to two photon states.

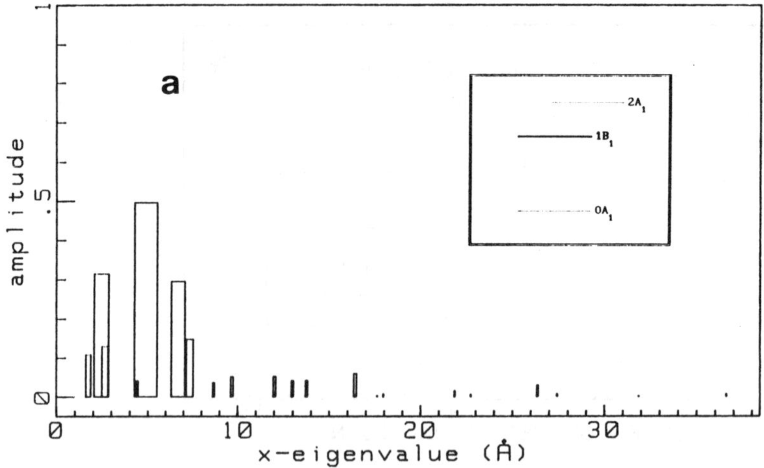

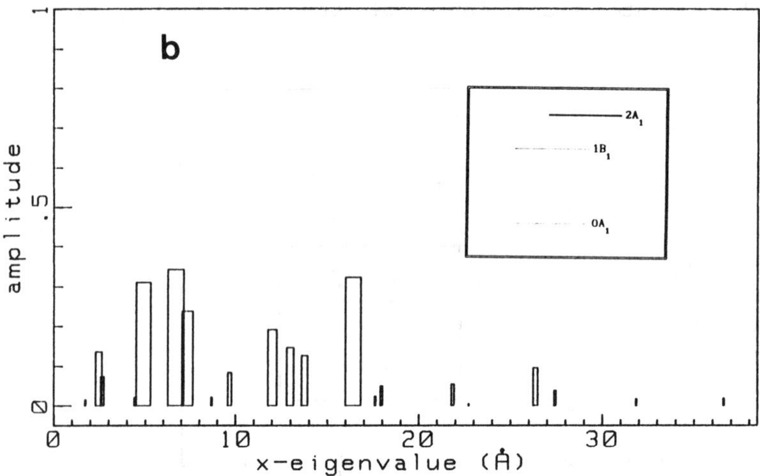

Figure 5 : Expansion of the one (a) and two (b) photons excitonic
wavefunction .
Each component is represented by a rectangle of height and width
proportional to the wave function amplitude and centered on the
corresponding x-eigenvalue.

4. THIRD ORDER SUSCEPTIBILITY (γ) CALCULATIONS

The standard time dependant perturbation theory allows a computation of γ from the transition dipole moments. For flat molecules the main contribution come from the three levels 0, 1, 2 and is given by the expression :

$$\gamma = \frac{1}{\varepsilon_o} \mu_{01}^2 \left[\mu_{12}^2 \ f(\omega) - \mu_{01}^2 \ g(\omega) \right] \tag{1}$$

where f and g are functions of the fundamental frequency ω and are given in ref [ ] for the different types of susceptibilities (harmonic generation, index variation, etc...). For $\omega = 0$ and third harmonic generation, we have :

$$\gamma(3\omega;\omega,\omega,\omega) = \frac{4}{\varepsilon_o} \frac{\mu_{01}^2}{E_1^2} \left[\frac{\mu_{12}^2}{E_2} - \frac{\mu_{01}^2}{E_1} \right] \tag{2}$$

In this expression the first term give a positive contribution. One expects that the hyperpolarizability per unit volume (γ/N) reaches a limiting value for infinite N. Since μ_{01}^2 varies as N and, as shown in fig. 4, μ_{12}^2 tends to a constant, this first term behaves properly. The second term which comes from virtual bleaching of the state 1 gives a negative contribution. One can see that this term varies as N^2. This lack of size-consistency is due to the neglect of doubly excited states. For very large molecules one must take into account the occurence of independant excitations on a single polymer chain. As a first approximation, one can renormalize this term by introducing a biexcited state built from two excitations assumed to be independant. Its energy is thus twice that of the monoexcited state 1.

If we assume that the first excitation fills a number N_o of repeat units in polymer chain of total length N, then simple statistical considerations show that the positive contribution due to the biexcited state exactly cancels the abnormalous increase of the bleaching term in eq. 3. The remaining contribution of bleaching writes :

$$\gamma_{bleach} = -\frac{4}{\varepsilon_o}\,\frac{\mu_{01}^4}{E_1^3}\,\frac{N_o}{N} \tag{3}$$

leading to a constant value of γ_{bleach}/N per infinite N.

An idea of N_o can be obtained from the extension of the excitonic wave function. Figure (6) shows that the exciton size reaches about 12 Å that is 3 times the repeat unit length. In addition fig. (3) shows that μ_{01}^2 varies as N for $N \geqslant 6$ which also means that two almost independant excitations can take place on the same molecule. A reasonable value of N_o would thus be $N_o = 3$, and for $N \geqslant 3$ we make use of the renormalized expression eq. (3).

The resulting value of hyperpolarizability for monomer unit γ/N (proportional to $\chi^{(3)}$) is plotted as a function of N in figure 7.

For large N (N = 16) the computed respective contributions of the positive and negative terms to γ/N are :

$$\begin{cases} \gamma_+/N = +2,1.\ 10^{-47} \qquad m^5/V^2 \\ \gamma_{bleach}/N = -0,6.\ 10^{-47} \qquad m^5/V^2 \end{cases}$$

One can notice that the main contribution is provided by the first term γ_+. For $N \geqslant 6$ the increase of γ/N with N comes only from the γ_+ term which saturates for $N \sim 16$.

The slower convergence with N of γ_+ term compared to γ_{bleach} is due to the larger extension of the state 2 wave function.

The computed γ value is :

$$\gamma/N = 1,5.\ 10^{-47} \qquad m^5/V^2$$

or, taking the local field factor = 1 as it must be for a cylindrical molecule [8],

$$\chi^{(3)} \simeq 1,4.\ 10^{-19} \qquad m^2/V^2 \ (= 1.0\ 10^{-11}\ esu)$$

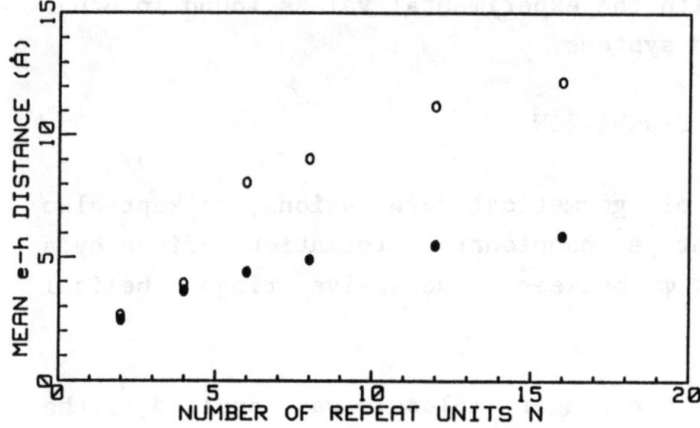

Figure 6 : Variation of the mean square electron-hole distance, $\sqrt{\overline{\langle x^2 \rangle}}$ with N for one (●) and two photon (○) state

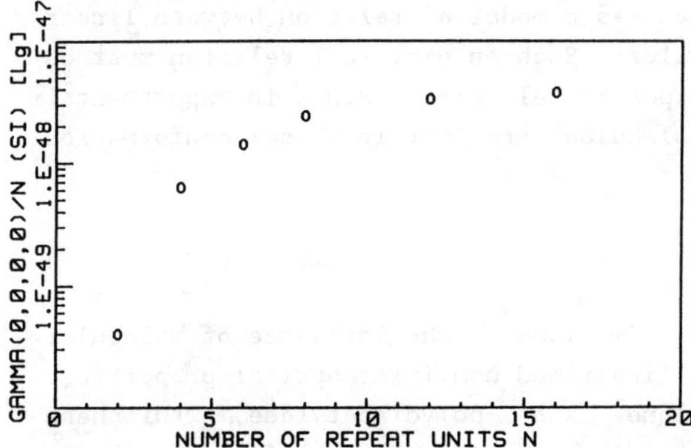

Figure 7 : Variation with N of the hyperpolarizability per monomer unit (γ/N) for the third harmonic generation at the limit $\omega = 0$. SI UNITS $\gamma_{esu} = 7,16.\ 10^{13}\ \gamma_{SI}$

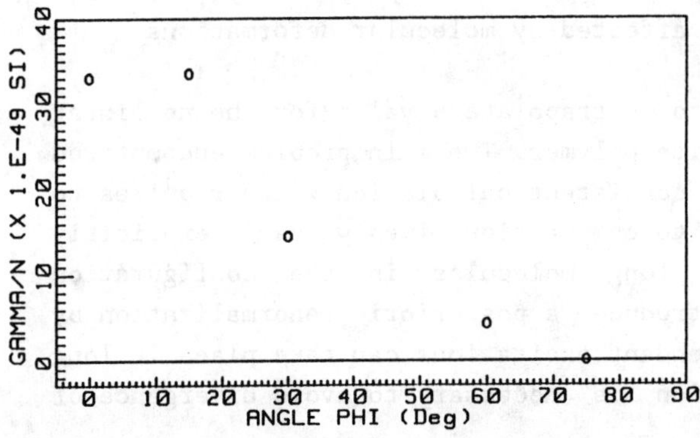

Figure 8 : Variation of γ/N (at zero frequency), for an helical polymer, as a function of φ angle defined in fig. 1

These values compare well with the experimental values found in other one dimensional π-conjugated systems.

5. INFLUENCE OF A BACKBONE DEFORMATION

To estimate the influence of geometical deformations, we kept also the possibility of a prototype nonplanar conformation defined by a constant rotation angle φ between successive rings (helical polymers).

The same procedure as for the planar molecule was applied to the calculation of γ/N as a function of the angle φ. The results for N = 12 are plotted on figure 8. The γ value is not much affected for angles less than 15° but decreases, dramatically for larger angles.

This calculation can be used as a model of relation between linear absorption spectra and γ values. Such an empirical relation must be used to derive γ from the experimental spectra since in experimental situations polythiophene molecules are not in planar conformation [3].

6. CONCLUSION

The CNDO procedure allows estimations of the influence of molecular length and deformations on linear and nonlinear optical properties. Contrarily to polyacetylene and polydiacetylene, thiophene derivatives can't be obtained in solution or thin films in planar conformation. We have shown that, while the nonlinear coefficient at zero frequency of the planar thiophene derivative, is roughly equivalent to that of polyacetylene or polydiacetylene, this nonlinearity is drastically, affected by molecular deformations.

We have been able also to extrapolate a value for the nonlinear susceptibility of the infinite polymer. The main problem encountered is to make use of a size consistent calculation which requires to consider biexcitations. Due to computation times we can't explicitly introduce them for such long molecules in the configuration interaction. But we can introduce a posteriori renormalization by assuming that totally independant excitations can take place in long molecules. Such a correction is necessary to avoid divergence of negative contribution of bleaching type terms.

The neglect of biexcited states in the CNDO procedure also produces a lack of accuracy in the excitation energies, especially in the case of the two-photon state. It is well known that this two photon state is located slightly below the one-photon state in the case of one dimensional π-electron systems as was experimentally shown for polyenes [9] or polydiacetylenes [10]. The same situation probably occurs in polythiophene on the contrary of the results of CNDO calculations.

From this viewpoint a further improvement would be to extrapolate to large molecules the corrections due to the explicit introduction of biexcited states possible only for small molecules.

REFERENCES

[1]   J.O. Morley, S.J. Lalama and A.F. Garito
         Phys. Rev. A, 20, (1979), 1179
         Journal of the Chem. Soc. Perkin Trans II 1987

[2]   J.R. Heflin, K.Y. Wong, O. Zamani-Khamiri and A.F. Garito
         Phys. Rev. B, 38 (1988) 1573
         B.M. Pierce 'Organic Materials for Nonlinear Optics'
         R.A. Hann and D. Bloor eds Royal Soc of Chemistry (N° 69)
         1988, p. 48

[3]   D. Fichou and F. Garnier - F. Charra, F. Kajzar and J.
         Messier - 'Organic Materials for Nonlinear Optics'
         R.A. Hann and D. Bloor eds Royal Soc of Chemistry (N°69)
         1988, p. 176

[4]   F. Kajzar, J. Messier and C. Sentein - R.L. Elsenbaumer and
         G.G. Miller
         Optical and Optoelectronic Applied Science and Engineering
         SAN DIEGO (August 6-11, 1989)

[5]   Quantum Chemistry Program Exchange (QCMPO29)

[6]   K. Nishimoto, N. Mataga, Z. Phys. Chem., 12, 335 (1957)

[7]   J. Orr and J. F. Ward, Mol. Phys. 20 (1971) 513

[8]   C. Cojan, G.P. Agrawal and C. Flytzanis
        Phys. Rev. B, 15 (1977) 909

[9]   B.E. Kohler, C. Spangler and C. Westerfield
        J. Chem. Phys. 89 (1988) 5422

[10]   P.A. Chollet, F. Kajzar and J. Messier
        Synthetic Metals, 18 (1987), 459

EXCITON RELAXATION IN PDA-4BCMU: FROM CRYSTALS TO FILMS

*‡M.J. NOWAK, *G.J. BLANCHARD, *G.L. BAKER, *S. ETEMAD, ‡Z.G. SOOS

Bellcore *
331 Newman Springs Rd.
Red Bank, N.J. 07701-7020
and
Department of Chemistry ‡
Princeton University
Princeton, N.J. 08544

ABSTRACT. We have investigated the role of disorder on the optical response of polydiacetylene 4BCMU using linear and time resolved absorption spectroscopy. In the highly ordered crystalline form, a sharp excitonic absorption is observed. On heating the crystals above the hydrogen-bond melting temperature (120 °C), the excitonic absorption blue-shifts and broadens. Further blue shifting and broadening is observed in the highly disordered spin-cast films. Time resolved absorption measurements show a fast ($\sim$10 ps) bleaching signal that is insensitive to the degree of disorder. However, in contrast to PDA-PTS crystals, a long-lived ($\sim$ 400 ps) photoresponse is observed only in the disordered samples.

I. Introduction

Conjugated polymers are a class of one-dimensional semiconductors with large on- and off-resonance optical nonlinearities.[1] [2] In exploring possible technological applications,[3] many workers have concentrated on the processable polydiacetylenes, from which high quality thin films may be cast or spun.[4] [5] The conjugated polymers are disordered in these amorphous films. In contrast, crystalline polydiacetylenes (PDAs) have ordered linear backbones that are well separated from neighboring chains. The polydiacetylene 4BCMU can be synthesized[6] in both crystalline and amorphous forms, allowing us to study the morphology-dependent nature of side-group to backbone coupling in a conjugated polymer using picosecond pump-probe spectroscopy.[7]

Poly(4BCMU) (Fig. 1) belongs to a family of closely related, soluble polydiacetylenes. The common structural elements in these polymers are urethane substituents spaced from the polymer backbone by one or more methylene (CH_2) groups, and an n-butoxycarbonyl group attached to the urethane that provides enhanced solubility. Poly(4BCMU), with four methylene units, has been widely studied particularly in the context of its solvato- and thermochromic transitions.[8] [9] [10] An important factor is the ability of the urethane groups to form hydrogen bonds with either an adjacent urethane on the same chain, or alternatively with a urethane of a neighboring chain. This ensemble of bonds forms a hydrogen-bond

421

J. L. Brédas and R. R. Chance (eds.), Conjugated Polymeric Materials:
Opportunities in Electronics, Optoelectronics, and Molecular Electronics, 421–427.
© 1990 *Kluwer Academic Publishers. Printed in the Netherlands.*

lattice with the tie points between chains acting as crosslinks, and intramolecular hydrogen bonds favoring the formation of a planar ribbon-like polymer conformation. As shown below in Fig. 1, the H-bond lattice plays an important role in maintaining a linear, stiff, long coherence length backbone.

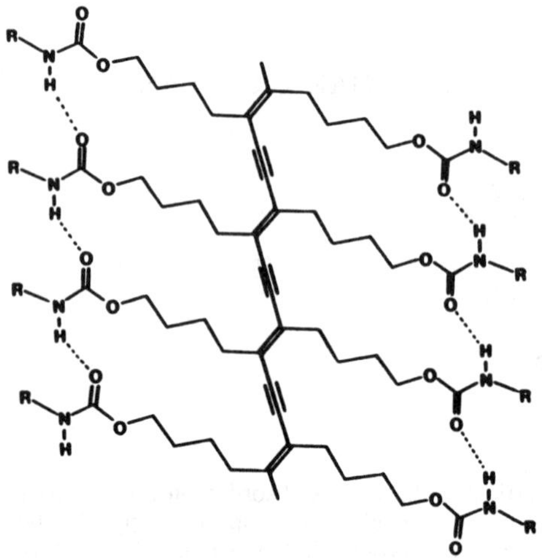

Fig. 1: Chemical structure of polydiacetylene-4BCMU.

In this paper we use low temperature picosecond pump-probe spectroscopy as a way to probe disorder in samples of 4BCMU ranging from crystalline to amorphous. Crystalline 4BCMU exhibits a sharp exciton in the linear absorption spectrum, with disorder increasing the exciton's energy through localization. Photoinduced spectra are a dynamic probe of that disorder, and in this work, we use time-resolved spectroscopy to study the effects of side-chain and backbone disorder.

II. Photoinduced Spectra

We have studied three samples of varying degrees of disorder. We designate as Sample A (Fig. 2) the pristine 4BCMU crystals from which the other samples are derived. 4BCMU monomer single crystals are exposed[6] to UV light until a semitransparent blue phase crystal is formed. Photopolymerization[11] yields highly ordered side-chains and backbones. Evidence for an H-bond lattice exists in the solvatochromic behavior[8][9][10] of the polymer in solvents of different polarity. Backbone disorder in the solid can be measured optically by taking the ratio of the absorbance parallel and perpendicular to the backbone. The anisotropy ratio of A exceeds 50 and indicates a relatively high degree of backbone alignment, consistent with a long coherence length and accounts for the low energy of the exciton absorption (Fig. 3). The sharpness of the A spectrum in Fig. 3 is comparable to the spectrum of high quality thin PTS crystals.

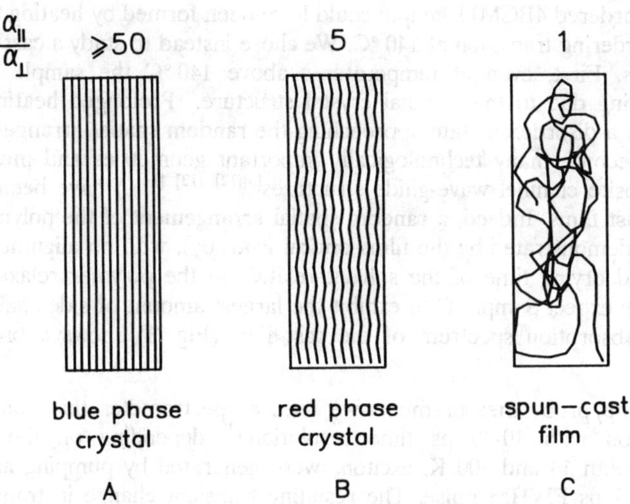

Fig. 2: Schematic diagram of three samples of PDA-4BCMU with varying amounts of disorder. The optical anisotropy ratio is given above.

Differential scanning calorimetry experiments of 4BCMU cast films have shown an endothermic transition at $\sim 120\,°C$ that corresponds to the melting of the hydrogen-bond lattice of the side-chains, with a second disordering transition appearing at $\sim 140\,°C$ associated with increased backbone disorder. Our sample B was formed by heating a pristine blue phase crystal to $120\,°C$. Its 300 K absorption spectrum is shown in Fig. 3 and shows a blue-shift of 0.45 eV and a broadening of the original exciton absorption. Along with the melting of the side-groups, a measured anisotropy of approximately 5 indicates a much lower degree of backbone linearity than in sample A.

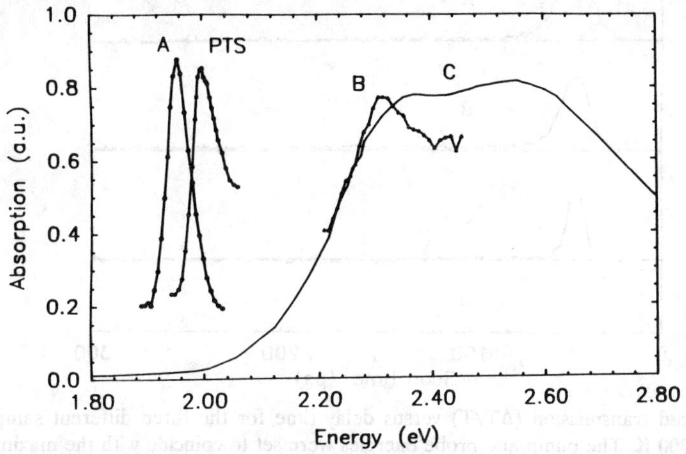

Fig. 3: The 300 K optical absorption spectra of the three samples of PDA-4BCMU. The optical spectrum of PDA-PTS is shown for comparison.

A highly disordered 4BCMU sample could have been formed by heating Sample B above the second disordering transition at 140 °C. We chose instead to study a cast film of 4BCMU for two reasons. First, even at temperatures above 140 °C the sample retains residual backbone ordering due to the original crystal structure. Prolonged heating above 140 °C eventually yields a disordered state approaching the random spatial arrangement of the cast film (Fig. 2). Second, many technologically important geometries and measurements (for example, composite channel wave-guide structures[5][12] [13] [14]) have been explored using 4BCMU spin-cast films. Indeed, a random spatial arrangement of the polymer exists in two dimensions (as demonstrated by the films optical isotropy), with no alignment of backbones due to the rapid drying time of the solvent relative to the polymer relaxation time. As a consequence, we expect Sample C to exhibit the largest amount of side-chain and backbone disorder. The absorption spectrum of the cast film (Fig. 3) shows a broad blue-shifted exciton peak.

For our pump/probe measurements we used a spectrometer that combines 2.7 cm^{-1} energy resolution with 10-20 ps time resolution,[7] depending on the wavelength. In experiments at both 14 and 300 K, excitons were generated by pumping at the absorption maxima with a 7 ps FWHM pulse. The resulting transient change in transmission, $\Delta T/T$, was monitored by a second 7 ps pulse at a variable time delay and at the same energy as the pump pulse. Both the pump and the probe beams were polarized parallel to the chain axis. All signals were linear in both pump and probe intensities. The high sensitivity of the spectrometer allows us to conduct experiments in the small signal limit and observe weak, long-lived signals typically unseen in conventional pump/probe measurements.

In Fig. 4, we present the 300 K on-resonance pump/probe results ($\omega_{pump} = \omega_{probe}$). In all three cases, the fast photoinduced bleaching signals are within the cross-correlation (10 ps) of our spectrometer at that particular wavelength, in agreement with previous work[15] on 3BCMU where the exciton lifetime has been estimated to be ~2 ps. Photobleaching is the expected result from partial saturation of the exciton absorption by a phase space filling mechanism.

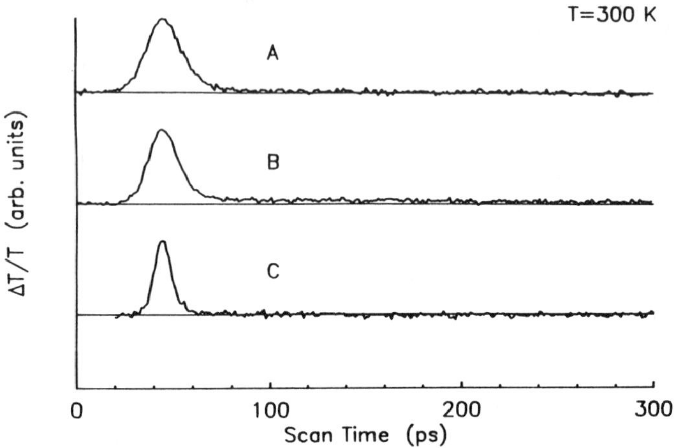

Fig. 4: Photoinduced transmission ($\Delta T/T$) versus delay time for the three different samples of PDA-4BCMU at 300 K. The pump and probe energies were set to coincide with the maximum of the excitonic absorption. The fast responses match the instrumental cross-correlation (not shown).

In Fig. 5 we present pump/probe results for all three samples at 14 K. The instrumental cross-correlation signal in Fig. 5a is slightly narrower than the photobleaching signal. The exciton lifetime at 14 K is comparable to the instrumental time resolution of 10 ps. Longer exciton lifetimes at low temperature has also been observed in 3BCMU.[15][16] As in the 300 K data, the fast photobleaching results are consistent with a phase space filling mechanism of saturation of the exciton absorption. In Samples B and C a long-lived response becomes apparent, with a decay time of ~ 400 ps in C. The magnitude of the long-lived response is essentially independent of the probe wavelength. At low temperatures, therefore, full ground state recovery does not occur until nanoseconds after photoexcitation. We associate the long-lived response with greater disorder, as discussed below. In crystalline PTS,[17] the long-lived signals at 14 K were almost an order of magnitude larger than in C, showed a characteristic rise-time of 35 ps and featured both photoinduced bleaching and absorption. The slow 4BCMU response, even in films, decays monotonically and is only a bleaching signal.

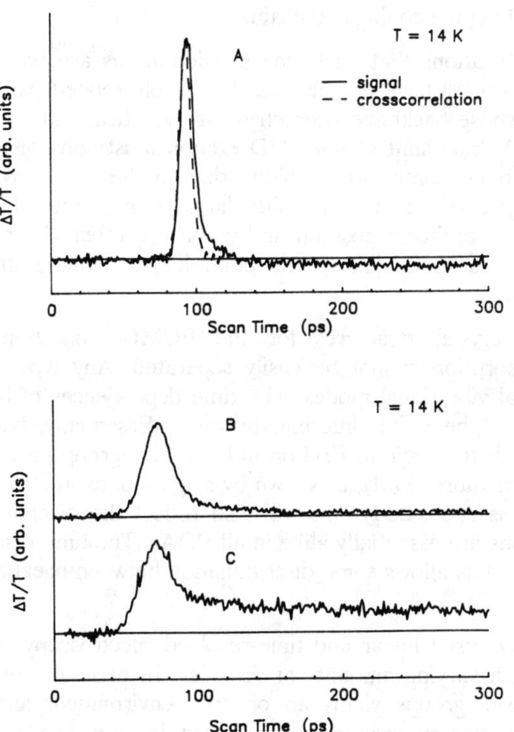

Fig. 5: Photoinduced transmission (ΔT/T) versus delay time for the three different samples of PDA-4BCMU at 14 K. The cross-correlation of the laser system is displayed along with spectrum A to draw attention to an exciton lifetime effect in the fast response at 14 K.

III. Discussion

Increasing disorder from sample A to C is associated with the slow response in pump-probe experiments. Both backbone disorder and side-group polarizabilities potentially affect the exciton absorption and localization. Backbone disorder in PDAs has been widely discussed.[18] Exciton to side-group coupling has recently been observed in PDA-PTS.[17] We suppose static (3-D backbone) disorder to control the extent of exciton delocalization and coupling to local modes, the relaxation of which primarily depends on the specific mode rather than on disorder.

Thermochromic and solvatochromic shifts in PDAs and in other conjugated polymers have generally been discussed in terms of disorder or strain along the conjugated backbone. A break in the conjugation, for example, leads to an ensemble of shorter segments whose excitation energies are higher. More realistic models invoke a wormlike chain and disordered transfer integrals, especially for the weak bonds with low barriers to rotation. While the connection between backbone conformation and excitation energies becomes complicated, disorder is always associated with blue shifts. A quantitative understanding is severely hampered by the absence of independent evidence for the nature or density of conformational defects along the conjugated chain.

Indeed, there are indications that backbone considerations are only part of the story. Sandman and coworkers[19] [20] have emphasised the closely related pair of crystals, PDA-DCH and PDA-THD, whose backbone geometries are identical within the available X-ray resolution. The 400 meV blue shift of the THD exciton must consequently reflect a side-group rather than backbone contribution. Shifts due to the local environment are well known in molecular crystals,[21] and such shifts have been proposed for PDA crystals. Although variations in the exciton's position and width are often observed, Enkelmann[22] finds little variation in the basic diacetylenic bond lengths among the most accurately determined PDA structures.

In the absence of crystal structures for the BCMUs, backbone and side-group contributions to the absorption cannot be easily separated. Any type of localization will enhance coupling to local vibrational modes. The time dependences of backbone and side-group contributions would be quite different, however. Essentially, backbone distortions decay instantaneously with the exciton. Exciton induced side-group reorientation,[17] on the other hand, may decay far more slowly, as shown by ring motions in PTS. Differences in the coupling between excitons and side-groups will also reflect the diversity possible in side-groups even if the excitons are essentially alike in all PDAs. The time resolution afforded by pump-probe experiments thus allows some discrimination between backbone and side-group contributions.

In conclusion, we have used linear and time-resolved spectroscopy to study the optical response of 4BCMU with varying amounts of disorder. In pristine samples of 4BCMU a network of H-bonded side-groups yields an ordered environment and linear backbone leading to a sharp, low energy exciton. An increase in disorder leads to a long-lived photoinduced bleaching response in low temperature experiments due to side-group relaxation. In contrast to the strong coupling and order of magnitude larger long-lived response observed in PDA-PTS, however, exciton coupling to side-groups is rather weak in PDA-4BCMU. Further studies of the photoinduced response in other PDA crystals and films are underway to elucidate side-group effects on excitons in conjugated polymers.

REFERENCES

1. **Nonlinear Optical Properties of Polymers**, Ed. A.J. Heeger, J. Orenstein and D.R. Ulrich, Material Research Society (1988).

2. W-S. Fann, S. Benson, J.M.J. Madey, S. Etemad, G.L. Baker and F. Kajzar, Phys. Rev. Lett. **62** 1492 (1988).

3. P.W. Smith, Bell Sys. Tech. J. **61** 1975 (1982).

4. P.D. Townsend, G.L. Baker, N.E. Schlotter, C.F. Klausner, S. Etemad, Appl. Phys. Lett. **53** 1782 (1988).

5. G.L. Baker, C.F. Klausner, J.A. Shelburne III, N.E. Schlotter, J.L. Jackel, P.D. Townsend, S. Etemad, Synth. Met. **28** 639 (1989).

6. G.N. Patel, Polymer Prepr., Am. Chem. Soc., Div. Polym. Chem., **19** (2), 154 (1978).

7. G.J. Blanchard, J. Chem. Phys. **87** 6802 (1987).

8. R.R. Chance, G.N. Patel and J.D. Witt, J. Chem. Phys. **71** 206 (1979).

9. G. Walters, P. Painter, P. Ika and H. Frisch, Macromolecules **19** 888 (1986).

10. M.F. Rubner, D.J. Sandman and C. Velazquez, Macromolecules **20** 1296 (1987).

11. A. Prock, M. Shand, R. Chance, Macromolecules **15** 238 (1982).

12. G.L. Baker, C.F. Klausner, U.S. Patent 4,824,522.

13. N.E. Schlotter, J.L. Jackel, P.D. Townsend and G.L. Baker, Appl. Phys. Lett. (submitted).

14. P. Townsend, J. Jackel, G. Baker, J. Shelburne III, S. Etemad, Appl. Phys. Lett. (Oct. 30, 1989).

15. M. Yoshizawa, M. Taiji, T. Kobayashi, invited paper to IEEE J.QE. Special Issue of Femtosecond Spectroscopy.

16. Kobayashi's group has also observed longer lifetimes in 4BCMU at low temperatures (private communication, M. Yoshizawa, 1989).

17. M.J. Nowak, G.J. Blanchard, G.L. Baker, S. Etemad and Z.G. Soos, Phys. Rev. Lett. (submitted).

18. **Polydiacetylenes**, NATO ASI Series No. 102, Ed. by D. Bloor and R. Chance, Nijhoff Publisher (1985).

19. D.J. Sandman and Y.J. Chen, Synth. Met. **28** D613 (1989).

20. M.E. Morrow, K.M. White, C.J. Eckhardt, and D.J. Sandman, Chem. Phys. Lett. **140** 263 (1987).

21. **Electronic Processes in Organic Crystals**, M. Pope and C. Swenberg, Clarendon Press (1982).

22. V. Enkelmann, Adv. Polym. Sci. **63** 91 (1984).

PHOTOINDUCED ABSORPTION AND NONLINEAR OPTICAL RESPONSE IN A POLYCONDENSED THIOPHENE-BASED POLYMER (PTT)

G. Ruani, A. J. Pal, R. Zamboni, C. Taliani

Istituto di Spettroscopia Molecolare, CNR, via de'Castagnoli 1, 40126 Bologna, Italy.

F. Kajzar

CEA-IRDI, DEIN-LPEM, Cen/Saclay, 91191 Gif-Sur-Yvette Cedex, France.

ABSTRACT Nonlinear optical properties of a new thiophene based conjugated polymer (polythieno(3,2-b)thiophene, hereafter referred to as PTT) have been studied by two kinds of nonlinear spectroscopies: Photoinduced Absorption (PA) and Third Harmonic Generation (THG). PTT is a novel highly nonlinear optical material comparable to polythiophene, and can be prepared directly as an amorphous free standing film. The electronic nonlinear response measured by means of THG gives an average value of resonant $\chi^{(3)}(-3\omega,\omega1,\omega1,\omega1) = 2\times10^{-11}$ e.s.u. in the 1.25-1.45 μm range of the fundamental laser wavelength. The oscillations observed in $\chi^{(3)}(-3\omega,\omega1,\omega1,\omega1)$ may be explained by three photon vibronic resonance enhancement with the $C=C$ stretching mode. Photoexcitation of electron-hole pairs in the neutral polymer generates long-lived bipolaron states within the semiconducting gap. We relate the variation of the refractive index due to photoexcitations to the cubic susceptibility. In this framework $\chi^{(3)}(-3\omega,\omega1,\omega1,\omega1)$ $= 1.3\times10^{-2}$ e.s.u. at the maximum of the low energy bipolaronic band.

J. L. Brédas and R. R. Chance (eds.), Conjugated Polymeric Materials:
Opportunities in Electronics, Optoelectronics, and Molecular Electronics, 429–441.
© 1990 *Kluwer Academic Publishers. Printed in the Netherlands.*

1. INTRODUCTION

It is generally well accepted that π electron conjugated polymers, in the semiconducting phase, exhibit high nonlinear optical properties with a fast response time. Up to now the most studied systems were topochemically polymerized polydiacetylenes (PDA). A substantial progress has been made in the understanding of the origin of nonlinear optical properties, and their frequency dispersion due to resonant contributions with excited electronic states[1]. Nevertheless the high crystallinity of these materials gives rise to a large light scattering[2] and results in substantial propagation losses making them unuseful for any applications in wave guiding thin film devices.

Therefore we have attempted to find other materials with better packing and giving amorphous thin films offering negligible light scattering for integrated optical applications.

Among these we focused our attention to conjugated polymers with non degenerate ground state based on fused thiophene rings which have been recently synthesized and characterized[3].

In this paper we report on a study of third order nonlinear optical properties study of polythieno(3,2-b)thiophene (PTT) by third harmonic generation (THG) as a function of incident laser wavelength and by photoinduced absorption spectroscopy (PA) in the IR range.

Third harmonic generation is a powerful technique giving not the only nonlinear susceptibility responsible for this process: $\chi^{(3)}(-3\omega;\omega,\omega,\omega)$, but also information on the electronic structure of the studied material.

Photoinduced absorption (PA) allows to study the same electronic states induced by photogeneration of charge carriers which have the same origin as those induced by chemical doping but without the possible additional distortion due to the Coulomb interaction with the dopant[4]. The breaking of the local symmetry due to this strong electron-phonon interaction allows totally symmetric modes (Raman active) to become active in IR and to be detected in photoinduced absorption[5]. The time-scale involved in the PA spectroscopy allows to investigate long-lived photoinduced species in steady-state conditions ($dn_i / dt = 0$, where n_i is the number of states of type i).

2. SAMPLE PREPARATION

Polymerization and thin film preparation were achieved by electrochemical oxidation of the monomer molecules in a two compartment cell at room temperature with platinum and ITO electrodes. The monomers were dissolved in suitable solvents (i.e. CH_2Cl_2, CH_3CN) with an electrolyte such as Bu_4NClO_4, $LiClO_4$ etc. The current density was kept constant during oxidation, at typically 0.1 mA/cm^2. Under these conditions the polymer grows at the anode forming a black film whose thickness may be controlled by varying the time of the process as well as the current density. The "as grown" polymer film was doped with anions provided by the electrolyte and was electrically conducting. Undoping was performed by short circuiting the electrodes for several hours and then by reversing the voltage for a few seconds[6,7].

Free standing films for THG measurements and optical absorption spectroscopy in the NIR-Vis-UV range were obtained by stripping the film from the electrode surface after swelling it with suitable solvents. Optical absorption spectra of a free standing film of PTT is shown in Fig. 1. The first strong electronic absorption, generally assigned to the lowest dipole allowed π-π^* transition, occurs at 2.75 eV in PTT.

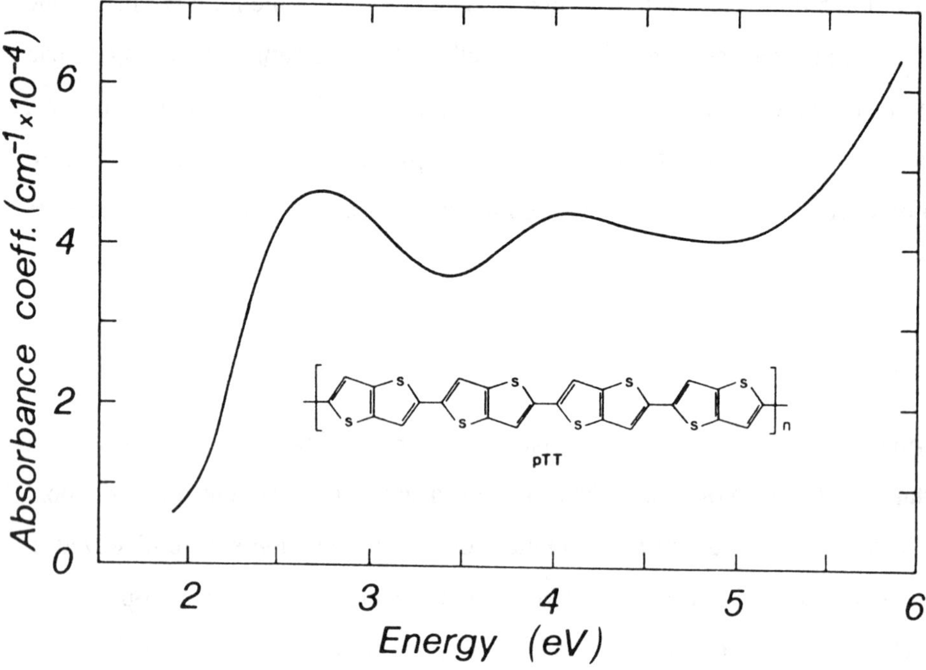

Fig. 1 - Room temperature optical absorption spectra of thin films of PTT. In the inset the idealized chemical structure of PTT is reported.

The morphology of thin films, tested by a scanning electron microscope (SEM), shows that these are rather amorphous. The surface is rather irregular showing the incipient tendency to develop nodules. The thicknesses of thin films used in THG experiments were in the range of 2500 Å.

Pressed pellets of homogeneous mixtures of the polymer in KBr were used for PA measurements. Special care was taken in grinding and pressing processes in order to achieve homogeneous and transparent pellets with good optical properties.

3. THIRD HARMONIC GENERATION

THG measurements were performed by transmission as a function of incident light wavelength using the technique described elsewhere[8]. Free standing films were deposited on silica substrates on the side facing the detector. All experiments were performed in vacuum in order to avoid environmental effects[9]. The sample was rotated along an axis perpendicular to the beam propagation direction and parallel to the incident light polarization. In such a geometry the harmonic intensity as a function of incident angle Θ is given by:

$$I_{3\omega}(\Theta) = \frac{64\pi^4}{c^2} \left(\frac{\chi^{(3)}}{\Delta\varepsilon}\right)_s^2 \left| e^{i(\psi_{3\omega}^P + \psi_{3\omega}^s)} \left[T_1 (e^{i\Delta\psi^s} - 1) + \rho T_2 e^{i\Delta\psi^s} (e^{i\Delta\psi^P} - 1) \right] \right|^2 I_\omega^3 \qquad (1)$$

where

$$\rho = \left(\frac{\chi^{(3)}}{\Delta\varepsilon}\right)_p \times \left(\frac{\chi^{(3)}}{\Delta\varepsilon}\right)_s^{-1} \qquad (2)$$

I_ω is the incident light intensity, $\Delta\varepsilon = \varepsilon_\omega - \varepsilon_{3\omega}$ is the dielectric constant dispersion, T_1 and T_2 are factors arising from transmission and boundary conditions[8,9]. The superscripts (or subscripts) p and s refer both to the polymer film and the substrate (silica), respectively. $\Delta\psi$'s in eqn. 1 are phase mismatches between fundamental (ω) and harmonic (3ω) frequencies

$$\Delta\psi = \psi_\omega - \psi_{3\omega} = 3\omega l (n_\omega \cos\theta_\omega - n_{3\omega} \cos\theta_{3\omega}) \qquad (3)$$

where $\theta_\omega(_{3\omega})$ are propagation angles at $\omega(3\omega)$ frequency, respectively, in a given medium and l is the polymer film thickness. We note here that the refractive index for polymer film is complex at harmonic frequency

$$n_{3\omega} = n_{3\omega}^\Gamma + ik_{3\omega} \qquad (4)$$

and the corresponding imaginary part for the PTT films calculated from optical
absorption is shown in Fig. 2.

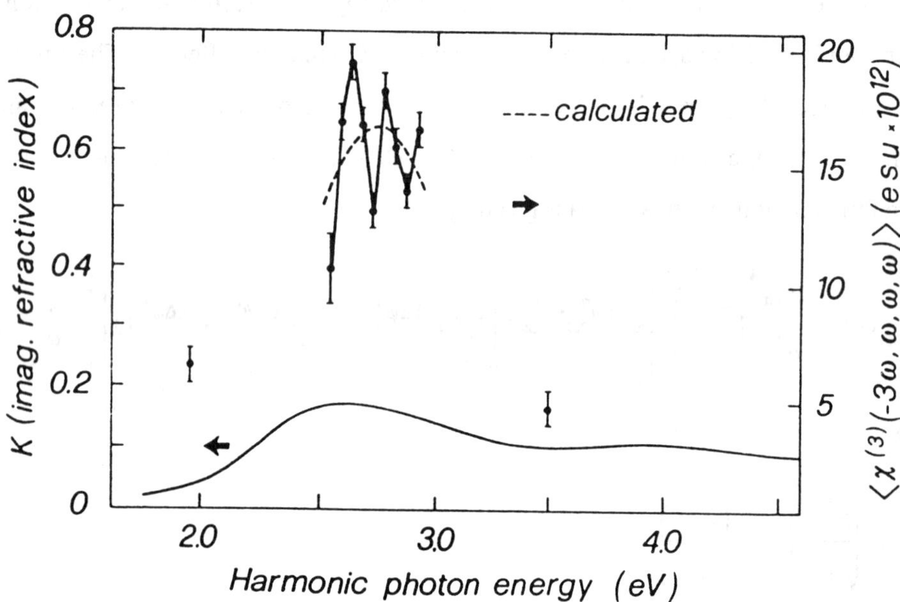

Fig. 2 - Energy dependence of average cubic susceptibility $<\chi^{(3)}>$ for PTT film.
Broken curve represents calculated values. Upper solid curve is a guide for
the eye showing the fine structure in $\chi^{(3)}$ spectrum. Lower solid curve represents the imaginary part of the refractive index.

The harmonic intensity from polymer film and substrate   was calibrated with
that from a 1 mm thick silica plate measured at the same geometry.

The measured average values of cubic susceptibility $\chi^{(3)}(-3\omega;\omega,\omega,\omega)$ are shown in
Fig. 2 for PTT. The averaging is done over all polymer chain distributions.

Assuming that the tensor component which is enhanced is parallel to the polymer chain direction (i.e. $\chi^{(3)}_{xxxx}$) we have

$$\chi^{(3)}_{xxxx} = <\chi^{(3)}(3\omega;\omega,\omega,\omega)> / <\cos^4\theta> \qquad (5)$$

where θ is the angle between polymer chain and incident light polarization. For a complete three dimensional chain disorder one gets:

$$<\cos^4\theta> = 1/5 \qquad (6)$$

For the case when all polymer chains are parallel to the substrate (disorder in two dimensions) one gets

$$<\cos^4\theta> = 3/8 \qquad (7)$$

One observes a strong resonance enhancement in the 1.25-1.45 μm fundamental wavelength range. This can be easily identified as a three photon resonance. In PTT the resonance is large. Assuming that the dominant term in $\chi^{(3)}(-3\omega;\omega,\omega,\omega)$ is responsible for three photon resonance and neglecting all other terms one obtains:

$$|\chi^{(3)}(-3\omega;\omega,\omega,\omega)| \propto A / [(E_{ng} - 3\omega)^2 + \Gamma^2]^{1/2} \qquad (8)$$

where A is a frequency and oscillator strength dependent parameter and Γ is the damping term.

By fitting eqn.(8) to experimental data one obtains for PTT a broad ($\Gamma = 2480$ cm^{-1}) resonance band located at 2.74 eV and visualized in Fig. 2 by a dashed line. The agreement of the fit with the experimental data is not so good. In fact the data shows a fine structure in χ dependence on harmonic photon energy. The spacing between neighbouring peaks is 1325 cm^{-1}. This value is very close to the strong collective $C=C$ excited states stretching vibration observed in the monomer[10] ($\nu_{C=C} = 1292$ cm^{-1}). Experimental accuracy ensured an excellent agreement and the structure can be interpreted as a three photon resonance with vibronic levels of the lowest excited singlet state. The measured three photon resonant values of average cubic susceptibility $\chi^{(3)}(-3\omega;\omega,\omega,\omega)$ is equal to 2×10^{-11} e.s.u.. Assuming a three dimensional disorder we derive an intrinsic $\chi^{(3)}(-3\omega;\omega,\omega,\omega)$ which is five times larger. Thus the measured $\chi^{(3)}$ values compete well with those obtained for other systems with comparable conjugation length like: polydiacetylene red form[2] and poly(3-butyl)thiophene[11] . The measured resonant $\chi^{(3)}(-3\omega;\omega,\omega,\omega)$ susceptibility is about three orders of magnitude smaller than that obtained by optical Kerr effect $(\chi^{(3)}(-\omega;\omega,-\omega,\omega))$ on the same thin films[12] , the last one corresponding to one photon resonance. The difference is fundamentally due to the different origin of these two susceptibilities. Whereas THG susceptibility is purely electronic related to a coherent process and connected with polarization of the electronic cloud, the Kerr susceptibility takes account also of other non-coherent processes such as population change, heating effects and may be dominated by them.

4. PHOTOINDUCED ABSORPTION

The photogeneration of electron-hole pairs by pumping the semiconducting one-dimensional material PTT above the gap cannot simply be considered as the generation of two, more or less interacting, quasi-particles in a rigid electronic band system. They will interact so strongly with the surrounding system creating, in the time scale of phonon interaction (10^{-13} sec) new electronic states. Most of these photoinduced species return to the ground state recombining non radiatively or, because of the confinement due to the non-degeneracy of the system, radiatively in about 10^{-9} sec (the peak of fluorescence for PTT is at 1.95 eV)[6]; the excited states, which don't recombine, evolve forming new electronic species i.e. bipolarons ($B++$ or B^{--}) which behave as long lived metastable states. In this view the photoinduced absorption response to cw photoinduced excitation probes only those metastable states that can reach a steady state condition under continuous generation of excitation i.e. the long live B^{++} and B^{--}.

As in the case of most of the conjugated polymers with non-degenerate ground state, in PTT also, photogeneration above the semiconducting π-π^* gap gives rise to the appearance of two new low energy electronic transitions and a bleaching of the π-π^* interband transition. This behaviour is accounted for by the formation of polaron/bipolaron states in the gap.

The PA spectra were obtained by using a modified Bruker FTIR interferometer (mod.IFS 88); a HeNe laser was used for photoexcitation at a photon energy of 1.96 eV. In order to minimize possible thermal effects, and to be sure that the signal was really due to photoexcitation, the power on the sample was kept below 70 mW/cm^2 and measurements at different laser power densities were carried out. At approx. 70 mW/cm^2 the PA saturates and the increase of the PA intensity departs from half of the square root dependence on the laser power, but the shape of the spectum is the same by going from 0.5 to 70mW/cm^2. The sample was kept at 80 K in a flow cryostat. A broad band MCT was used as a detector. Consecutive interferograms with laser on and off, with a period of about one second, were stored for a total number of about 4000 scans in order to achieve an appreciable S/N ratio.

The PA spectrum of PTT in the spectral range of 600-9000 cm^{-1} is shown in Fig. 3 (laser power 70 mW/cm^2, T = 80 K). The spectrum presents different photoinduced IR active vibration (IRAV) in the spectral range of 600-1600 cm^{-1} (see inset in Fig. 3) and a photoinduced electronic band at 0.44 eV.

IRAV bands show a pattern very similar, but better resolved, to that induced on PTT by chemical doping[6], indicating that the species which give rise to the appearance of these bands in IR by photoexcitation or chemical doping have a similar nature.

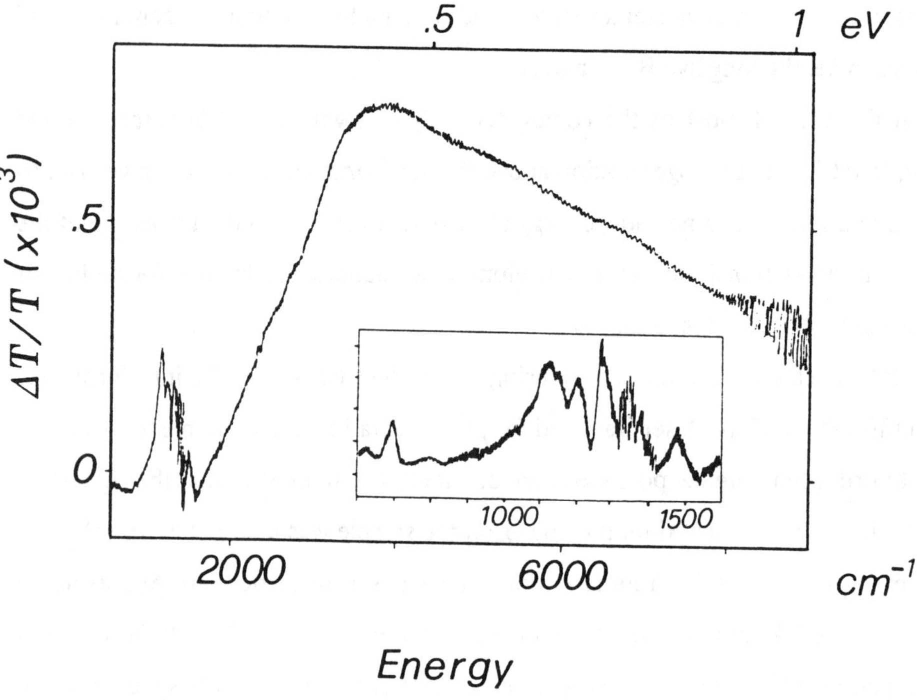

Fig. 3 - Photoinduced absorption spectrum of PTT in KBr pellet, at 80 K. In the inset a blow-up of the phonon spectral region is shown.

Also new electronic bands appear when PTT is chemically doped, but in this case the peak position of the lowest band is shifted considerably towards higher energy with respect to that obtained by photoexcitation (the lowest doping induced electronic band in the gap is at .85 eV, see Fig. 4). In fact even if the electronic species generated by chemical doping and photoexcitation have the same nature, the presence of the dopant, because of Coulomb interaction, modifies the intrinsic behaviour of the system and also the structure of the energy states.

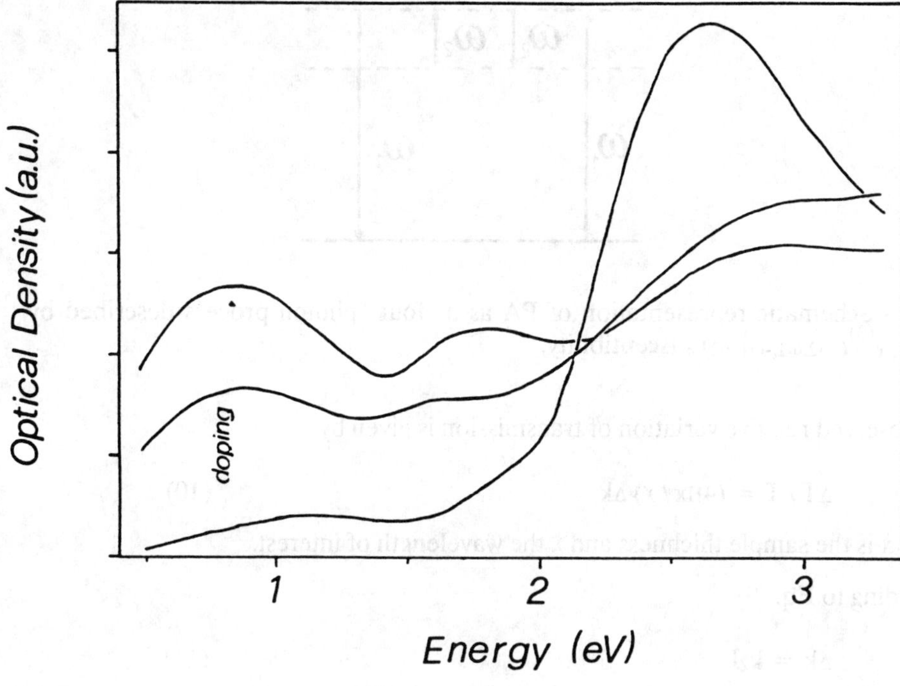

Fig. 4 - "In situ" optical absorption spectra of a PTT film at different doping levels. The arrow shows the direction of the increasing doping level.

The shift of the oscillator strength from $\Pi - \Pi^*$ transition towards the PA bands in the near IR gives rise, as it was proposed by Heeger et al.[14], to a variation of the refractive index with pump intensity I

$$n = n_0 + n_2 I = n_0 + (n_2^r + ik_2)I \qquad (9)$$

The process of PA may be described by a cubic susceptibility $\chi^{(3)}(-\omega_2, \omega_1, -\omega_1, \omega_2)$ where ω_1 and ω_2 are the pump and probe frequencies (see Fig. 5).

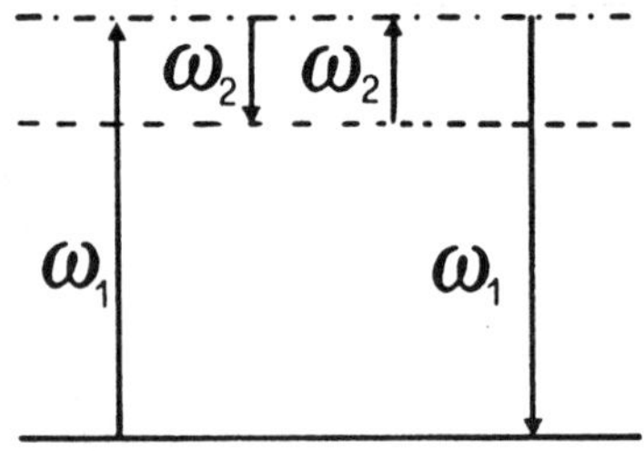

Fig. 5 - Schematic representation of PA as a four photon process described by $\chi^{(3)}(-\omega 2,\omega 1,-\omega 1,\omega 2)$ susceptibility.

The observed relative variation of transmission is given by

$$\Delta T / T = (4\Pi x / \lambda)\, \Delta k \tag{10}$$

where x is the sample thichness and λ the wavelength of interest.

According to Eq. (9)

$$\Delta k = k_2 I \tag{11}$$

In the present experiment $x = 10\ \mu m$, $\lambda = 2.48\ \mu m$ and $\Delta T / T = 0.8 \times 10^{-3}$ at saturation intensity (maximum probe intensity at which PA saturates[15]) $I_s = 75$ mW/cm^2. Putting these values into Eqs. (10) - (11) one obtains $\Delta k = 1.6 \times 10^{-5}$ and $k_2 = 2.3 \times 10^{-4}(cm^2/W)$. On the other hand $n_2 = 12\Pi\, \chi^{(3)}(-\omega 2,\omega 1,-\omega 1,\omega 2) / n_0^2 c$. Assuming $n_0 = 1.5$ one gets $Im\, \chi^{(3)}(-\omega 2,\omega 1,-\omega 1,\omega 2) = 1.3 \times 10^{-2}$ e.s.u. which is a very large value with however a slow response time in the order of ms.

In PTT, as in other conjugated polymers, nonlinear optical processes such as photoinduced absorption or third harmonic generation are related to the intrinsic instability of the system due to charge-phonon interactions.

ACKNOWLEDGEMENTS

The authors wish to thank Mr. S. Guerri for valuable technical contribution.

REFERENCES

1. Chollet P.A., Kajzar F. and Messier J. (1989), in Proceedings of International Symposium, Tokyo, July 1988, T. Kobayashi ed., Springer Verlag.
2. Chollet P.A., Kajzar F. and Messier J. (1988), in Nonlinear Optical and Electroactive Polymers, J.N. Prasad and D.R. Ulrich eds, Plenum Press, p. 121
3. Taliani C., Zamboni R., Danieli R., Ostoja P., Porzio W., Lazzaroni R., Phys. Scripta in press.
4. Kim Y.H., Hotta S., Heeger A. (1987), Phys. Rev. **B36**, 7486.
5. Male H.J. and Hicks J.H. (1986), Synth. Met. **13**, 7486.
6. Taliani C., Danieli R., Zamboni R., Ostoja P. and Porzio W. (1987), Synth. Met. **18**, 177.
7. Danieli R., Ostoja P., Tiecco P., Zamboni R. and Taliani C. (1986), J. Chem. Soc. Commun. 1476.
8. Kajzar F., Messier J.and Rosilio C. (1986), J. Appl. Phys. 60, 3040.
9. Kajzar F. and Messier J. (1985), Phys. Rev. **A32**, 2352.
10.Bertinelli F., Brillante A., Palmieri P. and Taliani C., (1977) J. Chem. Phys. **66**, 51.
11. Kajzar F., Messier J., Sentein C., Elsenbaumer R.L. and Miller G.G., Proceed. of 34th Annual SPIE Conference, San Diego, August 6-11, 1989 (in print).
12. Yang L., Dorsinville R., Wang Q.Z., Zou W.K., Ho P.P., Yang N.L., Alfano R.R., Zamboni R., Danieli R., Ruani G. and Taliani C., J. Opt. Soc. Am. **B** (in press).
13. Friend R.H., Bradley D.D.C. and Townsend P.D. (1987), J. Phys. **D 20**, 1367.
14. Heeger A.J., Moses D. and Sinclair M. (1986), Synth. Met. **15**, 95.
15. Kaneto K., Uesugi F. and Yoshino K. (1987), J. Phys. Soc. J. **56**, 3703.

STUDY OF SECOND HARMONIC GENERATION OF A HEMICYANINE DYE IN FLOATING
AND DEPOSITED ORGANIC MONOLAYERS

A. SCHEELEN, P. WINANT and A. PERSOONS
Department of Chemistry
Catholic University of Leuven
Celestijnenlaan 200D
B-3030 LEUVEN - Belgium

ABSTRACT. In this paper we describe second harmonic generation from a
hemicyanine dye incorporated in floating monolayers on an aqueous subphase and in Langmuir-Blodgett films. Values of the second order
susceptibilities are determined from SH-intensity vs. fundamental beam
intensity. Molecular hyperpolarisabilities and orientations are obtained and compared with previous values.

1. INTRODUCTION

Organic molecules offer a valid alternative for inorganic crystals in
applications in the field of non-linear optics due to their high
molecular hyperpolarizability [1]. A drawback of organic materials
for applications involving quadratic susceptibilities is the centrosymmetry of their crystalline structures. This centrosymmetry is
inherently absent in monolayers and suitably ordered multilayers.
   In this paper we report frequency doubling data from floating
and deposited monolayers of a substituted hemicyanine dye, MO, shown
in fig. 1, mixed with arachidic acid. The second order hyperpolarisability β of MO was estimated from intensities of both fundamental
and SH-signal and molecular orientations as derived from polarisation
ratios.

2. SECOND HARMONIC GENERATION OF FLOATING LAYERS

2.1. Experimental setup

Second harmonic generation of monolayers floating on water (SHGOW)
was measured in a reflection geometry, as shown in fig. 2. An injection seeded Nd:YAG laser (Spectra-Physics Mod. DCR-3) was used in
our studies. For beam attenuation a set of neutral density filters
was used (Schott NG-type). A part of the beam was sampled to measure
beam intensity. To eliminate visible light from the pump-flash
a RG-850 filter was mounted before the trough. The fundamental
beam, suitably polarized, was focussed on the water surface at an

J. L. Brédas and R. R. Chance (eds.), Conjugated Polymeric Materials:
Opportunities in Electronics, Optoelectronics, and Molecular Electronics, 443–449.
© 1990 *Kluwer Academic Publishers. Printed in the Netherlands.*

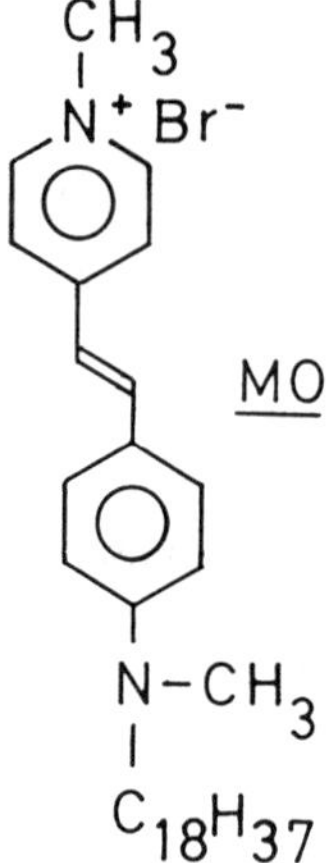

Figure 1. Structure of hemicyanine dye (MO).

angle of 45 degrees. A grounded RG1000 filter kept down reflections
from the glass of the trough. The SH-content in the reflected beam
was isolated by means of an IR-absorption filter (Schott KG-4) and
a 532 nm interference filter. A polarizer in front of the detecting
photomultiplier allowed the analysis of the polarisation of the
SH-signal, which was analyzed with a boxcar integrator. All measurements were carried out with a subphase of ultra pure water, (no
salt or buffer added) at a constant surface pressure of 20 nM/m,
controlled with a Wilhelmy balance and adjusted with a feedback
to the moving barrier of the Langmuir trough.

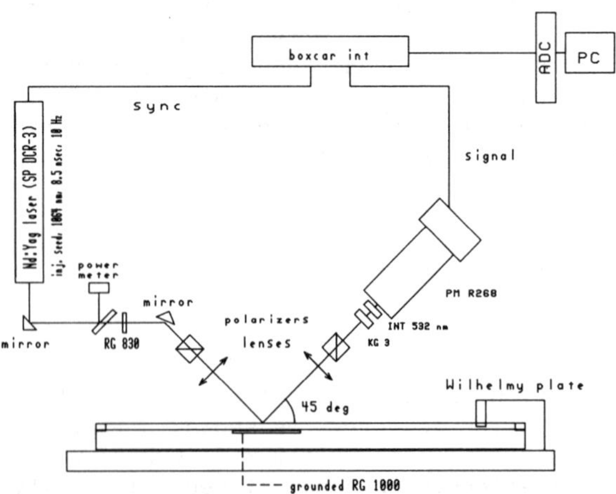

Figure 2. Experimental setup SHGOW.

LB-films were prepared according to standard practice [7].
A mixture of MO and arachidic acid in chloroform was dropped onto
an aqueous subphase (10 mM NaClO, pH 5.5). After evaporation of
the chloroform, the film was compressed and allowed to stabilize.
Transfer onto hydrophylic glass plates (Corning 7059) was carried
out at speeds of 30 mm^2/min. The slides were first covered with two
layers of pure arachidic acid before depositing the dye-arachidic
acid layer.

For the SHG measurements on LB-structures the pulses were focussed
onto the LB-film-coated (double-sided) substrates, which were mounted
on a rotation stage with its rotation axis vertical and perpendicular
to the incident 1064 nm beam. The second harmonic signal was measured,
after passing through an IR cut-off filter and a 532 nm interference
filter, by means of a photomultiplier and treated with gated electronics.

On rotating the sample in the incident beam, fringes are generated
as a result of the interference of the SH generated at the front side
of the sample with the SH generated at the back side of the sample
as shown in Fig. 3.

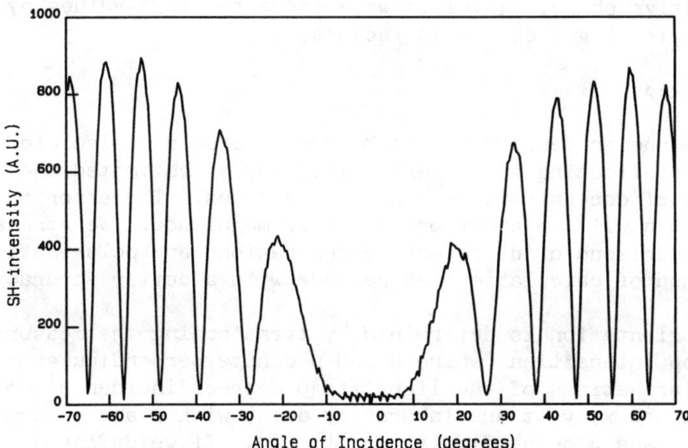

Figure 3.   SH intensity as a function of incidence angle for a double-
side LB-monolayer coated substrate; LB-film of a 90% MO - 10% AA layer.

2.2. Energy calculations

Incident intensity was measured with a power meter. We calculated
reflection and transmission losses on filters and lenses after the
power meter from polarisations, refraction indexes, transmission data
and geometric considerations. Detected energy was corrected for losses
from beam divergence and losses on filters and lenses (about 15%).
From the boxcar's pre-amplifier gain, gate width and sensivity,
PM load resistance and measured individual PM gain and spectral sensivity
curve we derived the signal/energy relations for the detected pulses.

These conversions result in the 'energy in' versus 'energy out' curves
as shown in fig. 4.

2.3. Hyperpolarisability calculations.

The nature (frequency doubled light) of the detected signal was verified by fitting the 'energy out' to a quadratic function of the 'energy in'. These fits have correlation coefficients as high as 0.995.
This, and the fact that only 532 nm is detected, enables us to identify
the detected signal positively as second harmonic generation.
  Quadratic susceptibilities of the layers were calculated from
'energy in' versus 'energy out' curves using : [2]

$$(\chi_s^{(2)})^2 = \frac{I_{2\omega} \, c^3 \, \epsilon_1 \, \epsilon_2^{1/2}}{32 \, \pi^3 \, \omega^2 \, \sec^2\theta \, (I_\omega)^2}$$

Here $\chi_s^{(2)}$ is the surface hyperpoliarsability. The ϵ-factor is estimated to be 2; the incidence angle is 45 degrees. $\chi_s^{(2)}$-values are converted to the usual bulk χ-values by multiplication with the length
(1 nm) of the active phase. χ-values were converted to β-values by
the relation (neglecting local field factors) :

$$\chi = N \langle\beta\rangle$$

where $\langle N \rangle$ is the number density of the MO-chromofores as calculated
from π-A-curves of floating monolayers. This way we estimated the
β's for a series of concentrations and polarisations. The error on
the absolute values will be about one order of magnitude. We stress
however that comparisons of different concentrations and polarisations
within this method of calculation can be made with a better accuracy
(about 20%).
  Molecular orientation is determined by transforming the measured
ss, sp, ps and pp intensities (where s and p denote perpendicular
and parallel polarisations of the light at 45 degrees incidence) to
$\chi_{zxx}^{(2)}$ and $\chi_{zzz}^{(2)}$ (where x, y, z are laboratory axis, with x and y forming
the water surface and z perpendicular to it) [3]. If we define

$$A = 2\chi_{zxx}^{(2)} / (\chi_{zzz}^{(2)} + 2\chi_{zxx}^{(2)})$$

then A is related to the molecular orientation as

$$A = \langle\sin^2\zeta \cos\zeta\rangle/\langle\cos\zeta\rangle$$

where ζ is the angle of the molecule with the surface normal. If
we assume a delta distribution for the molecular orientation ζ can
be determined as

$$\zeta = \sin^{-1} (A)^{1/2}$$

The second order polarisability β and the tilt angle θ between the molecular axis and the water surface are obtained from the results shown in fig. 4 as :

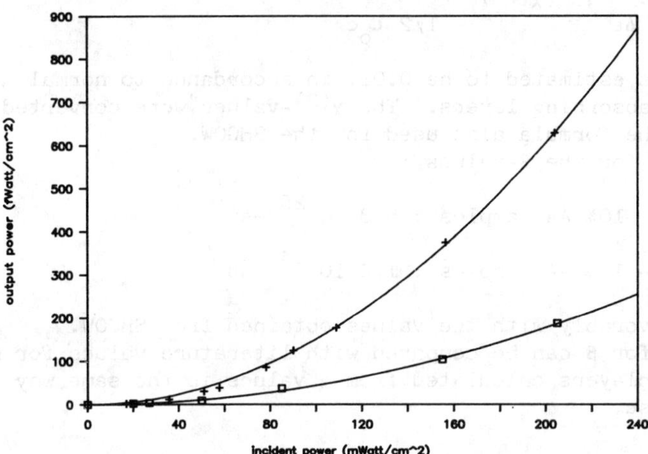

Figure 4.   SHGOW for MO : (+) MO 90%, AA 10%; (□) MO 10%, AA 90%.

90% MO, 10% AA : β = 5 10^{-25} esu

θ = 34 degrees

10% MO, 90% AA : β = 10 10^{-25} esu

θ = 35 degrees

These results clearly show the increasing efficiency per molecule for frequency doubling upon dilution. This is in agreement with previous results [4] and our reflection measurements [5,6]. We attribute this to antiparallel aggregation of molecules at higher concentrations. The angle found in SHGOW the chromophore in MO is making with the water surface is, within experimental error, equal to the result from reflection spectrometry where a value of about 30 degrees was found.

3. SECOND HARMONIC GENERATION ON LB-STRUCTURES

3.1. Hyperpolarisability calculation

To determine the quadratic susceptibility we measured the ratio of the intensity of the SH-signal to the intensity of the fundamental incident upon the monolayer. A gain curve was set up for

the photomultiplier to get an idea of the intensity
of the SH-signal. The intensity of the fundamental was measured with
a power meter. The ratio of the two intensities was correlated with
the following formula :

$$I_{2\omega} = \left[\frac{4\pi}{\Delta\epsilon}\right]^2 (\chi^{(2)})^2 \frac{(I_\omega)^2}{1/2\ \epsilon_o c}$$

The $\Delta\epsilon$-factor was estimated to be 0.01, in accordance to normal dispersion data of nonabsorbing layers. The $\chi^{(2)}$-values were converted
to β-values by the formula also used for the SHGOW.
We obtained for the β-values :

90% MO – 10% AA samples : $5.3\ 10^{-25}$ esu

10% MO – 10% AA samples : $1.4\ 10^{-24}$ esu

which compare favorably with the values obtained from SHGOW.
The values for β can be compared with literature values for measurements of monolayers calculated from χ values in the same way [8]
of about 10^{-25} esu.

4. CONCLUSION

We have shown that the chromophore in MO has a high second order polarisability. It is estimated to be about 10^{-25} esu in both floating
and deposited films. The increasing efficiency upon dilution is attributed to the effect of decreasing antiparallel aggregation within
one layer at lower concentrations. This is consistent with previous
reflection and absorption measurements. The orientation of the transition moment of the molecule versus the water surface is about 35 degrees, within experimental error equal to results obtained in reflection
spectrometry.

5. ACKNOWLEDGEMENTS

A. Scheelen and P. Winant are research assistants of the Belgian National Research Foundation. This work was supported by Nationale
Loterij and government frant GOA 87/91-109.

6. REFERENCES

[1] Williams, D.J., ed. (1983),"Nonlinear Optical Properties of Organic
    and Polymeric Materials", ACS symposium 233, American Chemical
    Society, Washington.
[2] Shen, Y.R. (1984), "The Principles of Nonlinear Optics", John
    Wiley, New York, Chpt. 25.
[3] Heinz, T.F., Tom, H.W.K. and Shen,Y.R. (1983), "Determination
    of molecular orientation of monolayer adsorbates by optical second
    harmonic generation", Phys.Rev. 1, 28, 1833.

[4] Schildkraut, J.S., Penner, T.L., Willand, C.S. and Ulman, A. (1988), "Absorption and second harmonic generation of monomer and aggregate hemicyanine dye in Langmuir-Blodgett films", Optics Letters, 13, 134.

[5] Winant, P., Scheelen, A. and Persoons, A. (1989), "Spectral properties and second harmonic geenration of hemicyanine dye in Langmuir Blodgett films", in J. Messier et al. (eds.), "Nonlinear Optical Effects in Organic Polymers", 219-224, Kluwer Academic Publishers, Dordrecht.

[6] Winant, P., Scheelen A. and Persoons, A. (1989), "Absorption, Reflection and Second Harmonic Generation study of Langmuir Blodgett films of a hemicyanine dye", in SPIE Vol. 1127 - nr. 25 - to be published.

[7] Kuhn, H., Mobius, D., and Bucher, H., (1972), "Spectrosocpy of monolayer assemblies", in Weissberger, A., and Rossiter, B.W., (eds.), "Techniques of Chemistry", vol. I, part IIIB, John Wiley.

[8] Allen, S., McLean, T.D., Gordon, P.F., Bothwell, B.D., Robin, P., Ledoux, I., (1988), "Properties of polyenic Langmuir Blodgett films", in SPIE vol. 971, "Nonlinear Optical Properties of Organic Materials", pag. 206.

NONLINEAR OPTICS IN SOLID SCHIFF BASES

E.HADJOUDIS[a],I.MOUSTAKALI-MAVRIDIS[a] AND J.ZYSS[b]
a) Institute of Physical Chemistry, N.R.C."Demokritos"
153 10 Aghia Paraskevi-Attiki, Greece.
b) Centre National d'Etudes des Telecommunications,
196 Av. Henri Ravera, F-92220 Bagneux, France.

ABSTRACT. The molecular and structural characteristics of
certain N-benzylideneanilines and N-Benzylideneaminopyridines are examined in order to find efficient organic crystalline materials for nonlinear optics. The nonlinear optical coefficients of the compounds, relative to a crystalline quartz standard were measured by the powder method.

1. INTRODUCTION

The concept of molecular engineering can be defined as a
predictive chemical action, at microscopic level, on the
physico-chemical property of interest. Crystal engineering,
is to designe molecules so as to guide their choice of
crystal structure and consequently induce the appearance of
certain properties in the solid state [1,2]. Presently we
are concerned with organic materials with large second harmonic generation (SHG). In order to find such materials we
undertook a systematic search on solid organic materials
upon which we have aquired experience over the years, namely Schiff bases [3] and therefore easier to account for the
above concepts.

The first compounds investigated, N-benzylideneanilines (I) and N-benzylideneaminopyridines (II) are aromatic
molecules substituted with an electron donating group in
the one aromatic ring and an electron accepting group on
the other. These molecules possess large dipole moments in
the group state and an intensive $\pi^* \leftarrow \pi$ electronic band in
the near ultraviolet range and they are therefore suitable
for S.H.G. measurements provided that their crystal
structure is non centrosymmetric [4,5]. It should be noted
that the magnitude of the induced dipole moment is affected
not only by the magnitude of the charge transfer between
the substituents but also by the conjugation, length and
planarity of the molecules [6]. The azomethine linkage is

J. L. Brédas and R. R. Chance (eds.), Conjugated Polymeric Materials:
Opportunities in Electronics, Optoelectronics, and Molecular Electronics, 451–456.

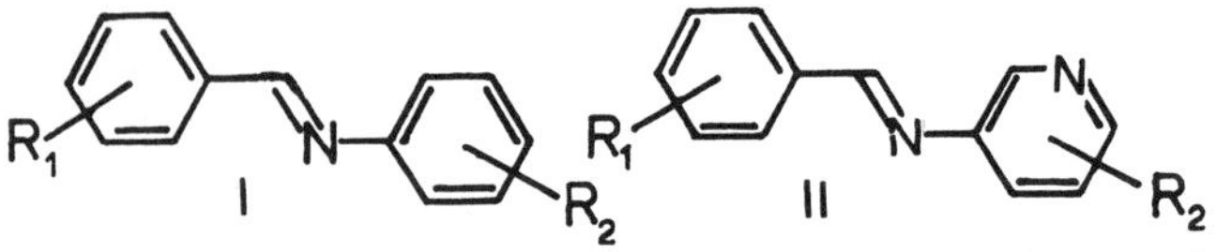

not planar and discrupts the conjugation. However the presence of hetero-nitrogen atoms in the aromatic amine have been shown to influence the planarity of the molecules [7, 8]. It should be emphasized that such nonlinear optical active species can be chemically attached to polymer chains allowing thus manipulation by various processing techniques eliminating some major inadequacies of crystalline inorganics [6].

2. EXPERIMENTAL

2.1. Preparation of compounds

The compounds were synthesized by direct condensation of the appropriate benzaldehyde with the appropriate amine or aminopyridine in ethanol, followed by repeated recrystallization from the same solvent if not otherwise stated. I.r., melting points and elemental analysis were utilized to establish the purity of the compounds.

2.2. Methods

The nonlinear optical coefficients of the compounds relative to a crystalline quartz standard was measured by the powder method [9]. The method permitts the rapid classification of compounds without the necessity of growing large single crystals.

3. RESULTS AND DISCUSSION

The first group of the prepared compounds and their S.H.G. relative to quartz is shown in Table 1 together with their color appearance in the crystalline state and their m.p.'s.

Table 1. N-benzylideneanilines

Compound I				S.H.G
R_1	R_2	color	m.p	$(I_{quartz}^{2\omega}/I_{quartz}^{2\omega})$
1. p-N(CH₃)₂	m-Cl	white-yellow	111	0.00
2. p-Cl	H	white-yellow	58	0.05
3. p-N(CH₃)₂	H	yellow	96.5	0.10
4. p-NO₂	H	yellow	91	0.20
5. p-N(CH₃)₂	p-Cl	gold-yellow	149	26.00
6. p-NO₂	p-Cl	yellow	122	110.00

An examination of the above Table shows some interesting features: Higher Second Harmonic generation values result when both **para**-positions are substituted with donor-acceptor groups presenting high resonance interaction (Compound 6). The importance of resonance (**para**-positions) is obvious from the comparison of compounds 1 and 5. In mono-substituted compounds (compounds 2,3 and 4) the intensity of the S.H.G. is small.

Futher, we decided to prepare more compounds substituted in both **para**-positions with donor/acceptor pairs in order to maximize the electron donor-acceptor interaction. These compounds are shown in Table 2.

Table 2. Para-substituted N-benzylideneanilines

| Compound I | | color | m.p.($^\circ$C) | S.H.G. |
R$_1$	R$_2$			(I^{2w}/Iquartz^{2w})
7. p-NO$_2$	p-OCH$_3$	dark yellow	133	$\sim$ 0
8. p-OCH$_3$	p-NO$_2$	yellow	126	$\sim$ 0
9. p-N(CH$_3$)$_2$	p-NO$_2$	orange	250	$\sim$ 0

Although it is expected that the above molecules will show a strong charge transfer effect, their S.H.G values relative to quartz are zero. This is probably due to the fact that we obtained centrosymmetric crystals. Indeed the crystal and molecular structures of compounds 7 and 9 have been determined. Compound 7 has been crystallized [10] in a centrosymetric space group, while compound 9 has been crystallized in four crystalline modifications [11,12] only one of which is non-centrosymmetric. For this last compound the molecular hyperpolarizability, b, has been reported [6] to be 23.4x10^{-30} esu at 1.9 μm. N-benzilideneanilines with p.p· substitutions show polymorphism in general. Thus the phenomenon has been observed in N-(4-nitrobenzylidene)-4´-methylaniline [13], N-(4-methylbenzylidene)-4´-methylaniline and N-(4-chlorobenzylidene)-4´-chloroaniline [14].

Our next step was to prepare compounds which have lower symmetry and possess an additional permanent dipole which affects the magnitude and the direction of the total molecular dipole moment without increasing the molecular breadth.This was achieved by inserting in the aniline ring an heteroatom (nitrogen)and the resulting compounds are shown in Table 3. We see that compound 11 presents a high S.H.G. value. The fact that this molecule exhibits second-harmonic generation makes one to suppose that the crystal structure is a priori acentric [13]. The molecular hyperpolarizability, b, of this compound measured in aceton was found to be 16x10^{-30} esu. The compound presents polymorphism. Thus it has been crystallized in two polymorphic modifications depending on the solvent and conditions of cry-

454

Table 3. _Para_-substituted N-Benzylideneaminopyridines.

| Compound II | | color | m.p. | S.H.G. |
R$_1$	R$_2$		^oC	($I^{2w}/I quartz^{2w}$)
10. p-CN	p-OCH$_3$	yellow	139	--
11. p-NO$_2$	p-OCH$_3$	yellow	140	~ 1000

stallization: (a) from ethanol or DMF crystallizes in platelets in the centrosymmetric space group P2$_1$/n with a unit cell α=11.080, b=3.376, c=14.364, A, β=112,56$_o$ with 4 molecules in the unit cell. We solved the structure of this compound up to a final consistency index 10 R=3.7%. (b) The second modification crystallizes in the form of elongated prisms from a mixture of ethanol or chlorofoprm in the non-centrosymetric space group, p2$_1$ with a unit cell α=3.8566, b=19.542, c=8.066, A, β=89.37° with 2 molecules in the unit cell [15]. This is the crystal form that shows the strong second harmonic generation effect (Table 3). The structure has been refined up to a final R=4.93%. The molecule has different conformations in the two crystal forms as it is shown in Fig.1 along with atom labelling. The orientation of the 2-methoxypyridyl ring with respect to the rest of the molecule is opposite indicating a free rotation of the ring. In both forms the methoxy group is coplanar to the pyridyl ring and faces the nitrogen atom, thus avoiding unfavorable steric interactions among the methyl and the ring hydrogen atoms.

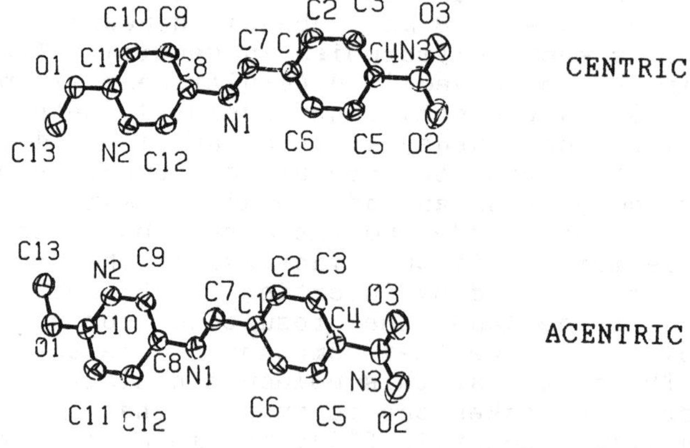

Fig.1 The two different conformations of the two crystal forms of 2-methoxy-N-(4-nitrobenzylidene)-5-pyridylamine.

    In   the   non-centrosymmetric modification the molecule
deviates considerably from planarity. If A and B are the
planes   of   phenyl and pyridyl rings respectively and C the
plane through atoms C1, C7, N1 and C8 then the angles between the normals to planes A, B and C are shown in Table 4.

Table 4. Dihedral angles (°) of the modifications of 2-
methoxy-N-(4-nitrobenzylidene)-5-pyridylamine [16].

	Centrosymmetric	Non-centrosymmetric
Between A and B	19.6	-35.7
Between A and C	3.6	4.8
Between B and C	22.7	-31.8

    Concerning   the bond lengths of the two modifications
and of N-benzylideneanilline, a shortening of the C(1)-C(7)
and N(1)-C(8) distances is observed and a lengthening of
the N(1)-C(7) distance as compared to N-benzylideneaniline.
This trend is attributed to the quinoid resonance forms
that result in intramolecular charge transfer between the
methoxy and nitro groups. It has been argued [17] that the
above mentioned lengths will depend also on the amount of
rotation of the aniline ring about the N(1)-C(8) bond. The
latter will determine the overlap between the lone pair
electrons of nitrogen atom and the π-system of the ring. In
the non-centrosymmetric modification the N(1)-C(8) bond
length has been descreased and the N(1)-C(7) has been
increased with respect to the centric one.
    This might reflect a stronger interaction between the
nitrogen atom lone pair and the pyridine π-system because
of the rotations of the latter.

4. REFERENCES

1. Chemla, D., Oudar, J.L. and Zyss, J.(1981) "Molecular
   engineering for modern optics", L´echo des RECHERCHES,
   47-60.
2. Badan, J., Hierle, R., Perigaud, A and Zyss, J. (1983)
   "Nonlinear   Organic Crystals: Theoretical Concepts,
   Materials, and Optical Properties", in D.J.   Williams
   (ed.), Nonlinear Optical Properties of Organic and
   Polymeric Materials,   ACS   Symposium Series   233,
   Washington D.C., pp. 81-107.
3. Hadjoudis, E., Vittorakis, M. and Moustakali-Mavridis,
   I.(1987) "Photochromism and Thermochromism of Schiff
   Bases in the Solid State and in Rigid Glasses",
   Tetrahedron 43, 1345-1360.
4. Nicoud, J.F.(1988) "Molecular and Crystal Engineering
   for Organic Nonlinear Optical Materials", Mol. Cryst.

Liq. Cryst. Inc. Nonlin. Opt., 257-268.

5. Filipenko, O.S., Shigorin, V.D., Ponomarev, V.I., Atovmyan, L.O., Safina, Z.Sh. and Tarnopol´skii, B.L. (1977) "Crystal structure and nonlinear optical properties of monoclinic p-nitro-p´-methylbenzylideneaniline", Sov. Phys. Crystallogr. 22, 305-309.

6. Leslie, T.M., Demartino, R.N., Won Choe, E., Khanarian, G., Haas, D., Nelson, G., Stamatoff, J.B., Stuetz, D.E., Yoon, H-N. (1987) "Development of Polymeric Nonlinear Optical Materials", Mol. Cryst. Liq. Cryst. 153, 451-477.

7. Moustakali-Mavridis, I. Hadjoudis, E. and Mavridis, A. (1978) "Crystal and Molecular Structure of Some Thermochromic Schiff Bases" Acta Cryst. B34, 3709-3715.

8. Moustakali-Mavridis, I., Hadjoudis, E. and Mavridis, A. (1980) "Crystal and Molecular Structure of Thernochromic Schiff Bases II", Acta Cryst. B36, 1126-1130.

9. Kurtz, S.K. and Perry, T.T. "A Powder Technique for the Evaluation of Nonlinear Optical Materials", J. Appl. Phys. 39, 3798-3813.

10. Meunier-Piret, J., Piret, P., Germain, G. and Van Meerssche, M. (1972) "Sructure Cristalline de la 4-Nitro-4´-Methoxy-N-Benzylideneaniline", Bull. Soc. Chim. Belges 81, 533-538.

11. Nakai, H., Shiro, M., Ezumi, K., Sakata, S. and Kubota, T. (1976) "The crystal and Molecular Structures of p-Nitrobenzylidene-p-dimethylaminoamiline and p-Dimethylaminobenzylidene-p-nitroaniline", Acta Cryst. B32, 1827-1833.

12. Nakai, H., Ezumi, K. and Shiro, M. (1981) "The Structures of Polymorphs of N-(p-Dimethylaminobenzylidena)-p-nitroaniline", Acta Cryst. B37, 193-197.

13. Ponomarev, V.I., Filipenko, O.S., Atovmyan, L.O., Grazhulene, S.S., Lempert, S.A. and Shigorin, V.D. (1977) "Polymorphism of p-nitro-p´-methylbenzylideneaniline", Sov. Phys. Crystallogr. 22, 223-225.

14. Bar, I. and Bernstein, J. (1982) "Conformational Polymorphism. 5. Crystal Energetics of an Isomorphic System Including Disorder", J. Phys. Chem. 88, 243-248.

15. Moustakali-Mavridis, I. and Hadjoudis, E. (1988) "Structure of 2-Methoxy-N-(4-nitrobenzylidene)-5-pyridylamine). II. Non-centrosymmetric Modification", Acta Cryst. C44, 1039-1041.

16. Moustakali-Mavridis, I., Terzis, A. and Hadjoudis, E. (1987) "Structure of 2-Methoxy-N-(4-nitrobenzylidene)-5-pyridylamine", Acta Cryst. C43, 1793-1796.

17. Burgi, H.B. and Dunitz, J.D. (1971) "Molecular Conformafation of Benzylideneanilines", Helv. Chim. Acta 54, 1255-1260.

TRIPLET EXCITON-POLARONS IN POLYDIACETYLENE SINGLE CRYSTALS

H. SIXL
Hoechst AG, Angewandte Physik,
Postfach 80 03 20
D-6230 Frankfurt am Main

and

W. RÜHLE
Physikalisches Institut
Universität Stuttgart
Pfaffenwaldring 57
D-7000 Stuttgart 80

ABSTRACT. Experimental evidence for mobile triplet-exciton-polarons in polydiacetylene single crystals is given by optically detected magnetic resonance experiments. The temperature dependence and the magnetic field dependence of the triplet state fine structure has been analyzed. The wavefunction of the π-electron triplet state is calculated on the basis of the precisely determined experimental fine structure tensor.

1. Introduction

Up to now polydiacetylene single crystals are the only known macroscopic polymer single crystals with perfectly arranged parallel polymer chains. The individual chains are kept at a distance of about 7 Å due to their large side groups. The polydiacetylene single crystals represent the only crystalline model system for conjugated polymers, thus opening the possiblity to evaluate the wavefunction of triplet-state elementary excitations appling Electron Spin Resonance (ESR) techniques. These triplet states are strongly related to charge carrying polarons and bipolarons.

The first observations of transient excitations were published by Orenstein, Etemad and Baker (1984) and by Hattori, Hayes and Bloor (1984). The triplet nature of these excitations was proven by Robins, Orenstein and Superfine (1986) applying magnetic fields and microwave transitions. Further information concerning the triplet kinetics was given by our pulsed ESR experiments after excimer laser excitation (Winter, Grupp, Mehring and Sixl (1987)).

J. L. Brédas and R. R. Chance (eds.), Conjugated Polymeric Materials:
Opportunities in Electronics, Optoelectronics, and Molecular Electronics, 457–470.
© 1990 Kluwer Academic Publishers. Printed in the Netherlands.

This contribution gives an exact characterization of triplet excited states on the conjugated polymer chains. Due to the low stationary concentration of triplet states conventional ESR is not applicable. The experimental method used in our experiments is the very sensitive optically detected magnetic resonance using the transient absorption as a detector of the resonances. The complete temperature dependency and the line shape analysis give evidence for a mobile triplet exciton-polaron state of the chains. The origin of the triplet state fine structure is based on a triplet diradical state. The triplet state wavefunction is related to the π-electron distribution on the carbene and dicarbene states observed during the solid state polymerization of diacetylene crystals (Sixl (1984), Sixl at al (1985), Kollmar et al (1987)).

2. Experimental

All experiments were performed with fully polymerized diacetylene single crystals of poly 2,4-hexadiin-1,6-diol-para-toluolsulfonate which has been partially deuterated. Only the CH_2-side groups of this polymer have been deuterated.

The monomer crystals were crystallized from solution at the university of Stuttgart by Tuffentsammer and Haas, split into 0.5 mm thick platelets in the (100)-plane and thermally fully polymerized at $60^{\circ}C$ for two days.

Figure 1 shows the structure of the two polymer chains A and B within the unit cell. R are the side groups starting with CD_2 between the chain and the toluol sufonate. The molecular z-axis is identical with the crystal b-axis. As shown by Robins and Orenstein (1986) the fine structure z_A- and z_B-axes deviate slightly from the crystallographic b-axis.

The polydiacetylene crystals show a transient absorption at 909 nm (Figure 2) due to triplet-triplet absorption which is obtained during optical excitation (360 nm) of the crystals as shown schematically in Figure 3. The ESR-transitions between the triplet sublevels in zero field and in a magnetic field (see later in Figure 5) are detected by the change of the transient absorption. Due to different selection rules for the decay of the triplet states a change of the triplet sublevel population (caused by microwave transitions) changes the total population of the excited triplet states and thus influences the intensity of the transient triplet-triplet absorption which is proportional to the total number of triplets.

ESR transitions were measured by a conventional X-band ESR spectrometer operating at 9.4 GHz. For a sensitive detection of the change of the transient absorption look-in modulation techniques were applied.

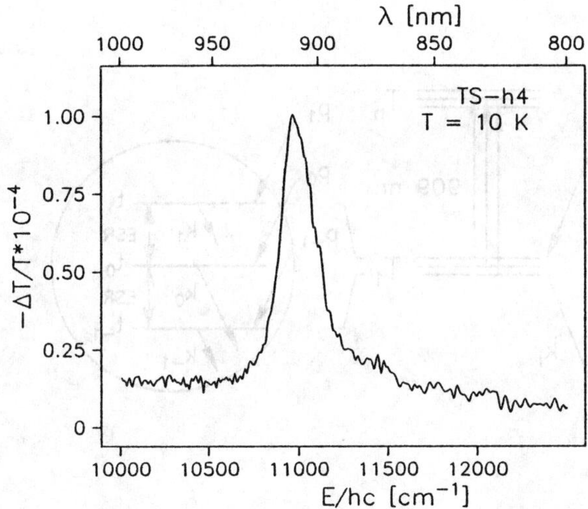

Figure 1. Orientation of the two chains A and B within the unit cell of diacetylene crystals. The side groups are denoted by R. The principal axes of the fine structure tensors are given z_A and z_B.

Figure 2. Transient absorption of a fully polymerized polydiacetylene crystal. The change in transmission $\Delta T/T$ is extremely small.

UV excitation was performed with a mercury high pressure arc (Osram HBO 500), a water filter and a UG 11 (Schott) filter. IR-exitation was done by a slide projector arc (24 V, 150 Watt) with filters RG 695 and 830, respectively.

3.   Experimental Results

3.1 MAGNETIC FIELD DEPENDENCE OF THE TRANSIENT ABSORPTION

In these experiments the transient absorption has been detected by modulation of the UV-excitation (75 Hz). As shown in Figure 4 in all situations there is a decrease of the absorption intensity as a function of the magnetic field. In addition in the directions $B||z$ and $B||x$ a resonance is observed due to antilevel crossing. The origin of these effects is the absence of spin lattice relaxation and a selective decay of the triplet sublevels as pointed out earlier (Winter et al (1987)).

In a magnetic field the zero-field spin functions become mixed, thus combining long lived states with short lived states. Due to the reduction of the lifetime the stationary concentration of the triplets is reduced, causing a decrease of the triplet-triplet absorption. At the antilevel crossing point the relevant spin states are perfectly mixed. This situation is shown in Figure 5 for a magnetic field parallel to the x-axis. The energy level diagram has been calculcated using the fine structure data of Winter et al. (1987).

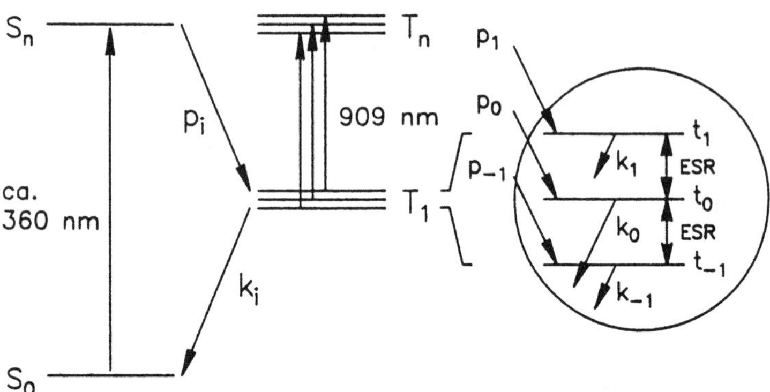

Figure 3.   Energy level diagram of the excitation pathways and the method of the optical detection of the magnetic resonance. The excitation is performed within the singlet states (S). The absorption is detected within the triplet states (T). In the circle the triplet manifold is enhanced in scale to show the relevant population (p), decay (k) and ESR-transition.

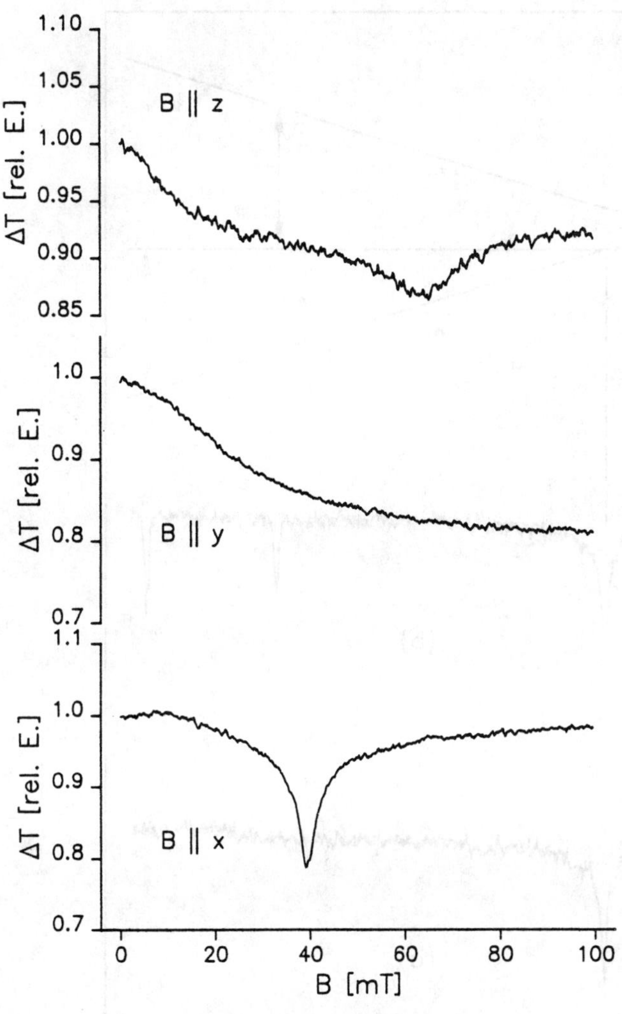

Figure 4.   Magnetic field dependence of the transient absorption in polydiacetylene at 10 K. To increase the signal to noise ratio the curves are averaged over 10 to 20 scans.

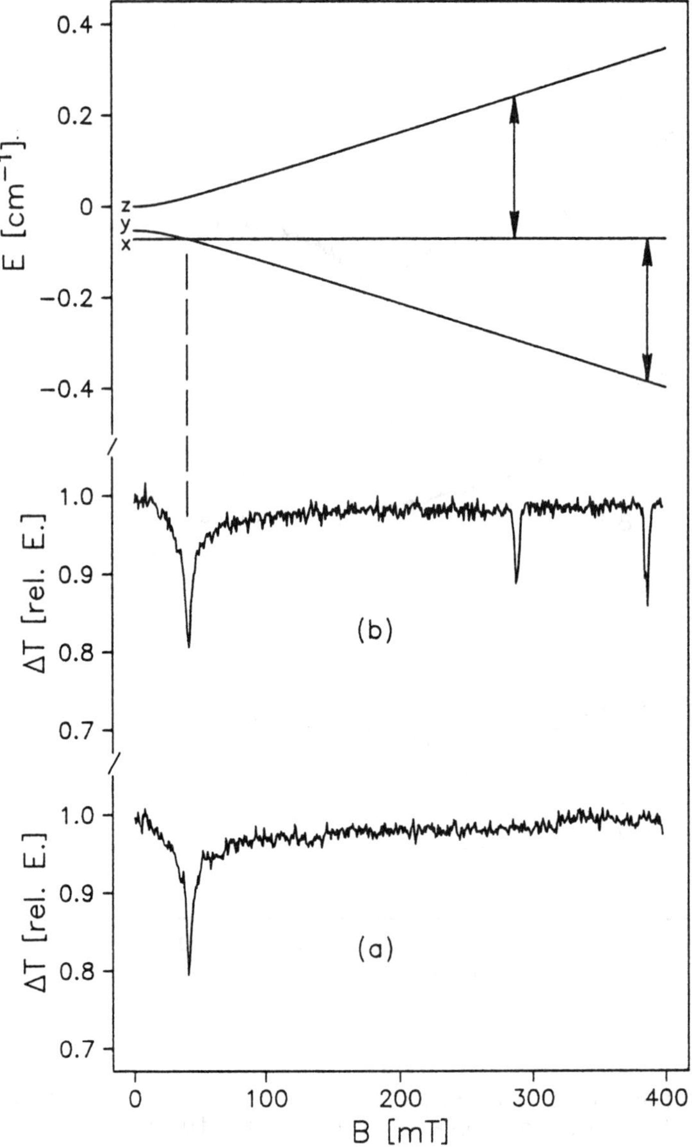

Figure 5.   Energy level diagram of the triplet state with magnetic field B parallel to the x-axis which is perpendicular to the zig-zag plane of the polymer chain. The characteristic signals in the (a) and (b) spectra of the triplet-triplet transient absorbtion are given by the level anticrossing and by the ESR transitions.

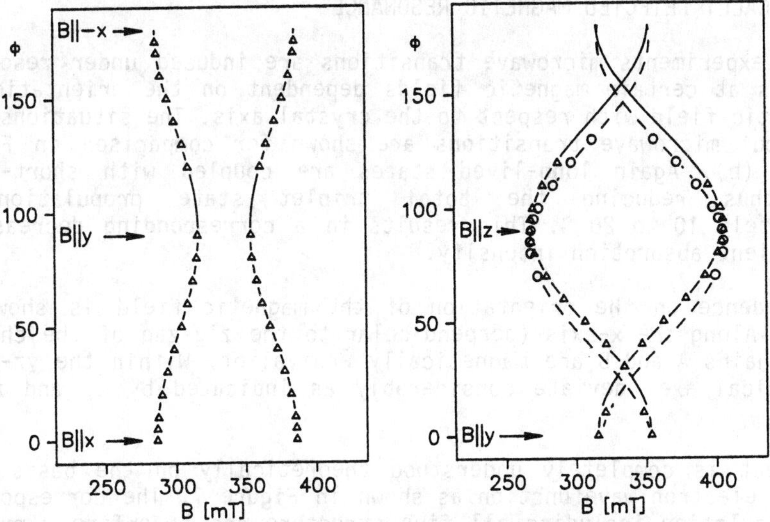

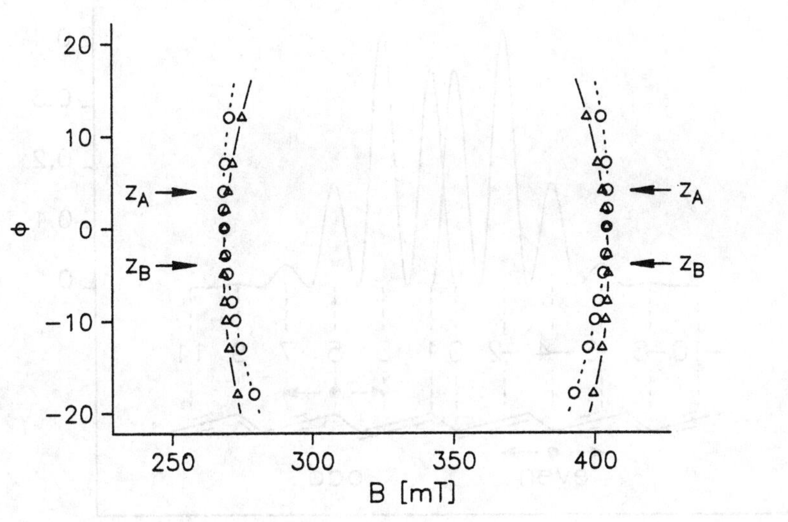

Figure 6.   Angular dependence of the resonance fields at 10 K. The calculated curves have been fitted to the experimental points. In x direction the A and B signals are identical. In y-direction the z_A and z_B axes deviate from the polymer chain direction (z = b see Figure 1) by an angle of $\pm$ 4 °.

3.2 OPTICALLY DETECTED MAGNETIC RESONANCE

In these experiments microwave transitions are induced under resonance
conditions at certain magnetic fields dependent on the orientation of
the magnetic field with respect to the crystal axis. The situations with
and without microwave transitions are shown for comparison in Figure
5(a) and (b). Again long-lived states are coupled with short-lived
states thus reducing the total triplet state propulation by
approximately 10 to 20 %. This results in a corresponding decrease of
the transient absorption intensity.

The dependence on the orientation of the magnetic field is shown in
Figure 6. Along the x-axis (perpendicular to the zig-zag of the chains)
the two chains A and B are magnetically equivalent. Within the yz-plane
the principal axes deviate considerably as indicated by z_A and z_B in
Figure 6.

This effect is completely understood theoretically on the basis of a
diradical electron wavefunction as shown in Figure 7. The corresponding
model calculation including all fine structure and hyperfine structure
data has been already published (Kollmar et al. (1988)).

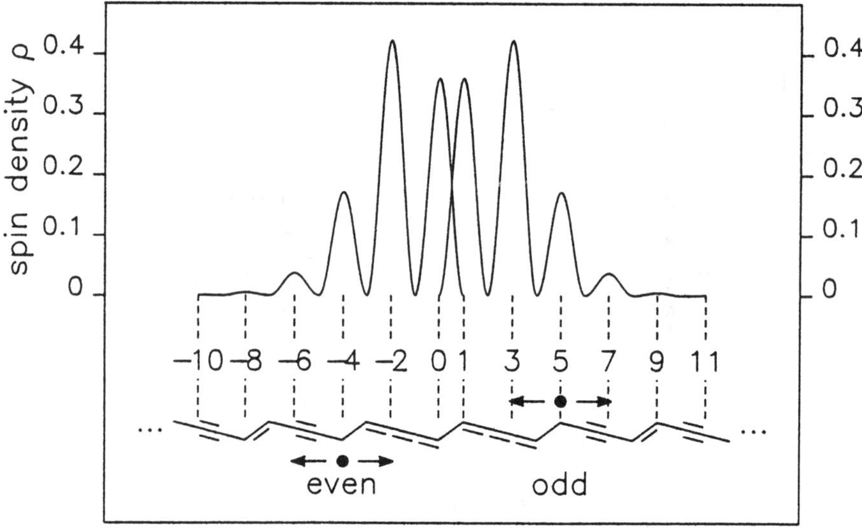

Figure 7. Triplet state electron wavefunction showing the spin density
distribution of the radical electron pair as calculated by Kollmar et
al. (1988) on the basis of the fine structure data using a configuration
model.

3.3 TRIPLET-ESR-LINE SPLITTING

For a sensitive detection of the optically detected magnetic resonance instead of the UV-modulation the microwave intensity was modulated during continuous UV and IR irradiation. Therefore with the exception of the ESR transition a constant base line independent of magnetic field effects is observed. The resulting ESR spectrum is shown in Figure 8. In this orientation the ESR lines are most intense. In addition the A and B chains of Figure 1 are magnetically equivalent. From the spectra it is obvious that there exists an additional substructure which has been partially resolved by spreading the scale as shown in the 6 K spectrum of Figure 9. The triplet ESR spectra can be detected up to 240 K as seen from Figure 9. The low field and high field transitions of the triplet state show a mirror symmetric substructure. The individual transitions are split into four lines at low temperatures. At 60 K the number of lines is reduced to two, which finally merge into one line at 200 K. This last transition is well known also from optical spectra, where a line splitting of the excition absorption is observed around 190 K. It is well known from the structural data of Enkelmann (1983) that the diacetylene polymer crystals undergo a phase transition between 180 and 200 K. Below the transition temperature the benzene rings of the toluene

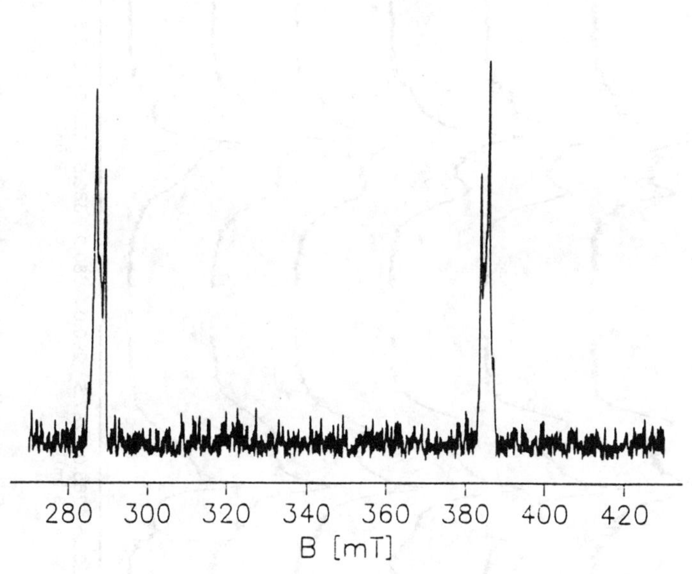

Figure 8.   Lock-in detected triplet ODMR spectrum using microwave modulation with magnetic field parallel to the x-axis and T = 10 K.

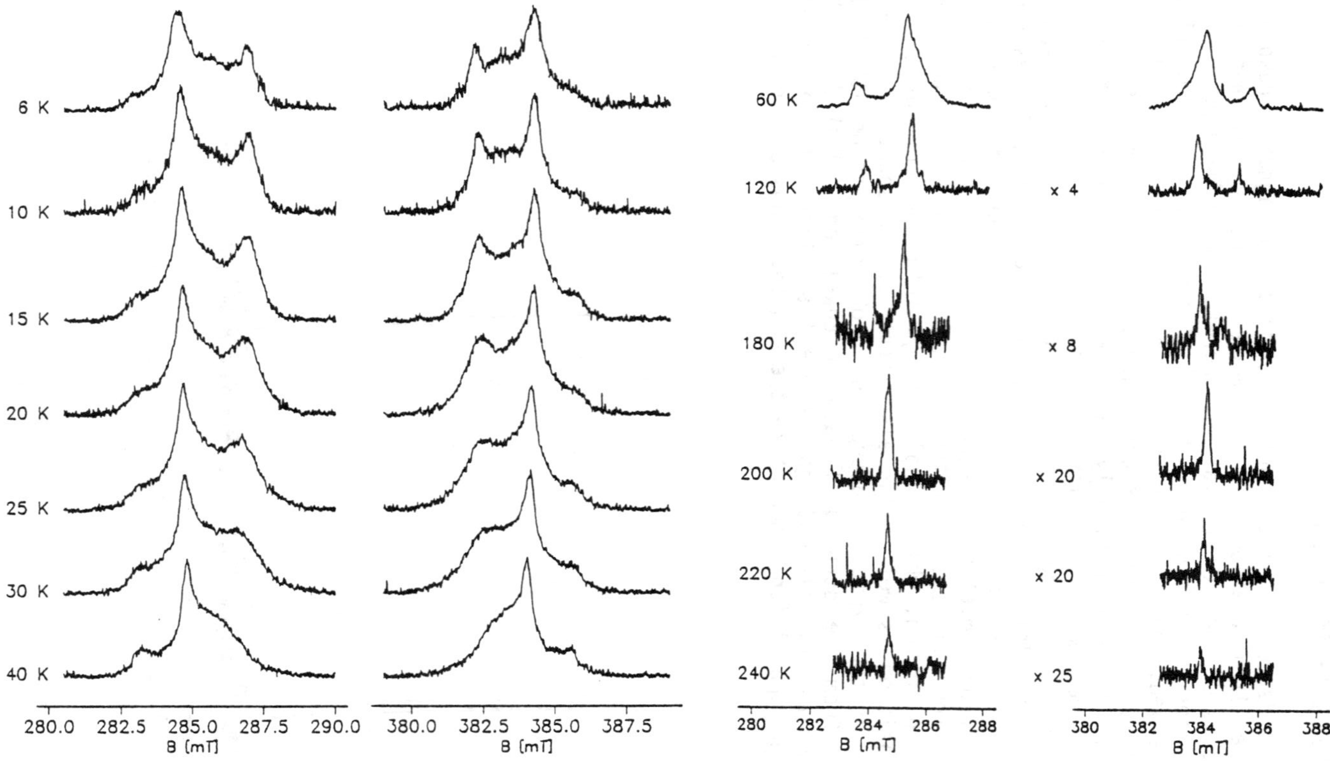

Figure 9. Temperature dependence of the triplet ODMR spectrum with low-field transition on the left and high-field transition on the right. The individual transitions show a four line spectrum at low temperature, a two line spectrum at 60 K and a one line spectrum above 200 K.

side groups shown in Figure 10 by Nowak (1989) are rotated by a certain angle in plus respectively minus direction. As a consequence the A and B chains of Figure 1 are surrounded by toluene rings rotated either in plus or minus direction. Therefore these chains feel a different interaction because in the plus situation the orbitals of the toluene ring become approximately orthogonal to the orbitals in the zig-zag backbone of the chains, whereas in the minus situation they become more aligned and therefore increase the interaction. The change in the D value of the triplet state is only about two percent originating from a two percent change in the average quadratic distance of the two radical electrons within the triplet state. Therefore this effects seems to be nicely understood. Unfortunately there is no clear explanation for the additional splitting of the lines below 60 K. Crystallographic investigations below 60 K would be very helpful.

3.4 DOUBLETS

Very surprisingly S = 1/2 states at g = 2 have been observed in the optically detected magnetic resonance experiments, which are present in all crystals. They are especially pronounced using FBS-crystals, where the toluene part of the side group is substituted by a fluorobenzene group. The doublet lines especially increase dramatically when switching off the UV-irradiation. This effect is shown in Figure 11. Another new observation is the fact that it is obviously possible to populate the triplet state with red light.

4.   Triplet State Population Pathways

In principle there are four routes to populate the triplet state:

Directly:	$S_0 \longrightarrow T_1$	possible
Intersystem Crossing:	$S_0 \longrightarrow S_1 \rightsquigarrow T_1$	possible
Fission:	$S_0 \longrightarrow S_i \rightsquigarrow 2T_1$	very efficient
Electron-hole excitation:	$S_0 \longrightarrow h^+ + e^- \rightsquigarrow T_1$	not probable

The fission process is spin allowed and consistent with all experimental observations, expecially with the excitation spectra and spin selection rules deduced from the polulation mechanisms ($p_x = p_y = p_z$). However, it is clear from the experiments that (with a lower efficiency) the direct route and the intersystem crossing route are present too although they are spin forbidden. The excitation route via electron-hole generation is not very probable because it leads to inconsistent selection rules for the population of the triplet state.

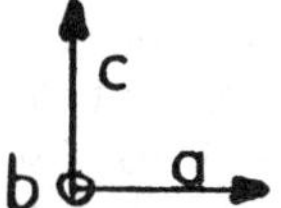
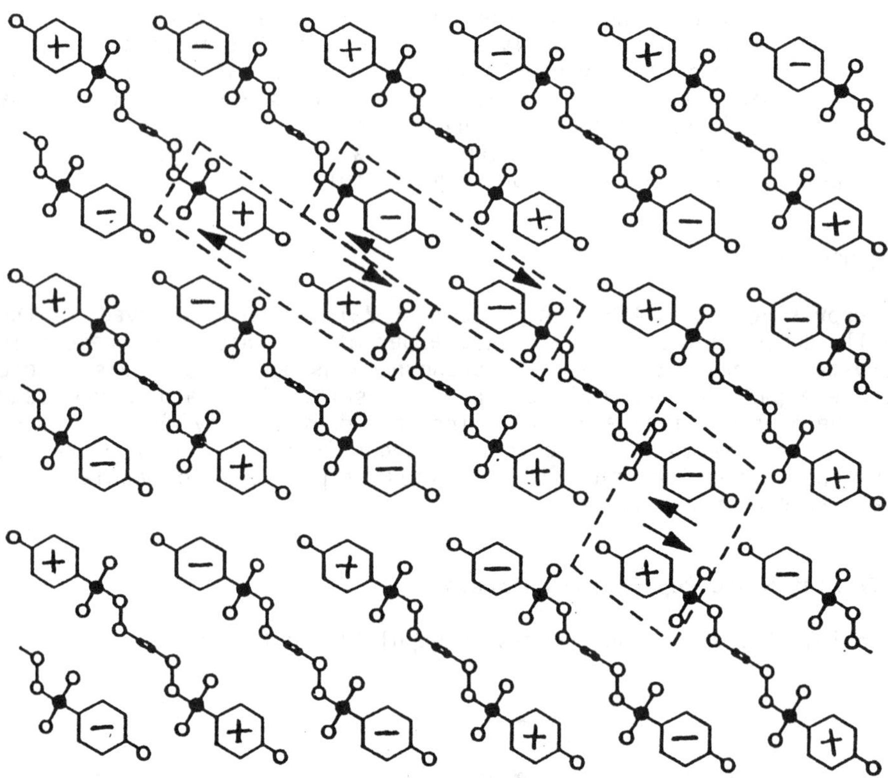

Figure 10. Crystallographic view along the polymer chains as worked out by M. Novak (1989). Above the phase transititon the toluene rings are almost parallel to the ac-plane. Below the phase transition the "plus" rings become approximately parallel whereas the "minus" rings are rotated away from the parallel orientation.

5. Motional Narrowing

The hyperfine interactions in polydiacetylene crystals are well-known. A model calculation of Kollmar et al. (1988) clearly shows that the experimentally determined line width of all triplet state spectra, which is 4 Gauss, is at least a factor of two smaller than the theoretically expected linewith of 10 G, calculated on the basis of the hyperfine data and the spin densities, which are deduced from the fine structure data.

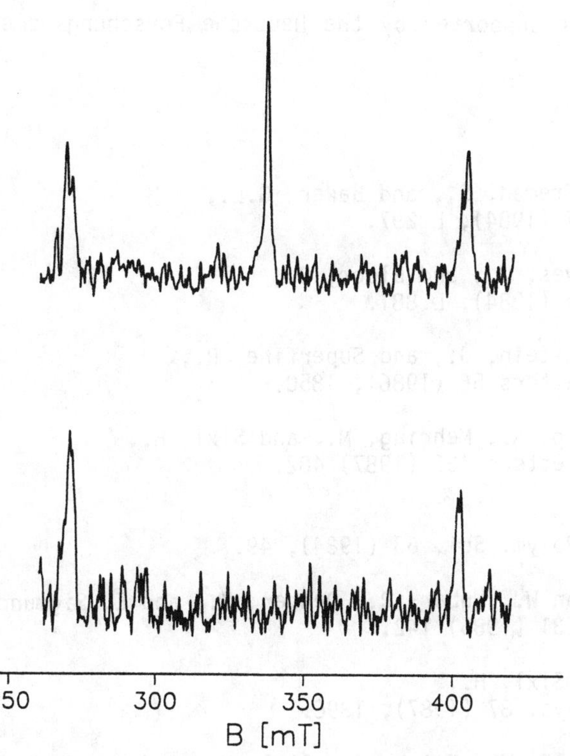

Figure 11. ODMR spectra using different excitation condition. The upper spectrum showing the doublet line is obtained without any UV contribution (the mercury arc has been switched off). Only the red light of the halogen lamp with wavelength larger than 695 nm is used. The lower spectrum is obtained with UV light.

In addition substitution of the CH_2 groups by CD_2 groups makes no significant effect of the line width. Therefore it seems to be obvious that the ESR lines are motionally narrowed. Nevertheless saturation effects clearly show that the lines are still inhomogeneously broadened, due to the fact that the triplet states are trapped on individual chains. No hopping between the chains is observed, otherwise the individual orientations of the A and B chains and the splitted lines A(+), A(-), B(+), B(-) at low temperature would be averaged.

Acknowledgements

The work has been supported by the Deutsche Forschungsgemeinschaft (SFB 329).

REFERENCES

Orenstein, J., Etemad, S., and Baker, G.L.,
    J. Phys. C 17 (1984), L 297.

Hattori, T., Hayes, W., and Bloor, D.,
    J. Phys. C 17 (1984), L 881.

Robins, L., Orenstein, J., and Superfine, R.,
    Phys. Rev. Letters 56 (1986), 1850.

Winter, M., Grupp, A., Mehring, M., and Sixl, H.,
    Chem. Phys. Letters 133 (1987) 482.

Sixl, H.,
    Advances in Polym. Sci. 63 (1984), 49.

Sixl, H., Neumann W., Huber, R., Denner, V., and E. Sigmund,
    Phys. Rev. B 31 (1985) 142.

Kollmar, C. and Sixl, H.,
    J. of Chem Phys. 87 (1987), 1396.

Kollmar, C., Rühle, W., Frick, J., Sixl, H., and von Schütz, J. U.,
    J. Chem. Phys.89 (1988), 55.

Enkelmann, V.,
    Makromol. Chem. 184 (1983), 1945.

Novak, M., this conference (1989).

Diffraction by Holographic Gratings in Diacetylene Crystals

Th. Vogtmann, H.-D. Bauer, Irene Müller, and M. Schwoerer

Physikalisches Institut and Bayreuther Institut für Makromolekülforschung (BIMF);
Universität Bayreuth, P.O. Box 101251, D-8580 Bayreuth, Federal Republic of Germany.

Abstract: The topochemical solid-state UV-photopolymerization of diacetylenes has been used to write holographic gratings in macroscopic TS6 single crystals. For visible light (633nm) these gratings act as thick volume phase gratings, showing efficiencies up to 63% and a small angular selectivity halfwidth down to 0.12°, which has been used to store 77 gratings simultaneously within one sample. The dependence of angular selectivity on sample thickness and UV-penetration depth for crystals with different polymer content is discussed and compared to simulations. Sensitivity, resolution, Q- and ρ-factors are given together with an analysis of the modulation amplitudes.

1. Introduction

Diacetylenes are the very example for compounds showing topochemical solid-state polymerization when being treated with pressure, heat or high energy radiation, e.g. X-rays or UV-light: once initiated, this reaction proceeds via a radical chain reaction,

$$n \; R - C \equiv C - C \equiv C - R \quad \xrightarrow{h\nu \; / \; kT} \quad \left[C = C = C = C \right]_n$$

the intermediate states of which have been studied in great detail [1-8]. Some compounds among this group, like TS6,

$$R = -CH_2 - SO_3 - \bigcirc - CH_3 \,,$$

the one treated here, can be grown from solution as macroscopic single crystals transformable quantitatively into polymer crystals. In TS6 all the polymer molecules are oriented with their backbones parallel to the b-axis of this monoclinic crystal. At room temperature the polymerization process of TS6 is a monomolecular reaction, but time-conversion curves, measured gravimetrically by solvent-extraction of unreacted monomer from partially polymerized crystals, don't show a first-order behavior at all: a flat region, the induction period, is followed by a fast increase of conversion, marking the onset of the so-called autocatalytic region. Apparently the main reason for this "s-shape" is the stepwise decrease of a 5% mismatch of monomer and polymer stacking distance in b-direction [9,10]. But as the activation energy has been

471

J. L. Brédas and R. R. Chance (eds.), Conjugated Polymeric Materials:
Opportunities in Electronics, Optoelectronics, and Molecular Electronics, 471–482.
© 1990 *Kluwer Academic Publishers. Printed in the Netherlands.*

found to be constant the kinetic chain length is regarded to be a function of conversion.

Richter et al. showed that it is possible to produce holographic gratings in TS6 diacetylene single crystals [11]. He used a frequency-doubled argon laser (λ_w=257nm) and obtained highly efficient surface phase gratings, due to the small penetration depth of this UV wavelength. We could optimize both the crystal quality and the experimental method to obtain holographic gratings in an efficient and reproducible way. The aim of our present work is to analyze the diffraction features of gratings obtained in this way in order to understand their diffraction properties.

2. Theory

2.1 General

Superposing two fully coherent plane waves in an isotropic medium (Fig.1) leads to an interference pattern with intensity

$$I(r) \sim E_1^2 + E_2^2 + 2E_1E_2\mathbf{s_1}\mathbf{s_2}\cos[(\mathbf{\sigma_1} - \mathbf{\sigma_2})r] \tag{2.1}$$

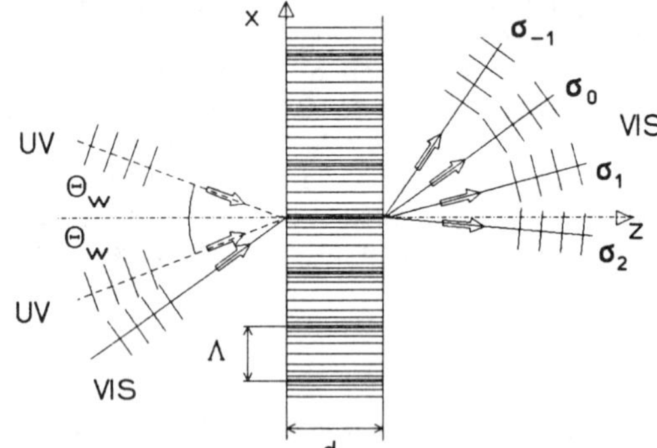

Fig. 1:
Writing and reading a holographic grating: two UV waves produce interference fringes of distance Λ in a medium of thickness d. In general the VIS beam can be diffracted in several diffraction orders.

where E_1, E_2 are the amplitudes, $\mathbf{s_1}$, $\mathbf{s_2}$ the polarization unit vectors and $\mathbf{\sigma_1}, \mathbf{\sigma_2}$ the propagation vectors of both waves, respectively. The resulting vector $\mathbf{\sigma_1} - \mathbf{\sigma_2}$ is the grating vector $\mathbf{K}$. For H-mode polarization ($\mathbf{s_1}$, $\mathbf{s_2}$ perpendicular to the plane of incidence, symmetric incidence) the polarization can be dropped, yielding a scalar grating. Here (2.1) reduces to

$$I(x) = 2I_0[1 + \varkappa\cos(Kx)] \tag{2.2}$$

$\varkappa$ is the contrast of the pattern, reaching its maximum value, 1, for E_1 = E_2. The spacing Λ of the interference fringes is given by the writing wavelength λ_w and the angle of incidence Θ_w:

$$\Lambda = \lambda_w/(2\sin\Theta_w) = 2\pi/K \tag{2.3}$$

For E-mode ($\mathbf{s_1}$, $\mathbf{s_2}$ in the plane of incidence, E_1 = E_2) we arrive at a vector grating:

$$I(x) = 2I_0[1+\cos(2\Theta_w)\cos(Kx)] \tag{2.4}$$

where Θ_w is the superposition angle taken inside the medium.

2.2 Coupled-wave approach

The general framework we used for analyzing our holographic gratings is the coupled-wave approach as it had been introduced by Kogelnik [12] and by Magnusson and Gaylord [13, 14]. The grating is assumed to be represented by a periodic variation of refractive index n and absorption coefficient α:

$$n(x) = n_0 + \sum_{h=1}^{\infty} [n_{ch}\cos(Khx) + n_{sh}\sin(Khx)]$$ (2.5)

$$\alpha(x) = \alpha_0 + \sum_{h=1}^{\infty} [\alpha_{ch}\cos(Khx) + \alpha_{sh}\sin(Khx)]$$

Inside the medium the amplitudes E_i of the diffracted waves (orders) i superpose to yield a total field

$$E(z) = \sum_{i=-m}^{m} s_i E_i(z)\exp(-j\sigma_i r)$$ (2.6)

where σ_i is the propagation vector, s_i the polarization unit vector of order i and $j = \sqrt{-1}$ (Fig. 1). The coupling of these waves is described by

$$\sigma_{\pm i} = \sigma_0 \pm iK$$ (2.7)

Putting (2.5-2.7) into the wave equation, Magnusson and Gaylord arrive at

$$c_i \frac{dE_i}{dz} + (\alpha_0 + j\vartheta_i) E_i + \frac{j}{2}\sum_{h=1}^{\infty} \{ E_{i-h}C_h s_{i-h}s_i + E_{i+h}D_h s_{i+h}s_i \} = 0$$ (2.8)

with

$$c_i = \cos\Theta - jK\cos\Phi/\beta_0 \qquad \vartheta_i = (\beta_0 - \sigma^2)/2\beta_0 \qquad \beta_0 = 2\pi n_0/\lambda$$

$$A_h = n_{ch} - jn_{sh} \qquad\qquad C_h = (2\pi/\lambda)A_h - jB_h$$

$$B_h = \alpha_{ch} - j\alpha_{sh} \qquad\qquad D_h = (2\pi/\lambda)A_h^* - jB_h^*$$

a system of differential equations they solved numerically with the help of a Runge-Kutta algorithm for several kinds of diffraction scenarios. Φ is the tilt angle between K and the z direction.

Assuming a pure volume sine grating with only one diffracted order present (Bragg case) this formalism can easily be simplified and one arrives at the case described by Kogelnik [12]: the diffraction efficiency of a transmission grating can now be given analytically as a function of the grating thickness coordinate z:

$$\eta(d) = E_1(d)E_1^*(d) = \left[\sin^2\left(\frac{\pi n_1 d}{\lambda\cos\Theta_0}\right) + \sinh^2\left(\frac{\alpha_1 d}{2\cos\Theta_0}\right)\right]\exp\left(-\frac{2\alpha_0 d}{\cos\Theta_0}\right)$$ (2.9)

In this relation the $\sin^2$-term describes a pure phase grating diffraction, whereas the $\sinh^2$-term describes an amplitude grating diffraction. The exponential term takes into account an overall absorption, Θ_0 is the Bragg angle for the readout wavelength λ, given by

$$\lambda_w/(2\sin\Theta_w) = \lambda/(2\sin\Theta_0)$$ (2.10)

The kind in which we write holographic gratings results in an even profile function and therefore the sin-terms in (2.5) can be neglected. Usually, the resulting K vector

474

points in x direction, therefore $\Phi = \pi/2$. With N Fourier coefficients taken into account (2.5) reads

$$n(x) = n_0 + \sum_{h=1}^{N} n_h \cos(hKx), \qquad \alpha(x) = \alpha_0 + \sum_{h=1}^{N} \alpha_h \cos(hKx) \qquad (2.11)$$

which, for a finite number of orders, m, can be written as a matrix equation:

$$\mathbf{S}'(z) = \mathbf{M}\,\mathbf{S}(z), \qquad (2.12)$$

where the vector $\mathbf{S}$ is given by

$$\mathbf{S}(z) = (E_m, E_{m-1}, \dots , E_0, E_{-1}, \dots , E_{-m})$$

and the matrix $\mathbf{M}$ by

$$M_{i,i} = -(\alpha_0 + j\vartheta_i)/\cos\Theta \qquad M_{i,i\pm h} = -(2\pi n_h/\lambda - j\alpha_h)\mathbf{S}_{i\pm h}\mathbf{S}_i/\cos\Theta$$

This differential equation can be solved:

$$\mathbf{S}(z) = \mathbf{S}(0)\exp(\mathbf{M}z) \qquad \mathbf{S}(0) = (\,0,\,0,\,\dots,\,E_0=1,\,0,\,\dots,\,0\,) \qquad (2.13)$$

Having found the eigenvalues ζ of the matrix $\mathbf{M}z$, the eigenvalues of $\exp(\mathbf{M}z)$ are given by $\exp(\zeta)$. These calculations can easily be implemented on a computer, providing efficiency values for a given grating in a very quick way, especially necessary for data fitting problems.

In the theory outlined above grating profiles are assumed not to vary with thickness z. This assumption, in general, does not meet the experimental situation, where the finite penetration depth l of the writing light rules the photoproduct distribution, which results in a substitution of $n_1 d$ and $\alpha_1 d$ in (2.9):

$$n_1 d \;\rightarrow\; \int_0^d n_1(x,z)dz \qquad \alpha_1 d \;\rightarrow\; \int_0^d \alpha_1(x,z)dz \qquad (2.14)$$

For a simple exponential decay of n_1 and α_1 with z n_1 and α_1 have to be substituted by their average values $\bar{n}_1$ and $\bar{\alpha}_1$. The matrix $\mathbf{M}$ in (2.12) also can be varied to take into account a finite penetration depth, which will not be discussed here.

The different refractive indices inside and outside the medium lead to a reflection of light at the surfaces of the sample, which can call for replacing the measured data for efficiency and transmission, η_{exp} and t_{exp}, by the "true" values

$$t = t_{exp}/(1-R)^2, \qquad \eta = \eta_{exp}/(1-R)^2 \qquad (2.15)$$

with the appropriate reflectivity value $R(\Theta,\lambda,\dots)$ for the experiment discussed.

3. Experimental

3.1 The sample

The recipe for growing TS6 single crystals has been described by Wegner et al. [15] and was optimized by Collet, Schott and Müller [1]. To get crystals of approx.

(1) A. Collet, M. Schott, and Irene Müller, private communication.

$1 \times 0.5 \times 0.2$ cm^3 and of good optical quality parameters like purification process, solvent, solvent evaporation velocity and temperature have been varied.

The crystals can be cleaved carefully with a razor blade parallel to the (100) surface, which is a natural growth plane including the polymer chain direction b. The platelets typically range from 20 to 500 µm in thickness. They are fixed on an aperture with a small portion of grease or heat conducting paste. Orientation of these samples is then done under a polarizing microscope with the help of the strong dichroism caused by a low polymer content present in even the very fresh crystals: Fig.2 shows a typical absorption spectrum of a fresh TS6 "monomer" crystal. Two main features are present: The strong monomer absorption in the UV region <350nm and the typical polymer band with its maximum around 570nm. This band grows in intensity and shifts towards longer wavelengths when further polymerization takes place. The reaction is regarded homogeneous, i.e. yielding a solid solution of polymer molecules in a monomer matrix.

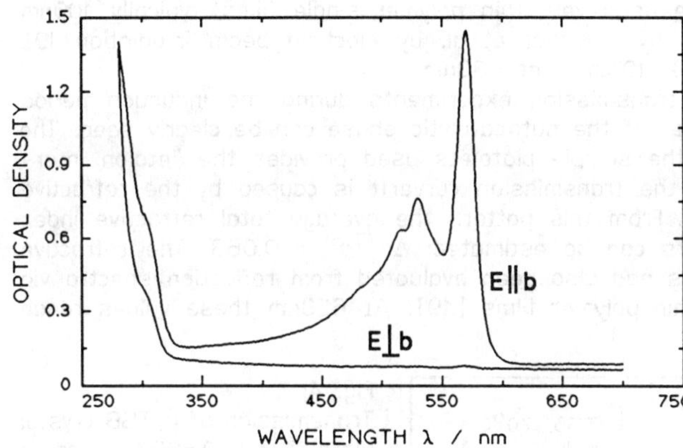

Fig. 2:
Absorption spectrum of a TS6 crystal platelet. Light travels perpendicularly to the (100) surface; polarization as indicated in the figure.

3.2 Setup

The setup we used is shown in Fig.3 and was described in more detail together with its advantages in previous papers [16, 17].

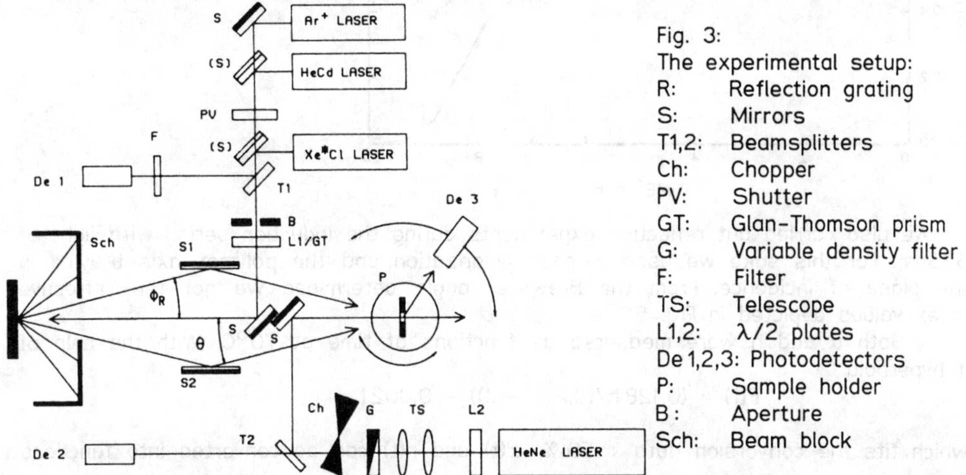

Fig. 3:
The experimental setup:
R: Reflection grating
S: Mirrors
T1,2:   Beamsplitters
Ch: Chopper
PV: Shutter
GT: Glan–Thomson prism
G: Variable density filter
F: Filter
TS: Telescope
L1,2:   λ/2 plates
De 1,2,3: Photodetectors
P: Sample holder
B: Aperture
Sch: Beam block

With a set of reflection gratings R we produced holographic gratings with Λ = 3.7, 3.3, 0.8 and 0.4 µm.

4. Results

4.1 Optical properties

During polymerization the optical properties n and α are known to change dramatically due to the production of molecules with extended π-electron systems. To analyze grating diffraction it is necessary to gain knowledge about these changes.

The absorption of TS6 "monomer" crystals in the visible region due to preformed polymer chains is small, typically $50 cm^{-1}$ (at 633nm, light polarized E||b). The pure polymer value cannot be measured from single crystals by transmission experiments. α values evaluated from reflection spectra via Kramers-Kronig analysis give $\alpha \approx 0.4 \times 10^4 cm^{-1}$ at 633nm [18]. We used very thin polymer single films, typically 100nm thick, which had been fabricated by Berrehar et al. by electron beam irradiation [19]. These films gave $\alpha = (2.4 \pm 0.4) \times 10^4 cm^{-1}$ at 633nm.

Furthermore, we carried out transmission experiments during the induction period of the reaction (Fig. 4). The onset of the autocatalytic phase can be clearly seen. The good optical quality of many of the sample platelets used provides the "etalon interference" pattern superposed to the transmission curve: it is caused by the refractive index change during the reaction. From this pattern the average total refractive index change during the first 7.5 hours can be estimated: $\Delta n^{(7.5)} \approx 0.063$. The refractive index of fully polymerized crystals had also been evaluated from reflection spectra via Kramers-Kronig [18] and using thin polymer films [19]. At 633nm these values range between 3 and 4.5.

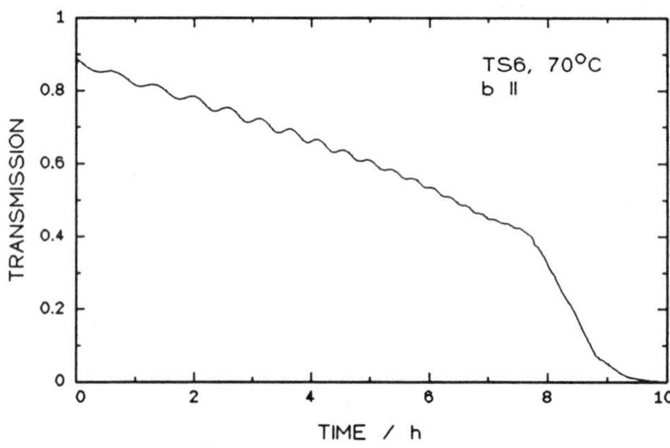

Fig. 4:
Transmission of a TS6 crystal platelet for λ=633nm as a function of time during thermal polymerization at 70°C.

We also carried out reflection experiments during the induction period with light of 633nm. For this sake we used E-mode polarization and the polymer axis b lying in the plane of incidence. From the Brewster angle determined we got the refractive index values depicted in Fig. 5.

Both α and n were measured as functions of time at 70°C. With the help of a hyperbola

$$P(t) = (0.128 h/(9.2 h - t)) - 0.0021$$

which fits the conversion data < 50 %, $\alpha(t)$ and $n(t)$ can be converted into functions

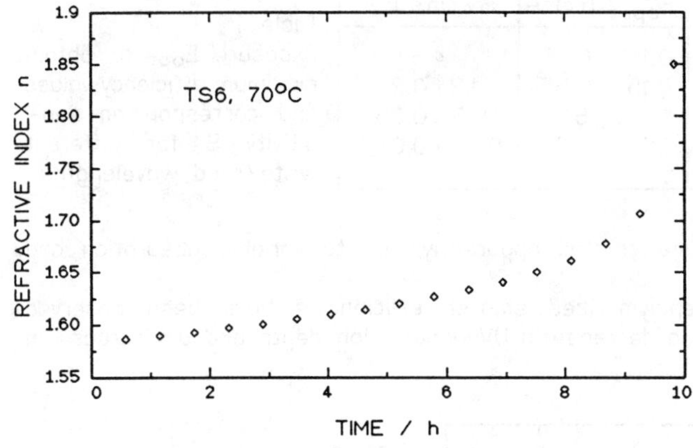

Fig. 5:
Refractive index of a TS6 crystal during thermal polymerization at 70°C, evaluated from Brewster angle measurements.

of polymer content P. Both are not linear at all, not even in the induction period, which is the only region accessible to our experiments.

4.2 Growth curves, sensitivity, efficiencies, and resolution

The exposure needed to reach maximum efficiency, E_{opt}, was determined from holographic growth curves (Fig.6) of the first diffraction order (higher orders hardly did appear). These measurements also provide the holographic sensitivity,

$$S = \sqrt{\eta}/E \qquad (4.1)$$

Values for E_{opt} and S are summarized in Table 1 for different writing wavelengths and for a UV-UV-experiment, too. The experiments were carried out with either H-mode of all polarizations and the b-axis parallel to these polarizations ("b∥ orientation") or E-mode and b lying in the plane of incidence ("b⊥ orientation").

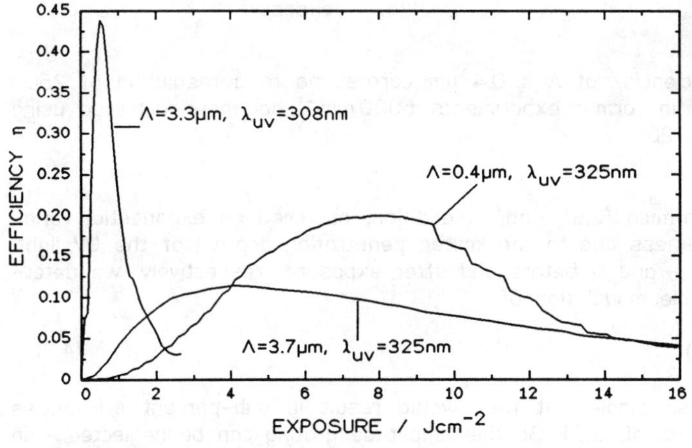

Fig. 6:
Holographic growth curves in TS6 samples for different UV wavelengths and different fringe spacings Λ.

Typical peak efficiencies range from 40 to 50% for Λ = 0.8 μm and 3.3 μm, with maximum values around 60 %. For Λ = 0.4 μm the η values ranged from 20 to 30% usually.

The reproducibility of η was not very good even when using samples of the same

λ_w / nm	λ_r / nm	E_{opt} / Jcm^{-2}	S / cm^{2}J^{-1}
257	633	0.1 ... 0.15	≈ 4.5
308	633	0.15 ... 0.65	1.2 ± 0.2
325	633	1.5 ... 6.0	0.15 ± 0.05
325	325	1.5 ... 6.0	0.03 ± 0.01

Table 1: Exposure E_{opt} to obtain maximum efficiency values and corresponding sensitivity S for different write/read wavelengths

thickness, cleaved from the same crystal, apparently due to variable absorption and crystal inhomogeneities.

With samples thermally prepolymerized smaller efficiencies have been observed (Fig. 7). This is caused by both a decrease in UV penetration depth and an increase in absorption at 633nm.

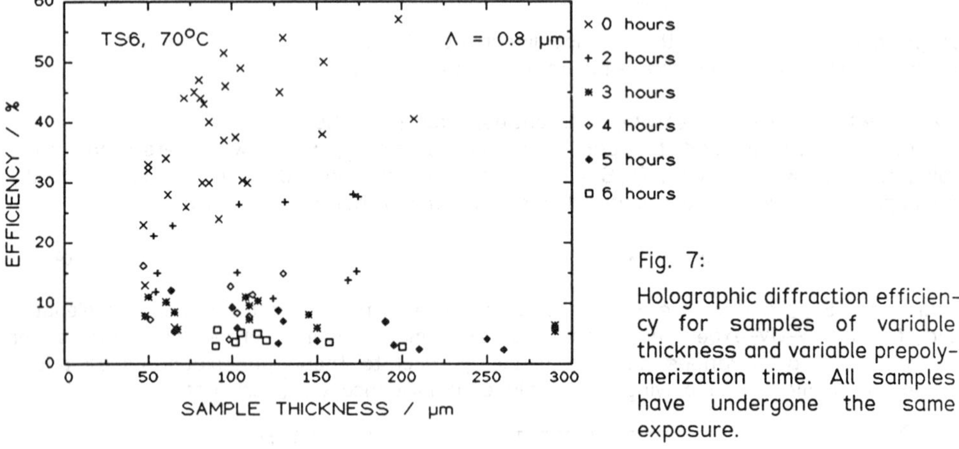

Fig. 7:

Holographic diffraction efficiency for samples of variable thickness and variable prepolymerization time. All samples have undergone the same exposure.

The still very high efficiencies at Λ = 0.4 μm correspond to a resolution of 2500 mm^{-1} with good contrast. In former experiments 5000 mm^{-1} had been achieved using electron beam lithography [20].

4.3 Modulation amplitudes

We used Kogelnik´s formula (2.9), and, in addition, assumed an exponential variation of n and α with thickness due to the limited penetration depth l of the UV light. From transmission values t_0 and t_1 before and after exposure, respectively, we determined the maximum possible modulation of α ,

$$\bar{\alpha}_1 = (1/d)\ln(t_0/t_1) \tag{4.2}$$

The values obtained are so small that they would result in sub-percent efficiencies when put in the sinh^2 -term of (2.9). So this amplitude grating can be neglected - in contrast to the overall absorption α_0,

$$\alpha_0 = -(1/d)\ln(t_1), \tag{4.3}$$

which cannot! In most cases a determination of n_1 using (2.9) could only be performed when taking into account reflection losses, which indicates that the efficiencies measured were close to the possible limit. The reflectivity R had been measured separately before.

Typical values for n_1 and optimum η reached 0.004 for fresh and 0.013 for pre-polymerized samples, respectively.

4.4 Q- factor and ρ-factor

Taking sample thickness as hologram thickness one can now readily calculate

$$Q = \frac{2\pi\lambda_0 d}{\Lambda^2 n_0}$$

for each grating. For $\Lambda = 3.3$ μm Q ranges from approx. 2.5 to 60, for 0.8 μm Q>800 could be reached. These values, together with the high efficiencies clearly indicate thick phase gratings.

Gratings with $\Lambda = 3.3$ μm and Q = 40 also showed a second diffraction order. This is obviously due to the fact that a highly modulated sine grating also can be the origin of higher diffraction orders, as was pointed out by Moharam and Young [21]. They introduced

$$\rho = \frac{\lambda_0^2}{\Lambda^2 n_0 n_1}$$

as a measure, whether a grating should show Bragg ($\rho \gtrsim 10$) or Raman–Nath behavior ($\rho \lesssim 1$). In the case mentioned, $\rho = 6$, lying between the two regimes.

4.5 Angular selectivity

The angular selectivity is the halfwidth $\Delta\Theta$ of the function $\eta(\Theta)$, where Θ is the angle of incidence. We found $\Delta\Theta$ to be a very well reproducible feature of our gratings showing an impressive dependence on sample thickness d for fresh crystals (Fig.8).

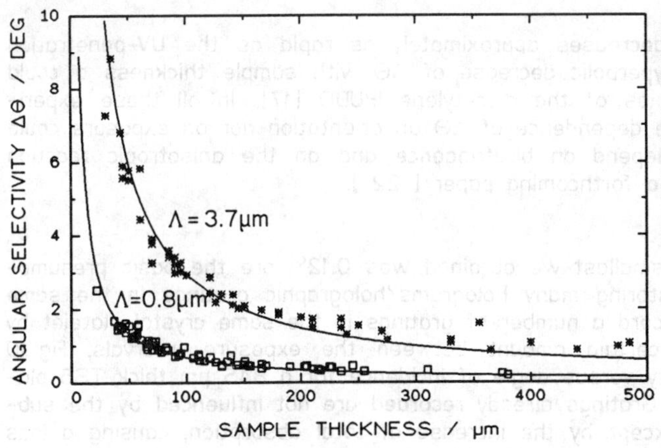

Fig. 8:
Angular selectivity $\Delta\Theta$ as a function of sample thickness d for fresh samples and two different grating distances Λ. For the fitted hyperbola curves see text.

Kogelnik only gives $\Delta\Theta \approx \Lambda/D$ as a rule of thumb, with the hologram thickness D. Magnusson and Gaylord do not give any analytical expression for this quantity. We implemented their coupled-wave approach in matrix form (2.13, 2.16) on a computer to simulate $\eta(\Theta)$ curves. We found a dependence

$$\Delta\Theta(D) = \chi n_0 \Lambda / D \qquad \text{with} \qquad \chi = (0.85 \pm 0.03)\,\text{rad}$$

for the first ascent period of each holographic growth curve. This hyperbola function describes the data in Fig. 8 very well. When using prepolymerized samples to write gratings, a remarkable deviation from this behavior can be observed (Fig. 9). At

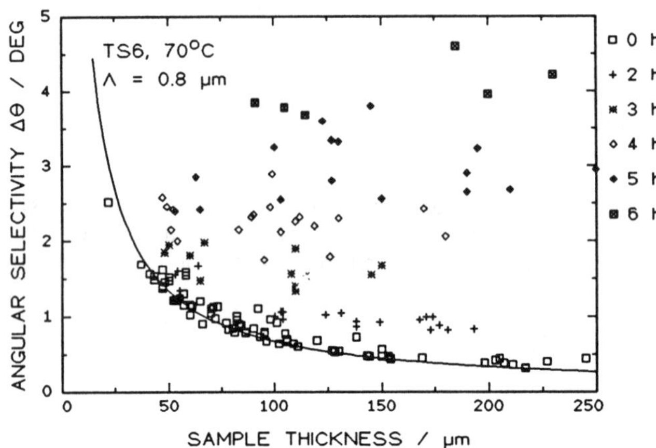

Fig. 9:
Angular selectivity $\Delta\Theta$ for holographic gratings of fringe distance $\Lambda = 0.8\,\mu m$ written in samples of different prepolymerization time.

high prepolymer levels the grating thickness obviously cannot be enlarged by taking thicker samples: the lower UV-penetration depth limits the maximum hologram thickness possible. In Table 2 the effective hologram thickness D is calculated for different prepolymerization times from the $\Delta\Theta$-limit taken from thick samples, $\Delta\Theta_\infty$, via

$$D = \chi n_0 \Lambda / \Delta\Theta_\infty$$

It can be seen that D decreases approximately as rapid as the UV-penetration depth, l, does. The typical hyperbolic decrease of $\Delta\Theta$ with sample thickness d could also be observed with samples of the diacetylene IPUDO [17]. In all these experiments neither any remarkable dependence of $\Delta\Theta$ on orientation nor on exposure could be observed. Effects, that depend on birefringence and on the anisotropic reaction kinetics will be discussed in a forthcoming paper [22].

4.6 Multiple exposure

Small values of $\Delta\Theta$, the smallest we obtained was 0.12°, are the basic presumption for the possibiliby of storing many holograms/holographic gratings in the same medium. We have tried to record a number of gratings in the same crystal platelet by rotating the sample for a certain amount between the exposure intervals. Fig. 10 shows the 1st order efficiency versus angle of incidence for a 845 µm thick TS6 platelet containing 77 gratings. Gratings already recorded are not influenced by the subsequent recording process except by the increase of total absorption, causing a loss of less than 10% in efficiency.

Table 2: Comparison of penetration depth l and hologram thickness D for different prepolymerization times.

Λ/μm	Prepolymerization time at 70°C / h	$\Delta\Theta_\infty$ / deg	l / μm	D/μm
0.8	0	0.25±0.05	420±50	250±50
	2	0.75±0.15	70±15	85±10
	3	1.5±0.3		42±8
	4	2.0±0.5	20±5	31±8
	5	2.7±0.5	16±5	23±5
	6	4.0±1.0		16±6
0.4	0	0.20±0.04	420±50	160±30
	2	0.5±0.1	70±15	63±10
	4	1.5±0.2	20±5	12±5
	6	2.7±0.3		12±3

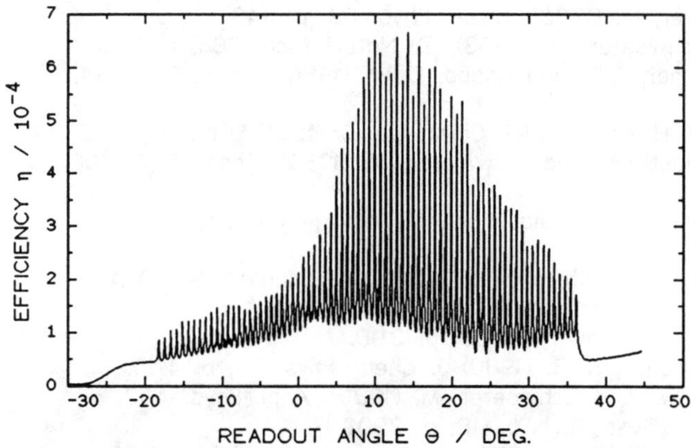

Fig. 10: Multigrating, consisting of 77 individual gratings, stored successively within the same sample. The figure shows the 1st order efficiency vs angle of incidence.

5. Summary

We showed that macroscopic diacetylene single crystals can be used to record high-efficiency holographic gratings using UV laser radiation of a wavelength with sufficient penetration depth. With a modified coupled-wave approach we analyzed the diffraction efficiencies that we observed when using a HeNe laser (633nm) for readout. The polymer conversion caused by the recording process is very small, though index modulations are in the range from 0.001 to 0.01. The absorption modulation is negligible.

Therefore the gratings obtained can be characterized as volume phase gratings with homogeneous absorption. The angular selectivity halfwidth could be shown to be very small in thick samples and was observed to increase with decreasing sample thickness or increasing prepolymerization time. The values obtained for fresh samples were in good agreement with numerical simulations based on the coupled-wave approach by Magnusson and Gaylord [18]. The values for prepolymerized thick samples give a good estimate for the effective hologram thickness achievable, which is approximately the UV penetration depth. Because of this excellent Bragg behavior we managed to store 77 gratings within the same sample. The resolution of our crystals is comparable to the best high resolution emulsions, but the holographic sensitivity is poor and hardly exceeds the one of photorefractive crystals.

The big variety of diacetylene compounds with some amongst, which are much more photosensitive and less thermally polymerizable, should provide noteworthy candidates for irreversible optical data storage, if crystal growth can be further optimized.

Acknowledgement

We thank G. Denninger, H. Hereth and G. Sauer for technical assistance and M. Schott, Paris, for providing the poly-TS6 thin films. This work was supported by the Deutsche Forschungsgemeinschaft, SFB 213/ B2.

References

1 Schwoerer, M., Huber, R. A., and Hartl, W. (1981), Chem. Phys. 55, p. 97.

2 Hartl, W., and Schwoerer, M. (1982), Chem. Phys. 69, p. 443.

3 Niederwald, H., and Schwoerer, M. (1983), Z. Naturforsch. 38a, p. 749.

4 Gross, H., Sixl, H., Fischer, S.F., and Knapp, E. W. (1984), Chem. Phys. 84, p. 321.

5 Neumann, W., and Sixl, H. (1984), Mol. Cryst. Liq. Cryst. 105, p. 41.

6 Müller-Nawrath, R., Angstl, R., and Schwoerer, M. (1986), Chem. Phys. 108, p. 121.

7 Cantow, H. J., editor (1984), "Polydiacetylenes", Adv. Polym. Sci. 63, Springer, Berlin.

8 Bloor, D., and Chance, R. R. editors (1985), "Polydiacetylenes", NATO ASI Series E 102, Nijhoff, Dordrecht.

9 Baughman, R.H. (1978), J. Chem. Phys. 68, p. 3110.

10 Baughman, R.H., and Chance, R. R. (1980) J. Chem. Phys. 73, p. 4113.

11 Richter, K.-H., Güttler, W., and Schwoerer, M. (1983), Appl. Phys. A32, p. 1.

12 Kogelnik, H. (1969), Bell Syst. Tech. J. 48, p. 2909.

13 Magnusson, R., and Gaylord, T. K. (1977), J. Opt. Soc. Am. 67, p. 1165.

14 Magnusson, R., and Gaylord, T. K. (1978), J. Opt. Soc. Am. 68, p. 1777.

15 Wegner, G. (1969), Z. Naturforsch. 24b, p. 824.

16 Kohler, Bryan E., Bauer, H.-D., Kohler, Bern E., Güttler, W., and Schwoerer, M. (1986), Chem. Phys. Lett. 125, p. 251.

17 Bauer, H.-D., Vogtmann, Th., Müller, I., and Schwoerer, M. (1989), Chem. Phys. 133, p. 303.

18 Bloor, D., and Preston, F. H. (1976), phys. stat. sol. a37, p.427.

19 Berrehar. J. Lapersonne-Mayer, C., and Schott, M. (1986), Appl. Phys. Lett. 48, p. 630.

20 Niederwald, H., Seidel, G., Güttler, W., and Schwoerer, M. (1984), J. Phys. Chem. 88, p. 1933.

21 Moharam, M. G., and Young, L. (1978), Appl. Opties 17, p. 1757.

22 Bauer, H.-D., Vogtmann, Th., Müller, I., and Schwoerer, M. (1989), Proceedings of the 4th International Conference on Unconventional Photoactive Solids, Mol. Cryst. Liq. Cryst., special issue, submitted for publication.

SCANNING TUNNELING MICROSCOPY
AT THE POLYMER–METAL INTERFACE

J.P. RABE and S. BUCHHOLZ

Max–Planck–Institut für Polymerforschung
Postfach 3148
D–6500 Mainz
West Germany

ABSTRACT. The Scanning Tunneling Microscope (STM) will be discussed as a tool to investigate the polymer–metal interface on the one hand and to address individual organic molecules on the other hand. Examples will be given for two classes of systems: (i) Single polymer molecules on a conducting substrate (alkylated cellulose on graphite) and (ii) two organic conductors (a radical cation salt and doped polyacetylene) interfaced and imaged by the tip of the STM.

1. INTRODUCTION

The Scanning Tunneling Microscope (STM) provides direct atomic resolution images of conducting surfaces [1]. More general, it may be viewed as a tool to interface a metal tip to single atoms, molecules or aggregates thereof. One may also look at it as a miniature version of a MIM or MIS device, allowing to perform electron spectroscopy on small sample areas down to the molecular length scale. Moreover, since the energy of the tunneling electrons is of the order of chemical binding energies or less, the electrons can either serve as a nondestructive probe or, alternatively, be used for chemical modification on an atomic scale. A review of the application of the STM in surface chemistry has been given recently [2].

483

J. L. Brédas and R. R. Chance (eds.), Conjugated Polymeric Materials:
Opportunities in Electronics, Optoelectronics, and Molecular Electronics, 483–493.
© 1990 *Kluwer Academic Publishers. Printed in the Netherlands.*

While the early STM work was at crystalline inorganic metal and semiconductor surfaces in ultrahigh vacuum, the method does, in principle, neither require any degree of order in the system under investigation nor vacuum conditions. This makes it a versatile tool for a wide range of materials in common ambients, provided they exhibit sufficient electron conductivity and limited molecular mobility at the surface.

For organic materials, in particular two types of systems are accessible for the STM: (i) Ultrathin films of nonconducting organics and (ii) bulk organic conductors. Important for a good understanding of experiments on thin films is a carefull characterization of the substrate material. While most of our experiments have been performed on highly oriented pyrolytic graphite (HOPG), also other substrates like gold and silver films, as well as MoS_2 and PbS have been investigated. In addition the ability to modify a given substrate material in a well controlled way may be of interest. We will demonstrate the local reactive etch of the basal plane of graphite in the STM. As an example for an organic adsorbate we will discuss imaging of individual molecules of alkylated cellulose, transferred onto graphite with the Langmuir–Blodgett technique. For organic conductors two examples are given, namely a radical cation salt, ß–$(BEDT–TTF)_2 I_3$, and oxidized polyacetylene films.

The results presented below were obtained with an STM described earlier [3]. It was operated at room temperature with Pt/Ir and W tips in air. The tunneling current was generally 2 nA. Both, 'Constant Current' and 'Variable Current' (or so called 'Constant Height') images were recorded. In the latter case the tip was scanned at 1 kHz in the x–direction and 40 Hz in the y–direction, corresponding to 10 images per second at a resolution of 100 lines. These current images were stored in real time on video tape. Further digital image processing was performed to remove some high frequency noise and to generate quasi 3–dimensional gray images.

2. ULTRATHIN ORGANIC FILMS

2.1. CONDUCTING SUBSTRATES

The most widely used substrate for STM imaging of organic adsorbates is graphite (HOPG), since it has a number of ideal properties for this purpose: It is atomically perfect and flat over distances of many 100 nm, and it does not form any insulating surface layer in common ambients. It is therefore possible to image HOPG in a broad variety of ambients with atomic resolution. However, its inertness causes the problem that also any deposited film will interact only weakly with the substrate. While in some cases the interaction turns out to be sufficient, it is nevertheless desirable to find alternative substrates. Besides other layer compounds noble metals have been suggested. However, gold and silver films evaporated at room temperature onto glass substrates are relatively rough for application in high resolution imaging of adsorbates. One way to vary the morphology of gold and silver films is the proper choice of the substrate and its temperature during evaporation. Fig. 1 shows STM images of gold and silver evaporated onto freshly cleaved mica at different substrate temperatures. It is apparent that the typical domain size inreases with the substrate temperature from about 5 nm at 100 K to several 100 nm at 650 K. In the high temperature case atomically flat terraces seperated by monoatomic steps can be identified on some islands. However, ridges of several nanometers depth always occur between the islands. Therefore, although these substrates exhibit atomically flat terraces up to 100 nm diameter, the ridges on a 100 nm scale may be a limitation for thin adsorbate film studies.

Noteworthy, gold surfaces which are flat to within about 0.1 nm on a micron scale can be prepared by melting a short wire into a ball within a gas flame and quenching it in air. Continuous tunneling in the STM for a some time appears to recrystallize it locally. Incidentally, surface melting and recrystallization of metallic glasses has been suggested before for information storage purposes [7].

Furthermore, we have imaged cleaved and fractured surfaces of MoS_2 and PbS at atomic resolution. However, the perfection and flatness is inferior to the above mentioned substrates.

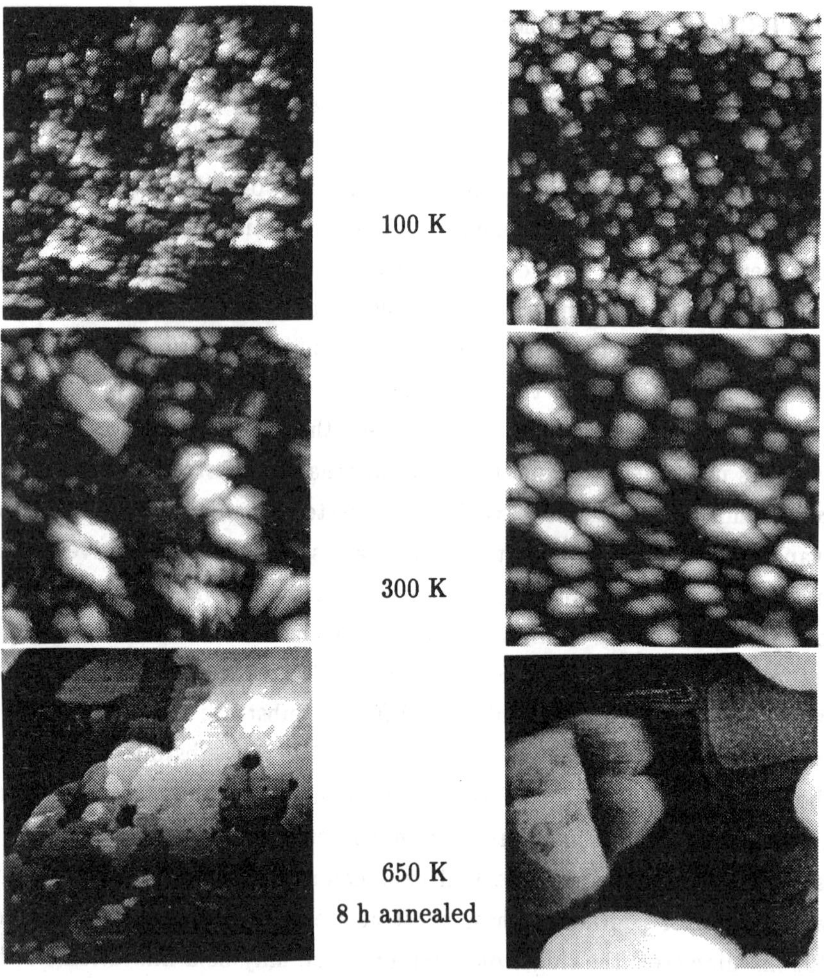

Fig. 1: STM images of gold (left column) and silver (right column), evaporated onto mica at different substrate temperatures. The image sizes are 250 nm x 250 nm. The brightness is proportional to the height with full contrast corresponding to about 3 nm for the samples prepared at 100 K and 300 K and 5 nm and 20 nm for gold and silver, respectively, evaporated at 650 K. On the high temperature images atomically flat terraces, separated by monoatomic steps can be identified.

2.2. SUBSTRATE MODIFICATION:
TIP INDUCED REACTIVE GRAPHITE ETCH

While the adhesion problem is particularly severe for single small molecules, it appears that ordered thin layers can be more readily imaged. For example, at the interface between HOPG and a number of liquid crystalline materials, images of highly ordered adsorbate layers have been obtained [4,5]. However, often one has to wait some time before they would appear, suggesting the possibility of a lack of nucleation sites.

One way to create such sites in situ is to etch the substrate surface locally with the STM. Indeed, we have demonstrated [6] that stable holes as small as one monolayer deep and a few nanometers in diameter can be etched into the basal plane of HOPG in a number of fluid ambients, including one of the above mentioned liquid crystalline materials, octylcyanobiphenyl. The reaction has been observed at negative tip bias between -2.5 V and -4V. Below a well defined threshold bias of -1.7 V high resolution images of the unperturbed HOPG are obtained, while at the threshold bias itself metastable adsorbate products with lifetimes on the order of 1 s are observed [6]. The similarity of these characteristics for several undried organic fluids on the one hand, and the inertness of dry toluene or helium atmospheres on the other hand indicate a common etch mechanism, possibly involving rest water in the fluids.

The modification of surfaces with the STM has been suggested as a means for information storage. The holes on HOPG may be also of interest in this respect since they can be produced highly reproducibly and are extremely stable. Moreover, HOPG is a very inert material, ideally suited for STM imaging.

2.3. ORGANIC ADSORBATES

Due to a lack of electronic conductivity of organic adsorbates on the one hand and their molecular mobility on the other hand, only few systems have been successfully imaged by STM to date, among them individual phthalocyanine molecules on copper [8] and a few polymers on graphite [3,9,10]. This lack of systematic experimental data contributes to an additional difficulty, namely the lack of a good understanding of the contrast mechanisms in STM imaging of organics.

In order to minimize both problems, mobility and conductivity, we have chosen to examine a ribbon polymer [6]. In this case even a weak interaction energy per monomer may sum sufficiently to permit an individual molecule to adsorb strongly enough for stable imaging. Moreover, if the diameter of the backbone is below 1 nm it is thin enough for a sizable tunnel current to pass through the molecule.

Laurylmethyl– and ethyl cellulose monolayers have been prepared on graphite by the Langmuir–Blodgett technique with a transfer ratio which is, however, smaller than unity [6,11]. Accordingly, with the STM (i) bare graphite, (ii) islands of a monolayer of mostly parallel rods, or (iii) single isolated polymer molecules were observed [6]. The imaged polymers are extended over many ten nanometers. Perfectly straight segments are displaced by characteristic kinks. The fact that both, laurylmethyl– and ethyl cellulose molecules are found to be highly extended is consistent with the lyotropic behaviour of many cellulose derivatives. However, it must be remembered that for the STM samples the polymers have been applied to graphite by the Langmuir–Blodgett technique, which involves particular surface– and orientation forces. The highly extended chain can, therefore, only be attributed to the adsorbed molecule.

3. ORGANIC CONDUCTORS

The charge transport in conjugated polymers is still a matter of debate. Part of the difficulty in obtaining reliable experimental data on the molecular conduction mechanism is the complicated morphology of the materials, which cause an ill defined contact between polymer molecules or bundles with the external circuitry. In this respect the STM may open new ways since it offers the possibility to address well defined parts of the sample. However, in order to test the concept on organic conductors, a well defined crystalline material will be discussed first.

3.1.   A RADICAL CATION SALT: ß–(BEDT–TTF)$_2$ I$_3$

The ß–form of the iodine salt of bis(ethylenedithio)tetrathiafulvalene is an organic conductor at room temperature and becomes superconducting below 1.6 K at ambient pressure. It crystallizes into a triclinic unit cell with well known parameters. In Fig. 2 STM images of the ab–plane of ß–(BEDT–TTF)$_2$ I$_3$ are displayed. The low resolution image (Fig. 2a) shows flat terraces and steps which are about 1.5 nm high, corresponding to a molecular monolayer. The stability of the image in air is not as good as in the examples discussed above, i.e. some material is scraped away during scanning. In fact, Fig. 2a shows a hole which had been formed during consecutive scanning. Clearly the damage becomes larger during scanning, indicating that it is not caused by pure chemical degradation, but that also the mechanical strength of the material comes into play.

On a crystalline surface high resolution images can still be obtained since the scraping procedure prepares fresh surfaces over again. Fig. 2b shows a high resolution image from the center of Fig. 2a, with the unit cell clearly visible. Also there is some more detailed structure which, however, is difficult to assign without any further consideration of the electronic structure of the material. A more detailed analysis is under way.

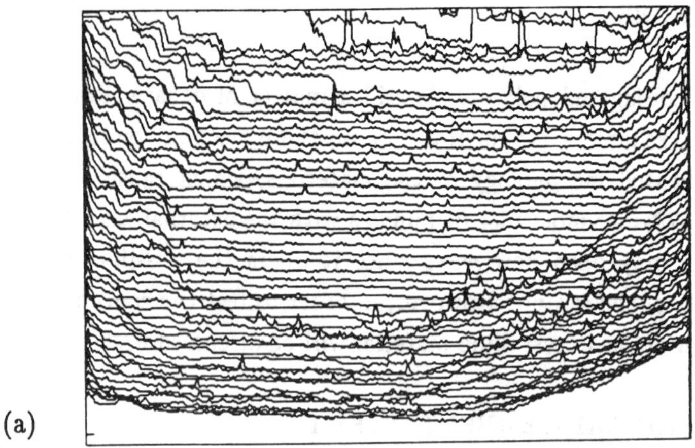

(a)

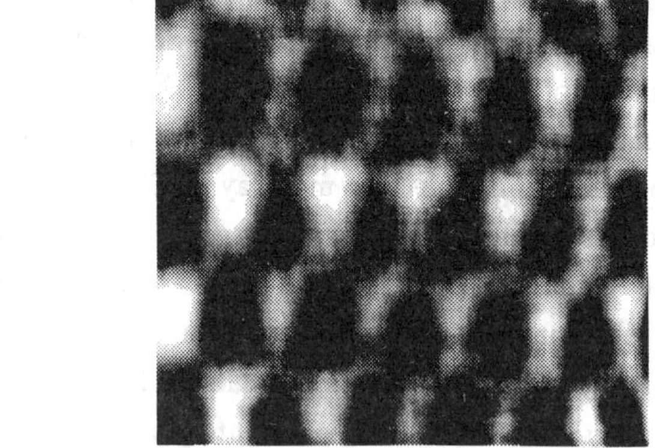

(b)

Fig. 2:

STM images of the ab–plane of ß–(BEDT–TTF)$_2$ I$_3$.

(a) Quasi 3–dimensional display of a 'Constant Current' image. The image size is 100 nm x 25 nm and the mark at the lower left corner corresponds to 1 nm in height.

(b) Top view 'Variable Current' image with the brightness proportional to the current. The image size is 3.5 nm x 3.2 nm.

3.2. POLYACETYLENE

Fig. 3 shows an STM image of an iodine doped polyacetylene film on a glass slide. After the initial tip approach an ongoing inward–creep of the tip was observed for about an hour, indicating that some compression of the film did occur. The image of Fig. 3 was obtained after the creep had stalled. Still, the reproducibility of small details in the image was not perfect from scan to scan. For the reasons given above the image cannot be considered as the image of the unperturbed surface but rather of a somewhat processed film. The image exhibits areas with parallel stripes, 5 nm to 10 nm in width, which may be bundles of polymers, indicating that the STM provides a means to address single polyacetylene fibrils. However, more experiments on better characterized samples are necessary to substantiate this and, if possible, make use of it for some local spectroscopy.

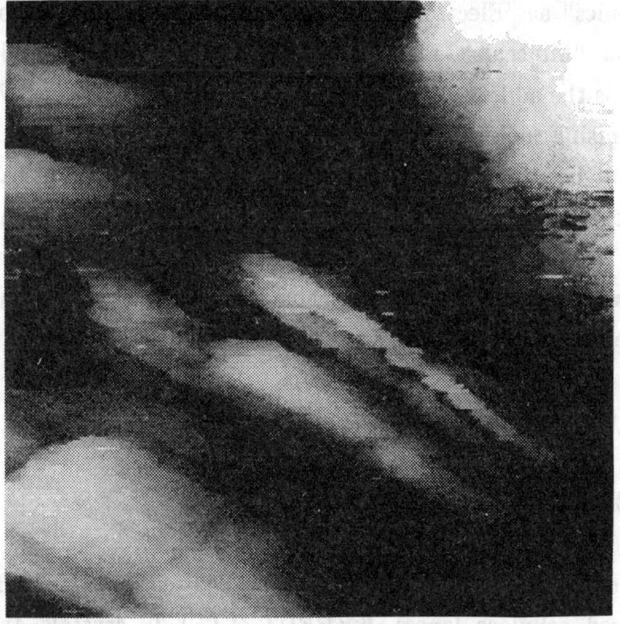

Fig. 3:
STM image of iodine doped polyacetylene. The image size is 250 nm x 250 nm. The brightness is proportional to the height with full contrast corresponding to about 20 nm in height.

492

4. CONCLUSIONS

The results given above have demonstrated the ability of the STM to image single organic molecules and to chemically modify the surface of a molecular solid on a nanometer scale. This means that individual molecules can be addressed by the metallic STM tip, or, in other words, interfaced to the external electronic circuitry. It also shows that organic materials can be chemically altered on a nanometer scale. What still remains to be shown is that any useful spectroscopic information like, e.g., I–V–curves or inverse photoemission spectra can be obtained for organic materials much like for inorganic metals and semiconductors.

With this the idea [12,13] of building molecular scale devices and manipulating on a nanometer scale becomes more of a reality. Indeed, in the development of "Molecular Electronics" as "Electronics on a Molecular Level" (cf. Report of the Working Group on "Molecular Electronic Prospects", these Proceedings) an important problem is the addressing of any molecular wires or switches. While the STM is a very promising tool in the development of such systems, it is a different issue whether it will be very useful in any actual device, since it is based on a fairly slow, i.e. mechanical concept. On the other hand, STMs with dimensions of 1000 x 200 x 8 μm^3 have been made by planar microfabrication techniques [14], indicating that some parallel processing scheme appears feasable, which may alleviate also this problem.

ACKNOWLEDGEMENT

The authors are very grateful to A.M. Ritcey, D. Schweitzer and J. Halim for providing derivatized cellulose layers, ß–(BEDT–TTF)$_2$ I$_3$ crystals and doped polyacetylene films, respectively. The work has been supported by the Bundesministerium für Forschung und Technologie under the title "Ultrathin Polymer Layers" 03M4008E9 and by the European Science Foundation (Additional Activity: Chemistry and Physics of Polymer Surfaces and Interfaces).

5. REFERENCES

[1]   G. Binnig, H. Rohrer, C. Gerber, E. Weibel,
       Phys. Rev. Lett. 50 (1983) 120.

[2]   J.P. Rabe,
       Angew. Chem. Int. Ed. 28 (1989) 117 and 1127.

[3]   J.P. Rabe, M. Sano, D. Batchelder, A.A. Kalatchev,
       J. Microscopy (Oxford) 152 (1988) 573.

[4]   J.S. Foster, J.E. Frommer,
       Nature 333 (1988) 542.

[5]   J.K. Spong, H.A. Mizes, L.J. LaComb Jr., M.M. Dovek, J.E. Frommer,
       J.S. Foster,
       Nature 338 (1989) 137.

[6]   J.P. Rabe, S. Buchholz, A.M. Ritcey,
       J. Vac. Sci. Techn. A, Jan./Feb. 1990.

[7]   U. Staufer, R. Wiesendanger, L. Eng, L. Rosenthaler, H.–R. Hidber,
       H.–J. Güntherodt,
       J. Vac. Sci. Technol. A 6 (1988) 537.

[8]   P.H. Lippel, R.J. Wilson, M.D. Miller, C. Wöll, S. Chiang,
       Phys. Rev. Lett. 62 (1989) 171.

[9]   T.R. Albrecht, M.M. Dovek, C.A. Lang, P. Grütter, C.F. Quate,
       S.W.J. Kuan, C.W. Frank, R.F.W. Pease,
       J. Appl. Phys. 64 (1988) 1178.

[10]   R. Yang, K.M. Dalsin, D.F. Evans, L. Christensen, W.A. Hendrickson,
       J. Phys. Chem. 93, 511 (1989).

[11]   A.M. Ritcey and G.Wenz (to be published) and
       A.M. Ritcey and G.Wegner (to be published).

[12]   R.P. Feynman,
       Engin. Sci. (Febr. 1960) p.22.

[13]   C. Schneiker, S. Hamaroff, M. Voelker, J. He, E. Dereniak, R. McCuskey,
       J. Microscopy 152 (1988) 585.

[14]   S. Akamine, T.R. Albrecht, M.J. Zdeblick, C.F. Quate,
       IEEE Electron Devices Letters (to be published).

NLO COEFFICIENTS OF POLYENES: SIZE AND ALTERNATION DEPENDENCE

Z.G. Scos, G.W. Hayden, and P.C.M. McWilliams
Department of Chemistry
Princeton University
Princeton, N.J. 08544-1009 U.S.A.

ABSTRACT. The static polarizability and second hyperpolarizability
of N-site polyenes are found for noninteracting (Hückel) and
interacting (Pariser-Parr-Pople) chains with alternating transfer
integrals $t(1 \pm \delta)$. Saturation with increasing N is shown to depend
on $\delta > 0$ and contributions from unlinked clusters are shown to cancel.
Reduced NLO coefficients are introduced to compare electron-electron
correlations in quantum cell models with identical optical gaps E_g,
rather than identical δ.

I. INTRODUCTION

The nonlinear optical (NLO) response of conjugated polymers is
large, fast, and frequency dependent [1]. Enhanced NLO response is
due to electron delocalization along the polymer, which reduces
excitation energies and increases transition moments. As discussed
throughout the workshop, conjugated polymers are excellent candidates
for a variety of important applications. Many materials issues
remain under consideration. A major challenge for theorists is to
connect the observed NLO responses of molecular and polymeric systems
with their electronic structure. Excited electronic states, in
particular, must be treated as accurately as possible.

Theoretical studies of NLO properties fall into two broad
classes. Quantum chemical approaches, whether semiempirical or ab
initio, increasingly treat all electrons in a molecule, or at least
all valence electrons. Solid-state models are typically restricted
to π-electrons and deal with larger systems. Molecular and model
calculations emphasize different aspects of the problem. The
polarizability of small hydrocarbons generally increases with larger,
more flexible bases that describe the perturbed orbitals more
accurately [2]. Quantum chemical results using molecular orbitals
(MOs) converge from below. Small molecular size mitigates the
effects of charge correlations in virtual states. In π-electron
models, by contrast, the atomic orbitals are fixed by hypothesis and

495

J. L. Brédas and R. R. Chance (eds.), Conjugated Polymeric Materials:
Opportunities in Electronics, Optoelectronics, and Molecular Electronics, 495–508.
© 1990 Kluwer Academic Publishers. Printed in the Netherlands.

496

only their coefficients change in electric fields. MO or band
results now overestimate the response by neglecting charge
correlations in extended systems [3].

The present discussion is restricted to quantum cell models.
Frontier orbitals are parametrized to various solid-state
properties. Such models are inherently finite dimensional and
amenable to exact analysis for noninteracting electrons. They also
have long histories, starting with Hückel theory for conjugated
molecules and tight-binding models for simple metals [4]. While
π-electron models are generally meant for conjugated polymers, very
similar treatments apply to σ-conjugation in polysilane (PS) polymers
and to ion-radical and charge-transfer organic solids.

The proper choice of π-electron models has also been extensively
discussed. The Su-Schrieffer-Heeger [5] approach to polyacetylene
(PA) involves electron-phonon coupling and noninteracting π-electrons,
as in Hückel theory. A general and unified picture is achieved,
provided that new parameters are used for each polymer and that some
correlation corrections are added on. Hubbard or Pariser-Parr-Pople
(PPP) models, by contrast, deal with interacting π-electrons. PPP
theory is preferred for conjugated molecules since most low-lying
states are reasonably given by a common set of parameters [4]. Such
transferability is essential for models that attempt quantitative
predictions. Models with interacting π-electrons are naturally more
difficult to apply to polymers.

The relevant quantum cell models for π-electrons are typically
based on a $2p_z$ orbital at each carbon. Since transfer integrals and
other parameters are defined rather than evaluated, however, any
frontier orbital may be used. Two sp^3 hybrids are natural for
σ-bonding along the PS backbone. A finite basis necessarily leads to
a finite number of excited states. Diagrammatic valence-bond (DVB)
techniques [6] yield exact static and dynamic NLO coefficients by
including all virtual states for systems with $N \leq 12$ orbitals. PPP,
Hubbard, or any other spin-independent interactions may be postulated.
Such models use the zero-differential-overlap (ZDO) approximation
[7], which is convenient rather than necessary for DVB theory.

Coulomb interactions among ionic C^+ and C^- sites are given in
PPP theory by the Ohno formula [8],

$$V(R) = e^2/(R_0^2 + R^2)^{1/2} \tag{1}$$

The on-site repulsion of two electrons in a $2p_z$ orbital in C^- is $V(0)$
= U = 11.26 ev, a value taken from atomic data. This corresponds to
R_0 = 1.28 A. No adjustable parameter is then needed in PPP theory
once the geometry of a hydrocarbon has been specified. The larger
sp^3(Si) orbitals suggests a smaller $V(0)$, which is consistent with
lower ionization potential for Si, but such parametrization has not
been tested. Adjustable on-site U's are used in Hubbard models,

partly to follow correlation effects and partly to model many types of frontier orbitals.

Finite polyenes and PA have long been prototypical of conjugated systems and will remain so for NLO properties. We discuss here the polarizability and NLO response of quantum cell models of the simplest polymer, all-trans polyacetylene (PA). Finite polyenes in Fig. 1 have alternating transfer integrals $t(1 \pm \delta)$. We consider the dependence of NLO coefficients on the polyene length N, on the alternation δ, and on Coulomb interactions $V(R)$ among π-electrons. We also suggest that comparisons are more appropriate between models with identical observables than with identical microscopic parameters. For instance, Hückel and PPP comparisons for PA should involve parameters leading to the same optical gap, E_g.

II. REDUCED NLO COEFFICIENTS: STATIC POLARIZABILITY

Hückel models for conjugated molecules were parametrized to the lowest dipole-allowed excitation. The corresponding quantity for conjugated polymers is the optical gap $E_g = 4t\delta$ for noninteracting electrons. Since NLO coefficients depend on the optical gap, correlation contributions to E_g in Hubbard or PPP models naturally reduce the response. The relevant experimental comparisons are among models that have comparable E_g, even if this entails different alternation or other microscopic parameters. Extensive tuning or reparametrization of models is premature at the present, early stages of NLO applications. Reduced NLO coefficients offer a very simple method for comparing models with different E_g.

We define dimensionless reduced NLO coefficients that already contain the optical gap E_g and the mean bond length, a. The procedure is elementary: every transition moment $\langle R|\mu_x|S\rangle$ is scaled as ea, every excitation energy as E_g, and every frequency as $f = \hbar\omega/E_g$. A pth-order susceptibility contains p+1 transition moments and p frequency-dependent energy denominators. The linear response is the polarizability $\alpha_{ij}(\omega)$, with i, j = x, y, z.

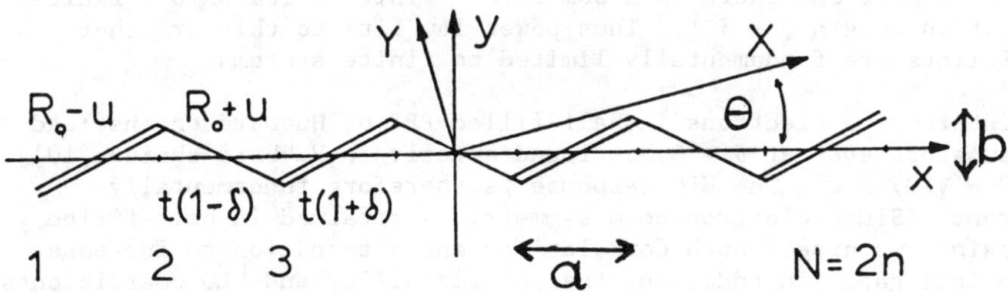

Fig. 1. Schematic representation of N-site *trans* polyenes with bond lengths $R_0 \pm u$, transfer integrals $t(1 \pm \delta)$, and principal axes X, Y for the static polarizability.

498

The reduced static polarizability per site, for example, is

$$\tilde{\alpha}_{ij}(0) = \alpha_{ij}(0) \, E_g \, / \, N(ea)^2 \tag{2}$$

$$= \sum_P 2E_g \langle G|\tilde{\mu}_i|P\rangle\langle P|\tilde{\mu}_i|G\rangle/N(E_P-E_G)$$

The sum is over the excited states P of H, with excitation energy
E_P - E_G from the ground state. The reduced dipole displacement
operators are

$$\tilde{\mu}_i = (\mu_i - \langle G|\mu_i|G\rangle) \, /er_i \tag{3}$$

with r_x = a and r_y = b in Fig. 1. The dipole operator μ is a
site-diagonal one-electron operator in the ZDO approximation used in
quantum cell models. There is no contribution in (2) from P = G. In
the special case of *trans* polyenes, i, j are in the molecular plane
in Fig. 1 and there is no ground-state dipole moment. Dipole allowed
transitions connect the singlet $1A_g$ ground state to mB_u singlets.
The choice of H then completely specifies the static polarizability
of the N-site polyene in Fig. 1. Approximations for the eigenstates
G and P result in approximate polarizabilities.

Hameka and coworkers [9] studied the static polarizability of
trans polyenes in the usual Hückel case of δ = 0 in Fig. 1. They
find $\alpha_{xx}(0)$ to increase as N^a, with a = 5.3. Now $\alpha_{xx}(0)/N$ diverges
in the limit of an infinite chain. The energy gap in (2) vanishes
and perturbation theory breaks down for a half-filled metallic band.
In the molecules of interest, however, finite-size gaps are
sufficiently large to make detailed variations of t(R) with R
unimportant. Such bond-length alternation in the infinite Hückel
chain is crucial, as it leads to a semiconductor with optical gap E_g
= 4tδ, so that perturbation theory converges. Electron-phonon
contributions lead to a Peierls instability in one-dimensional
systems and qualitatively alter the physics. Now the polarizability
per site becomes independent of N for large N, since evaluation of
(2) amounts to summing over excited states with a single electron-
hole (e-h) pair and there is a sum rule. Finite δ leads to a finite
correlation length $\xi \sim \delta^{-1}$. Thus power-law fits to this or other
coefficients are fundamentally limited to finite systems.

Interacting electrons in half-filled PPP or Hubbard chains lead
to insulators even at δ = 0, as found exactly in Hubbard chains [10]
with U = V(0) > 0. The NLO response is therefore fundamentally
different. Since electron-hole symmetry is retained in half-filled
PPP chains, we expect both correlations and alternation to increase
the optical gap. In addition, the polarizability and NLO coefficients
depend on the excited states and the ordering of 1^1B_u and 2^1A_g states,
which is not properly predicted by Hückel models or by one-electron
approximations to PPP models. Exact PPP solutions [11] with traditional molecular parameters provide the proper ordering and generally
fit molecular data. Such solutions are readily accessible to

polyenes with N ≤ 12 carbons. We limit the present discussion to
exact Hückel and PPP models with alternation δ and bandwidth 4t =
9.6ev.

The principal axes XY of $\alpha_{ij}(0)$ depend on the alternation δ in
Fig. 1. They are along the double bonds in the dimer limit $\delta = 1$ and
along the molecular axes xy for $\delta = 0$. The inverse of the static
polarizability per site is shown in Fig. 2 for PPP models of all-
trans polyenes [12]. The standard parameters have previously been
used for excitation energies and other properties. The N^{-1} plot is
sufficiently smooth to allow linear extrapolation to the infinite
chain. In addition to the ground-state polarizability of 7.4 A^3, we
show extrapolations in Fig. 2 for the 2^1A_g and 1^1B_u state. They are
obtained by replacing G in (2) with the desired states. Since $2A_g$ is
below $1B_u$ for these parameters, there is a finite gap and the $2A_g$

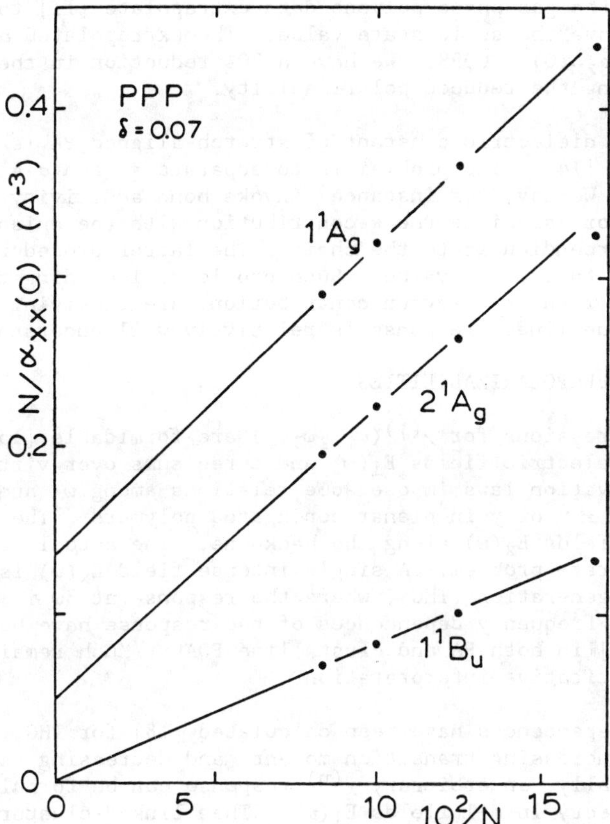

Fig. 2.   Inverse of the static polarizability per site, $N/\alpha_{xx}(0)$, of
the 1^1A_g, 2^1A_g, and 1^1B_u states of PPP models for N ≤ 12 site *trans*
polyenes.

polarizability per site remains finite. But the $1B_u$ polarizability apparently diverges for $N \to \infty$. We note that (2) has vanishing energy denominators in this case, since the A_g states form a continuum above the alternation gap to $2A_g$.

The polarizability along the PA chain is an order of magnitude larger in Hückel theory [13] at $\delta = 0.07$. But larger $\delta = 0.18$ is required if the entire gap $E_g = 4t\delta = 1.8$ ev is attributed to alternation and the bandwidth is $4t = 10$ ev. We contrast instead models with equal E_g by considering the reduced polarizability per site. The Hückel result for $\delta \ll 1$ is particularly simple,

$$\tilde{\alpha}_{xx}(0) = 2(3\pi\delta)^{-1} \tag{4}$$

and gives 1.19 at $\delta = 0.18$. The PPP polarizability in Fig. 2 is low partly because the gas-phase polyene data extrapolate [11] to $E_g = 2.8$ ev, well above the solid-state value. The extrapolated $\alpha_{xx}(0)$ and E_g lead to $\tilde{\alpha}_{xx}(0) = 0.98$. We have a 20% reduction in the linear response on using the reduced polarizability.

The static dielectric constant of stretch-aligned PA is known to better than 20% [14]. The problem is to separate π- and σ-electron contributions. We may, for instance, invoke bond additivity for the σ-contribution or associate the π-contribution with the anisotropy parallel and perpendicular to the chain. The latter procedure leads to $\tilde{\alpha}_{xx}(0)$ close to the PPP value. Such problems, including the identification of the π-electron contribution, are receiving more attention and the linear response is relatively well understood.

III. SECOND HYPERPOLARIZABILITIES

Formal expressions for $\chi^{(3)}(\omega_1 \omega_2 \omega_3 \omega_4)$ are formidable [15]. There are four electric fields $E_i(\omega)$ and three sums over virtual states. Conservation laws impose some relations among ω_j and polarizations $i = x$ or y in planar conjugated polymers. The dominant component has fields $E_x(\omega)$ along the backbone. The actual or local field is a separate problem. A single intense field $E_x(\omega)$ is used in third harmonic generation (THG), where the response at 3ω is sought. The angular and frequency dependences of the response have been of interest [16,17] in both PA and crystalline PDAs. Much remains to be done for a quantitative interpretation.

Strong N-dependences have been calculated [18] for THG, as expected from increasing transition moments and decreasing excitation energies. Formally, an arbitrary $\chi^{(n)}$ response can be formulated as perturbation theory in $n+1$ fields $E_i(\omega)$. Then linked-cluster expansions demonstrate quite generally the extensive nature of THG or other $\chi^{(n)}$. Unlinked diagrams with quadratic or higher asymptotic N-dependences must cancel exactly. It is advantageous to discard such contributions at the outset.

To illustrate the cancellation of unlinked diagrams, we consider THG with $E_x(\omega)$ along the backbone and adopt a single-particle picture in which the ground-state $|G>$ has as filled valence band. The intermediate states $|r>$ and $|t>$ contain one electron-hole pair at energies ϵ_r and ϵ_t, respectively, relative to E_G. Each sum over e-h pairs goes as N in half-filled models, as may be shown for the linear response. Three unlinked processes based on two e-h pairs are sketched in Fig. 3. Each goes as N^2 asymptotically. The first involves sequential creation and annihilation; the third virtual state $|s>$ is then the ground state $|G>$. The second two have two e-h pairs in $|s>$ and $\epsilon_s = \epsilon_t + \epsilon_r$ for independent electrons; they differ in the order of annihilating the pairs, with $\epsilon_t = \epsilon_r$ in Fig. 3b and $\epsilon_t \neq \epsilon_r$ in Fig. 3c. Since the identities of the e-h pairs are preserved, there is a common factor of $\mu_{Gr}^2 \mu_{Gt}^2$ for all terms in Fig. 3. The sums over virtual states $|t>$ and $|r>$ are carried out after combining these processes.

The energy denominators for THG go as $D_1(\omega) + D_1(-\omega) + D_2(\omega) + D_2(-\omega)$, with

$$D_1^{-1}(\omega) = (\epsilon_r + 3\omega)(\epsilon_s + 2\omega)(\epsilon_t + \omega)$$

$$D_2^{-1}(\omega) = (\epsilon_r + \omega)(\epsilon_s + 2\omega)(\epsilon_t - \omega)$$

$$(5)$$

and ϵ_s specified in Fig. 3. We obtain four denominators for each process in Fig. 3. Direct addition shows cancellation between $\epsilon_s = 0$ terms in Fig. 3a and $\epsilon_s = \epsilon_r + \epsilon_t$ terms in Fig. 3b and c [19]. The relations hold for complex ϵ that include excited-state lifetimes. The combined denominators reduce to an expression in $\epsilon_r - \epsilon_t$ that vanishes on summing over r and t. The only nonvanishing contribution

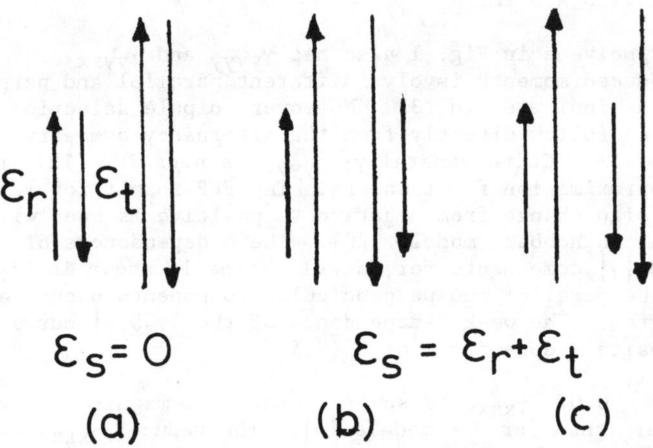

Fig. 3. Decoupled electron-hole pairs at ϵ_r and ϵ_t in one-electron treatments of THG, along with the excitation energy of the intermediate state $|s>$ of the same symmetry as the ground state $|G>$.

from Fig. 3 involves a double excitation, with two e-h pairs in the same orbitals, when the order of annihilation is irrelevant. But this goes as N for N orbitals.

The sufficient conditions for the cancellation are (i) $\epsilon_s = \epsilon_r + \epsilon_t$ and (ii) common transition moments for the processes in Fig. 3. In systems with a finite gap E_g between $|G\rangle$ and any other excitation, we have a finite correlation length ξ. The Hückel approximation for alternating chains leads to $\xi \sim \delta^{-1}$. Excitations localized in different regions of a large system then satisfy both (i) and (ii). Such considerations show $\chi^{(n)}$ to be extensive even in interacting systems, provided that ξ is finite, but are of limited value in simplifying the sums over excited states.

The length dependence of THG in alternating Hückel chains is shown in Fig. 4. The reduced second hyperpolarizability per site is $\gamma_{xxxx}(-3\omega,\omega,\omega,\omega)\ E_g^3/N(ea)^4$. Both static ($\omega = 0$) and dynamic ($\omega = E_g/4\hbar$) coefficients are shown. No lifetime corrections were used far from resonance. As expected, γ/N saturates for $\delta \neq 0$, when $\xi \sim \delta^{-1}$ is finite, and the saturation occurs at smaller N with increasing δ. The N-dependence of the regular ($\delta = 0$) chain, which does not saturate, agrees with the exponent [9] of 5.3 when the N^{-3} dependence of E_g^3 is taken into account.

Since the gap $E_g(N,\delta)$ increases with δ, γ/N is expected to decrease and this behavior is seen in Fig. 4. The anomalous behavior at $\delta = 0$ was carefully checked and apparently reflects the combined behavior of many transition moments. The $\delta = 0$ curve remains below $\delta = 0.07$ for small N even without the E_g^3 factor of the reduced second hyperpolarizability. By contrast, γ_{xxxx} decreases monotonically with δ in PPP chains with $N \leq 12$.

The linear polyene in Fig. 1 also has γ_{yyyy} and γ_{yyxx} components. Reduced moments involve different parallel and perpendicular lengths, as indicated in (3). Moreover, dipole selection rules for a field $E_y(\omega)$ follow directly from the alternancy symmetry of the Hückel model for PA. Quite generally, γ_{yyyy} is *negative* [12] in a one-electron approximation for *trans* PA. The PPP result for γ_{yyyy} is positive and a sign change from negative to positive is seen with increasing $U/|t|$ in Hubbard models [20]. The N dependences of various reduced $|\gamma|$ components for Hückel chains is shown in Fig. 5. Saturation of the parallel and perpendicular components occurs at comparable lengths. The weak N-dependence of the $|\gamma_{xxyy}|$ curve is due to the opposite N-dependence of $E_g(N,\delta)$.

At fixed $\delta = 0.07$, γ_{xxxx} is several orders of magnitude higher for Hückel models than for PPP models [3]. The reduced γ_{xxxx} per site at $\hbar\omega = E_g/4$ in Table 1, by contrast, differ by only factors of unity, mainly because the Hückel value of E_g is about half of the PPP gap. We may also tune to the PPP gap by defining an effective alternation δ_e for Hückel chains. The resulting Hückel γ_{xxxx} in

Table 1 are virtually equal with the PPP values, although their
N-dependences are slightly different. Reduced NLO coefficients
provide useful information about models with different E_g. Tuning
models to the same E_g is also attractive. Such procedures fail,
however, for transverse components like γ_{yyyy} that change sign in
interacting systems. The advantages and shortcomings of reduced NLO
coefficients require further study.

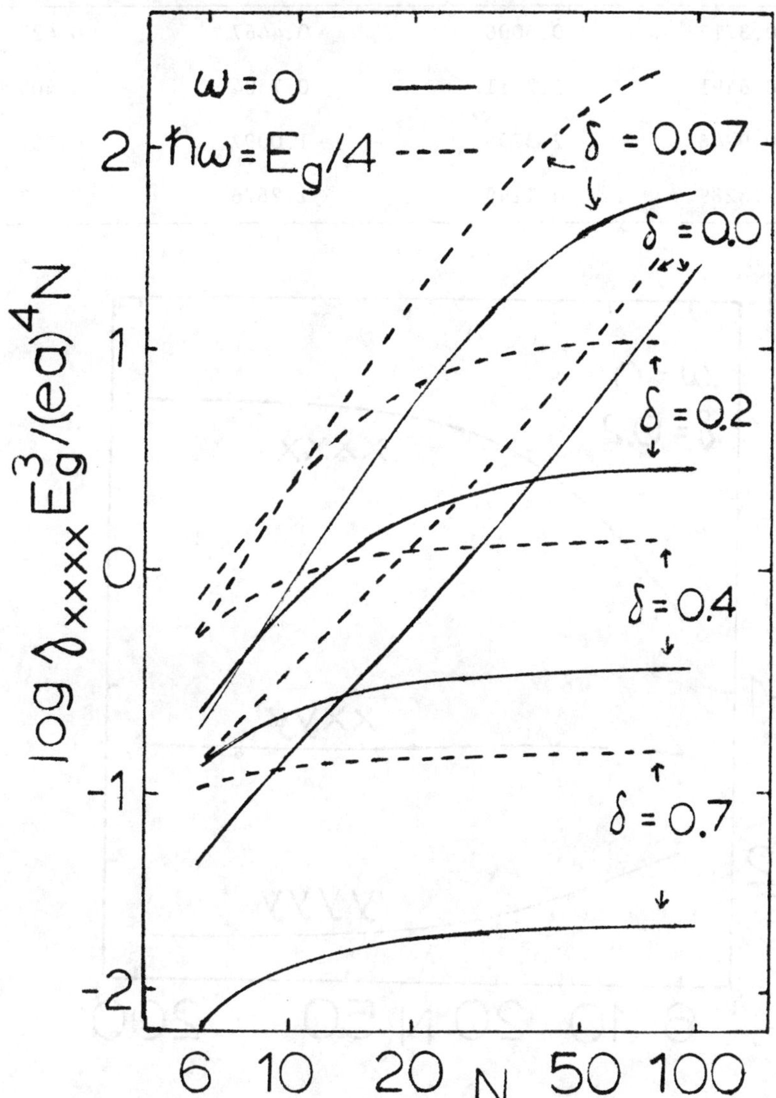

Fig. 4. Reduced static ($\omega = 0$) and dynamic ($\hbar\omega = E_g/4$) THG
coefficient parallel to N-site Hückel chains with alternation δ.

504

Table 1. Reduced γ_{xxxx} $(-3\omega,\omega,\omega,\omega)E_g^3/N(ea)^4$ at $\hbar\omega = E_g/4$
for N-site PPP and Hückel chains with $\delta = 0.07$ and for Hückel chains
with effective alternation δ_e to match the PPP value for E_g. The
bandwidth is fixed at $4t = 9.6ev$.

N	PPP(δ=0.07)	Hückel(δ=0.07)	Hückel(δ=δ_e)	δ_e
6	0.3717	0.5006	0.4467	0.427
8	0.6491	1.2111	0.7304	0.403
10	0.9746	2.3796	1.0097	0.386
12	1.3289	4.1149	1.2676	0.373

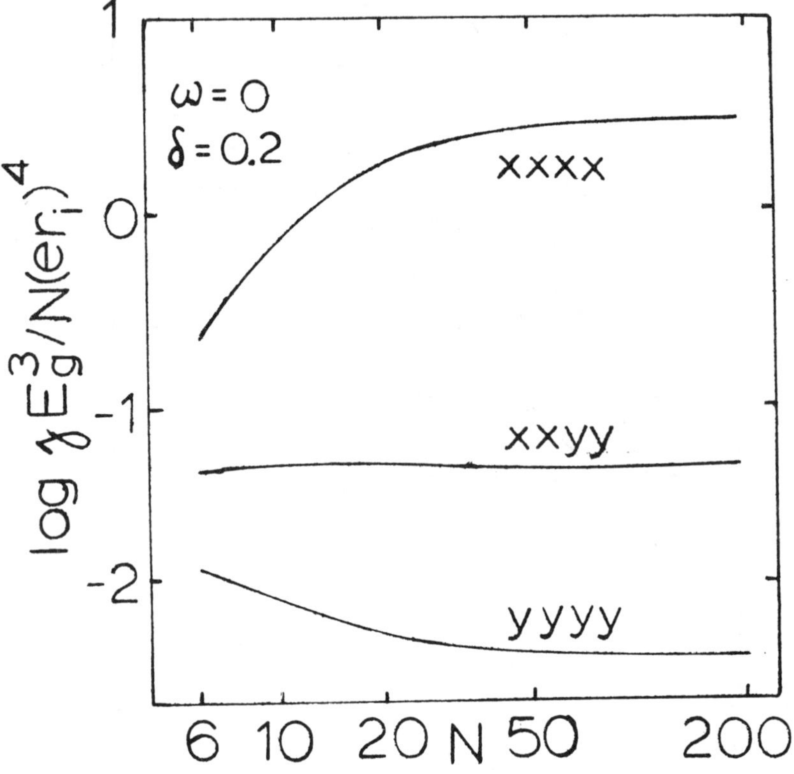

Fig. 5. Reduced THG coefficients γ_{xxxx}, γ_{xxyy}, and γ_{yyyy} for N-site
Hückel chains with alternation $\delta = 0.20$.

IV. DISCUSSION

The polarizability and second hyperpolarizability of finite
polyenes depend on both the length N and alternation δ. The limiting
static polarizability per site at large N is shown in Fig. 2 for
interacting π-electrons in PPP models with $\delta = 0.07$. Longer chains
of $N \sim 10^2$ were required in Figs. 4 and 5 to demonstrate saturation
for the second hyperpolarizabilities of noninteracting π-electrons in
Hückel chains. Proper extrapolations remain an important topic for
modeling conducting polymers.

The various γ/N components saturate around $N \sim \delta^{-1}$ in finite
Hückel chains. The effective conjugation length ξ discussed by
Agrawal <u>et al</u>. [21] in infinite chains also reduces to δ^{-1}. The
correlation length ξ cannot exceed the actual length N. Thus closely
similar ideas flow from either finite or infinite Hückel chains with
alternation δ. The proportionality constant between ξ and δ^{-1}
probably reflects the boundary conditions, but is not important here.

Power-law fits to NLO coefficients are useful for and restricted
to finite systems, to molecules rather than polymers. Even for small
N, the THG coefficient in Fig. 4 depends on δ, with the power-law
exponent decreasing with increasing δ. Power-law fits to molecular
series are encouraging but not compelling unless δ is independently
known in Hückel theory. Table 1 shows that an effective δ that gives
the E_g of a correlated system goes a long way towards giving the
proper γ_{xxxx} at $\hbar\omega = E_g/4$. Such an attractive phenomenological
approach must be used with care, however, since different δ's are
unphysical for the same system. The low-energy features of PA are
discussed [5] in SSH theory with a bandwidth $W = E_g/\delta = 10ev$.
However, the static γ_{xxxx} coefficient, which goes as δ^{-6}, has been
fit to experiment [22] for $W = E_g/\delta = 5ev$. As the magnitude and
frequency dependences of NLO coefficients become better known, the
choice of effective δ's will become more restricted.

Current $\chi^{(3)}$ results for polyenes and for conjugated polymers
like PA or polydiacetylanes (PDAs) show the power-law behavior of the
former to match the latter at surprisingly low N, around $N \sim 30$. The
available polymers may of course be limited by defects, so that their
intrinsic NLO response may be considerably higher. The molecular
response may be significantly increased by local field
contributions. Local fields are not enhanced along the PDA backbone
in the dipole approximation [23]. Such experimental considerations
may account for the low crossover between molecules and polymers.

The observed crossover around $N \sim 30$ may be rationalized
theoretically. The THG crossover between molecules and polymers in
alternating Hückel chains is the saturation around $\xi \sim N \sim \delta^{-1}$ in
Fig. 4. The longest possible correlation lengths are expected for
noninteracting electrons in perfectly all-*trans* polyenes. Conformational defects are expected to reduce ξ, as has been widely discussed

506

for many conjugated polyenes. We retain for simplicity all-
trans chains without such defects.

We suggest instead that electron-electron interactions can also
reduce ξ, below the Hückel value, with the correlation length
associated with e-h separation of particular interest for electric
dipole transitions. The effective δ in Hückel models of PA is around
$\delta \sim 0.20$, based on $W \sim 10ev$ and $E_g \sim 1.8ev$, rather than $\delta = 0.07$ in
PPP models of polyenes. The crossover to constant γ_{xxxx}/N for Hückel
chains in Fig. 4 occurs around $N \sim 30$ for $\delta = 0.20$. A significant
e-e contribution to the effective alternation may also rationalize
relatively small differences, of an order of magnitude, among
conjugated polymers with different bond-length patterns. The e-e
contributions remain essentially the same and largely fix ξ.
Accurate NLO results for both interacting and noninteracting models
should allow the testing of such hypotheses.

Reduced NLO coefficients offer better comparisons among
theoretical models by automatically including some E_g effects.
Furthermore, resonances at $f = n\hbar\omega/E_g$ with $n = 2$ (two-photon) or 3
(three-photon) are automatically referenced to the optical gap. We
anticipate improved extrapolations for the dynamic response at
constant f rather than constant $\hbar\omega$. The tuning of model Hamiltonians
to give identical E_g also raises both conceptual and computational
questions. Initial efforts naturally center on thoroughly understanding the NLO properties of the quantum cell models that describe
the frontier electrons in conjugated polymers.

One of us (Z.G.S.) thanks R. Silbey for a stimulating discussion
of correlation lengths and S. Etemad for comments about third
harmonic generation.

REFERENCES

1. P.N. Prasad and D.R. Ulrich, eds. Nonlinear Optical and
    Electroactive Polymers, (Plenum, New York, 1988); A.J. Heeger,
    J. Orenstein, and D.R. Ulrich, eds., Optical Properties of
    Polymers, Proceedings of MRS Symposium <u>109</u> (1988).

2. J.M. André, C. Barbier, V. Bodart, and J. Delhalle, in Nonlinear
    Optical Properties of Organic Molecules and Crystals, Vol. 2
    (eds. D.S. Chemla and J. Zyss, Academic Press, New York, 1987
    (p. 137-158. These two volumes provide a comprehensive
    introduction to NLO properties of organic solids.

3. Z.G. Soos and S. Ramasesha, J. Chem. Phys. <u>90</u>, 1067 (1989); S.
    Ramasesha and Z.G. Soos, Chem. Phys. Lett. <u>153</u>, 171 (1988).

4. Z.G. Soos and G.W. Hayden, in Electroresponsive Molecular and
    Polymeric Systems (ed. T.A. Skotheim, Marcel Dekker, New York,
    1988) p. 197.

5. A.J. Heeger, S. Kivelson, J.R. Schrieffer, and W.P. Su, Rev. Mod. Phys. $\underline{60}$, 781 (1988).

6. S. Mazumdar and Z.G. Soos, Synth. Met. $\underline{1}$, 77 (1979); S.R. Bondeson and Z.G. Soos, Phys. Rev. $\underline{B22}$, 179 3 (1980); S. Ramasesha and Z.G. Soos, Int. J. Quantum Chem. $\underline{25}$, 1003 (1984); J. Chem. Phys. $\underline{80}$, 3278 (1984); see also ref. 4.

7. L. Salem, The Molecular Orbital Theory of Conjugated Systems, (Benjamin, New York, 1966), p. 56-60.

8. K. Ohno, Theor. Chim. Acta $\underline{2}$, 219 (1964).

9. H.F. Hameka, J. Chem. Phys. $\underline{67}$, 2935 (1977); E.F. McIntyre and H.F. Hameka, $\underline{ibid}$, $\underline{68}$, 3481 (1978).

10. E.H. Lieb and F.Y. Wu, Phys. Rev. Lett., $\underline{20}$, 1445 (1968).

11. Z.G. Soos and S. Ramasesha, Phys. Rev. $\underline{B29}$, 5410 (1984).

12. Z.G. Soos and G.W. Hayden, Phys. Rev. $\underline{B40}$, 3081 (1989).

13. C. Cojan, G.P. Agrawal, and C. Flytzanis, Phys. Rev. $\underline{B15}$, 909 (1977).

14. G. Leising, Phys. Rev. $\underline{B38}$, 10313 (1988).

15. D.C. Hanna, M.A. Yuratich and D. Cotter, "Nonlinear Optics of Free Atoms and Molecules" (Springer-Verlag, Berlin, 1979); P.W. Langhoff, S.T. Epstein, and M. Karplus, Rev. Mod. Phys. $\underline{44}$ (1972).

16. W.S. Fann, S. Benson, J. Madey, S. Etemad, G.L. Baker, and F. Kajzar, Phys. Rev. Lett. $\underline{62}$, 1492 (1989); G. Leising, Phys. Rev. B.$\underline{39}$, 3701 (1989); M. Sinclair, D. McBranch, D. Moses and A.J. Heeger, Synth. Met. $\underline{28}$, D645 (1989).

17. S. Etemad, G.L. Baker, D. Jaye, F. Kajzar, and J. Messier, Proc. SPIE $\underline{682}$, 44 (1987); G.M. Carter, Y. J. Chen, M.F. Rubner, D.J. Sandman, M.K. Thakur, and S.K. Tripathy, in Ref. 2, p 85-120 (1987).

18. C.P. De Melo and R. Silbey, J. Chem. Phys. $\underline{88}$, 2558, 2567 (1988) and references therein; J.R. Heflin, K.Y. Wong, O. Zamani-Khamiri, and A.F. Garito, Phys. Rev. $\underline{B38}$, 1573 (1988).

19. P.C.M. McWilliams and Z.G. Soos, unpublished results.

20. Z.G. Soos, S. Ramasesha, and G.W. Hayden, NATO Advanced Research Workshop, Torino, Italy (1988) (eds. D. Campbell and D. Baeriswyl, Plenum New York) in press.

21. G.P. Agrawal, C. Cojan, and C. Flytzanis, Phys. Rev. $\underline{B17}$, 776 (1978).

22. W. Wu and S. Kivelson, Synth. Met. $\underline{28}$, D575 (1989).

23. M. Hurst and R.W. Munn, Chem. Phys. $\underline{127}$, 1 (1988).

ELECTRONIC STRUCTURE AND STATIC ELECTRIC DIPOLE POLARIZABILITY OF ACETYLENIC ANALOGS OF CARBOCYANINES

V.P. Bodart [a], J. Delhalle, J.M. André
Laboratoire de Chimie Théorique Appliquée
Facultés Universitaires Notre-Dame de la Paix
61, rue de Bruxelles B-5000 Namur
Belgium

ABSTRACT. The equilibrium geometry and electric polarizability of a model acetylenic cyanine are compared with those of three isoelectronic molecules and the corresponding carbocyanine. The carbocyanine is the most polarizable system of the series.

1. INTRODUCTION

Linear and nonlinear optical properties of organic polymers are presently under intense investigation in view of technological applications [1-3]; systems to be designed are those with high electric responses. Progress in optoelectronics is conditioned by substantial improvements of several characteristics of the optical materials. Among the most important ones is the magnitude of the susceptibility $\chi^{(n)}$. Many π-conjugated organic materials show fast response times and large optical nonlinearity [1,2], but their largest third-order nonlinear-optical susceptibility is still too small to permit these materials to be used in actual applications.

A first and minimal condition for materials to be of potential interest for optoelectronics and nonlinear optics is certainly to have high electric responses of their atomic and molecular constituents to external field perturbations. Quantum chemistry calculations can provide a convenient and fast way to estimate the microscopic electric responses of model systems (molecules and oligomers) as a function of their chemical nature, molecular structure, organization patterns at the molecular level, chain length, and other factors that can influence the ultimate performances of actual materials [4].

Substantial electronic polarizations have always been obtained in systems with valence electrons highly delocalized over large distances as obtained in conjugated oligomers and polymers [3]. In designing new conjugated chains, it is important to control the molecular organization over distances at least corresponding to the so-called "delocalization length" (note that the concept of "delocalization length" itself is largely an intuitive concept as discussed in the contributions to these proceedings by Silbey and Soos).

---

[a] Now permanently at : SOLVAY Cie, Laboratoire Central,
310, rue de Ransbeek, B-1120 Bruxelles (Belgium).

J. L. Brédas and R. R. Chance (eds.), Conjugated Polymeric Materials:
Opportunities in Electronics, Optoelectronics, and Molecular Electronics, 509–516.
© 1990 *Kluwer Academic Publishers. Printed in the Netherlands.*

Local structural defects, *e.g.* conformational twists, kinks, etc. can significantly reduce this delocalization length and, unfortunately, the longer the chain the more difficult it is to enforce the desired structural and electronic homogeneity [5].

In spite of already reported computations on the 2nd and 3rd molecular hyperpolarizabilities of conjugated chains [6,7], the present state of quantitative predictions of these properties is still in its infancy. It suffices to refer to recent works on the Li atom [8] and on the Ne atom [9] to measure the magnitude of the difficulties to be overcome (errors due to basis set limitations, importance of the correlation corrections in large size systems, etc.). The static electric dipole polarizability, comparatively less tricky to evaluate, describes the polarization linear in the external field. In a first approximation, it can serve as a reference property to rank a series of related conjugated molecules and chains regarding the efficiency in delocalizing their π-electrons. It is the point of view adopted in this contribution.

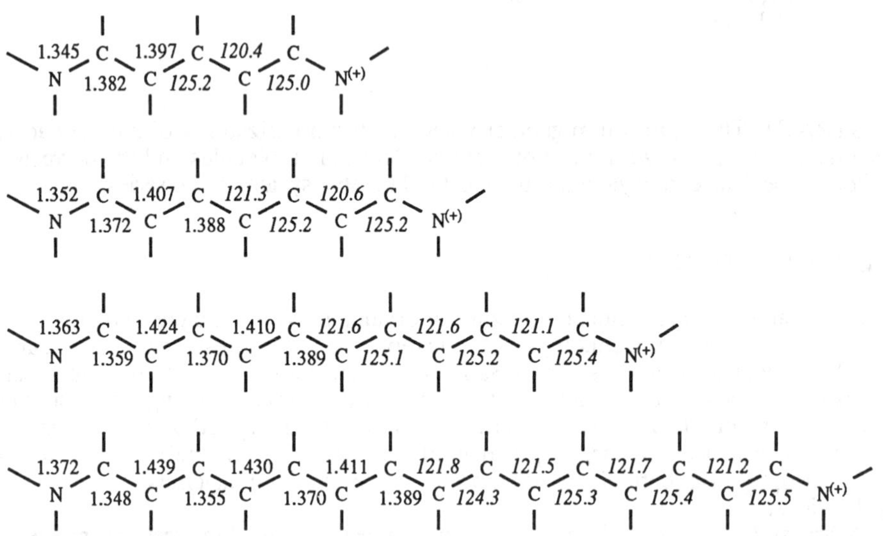

Figure 1. Bond lengths (in Å) and bond angles (in degrees) of four carbocyanines of increasing length $H_2N(-CH=CH)_n-CH=N^+H_2$ (n = 2, 3, 5 and 7) . The geometries have been optimized at the STO-3G level.

Carbocyanines, $H_2N(-CH=CH)_n-CH=N^+H_2$ are characterized by a highly delocalized conjugated pathway evidenced both by a low degree of alternation between double and single bonds and a high polarizability of their backbones [10-12]. This idea has been already developed in the late 40's by Kuhn [13-18]. It was his pioneering contribution to relate the nature of the first $\pi-\pi^*$ transition to the concept of bond alternation in conjugated chains particularly in the even-atom polyene and in the odd-atom polymethine chains. Indeed, Kuhn showed that in the series of polymethine dyes, the bond lengths between all the carbon atoms are equal due to a resonance balance between equivalent extreme forms:

$$>N\text{-}(CH=CH)_n\text{-}CH=N^+< \quad \leftrightarrow \quad >N^+=CH\text{-}(CH=CH)_n\text{-}N<$$

All carbon-carbon bonds in the skeleton have 50% of double bond character. This fact was later confirmed by X-ray diffraction studies. A simple FE model calculation shows that there is no energy gap between the valence and conduction bands and the limit of the first UV-visible transition for an infinite chain is zero. On the opposite, when the same approach is used in the case of polyenes (polyacetylenes), a much poorer agreement with experiment is obtained. In order to reproduce the experimental data, Kuhn had to include a sinusoidal perturbative potential so that the electronic distribution corresponds to alternating single and double bonds. In this case, the extrapolation of the UV spectrum tends to a non-zero energy gap for the infinite chain. This question has been investigated in more detail [19] by plotting the FE transition energies of the series of oligomers $>(CH=CH)_n<$ (for n=2,3,...6) and $>N-(CH=CH)_n-CH=N^+$ (for n=1,2,...5) as compared to the experimental values taking into account the bond end lengths as adjustable parameters. In the polymethine series, the regression line has its y-intercept at the origin of the axis. As expected, it corresponds to a zero energy gap, i.e., to a metallic character. In the polyene case, however, the y-intercept is of the order of 1.8 eV, a value in close agreement with the experimental energy gap of polymeric polyacetylene. This discussion is of crucial importance in the present discussion since in the Unsöld approximation, polarizabilities are direclty related to first electronic transitions. In this way, carbocyanines are to be significantly more polarizable than polyacetylenes. The question of bond alternation is thus quantitatively illustrated in Figure 1 where the relevant geometrical parameters (bond distances and bond angles calculated [20] at the STO-3G level [21]) are given for a series of model carbocyanines of increasing length (n = 2, 3, 5 and 7). The absolute values of STO-3G geometrical parameters are known to depart somewhat from experiment, but the structural trends in a series of related molecules are usually predicted right [22]. For instance, the bond alternation degree in polyenes is known to be exaggerated [23-26], but the tendencies are correctly reproduced.

Note that bond alternation is quickly restored as the carbocyanine chain length increases. The alternation degree of carbocyanines is classically formalized in terms of specific symmetric resonance structures which are chemically induced through the interplay of the $-NR_1R_2$ ($=N^+R_1R_2$) and ($-NR_3R_4$) $=N^+R_3R_4$ end groups; the corresponding resonant structures are shown in Figure 2a.

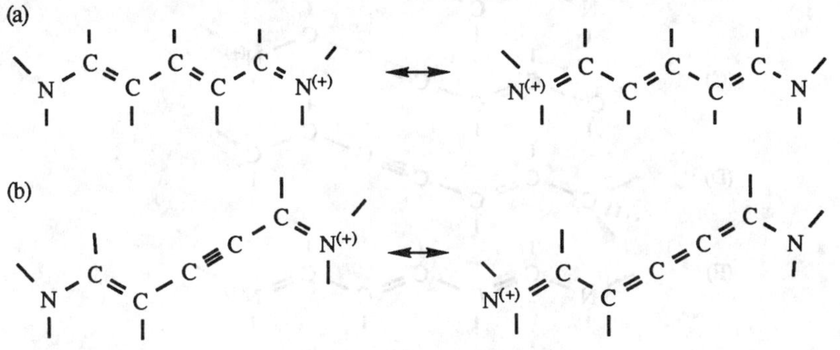

Figure 2. Resonance structures invoked for: (a) classical carbocyanines and (b) their acetylenic analogs.

Since the electronic polarization, in addition to the delocalization length, directly depends upon the number of active valence electrons per repeat unit, we address in this

512

paper the possibility of chemically deconfining π-electron rich moieties such as the -C≡C- triple bond to increase the number of electrons participating to the electronic response. The hope is that, by inserting a triple bond in a carbocyanine backbone, the -C≡C- distance will increase and lead to enhanced the polarizability; the resonance forms for acetylenic carbocyanines are shown Figure 2b.

2. MODEL MOLECULES AND METHODOLOGY

To our knowledge, the first synthesis of an acetylenic cyanine has been made by Mee *et al.* [27-29]. They noted an hypsochromic shift of the first optical transition as compared to the carbocyanine analogs (*i.e.* containing the same number of carbon atoms). They explained this shift on the basis of the asymmetric electron distribution of the acetylenic cyanines. This can also be related to the resonance structures shown in Figure 2b which are not equivalent as opposed to the carbocyanine case (Figure 2a). In Figure 2b, one of the resonance structure is reminiscent of the monomer unit of polydiacetylene, while the second is typical of the butatrienic form. Butatriene belongs to the family of cumulenes which seem to be the most polarizable of the conjugated chains investigated so far [7,26, 30-32]. It is thus interesting to determine which of these two limiting forms is dominant in the representation of the actual structure. A series of five molecules, denoted A, B, C, D and E and shown in Figure 3 has been considered. Molecule A is the simplest of the acetylenic cyanines, molecules B to D are isoelectronic to A and differ only by the nature of their chain ends. The general formula is R_1-(CH=CH)-C≡C-CH=R_2. For comparison purposes, the results for carbocyanine E, H_2N-CH=CH-CH=CH-CH=N^+H_2, have been added to the series.

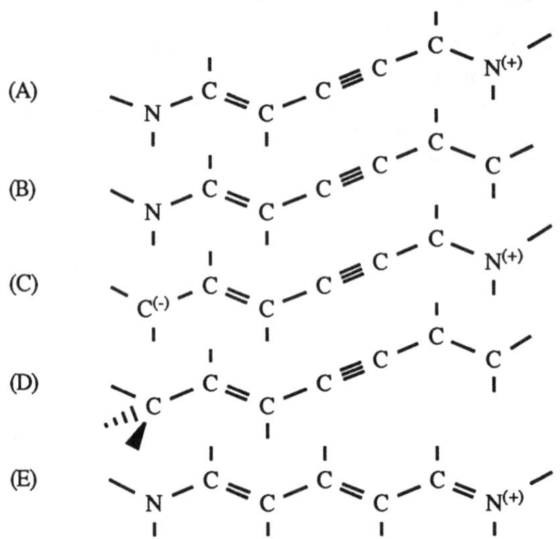

Figure 3. Molecular structure of the series of related molecules (A to E).

Both the equilibrium geometry and polarizability have been computed for these molecules. Equilibrium structures have been computed with the Gaussian-82 series of program [33], adapted for an FPS-164 attached to an IBM-4341 computer. Full geometry optimizations were carried out with the minimal STO-3G basis set [21] on the model systems kept planar. The Fletcher-Powell procedure was used to minimize the forces on the nuclei and the standard threshold conditions of the Gaussian-82 program have been kept , i.e. 10^{-10} a.u. for the two-electron integral cutoff, 10^{-9} for the requested convergence on the density matrice elements and 5.10^{-4} hartree bohr^{-1} as the minimum residual forces on the cartesian components. The optimized geometries have been used as input for static dipole polarizability calculations. The average polarizability α ($\alpha = [\alpha_{xx} + \alpha_{yy} + \alpha_{zz}]/3$) has been computed with the so-called finite-field method originally proposed by Cohen and Roothaan [34]; it is expressed in atomic units (1 a.u. of polarizability $= 1.6488 \times 10^{-41}$ C^{2}m^{2}J^{-1}). As in the case of geometry, the STO-3G predictions quantitatively depart from the experimental values, but they are qualitatively correct [35,36].

3. GEOMETRIC STRUCTURE AND ELECTRIC DIPOLE POLARIZABILITY

Equilibrium geometry parameters are listed in Table 1 according to the convention shown in Figure 4.

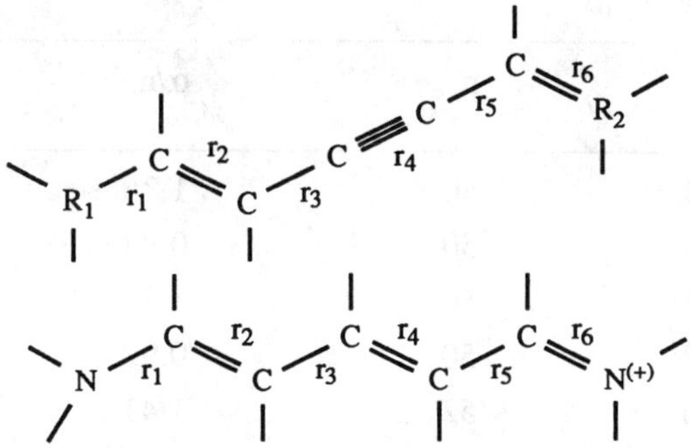

Figure 4. Numbering system used to denote the bonds in molecules A to E.

TABLE 1. STO-3G optimized bond lengths (in Å) for the five molecules (A to E). Experimental data have been added in the case of molecule A.

	R_1	R_2	r_1	r_2	r_3	r_4	r_5	r_6
A	NH_2	NH_2^+	1.351	1.373	1.389	1.208	1.371	1.335
exp.			1.370	1.386	1.376	1.212	1.394	1.356
B	NH_2	CH_2	1.395	1.329	1.444	1.178	1.451	1.316
C	CH_2	NH_2	1.315	1.483	1.309	1.246	1.308	1.400
D	CH_3	CH_2	1.519	1.319	1.452	1.177	1.453	1.316
E	NH_2	NH_2^+	1.345	1.382	1.397	1.397	1.382	1.345

The results show that the carbon framework , -(CH=CH)-C≡C-CH=, common to all structures is quite sensitive to the nature of the chain ends. Note first the satisfactory agreement between experimental and theoretical predictions for (A). Molecule A has a structure intermediate between acetylenic and butatrienic forms. Only r_4 is substantially shorter (1.208 Å) than the other CC bonds in the molecule; a typical -C≡C- bond length at the STO-3G level is 1.170 Å [22]. This indicates a net increase of the -C≡C- bond length due to its incorporation in a cyanine type structure. Molecules B and D exhibit values typical of an acetylenic bond length and, from our previous experience [25,37], it is anticipated that their electric polarizability will be smaller than for molecule A. Molecule C has a clear butatrienic configuration which originates from the constraining effect of the =CH_2 group on the left side of the structure.

Absolute value of the average polarizability for molecules A to E are given in Table 2. However, to make a fair comparison between the five systems, it is more appropriate to consider the average polarizability divided by the total number of electrons in each molecule.

TABLE 2. Average polarizability α (in a.u.) and average polarizability divided by the number of electrons n_e in the molecule (α/n_e) .

	α	n_e	α/n_e
A	68.25	50	1.36
B	46.36	50	0.93
C	63.67	50	1.27
D	46.07	50	0.92
E	73.30	52	1.41

The largest polarizability is obtained for carbocyanine E, followed by the acetylenic cyanine A which has a geometrical structure intermediate between the alternating acetylenic structure and the more regular butatrienic form. Then comes molecule C which includes a butatrienic structure in its skeleton. Finally molecules B and D have the smallest values of polarizability in good agreement with the fact that the -C≡C- triple bond has not been affected.

Thus, as hoped at the beginning of this study, the acetylenic cyanine exhibits a net polarizability enhancement over isoelectronic molecules B and D for which the r_4 distance is basically a -C≡C- triple bond of which the 4 π-electrons remain strongly confined. The most polarizable system is still the classical carbocyanine E. Also interesting to note is the fact that even though molecule C incorporates a butatrienic structure, it is still less polarizable than A. This is due to the fact that delocalization in A takes place over the entire molecule while in C, the electronic delocalization is interrupted at r_2 (1.483 Å = the longest CC bond distance found in the molecules A to E).

4. CONCLUSIONS

The largest polarizability is noted for the classical carbocyanine E which also has the maximum equalization of bond distances and thus the best chances for electronic delocalization of the structures considered in this work. The acetylenic analog (A), which has a structure intermediate between the vinylacetylene and butatriene forms, has the next largest polarizability. Compounds (C), which exhibits a butatriene-like structure known to be a highly polarizable fragment, has nonetheless an average polarizability smaller than (E) and (A). This is because of the large r_2 distance which, as pointed out in an early theoretical analysis by Flytzanis, disrupts the actual delocalization and limits its extent to a smaller portion of the molecular framework.

The results on the triply bonded moiety ($-C{\equiv}C-$) are somewhat deceptive, but other fragments could be more suitable. Work along these lines is in progress.

Acknowledgments. J.M.A and J.D. thank Prof. J.L. Brédas and Dr. R. Chance for their generous hospitality during this very stimulating workshop. The authors also acknowledge with appreciation the support of this work under the ESPRIT II - EEC Contract No. 2284 on "Optoelectronics with Active Organic Molecules", and the National Fund for Scientific Research (Belgium), IBM Belgium and the Facultés Universitaires Notre-Dame de la Paix (FNDP) for the use of the Namur Scientific Computing Facility.

REFERENCES

1.   Chemla D. and Zyss J. (eds.) (1987) 'Nonlinear optical properties of organic molecules and crystals', Academic Press, New York. vols. 1 and 2.
2.   Messier J., Kajzar F., Prasad P. and Ulrich D. (eds.) (1989) 'Nonlinear optical effects in organic polymers', Kluwer Academic Publishers, NATO ASI Series E vol.   162.
3.   Flytzanis Ch. (1987) 'Dimensionality effects and scaling laws in nonlinear optical susceptibilities', in Chemla D. and Zyss J. (eds.) 'Nonlinear optical properties of organic molecules and crystals', Academic Press, New York. vol. 2, pp. 121-135.
4.   Delhalle J., Dory M., Fripiat J.G., André J.M. (1989) 'Theoretical design of organic molecules and polymers for optoelectronics' in Messier J., Kajzar F., Prasad P. and Ulrich D. (eds.) 'Nonlinear optical effects in organic polymers', Kluwer Academic Publishers, NATO ASI Series E vol. 162, pp. 13-28.
5.   André J.M., Barbier C., Bodart V.P. and Delhalle J. (1987) 'Trends in calculations of polarizabilities and hyperpolarizabilities of long molecules', in Chemla D. and Zyss J. (eds.) 'Nonlinear optical properties of organic molecules and crystals', Academic Press, New York. vol. 2, pp. 137-158.
6.   Hurst G.J.B. , Dupuis M. and Clementi E. (1988) J. Chem. Phys. 89, 385.
7.   Chopra P., Carlacci L., King H.F. and Prasad P. (1989) '*Ab initio* calculations of polarizabilities and hyperpolarizabilities in organic molecules with extended pi-electron conjugation', J. Phys. Chem., in press.
8.   Maroulis G. and Thakkar A. (1989) 'Static hyperpolarizabilities and polarizabilities of Li', J. Phys. B, in press.
9.   Shelton D.P. (1989) Phys. Rev. Lett. 62, 2660.
10.   Smith D.L. and Luss H.R. (1972) Acta Cryst. B28, 2793.
11.   Smith D.L. and Luss H.R. (1975) Acta Cryst. B31, 402.
12.   Bigelow R.W. and Freund H.J. (1986) Chem. Phys. 107, 159.
13.   Kuhn H.(1948) Chimia Aarau 2, 1.
14.   Kuhn H.(1948) J.Chem.Phys. 16, 840.

516

15.	Kuhn H.(1948) Helv.Chim.Acta 31, 1441.
16.	Kuhn H.(1949) J.Chem.Phys. 17, 1198.
17.	Kuhn H.(1955) Chimia Aarau 9, 237.
18.	Kuhn H.(1959) Angew.Chem. 71, 93.
19.	André J.M. (1965) 'Etude comparative des méthodes de Hückel et de l'électron libre', Ms.Sc Thesis, Université Catholique de Louvain, Leuven-Belgium.
20.	Bodart V.P. (1987) 'Calcul de polarisabilités et hyperpolarisabilités de chaînes organiques conjuguées : étude formelle et application en optique non linéaire', Ph.D.Thesis, Facultés Universitaires Notre-Dame de la Paix, Namur-Belgium, pp. 241 - 253.
21.	Hehre W.J., Stewart R.F. and Pople J.A. (1969) J. Chem.Phys. 51, 2657.
22.	W.J. Hehre W.J., Radom P., Schleyer P. von Ragué and Pople J.A. (1986) '*Ab initio* Molecular Orbital Theory', Wiley, New York, and references therein.
23.	Bock C.W. and Trachtman M.J. (1984) J. Mol. Struct. (Theochem) 1, 109.
24.	Brédas J.L., Heeger A.J. and Wudl F. (1986) J. Chem. Phys. 85, 4673.
25.	Bodart V.P., Delhalle J., André J.M. and Zyss J. (1985) Can. J. Chem. 63, 1631.
26.	Bodart V.P., Delhalle J., Dory M., Fripiat J.G. and André J.M. (1987) J. Opt.Soc.Am. B4, 1047.
27.	Mee J.D. (1974) J. Am. Chem. Soc. 96, 4712 (1974).
28.	Mee J.D. (1977) J. Org. Chem. 42, 1035.
29.	Mee J.D. and Sturmer D.M. (1977) J. Org. Chem. 42, 1041.
30.	Delhalle J., Bodart V.P., Dory M., André J.M. and Zyss J. (1986) Int. J. Quantum Chem. Symp. 19, 313.
31.	Baratan D.N., Onuchic J.N. and Perry J.W. (1987) J. Phys. Chem. 91, 2696.
32.	Kirtman B. and Hasan M. (1989) Chem. Phys. Lett. 157, 123.
33.	Binkley J.S. , Frisch M., K. Raghavachari, DeFrees D.J., Schlegel H.B., Whiteside R.A., Fluder E., Seeger R. and Pople J.A. (1983) GAUSSIAN 82, Carnegie-Mellon University.
34.	Cohen H.D. and Roothaan C.C.J. (1985), J. Chem. Phys. 43, 534.
35.	Chablo A. and Hinchliffe A. (1980) Chem. Phys. Lett. 72, 149.
36.	Dory M., Beudels L., Delhalle J., André J.M. and Dupuis M., to be published.
37.	Bodart V.P., Delhalle J., André J.M. and Zyss J. (1985) ' Structural dependence of the longitudinal electric polarizability of finite polyene chains : an *ab initio* study' in Bloor D. and Chance R. (eds.) 'Polydiacetylenes : synthesis, structure and electric properties', Martinus Nijhof Publishers, Dordrecht, pp. 125-133.

MACROCYCLES AS MOLECULAR UNITS TO BUILD UP ELECTRORESPONSIVE MATERIALS: A COMPARATIVE THEORETICAL INVESTIGATION OF THE ELECTRONIC AND OPTICAL PROPERTIES OF PHTHALOCYANINE AND RELATED SYSTEMS

E. Ortí
Departamento Química Física, Universitat de València,
Dr. Moliner 50, 46100 Burjassot - Valencia (Spain).

J.L. Brédas
Service de Chimie des Matériaux Nouveaux, Département des
Matériaux et Procédés, Université de Mons, B-7000 Mons (Belgium).

1. Introduction

Besides their classical and well-known applications as, e.g., dyes, pigments, catalysts, etc. [1], phthalocyanines (Pc's) have been of particular interest in many fields of basic and applied research concerning solar cells [2], photosensitizers [3], low dimensional metals [4], gas sensors [5], electrochromism [6], Langmuir-Blodgett (LB) films [7], and nonlinear optics [8]. As an example, gas sensors [9], electronic devices [10], laser recording materials [11], and photovoltaic cells [12] can now be built from LB films of phthalocyanines. This great variety of interesting applications results from a number of unique properties that phthalocyanines exhibit. On the one hand, they show an exceptional thermal and chemical stability (undergoing no noticeable degradation in air up to 400-500° C) and a very large chemical versatility (more than 70 different metallophthalocyanines (MPc's) can be obtained). On the other hand, phthalocyanines present remarkable optical (intense absorptions in the visible), electrical (semiconducting and photoconducting), and magnetic properties that can be easily tuned.

The optical and electrical applications mentioned above for phthalocyanine-based materials are directly determined by the electronic and structural characteristics of phthalocyanine molecules. Phthalocyanines basically correspond to macrocyclic planar

517

J. L. Brédas and R. R. Chance (eds.), Conjugated Polymeric Materials:
Opportunities in Electronics, Optoelectronics, and Molecular Electronics, 517–530.
© 1990 *Kluwer Academic Publishers. Printed in the Netherlands.*

compounds with a very large aromatic π-electron system delocalized over a carbon-nitrogen backbone of 40 atoms. Furthermore, they present crystal structures where molecules are packed in columnar stacks providing a pathway for electron delocalization through the π-π overlap between adjacent macrocycles [13]. Thus, when phthalocyanines are cocrystallized with an oxidant agent like I_2 to yield partially oxidized stoichiometric $M(Pc^{+0.33})I^{-0.33}$ compounds, electrons are extracted from the HOMO (highest occupied molecular orbital) or valence band delocalized along the columnar stack and conductivities as high as 1000 S/cm are obtained parallel to the stacking direction [4a,b]. The material $M(Pc)I$ is called an "organic or molecular metal" since the temperature dependence of the conductivity is metal-like down to very low temperatures [4a]. The same kind of mechanism acts for phthalocyanine sensors. When a gaseous molecule, such as NO_2, is chemisorbed onto the MPc surface, a charge transfer occurs enhancing the surface conductivity by greatly increasing the number of charge carriers (which for phthalocyanines are holes). The charge transfer is possible as the ionization potential of MPc compounds is low enough that a redox reaction readily occurs, removing an electron from the ring system.

It is to be noted that, unlike the situation in tetracyanoplatinate salts where electrons move through the chain of Pt atoms, the presence of metal atoms is not a requirement to achieve high conductivities in phthalocyanines, since, as mentioned above, the interaction between π-molecular orbitals on adjacent macrocycles can provide the electronic pathway. Partially oxidized crystals of metal-free phthalocyanine, H_2Pc, show conductivities of 700 S/cm at room temperature and greater than 3500 S/cm at 1.5 K [14]. The presence of metal atoms is however a requirement to obtain cofacially joined metallophthalocyanine polymers $[M(Pc)L]_n$ where the metallomacrocycles are linked together by bisaxially metal bonded bridging ligands, L [4c,15].

The optical properties of phthalocyanines are very similar to those of the closely related square planar porphyrins and are mainly characterized by an intense electronic absorption band in the visible (Q band) and another broad absorption band in the near ultraviolet (B or Soret band) [16]. These absorption bands are associated with singlet π-π^* transitions involving primarily the highest two occupied and the lowest two unoccupied molecular orbitals. This constitutes the basis for the so-called four-orbital model developed by Gouterman et al. [17] to interpret the porphine optical spectra.

It is therefore clearly established that the most important electronic and optical properties of phthalocyanine-based materials are determined by the upper occupied and lower unoccupied energy levels of the macrocyclic system. Furthermore, since the interactions between phthalocyanine molecules in the crystals are mostly of van der Waals type, the electronic and structural characteristics of the molecules are preserved in the crystalline state. Thus, the determination of the electronic structure of the isolated molecular units constitutes a crucial step towards the understanding of the electronic and optical properties of phthalocyanine-based materials.

In this contribution, we discuss and review the electronic and optical properties of phthalocyanine-based molecular building units used to obtain electroresponsive materials. These units are:

    i) Basic macrocyclic units: Metal free phthalocyanine (H_2Pc) and tetrabenzoporphyrin (H_2TBP).

    ii) Phthalocyanine radicals: Lithium phthalocyanine (LiPc) and lutetium diphthalocyanine ($LuPc_2$).

    iii) Extended phthalocyanine macrocycles: 2,3-naphthalocyanine ($2,3-H_2Nc$), 1,2-naphthalocyanine ($1,2-H_2Nc$), and 9,10-phenanthrenocyanine (H_2Phc).

To investigate the electronic structure of all of these molecular systems, we make use of the nonempirical valence effective Hamiltonian (VEH) method. A brief outline of our theoretical approach is presented in Sect. 2.

2. Theoretical Approach

All the calculations presented and discussed in this work have been performed in the framework of the quantum-chemical VEH pseudopotential technique [18,19]. The VEH method takes only into account the valence electrons and is based on the use of an effective Fock Hamiltonian parameterized to reproduce the results of ab initio calculations without performing any self-consistent-field (SCF) process or calculating any bielectronic integral. In this way, the VEH method constitutes an especially useful tool to deal with very large molecular systems since it yields one-electron energies of ab initio double-zeta quality at a reasonable computer cost.

The suitability of the VEH approach for describing the electronic structure of phthalocyanine-type macrocycles is supported by previous works on metal-free phthalocyanine [20,21] and porphine [22] systems. An excellent agreement is found between the calculated VEH electronic structures and the experimental photoemission (UPS or XPS) data both for gas and solid phases. It is also our experience that the VEH calculations provide good estimates for the energies of the lower energy optical transitions. This feature is due to the fact that the VEH parameterization is not contaminated by any information coming from the unoccupied Hartree-Fock molecular orbitals [24]. However, in order to discuss the optical properties of the macrocycles investigated in this work, two features should be considered in relation to the theoretical

520

approach:

(i) The VEH method yields, as any other ab initio technique, two wide a valence
structure and a contraction of the energy scale for the occupied valence levels
becomes necessary to obtain the best correlation with experimental photoemission
data. A contraction factor of 1.3 has been consistently used as discussed in detail in
Ref. 20.

(ii) The VEH method predicts a HOMO-LUMO energy gap of only 1.22 eV for the
phthalocyanine molecule, underestimating by 0.59 eV the energy of the first optical
transition (1.81 eV) reported from vapor absorption spectra for H_2Pc [23].

Therefore, the energies of electron transitions from the occupied to the unoccupied levels
are calculated as one-electron energy differences after contracting by a factor of 1.3 the
energy scale for the occupied levels and rigidly shifting by 0.59 eV to higher energies the
unoccupied levels. These adjustments are systematically used in all the phthalocyanine-
based molecular systems investigated in this work in order to provide a fully consistent
approach than can result in meaningful comparisons. The tetrabenzoporphyrin molecule
constitutes a different system, as discussed later on, and the 0.59 eV shift is not
necessary in this case.

Calculations have been performed by using the VEH parameters previously reported
for the nitrogen atoms [24] and those recently obtained for the carbon and hydrogen
atoms [25]. Molecular geometries have been obtained from crystallographic data, when
available, or constructed from the neutron diffraction structure reported for H_2Pc [26].

3. Basic Macrocyclic Units

As a first step, we present the electronic structure of the metal-free phthalocyanine
(H_2Pc) and tetrabenzoporphyrin (H_2TBP) macrocycles. Although phthalocyanines have
been more extensively studied, tetrabenzoporphyrins give rise to low dimensional
molecular metals and polymers which exhibit conductivities similar to those reported for
phthalocyanine-based compounds [4b,c]. Indeed, the synthesis has been reported for a
mixed polymer $(GeTBPcO)_n$ containing phthalocyaninatogermanium (GePc) and
tetrabenzoporphyrinatogermanium (GeTBP) subunits and combining the different
electronic properties of both macrocyclic units [27].

As can be seen from Fig. 1a, phthalocyanine and tetrabenzoporphyrin macrocycles
only differ by the units linking the isoindole moieties. The four bridging aza nitrogen
atoms of phthalocyanine are replaced by four methine carbon atoms in
tetrabenzoporphyrin. The molecular geometry employed for H_2Pc is based on the C_{2h}
structure reported by Hoskins et al. [26] averaged to a D_{2h} symmetry as explained in

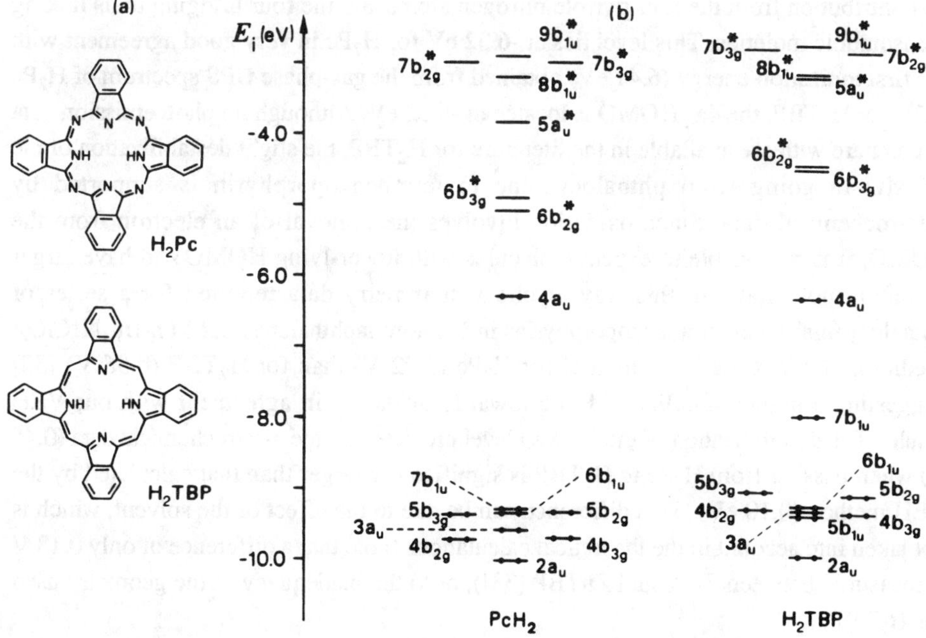

Figure 1. (a) Molecular structures of metal-free phthalocyanine, H$_2$Pc (top), and metal-free tetrabenzoporphyrin, H$_2$TBP (bottom). (b) VEH one-electron energies (E$_i$) and symmetries of the upper occupied and lower unoccupied molecular orbitals calculated for H$_2$Pc and H$_2$TBP.

previous works [20,28]. Since no structural data are available for metal-free tetrabenzoporphyrin, the input geometry for H$_2$TBP is obtained by linking the isoindole moieties of H$_2$Pc by C-C bonds of about 1.38 Å taken from the X-ray structure of porphine [29]. This assumption is supported by the almost identical molecular structures reported for Ni(TBP)I [30] and Ni(Pc)I [31] compounds.

The VEH one-electron energy level distributions obtained for H$_2$Pc and H$_2$TBP are schematically displayed in Fig. 1b. Only the highest occupied and lowest unoccupied molecular orbitals are included in Fig. 1b and they have been classified according to the D$_{2h}$ point group of symmetry. All the orbitals in Fig. 1b are of π-nature.

The VEH molecular orbital distributions calculated for H$_2$Pc and H$_2$TBP are very similar (Fig. 1b). For both molecules, the HOMO corresponds to the 4a$_u$ level which has no contribution from the four pyrrole nitrogen atoms and the four bridging units linking the isoindole moieties. This level lies at -6.32 eV for H_2Pc in very good agreement with the first ionization energy (6.41 eV) obtained from the gas-phase UPS spectrum of H_2Pc [32]. For H_2TBP, the $4a_u$ HOMO is located at -6.22 eV. Although no photoemission data to compare with are available in the literature for H_2TBP, the slight destabilization of the HOMO in going from phthalocyanine to tetrabenzoporphyrin is supported by electrochemical data. Since oxidation involves the removal of an electron from the HOMO, it is reasonable to expect molecules with lower-lying HOMO's to have larger oxidation potentials. In this way, cyclic voltammetry data reported for a series of phthalocyanines and tetrabenzoporphyrins in 1-chloronaphthalene / 0.2 M $n\text{-}Bu_4N^+ClO_4^-$ predict a larger oxidation potential for H_2Pc (1.02 V) than for H_2TBP (0.56 V) [33] suggesting a higher stability of H_2Pc towards oxidation in agreement with our VEH results. The destabilization of the HOMO level predicted by the electrochemical data (0.46 V) when passing from H_2Pc to H_2TBP is significantly larger than that calculated by the VEH method (0.10 eV). This difference can be due to the effect of the solvent, which is not taken into account in the theoretical calculations (note that a difference of only 0.18 V is measured between ZnPc and ZnTBP [33]), or to the inadequacy of the geometry used for H_2TBP.

The lowest unoccupied molecular orbital (LUMO) corresponds to a pair of almost degenerate $6b_{2g}^*\text{-}6b_{3g}^*$ molecular orbitals for both phthalocyanine and tetrabenzo-porphyrin. As discussed previously [34], the HOMO-LUMO electronic transitions are to be correlated with the double-peak Q absorption band observed in the visible. The VEH method predicts a HOMO-LUMO energy gap of only 1.22 eV for phthalocyanine, underestimating by 0.59 eV the energy of the first optical transition (1.81 eV) reported for H_2Pc vapors [23]. On the contrary, the VEH method calculates a HOMO-LUMO energy gap of 1.82 eV for tetrabenzoporphyrin in excellent agreement with the energy of the Q band (1.86 eV) measured for H_2TBP in 1-chloronaphthalene [33]. A similar energy value could be expected for the Q band of H_2TBP in the gas phase, since the energy of this band for H_2Pc slightly changes from the gas phase (1.81 eV) [23] to 1-chloronaphthalene solution (1.87 eV) [33]. This is also the case for ZnTBP where the Q band appears at 1.98 and 1.97 eV in the gas phase [35] and 1-chloronaphthalene solution [33], respectively. The failure of the VEH method in estimating the HOMO-LUMO energy gap in H_2Pc can be explained by the fact that the parameterization for the nitrogen atom [24] was based on pyrrole and dimethylamine molecules without taking into account any chemical environment related to the aza nitrogen atoms linking the isoindole moieties in H_2Pc. These aza nitrogens contribute a great deal to the $6b_{2g}^*\text{-}6b_{3g}^*$ molecular orbitals. To correct this shortcoming of the VEH method, a shift of 0.59 eV to higher energies of the unoccupied levels is systematically used for all the systems studied in this work which derive from the phthalocyanine macrocycle.

Despite the great similarity observed between the VEH electronic structures calculated for phthalocyanine and tetrabenzoporphyrin, there is an important difference concerning the location of the second HOMO (see Fig. 1b). While for H_2Pc the $7b_{1u}$ level is imbedded in a group of very close-lying orbitals separated by a gap of 3.00 eV from the $4a_u$ HOMO, for H_2TBP it lies alone 1.66 eV below the $4a_u$ HOMO. This explains the differences observed for the B or Soret band in the absorption spectra of phthalocyanines and tetrabenzoporphyrins. Phthalocyanines present a broad band centered around 320-340 nm [23]. This band can be correlated with the series of very close-lying electronic transitions centered around the $7b_{1u} \rightarrow 6b_{2g}$*, $6b_{3g}$* excitations starting at 3.90 eV (317 nm) for H_2Pc. On the contrary, tetrabenzoporphyrins show a very sharp Soret structure observed for ZnTBP at 405 nm [35,36]. The sharpness of this structure is due to the fact that here the B band only results from the excitation of an electron from the isolated $7b_{1u}$ level to the $6b_{2g}$*-$6b_{3g}$* LUMO's. These transitions are calculated to be at 3.09 eV (400 nm) for H_2TBP, in excellent agreement with the experimental value measured for ZnTBP.

4. Phthalocyanine Radicals

We discuss in this section the electronic structures of lithium phthalocyanine (LiPc) and lutetium diphthalocyanine (LuPc$_2$) compounds. The importance of these systems in the context of organic semiconductors is determined by the fact that they are claimed to be the first molecular semiconductors reported to date [37-39]. Room temperature conductivities as high as 2×10^{-3} and 6×10^{-5} S/cm have been respectively measured for single crystals of LiPc [38] and LuPc$_2$ [40] without any external doping. These conductivities are to be compared with the low values ($<10^{-12}$ S/cm) reported for single crystals of H_2Pc or common metallophthalocyanines [13].

The VEH calculations for LiPc and LuPc$_2$ were performed on the basis of their respective X-ray crystallographic structures [41,42] without taking explicitly into account the lithium or lutetium atoms. As previously discussed [34], the role of the metals in these cases is fairly well represented simply by considering their influence on the geometry of the phthalocyanine rings. Fig. 2 displays the resulting VEH molecular orbital distributions for the LiPc and LuPc$_2$ compounds. The electronic structure calculated for H_2Pc is included for the sake of comparison. All the orbitals depicted in Fig. 2 are of π-type.

As can be seen from Fig. 2, the molecular orbital distribution of the LiPc model is

524

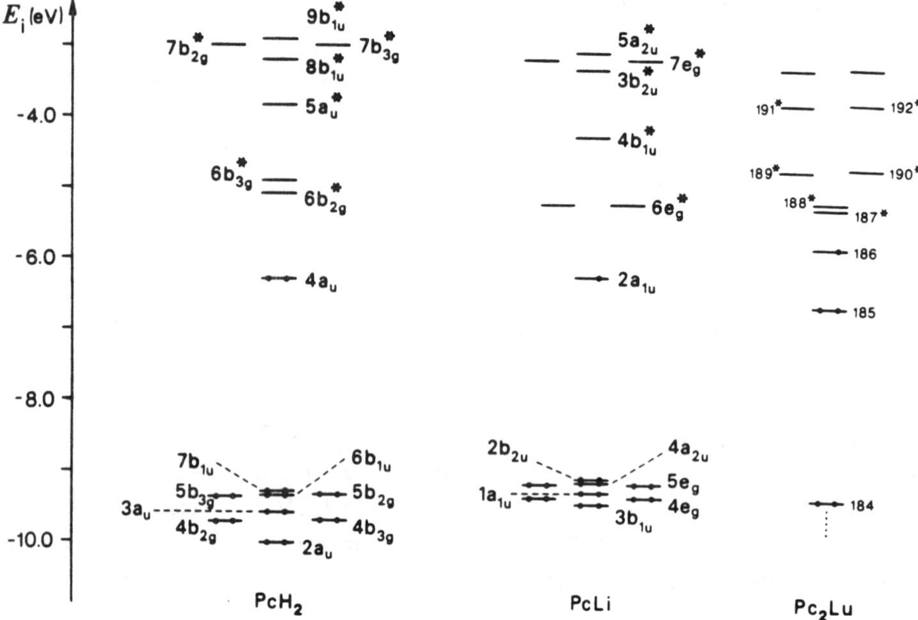

Figure 2. VEH one-electron energies (E_i) of the upper occupied and lower unoccupied molecular orbitals calculated for H_2Pc and the LiPc - LuPc$_2$ model systems. MO's of H_2Pc and LiPc are classified according to D_{2h} and D_{4h} symmetries, respectively.

very similar to that obtained for H_2Pc. The main difference between LiPc and H_2Pc is that for the former the $2a_{1u}$ HOMO is occupied by a single electron due to the radical nature of the phthalocyanine ring. (The Li atom gives a single electron to the Pc ring which becomes a stable monoanion radical, Pc$^-$). As a consequence, additional electronic transitions from lower occupied levels to the $2a_{1u}$ HOMO take place in LiPc. Note that the $6b_{2g}$*-$6b_{3g}$* MO's of H_2Pc evolve to give rise in LiPc to the degenerate $6e_g$* LUMO. As a result, the HOMO $\rightarrow$ LUMO transition is no longer split into two peaks.

The experimental optical spectrum of LiPc [43] shows well defined absorptions at 1.53 eV (810 nm), 2.62 eV (473 nm), 2.90 eV (427 nm), and 3.35 eV (370 nm) which can be assigned on the basis of our theoretical calculations. The absorption band at 1.53 eV results from the electronic HOMO ($2a_{1u}$) $\rightarrow$ LUMO ($6e_g$*) transition calculated at 1.62 eV and can therefore be correlated with the Q band observed in H_2Pc. The band observed at 3.35 eV can be attributed to the electronic transitions from the low-lying levels ($2b_{2u}$,

$4a_{2u}$,....) to the $6e_g^*$ LUMO and corresponds to the B or Soret band of H_2Pc. Finally, the intermediate absorptions bands at 2.62 and 2.90 eV originate in the $5e_g$, $4e_g \rightarrow 2a_{1u}$ and $3e_g \rightarrow 2a_{1u}$ electronic transitions calculated at 2.42 and 3.01 eV, respectively. These two bands are not present in H_2Pc because they imply the promotion of an electron to the half-filled HOMO level which is doubly occupied in H_2Pc.

An overall splitting of the molecular orbitals is observed in Fig. 2 when passing from the monomeric phthalocyanines to the $LuPc_2$ molecule, due to the interaction between the two Pc rings. Although these rings are severely distorted from planarity, the mean distance separating the planes formed by the four isoindole nitrogens of both macrocycles is very short (2.69 Å). As a consequence, a large splitting (0.83 eV) is obtained for the MO's 185 and 186 resulting from the doubling of the π-HOMO of the monomers. The calculated splitting is in very good agreement with the absorption band observed at 0.90 eV in the near infrared electronic spectrum of $LuPc_2$ in dichloromethane solution [44]. This absorption band therefore corresponds to the promotion of an electron from the MO number 185 to the half-occupied HOMO (see Fig. 2) and can be considered as an intramolecular charge transfer between the two phthalocyanine rings since it is polarized following the axis perpendicular to the planes of the rings. The $LuPc_2$ molecule shows another absorption band in the near infrared region at 1.37 eV [44], which can be assigned to the rather weak (oscillator strength $\approx$ 0.2) $186 \rightarrow 187^*,188^*$ HOMO-LUMO transitions calculated to be around 1.20 eV.

Two absorption bands are observed for $LuPc_2$ in the visible region at 1.88 and 2.70 eV [44,45]. The former corresponds almost exactly to the center of gravity of the $186 \rightarrow 189^*$, 190^* and $185 \rightarrow 187^*$, 188^* theoretical transitions (1.87 eV) and can be clearly correlated with the strong Q absorption band observed for the monomers. As for LiPc, the band at 2.70 eV can be attributed to electronic transitions from deeper levels to the half-occupied HOMO.

In summary, the VEH calculations allow for a complete assignment of the optical features observed for the LiPc and $LuPc_2$ molecular semiconductors. For these compounds, it is found that the major factor influencing the evolution of the optical transitions is not the electronic structure of the metal but rather the geometric structure of the phthalocyanine rings.

5. Extended Phthalocyanine Macrocycles

Since the electrical conductivity of phthalocyanine crystals strongly depends on the orientation and spacing of phthalocyanine rings in the molecular stacks, more effective π-

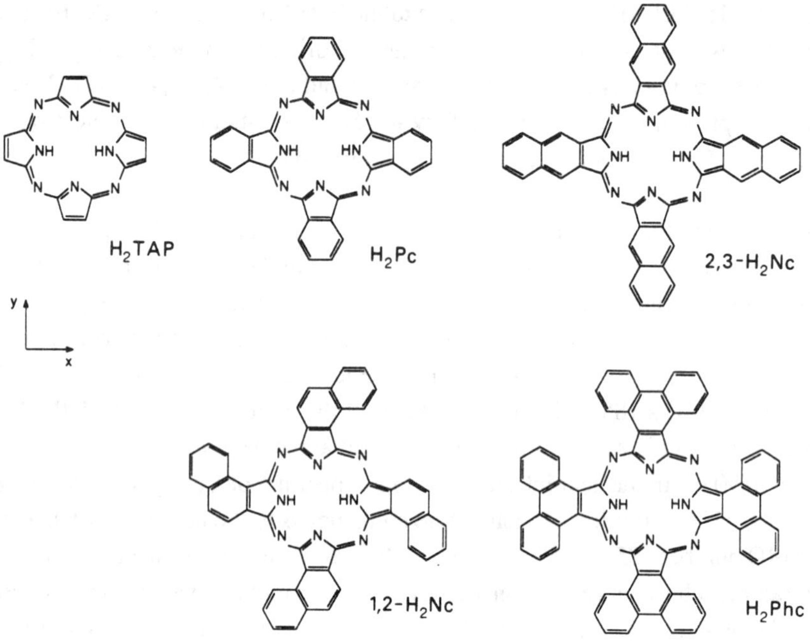

526

Figure 3. Molecular structures of tetraazaporphyrin (H₂TAP), phthalocyanine (H₂Pc), 2,3-naphthalocyanine (2,3-H₂Nc), 1,2-naphthalocyanine (1,2-H₂Nc), and 9,10-phenanthrenocyanine (H₂Phc).

interactions and, hence, higher conductivities could be expected if the conjugated system of the phthalocyanine macrocycle is extended. In this section, we analyze the electronic structure of the extended phthalocyanines depicted in Fig. 3 in order to investigate how the extension of the aromatic structure affects those molecular electronic properties more closely related to electrical conductivity. The macrocycles studied are 2,3-naphthalocyanine (2,3-H₂Nc), 1,2-naphthalocyanine (1,2-H₂Nc), and 9,10-phenanthrenocyanine (H₂Phc). Note that 1,2-H₂Nc and H₂Phc can be regarded as derivatives of phthalocyanine with angular annellation of the additional benzene rings, while 2,3-H₂Nc presents a linear annellation of the benzene rings.

The VEH one-electron energy level distributions obtained for the five molecular systems sketched in Fig. 3 are displayed in Fig. 4. The electronic structures of the reference macrocycles tetraazaporphyrin (H₂TAP) and phthalocyanine are included for the sake of comparison. All the molecular orbitals in Fig. 4 are of π-nature and they have

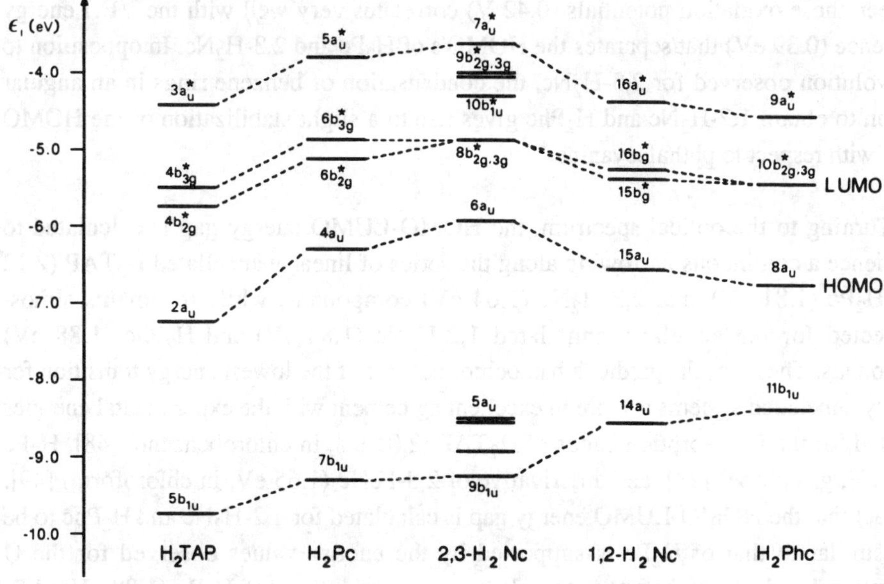

Figure 4. VEH one-electron energies (E_i) and symmetries of the upper occupied and lower unoccupied molecular orbitals calculated for H_2TAP, H_2Pc, 2,3-H_2Nc, 1,2-H_2Nc, and H_2Phc.

been classified acording to the D_{2h} symmetry except for 1,2-naphthalocyanine for which a C_{2h} geometrical structure was assumed [46]. The D_{2h} geometries of tetraazaporphyrin, 2,3-naphthalocyanine, and phenanthrenocyanine were built from the geometrical structure reported for H_2Pc [26].

Fig. 4 illustrates that the HOMO level, which in all cases corresponds to a level of a_u symmetry, undergoes an energy destabilization in going from tetraazaporphyrin (-7.27 eV) to phthalocyanine (-6.32 eV) and from phthalocyanine to 2,3-naphthalocyanine (-5.93 eV). This trend indicates that the extension of the conjugated system by annellation of benzene rings following the axis of the pyrrole units (linear annellation) produces a continuous destabilization of the highest occupied molecular orbital and, as a consequence, electrons are more easily extracted along the series H_2TAP, H_2Pc, 2,3-H_2Nc. This result is supported by the larger oxidation potentials measured for phthalocyanine compounds compared to those observed for 2,3-naphthalocyanines. For exemple, oxidation potentials of 1.00 and 0.58 V vs SCE have been measured in CH_2Cl_2 for $SiPc(OR)_2$ and 2,3-$SiNc(OR)_2$ compounds, respectively [47]. The difference between these oxidation potentials (0.42 V) correlates very well with the VEH energy difference (0.39 eV) that separates the HOMO's of H_2Pc and $2,3-H_2Nc$. In opposition to the evolution observed for $2,3-H_2Nc$, the condensation of benzene rings in an angular fashion to obtain $1,2-H_2Nc$ and H_2Phc gives rise to a slight stabilization of the HOMO levels with respect to phthalocyanine.

Turning to the optical spectrum, the HOMO-LUMO energy gap is calculated to experience a continuous narrowing along the series of linearly annellated H_2TAP (2.12 eV), H_2Pc (1.81 eV), and $2,3-H_2Nc$ (1.64 eV) compounds, while it remains almost unaffected for the angularly annellated $1,2-H_2Nc$ (1.84 eV) and H_2Phc (1.88 eV) compounds. These results predict a bathochromic shift of the lowest energy transition for linearly annellated systems and are in excellent agreement with the experimental energies reported for the Q absorption bands of H_2TAP (2.01 eV, in chlorobenzene) [48], H_2Pc (1.81 eV, gas phase) [23], and a derivative of $2,3-FeNc$ (1.65 eV, in chloroform) [49]. The fact that the HOMO-LUMO energy gap is calculated for $1,2-H_2Nc$ and H_2Phc to be very similar to that of H_2Pc is supported by the energy values observed for the Q absorption band of octahedrally complexed iron derivatives of H_2Pc (1.88 eV), $1,2-H_2Nc$ (1.90 eV), and H_2Phc (1.89 eV) [46].

In summary, the VEH results show that, for phthalocyanine-type macrocycles, linear annellation leads to a more effective conjugation than angular annellation in the macrocyclic system. Lower oxidation potentials and narrower energy gaps are thus predicted for linearly annellated than for angularly annellated phthalocyanines and, as a result, better conductivity properties are to be expected for crystals and polymers derived from the former. This is consistent with the higher electrical conductivities reported for $2,3-FeNc$ (4×10^{-5} S/cm) and the corresponding diisocyano-bridged (dib) polymer $[2,3-FeNc(dib)]_n$ (2×10^{-3} S/cm). These conductivities, obtained without any intentional external doping [50], can be compared to those of FePc (4×10^{-9} S/cm) and $1,2-FeNc$ (4×10^{-9} S/cm) [49]. Preliminary VEH calculations on more extended linearly annellated phthalocyanines suggest the possibility of obtaining crystals and polymers with very small bandgaps.

Acknowledgements

We thank the CIUV (Centro de Informática de la Universitat de València) and the University of Mons Computer Center for use of computing facilities. EO is indebted to the University of Mons for its hospitality and financial support. This work has been partially supported by the DGICYT Project PS88-0112 and the Project No 683.2/89 from the University of Valencia. Finally, the authors are indebted to Professor M. Hanack for communicating unpublished data and acknowledge stimulating discussions with Dr. C. Clarisse.

References

1. For recent reviews, see A.B.P. Lever, M.R. Hempstead, C.C. Leznoff, W. Liu, M. Melnik, W.A. Nevin, and P. Seymour, Pure Appl. Chem. 58, 1467 (1986); K. Kasuga and M. Tsutsui, Coord. Chem. Rev. 32, 67 (1980).
2. R.O. Loutfy and J.H. Sharp, J. Chem. Phys. 71, 1211 (1979); C.W. Tang, Appl. Phys. Lett. 48, 183 (1986).
3. M. Kato, Y. Nishioka, K. Kaifu, K. Kawamura, and S. Ohno, Appl. Phys. Lett. 46, 196 (1985).
4. For recent reviews, see (a) T.J. Marks, Science 227, 881 (1985); (b) B.M. Hoffman and J.A. Ibers, Acc. Chem. Res. 16, 15 (1983); S.M. Palmer, J.L. Stanton, J. Martinsen, M.Y. Ogawa, W.B. Hener, S.E. Van Wallendael, B.M. Hoffman, and J.A. Ibers, Mol. Cryst. Liq. Cryst. 125, 1 (1985); (c) M. Hanack, A. Datz, R. Fay, K. Fischer, U. Keppeler, J. Koch, J. Metz, M. Mezger, O. Schneider, and H.J. Schulze in "Handbook of Conducting Polymers", edited by T. A. Skotheim, (Marcel Dekker, New York, 1986), Vol 1, Chap. 5, p. 133.
5. R.A. Collins and K.A. Mohamed, J. Phys. D 21, 154 (1988); T.A. Temofonte and K.F. Schoch, J. Appl. Phys. 65, 1350 (1989).
6. H. Yamamoto, T. Sugiyama, and M. Tanaka, Jpn. J. Appl. Phys. 24, L305 (1985).
7. S. Baker, M.C. Petty, G.G. Roberts, and M.V. Twigg, Thin Solid Films 99, 53 (1983); M.J. Cook, A.J. Dunn, M.F. Daniel, R.C.O. Hart, R.M. Richardson, and S.J. Roser, ibid. 159, 395 (1988); S. Palacin, P. Lesieur, I. Stefanelli, and A. Barraud, ibid. 159, 83 (1988).
8. J. Simon, P. Bassoul, and S. Norvez, New J. Chem. 13, 13 (1989).
9. W.R. Barger, H. Wohltjen, and A.W. Snow, Transductors 85, (IEEE, New York, 1985), p. 410.
10. G.G. Roberts, M.C. Petty, S. Baker, M.T. Fowler, and N.J. Thomas, Thin Solid Films 132, 113 (1985).
11. Y. Nishimura, H. Kawada, M. Haruta, Y. Hirai, N. Mochizuki, and T. Nakagiri, Jpn Patent 86 37475, 22nd February 1986.
12. M. Yonegama, M. Sugi, M. Saito, K. Ikagami, S. Kuroda, and S. Iizima, Jpn. J. Appl. Phys. 25, 961 (1986).
13. J. Simon and J.-J. André, "Molecular Semiconductors", (Springer, Berlin, 1985), Chap III.
14. T. Inabe, T.J. Marks, R.L. Burton, J.W. Lyding, W.J. McCarthy, C.R. Kannewurf, G.M. Reisner, and F.H. Herbstein, Solid State Commun. 54, 501 (1985).
15. T. Inabe, J.G. Gaudiello, M.K. Moguel, J.W. Lyding, R.L. Burton, W.J. McCarthy, C.R. Kannewurf, and T.J. Marks, J. Am. Chem. Soc. 108, 7595 (1986); J.G. Gaudiello, G.E. Kellogg, S.M. Tetrick, and T.J. Marks, ibid. 111, 5259 (1989).
16. A.B.P. Lever, Adv. Inorg. Chem. Radiochem. 7, 115 (1965); D. Eastwood, L. Edwards, M. Gouterman, and J. Steinfeld, J. Mol. Spectrosc. 20, 381 (1966).

530

17. M. Gouterman, G.H. Wagnière, and L.C. Snyder, J. Mol. Spectrosc. 11, 108 (1963); C. Weiss, H. Kobayashi, and M. Gouterman, *ibid.* 16, 415 (1965).
18. G. Nicolas and Ph. Durand, J. Chem. Phys. 70, 2020 (1979); 72, 453 (1980).
19. J.M. André, L.A. Burke, J. Delhalle, G. Nicolas, and Ph. Durand, Int. J. Quantum Chem. Symp. 13, 283 (1979); J.L. Brédas, R.R. Chance, R. Silbey, G. Nicolas, and Ph. Durand, J. Chem. Phys. 75, 255 (1981).
20. E. Ortí and J.L. Brédas, J. Chem. Phys. 89, 1009 (1988).
21. E. Ortí and J.L. Brédas, Synth. Met. 29, F115 (1989).
22. E. Ortí and J.L. Brédas, Chem. Phys. Lett., in press.
23. L. Edwards and M. Gouterman, J. Mol. Spectrosc. 33, 292 (1970).
24. J.L. Brédas, B. Thémans, and J.M. André, J. Chem. Phys. 78, 6137 (1983).
25. B. Thémans, J.M. André, and J.L. Brédas, to be published.
26. B.F. Hoskins, S.A. Mason, and J.C.B. White, J. Chem. Soc., Chem. Commun., 554 (1969).
27. M. Hanack and T. Zipplies, Synth. Met. 25, 341 (1988).
28. E. Ortí, M.C. Piqueras, R. Crespo, and J.L. Brédas, submitted for publication.
29. B.M.L. Chen and A. Tulinsky, J. Am. Chem. Soc. 94, 4144 (1972).
30. J. Martinsen, J.L. Pace, T.E. Phillips, B.M. Hoffman, and J.A. Ibers, J. Am. Chem. Soc. 104, 83 (1982).
31. C.J. Schramm, R.P. Scaringe, D.R. Stojakovic, B.M. Hoffman, J.A. Ibers, and T.J. Marks, J. Am. Chem. Soc. 102, 6702 (1980).
32. J. Berkowitz, J. Chem. Phys. 70, 2819 (1979).
33. R. Behnisch, B. Speiser, M. Hanack, and G.E. Kellogg, Inorg. Chem., in press.
34. E. Ortí, J.L. Brédas, and C. Clarisse, J. Chem. Phys., in press.
35. L. Edwards, M. Gouterman, and C.B. Rose, J. Am. Chem. Soc. 98, 7638 (1976).
36. L. Bajema, M. Gouterman, and C.B. Rose, J. Mol. Spectrosc. 39, 421 (1971).
37. M. Maitrot, G. Guillaud, B. Boudjema, J.-J. André, H. Strzelecka, J. Simon, and R. Even, Chem. Phys. Lett. 133, 59 (1987).
38. Ph. Turek, P. Petit, J.-J. André, J. Simon, R. Even, B. Boudjema, G. Guillaud, and M. Maitrot, J. Am. Chem. Soc. 109, 5119 (1987).
39. P. Turek, P. Petit, J.-J. André, J. Simon, R. Even, B. Boudjema, G. Guillaud, and M. Maitrot, Mol. Cryst. Liq. Cryst. 161, 323 (1988).
40. P. Petit, K. Holczer, and J.-J. André, J. Phys. 48, 1363 (1987).
41. H. Sugimoto, M. Mori, H. Masuda, and T Taga, J. Chem. Soc. Chem. Commun., 962 (1986).
42. A. De Cian, M. Moussavi, J. Fischer, and R. Weiss, Inorg. Chem. 24, 3162 (1985).
43. Ph. Turek, J.-J. André, A. Giraudeau, and J. Simon, Chem. Phys. Lett. 134, 471 (1987).
44. D. Markovitsi, T.-H. Tran-Thi, R. Even, and J. Simon, Chem. Phys. Lett. 137, 107 (1987).
45. C. Clarisse and M.T. Riou, Inorg. Chim. Acta 130, 139 (1987).
46. M. Hanack, G. Renz, J. Strähle, and S. Schmid, Chem. Ber. 121, 1479 (1988).
47. B.L. Wheeler, G. Nagasubramanian, A.J. Bard, L.A. Schechtman, D.R. Dininny, and M.E. Kenney, J. Am. Chem. Soc. 106, 7404 (1984).
48. R.P. Linstead and M. Whalley, J. Chem. Soc., 4839 (1952).
49. M. Hanack and S. Deger, Isr. J. Chem. 27, 347 (1986).
50. S. Deger and M. Hanack, Synth. Met. 13, 319 (1986).

CONTROL OF INTRAMOLECULAR INTERFERENCES THROUGH BENZENE AND CYCLOPHANE USING DONOR AND ACCEPTOR GROUPS

P. Sautet [+] *and* *C. Joachim**

[+] Laboratoire de Chimie Théorique
Ecole Normale Supérieure de Lyson
69364 Lyon Cedex (France)
* Molecular Electronic Group
CEMES-LOE/CNRS
29, rue Jeanne Marvig - 31055 Toulouse Cedex (France)

ABSTRACT. The elastic scattering quantum chemistry method is used to calculate electron transmission through donor-acceptor substituted benzene and cyclophane embedded in a $(CH)_x$ infinit chain. Through benzene interferences are controlled by these substituants either directly grafted on the ring or grafted on the other cyclophane ring. The cyclophane structure open a new way in the control of intramolecular interference due to the spatial separation in this molecule between the tunnelling channels and the polarization groups.

1. INTRODUCTION

Electron interference phenomena have been studied recently in solid state physics [1] using for example mesoscopic systems [2]. Because these interferences come from the wave character of the electron, they must also appear at a scale lower than the mesoscopic one. At the molecular scale for example, an electronic "interferometer" can be obtained using a single molecular impurity (with a donor (D) or acceptor (A) character) grafted on a conducting polymeric chain [3].

This impurity introduces new electronic states in the impurity + conducting chain system which must be located in or near the energy bands of the conducting chain to be active. Therefore, when an electron coming from this chain is scattered by the impurity, it can tunnel through the conducting chain electronic levels, through the impurity electron levels or through both. The main difference between these tunnelling channels is the phase shift of the tunnelling electron wave passing through the impurity compared to the one passing through the conducting chain [4]. If the electron wave propagation in the neighbourhood of the impurity is elastic, electronic interferences occur between the part of the electron wave passing through the impurity and the part passing through the conducting chain.

531

J. L. Brédas and R. R. Chance (eds.), Conjugated Polymeric Materials:
Opportunities in Electronics, Optoelectronics, and Molecular Electronics, 531–543.
© 1990 *Kluwer Academic Publishers. Printed in the Netherlands.*

The interference patterns so obtained can be predicted by calculating the electronic transmission T(E) through the impurity + conducting chain system as a function of the energy E of the incident electron [5]. It can be measured from the elastic conductance of the system like for mesoscopic systems [6]. T(E) is related to this conductance by the Landauer formula [7] or its multichannel extension [8].

To improve the efficiency of such an interferometer, we have studied T(E) for a benzene embedded in a polyacetylene infinit chain [9]. The benzene has the same shape than the well known normal mesoscopic metallic loop where the Aharonov-Bohm effect have been detected. Therefore, even if its diameter is much more smaller, interferences exist through such a subnanoscopic loop for an electron plane wave coming from the conducting chain and divided in two parts by the ring. The two parts interfer after passing through the benzene [9].

But an asymmetry is required between the two electronic tunnelling paths through the benzene to get an interference. Examples of such asymmetry are chemically controlable [10] or photo sensible donor-acceptor groups [11] grafted only on one side of the benzene which actively control the interference pattern.

Interest in the benzene compared to a single impurity is that interferences are clearly topologic in nature : the electron plane wave passes through a different number of C-C bonds in each part of the benzene depending on the connection site chosen between the benzene and the conducting chains. Therefore, when a Donor or an Acceptor group is grafted on one side of the benzene only, the interesting effect is the polarization of the π electrons system of the benzene. This polarization induces a shift of the intereference pattern. This effect is formally equivalent to the electric Aharonov-Bohm effect in mesoscopic normal metallic loops [12].

The problem with the benzene approach to molecular interferences is that donor and acceptor introduce not only interference pattern shifts but also new interferences which in general completly transform the original one. This is due in part to the electronic coupling between D, A and the benzene π electron system which must be high enough to shift the original interference pattern. It is therefore important to find a way to shift the interference through the benzene without introducing new large interferences.

One solution may be to insert saturated bonds between the donor (the acceptor) and the benzene. In this case, the π electron system of the two will be uncoupled in first approximation. It will result very narrow new interference holes but also a drastric decrease of the shift effect which is not very suitable. An other solution is to use the privilegied cyclophane structure where two benzenes are tighted co-planar. One ring is connected to the conducting chains and the other is used to attach the donor and the acceptor groups $\underline{1}$.

$\underline{1}$

This paper presents results on the study of such a system compared to the benzene and to the substituted benzene. The paper is divided as follow. The ESQC technique used to calculate at a quantum chemistry level T(E) is shortly described section 2 together with a discussion on the relation between T(E) and the elastic conductance through a molecule. Results on the interference pattern shift through the benzene by donor and acceptor groups are analyzed section 3. Section 4 presents results on cyclophane and susbstituted cyclophane together with a comparison with through benzene interference. Extensions of this work are discussed in conclusion.

2. ELECTRON TRANSMISSION THROUGH A SPACER AND CONDUCTANCE

Let us consider a regular polymer chain where a molecule (the spacer), different from the monomer of the polymer, has been inserted. In the regular part of this polymer, each electron experiments far form the spacer a periodic potential. The wave function of a delocalized electron on the polymer is a "Bloch wave" made of a linear combination of elementary Bloch functions within each energy band of the polymer.

But the translational invariance all along this polymer is broken by the spacer because the potentiel energy of an electron on or near the spacer is different from the one encountered far from this spacer. Therefore, an electron at the Fermi energy E_F coming from a conducting band of the regular polymer and attempting to reach the other regular part after the spacer must tunnel through this spacer. In this elastic scattering process, a part of the electron incident Bloch wave passes through the spacer while the other part is reflected back to the regular chain. The transparency of such a spacer to an electron flux is usually characterized by the transmission coefficient T(E), ratio between the transmitted and the incident electron flux.

At first, it may seems that a conjugated spacer keeps the conjugated nature of the polymer. In this case, electron delocalization around the spacer will be high and so will be T(E). A non-conjugated spacer must localized the Bloch wave leading to a low T(E). However, such analysis does not consider the energy dependence of T(E). To get a clear picture of

534

this dependence, the detailled electronic structure of the spacer and of the polymer must be taken into account. The Elastic Scattering Quantum Chemistry (ESQC) method have been implemented exactly to get a quantum chemistry description of the polymer-spacer-polymer system and to provide $T(E)$ from the standard scattering matrix technique used for mesoscopic systems [8].

In ESQC, each molecular orbital ϕ_m of the spacer calculated with a quantum chemistry program (Extended Huckel for example) is considered as a possible tunnelling channel for the electrons provided by the polymer. Such an orbital located within an energy band of the polymer and overlapping well with the adjacent monomers orbitals of the polymer gives a large transmission coefficient $t_m(E)$ through the channel opens by ϕ_m. Molecular orbitals out of the bands or with a small overlap lead to a small $t_m(E)$.

The overall $T(E)$ is calculated in ESQC from to superposition of all these individual channel contributions. But this "mixing step" is not additive. One cannot set $T(E) = \sum t_m(E)$ because each molecular level shifts the phase of the tunnelling electron wave using the corresponding channel. Interference effects appear when two or more channels are considered together. ESQC combined all these channel effects by calculating a multi-channel scattering matrix. Moreover, a qualitative rule to predict the nature of these interferences between two channels (destructive or constructive) has also been presented [9].

Next to the $T(E)$ calculation is the relation of this calculation with experimental results. When an electron flux is kept through the polymer by a low bias voltage applied to the polymer far from the spacer, only Fermi energy electrons are provided to the spacer. At low temperature, the mean free path of an electron in the polymer becomes greater than the effective tunnelling length of the spacer. In this case, the relation between the voltage V applied to the polymer and the current I stabilized through the polymer-spacer-polymer junctions is given by [8] :

$$I = \frac{e^2}{\pi\hbar} T(E_F)\, V \tag{1}$$

Therefore, a measurement of the sample conductance $G = e^2/\pi\hbar\, T(E_F)$ provides $T(E_F)$. This conductance is not equal to the "spacer conductance", but to the polymer-spacer-polymer conductance because in (1), the voltage V is applied to the polymer and not to the spacer.

Another definition of the conductance uses the difference between the quasi-chemical potential taken as close as possible to the spacer to approach the potential difference near the spacer. With this difference instead of V and following the Landauer formula [7,8], the "spacer conductance" can be written [13] :

$$G = \frac{e^2}{\pi\hbar} \frac{T(E_F)}{1 - T(E_F)} \tag{2}$$

This leads to a better electrical characterization of the spacer even if the exact position to measure the quasi-chemical potential on the polymer near the spacer is not clearly defined.

In the following, only the T(E) variation will be discussed. Those interested by the conductance more than by T(E) have to use (1) or (2) to evaluate the conductance according to the Fermi level position of the chosen polymer and to the two probes, four probes style of the measurement.

3. INTERFERENCES THROUGH A SUBSTITUTED BENZENE

Tunnelling through a benzene ring produces electron interferences which are topological in nature [9]. The shape of these interferences depend on how the ring is connected to the conducting chain which provides and collects the tunnelling electrons. If the chain is a polyacetylene, three geometries can be chosen for the insertion of a benzene : the para, meta or ortho connexion. In each case, the main contribution to T(E) comes from the ϕ_2 and ϕ_3 (ϕ_2^*, ϕ_3^*) benzene molecular orbitals located within the $(CH)_x$ π and π^* band respectively (see figure 1 for the description of these orbitals). The outer most orbitals ϕ_1 and ϕ_1^* are of secondary importance [9].

The way a change in the connexion shifts the interference patterns has already been discussed [9]. To study next how a donor and an acceptor substitution control the through benzene interference, an initial geometry has to be chosen among the three possible connexions. The meta one is used in the following to be able to compare results on benzene with results on cyclophane. We have already demonstrated that in a naked benzene meta-connected to a $(CH)_x$, the ϕ_2 benzene molecular orbital creates the main tunnelling channel for electrons coming from the π bands [9]. ϕ_3 opens a tunnelling channel too but less coupled to the $(CH)_x$ molecular orbitals than the ϕ_2 one. Moreover, tunnelling through this channel leads to a transmitted wave with a phase different from the one of the ϕ_2 channel. The T(E) hole in the π band is a result of a destructive inteference between these ϕ_2 and ϕ_3 channels as shown for the through benzene T(E) figure 2-a. The same phenomena occurs in the π^* band. Therefore, to shift these interference holes, we have to play with the energy position of the ϕ_3 and ϕ_3^* orbitals in a way that they can be polarized by substituants grafted on the benzene. Let us graft these substituants as shown below $\underline{2}$ where A stands for acceptor and B for donor.

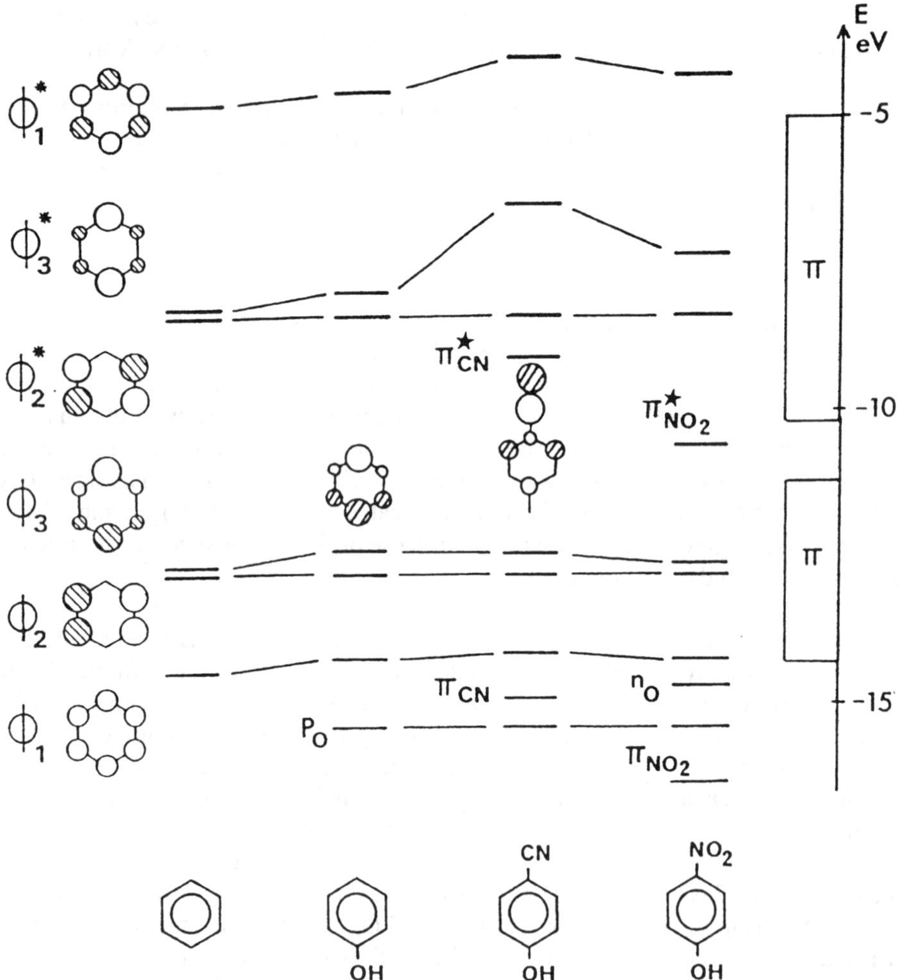

Fig. 1 - Energy and symmetry of the benzene and substituted benzene molecular orbitals compared to the position of the $(CH)_x$ π and π^* band. The genealogy of each molecular level is also shown.

The molecular orbitals of the D-benzene-A are shown figure 1 for D = OH and A = CN, NO_2 together with the naked benzene orbitals and the energy position of the $(CH)_x$ π and π^* bands. The π benzene orbitals are polarized by the donor and acceptor groups. But the ϕ_2 and ϕ_2^* benzene orbitals are not affected by this substitution since they have node on the carbon atoms where the D and A groups are grafted. On the contrary, ϕ_3 and $\phi_3{}^*$ orbital energies increase due to the donor effect of the Pz oxygene orbital of the OH group. Let us now described in detail the OH, CN and NO_2 work on the benzene interference patterns. The T(E) for each of the substituted benzene spacer studied are presented figure 2. For the OH substituant, the transmission holes in both bands are shifted up compared to the naked benzene. Moreover, since not only the energy but also the shape of ϕ_3 is changed by an OH substitution, there is a large difference between the T(E) obtained with an ortho substitution (figure 2-b1). and a meta substitution (figure 2-b2).

To compensate the OH group effect, acceptor groups can be substituted on the benzene ring opposite to the OH group. For a CN group, the main effect appears in the $(CH)_X$ π^* band. It is the $\pi^*{}_{CN}$ LUMO on the CN fragment which interacts with the $\phi_3{}^*$ benzene orbital and strongly destabilizes this level. The destructive interference controlled by $\phi_3{}^*$ is hence shifted up from the middle of the π^* band to its top. But it is not the only contribution of the $\pi^*{}_{CN}$ orbital. Since the corresponding level is in the energy range of the $(CH)_X\pi^*$ band, it introduced a new destructive interference hole at the bottom of this band. The width of this hole is small because the $\pi^*{}_{CN}$ orbital has small coefficients in the meta position with respect to the CN group (figure 2-c1). For the case figure 2-c2, the overlap between the (CH)x molecular orbitals and the $\pi^*{}_{CN}$ is bigger. The new destructive interference introduced by the $\pi^*{}_{CN}$ is broader than with a CN in the meta position. Moreover, the ϕ_3 destabilization induced by $\pi^*{}_{CN}$ is very weak in this case.

If the CN group is remplaced by a better acceptor group like : NO_2, the first vacant $\pi^*{}_{NO2}$ orbital is lower in energy than the $\pi^*{}_{CN}$ one and is located in the $(CH)_X$ band gap between the π and π^* band. Therefore, no new destructive interference hole is added to the benzene interference pattern and the $\phi_3{}^*$ benzene orbital is less destabilized compared to the destabilization induced by the CN group. In the $(CH)_X$ π band, the ϕ_3 orbital is slightly stabilized by its interaction with the $\pi^*{}_{N02}$ orbital. But its polarisation toward the donor OH group increases due to this strong acceptor. Therefore, when the $(CH)_X$ chains are connected far from the donor (figure 2-d2) the interference through the ϕ_3 channel becomes narrow and a high hump appears at the top of this band.

The main effect of an asymmetric polarization of the π benzene electron by donor and acceptor groups is a shift in energy of the position of the T(E) interference holes This is a good way to bring one of these holes in a well defined energy zone (near the Fermi energy) to be detected experimentally. Moreover, a systematic study of the relation between the strength of the donor-acceptor dipolar moment on the benzene and the interference hole shift is clearly needed to compare this phenomena with the electric Aharonov-Bohm effect [12]. At the molecular scale, the advantage compared to the mesoscopic scale is that there is no-more problems for the penetration of the polarization field through the loop. Donor and acceptor groups do not only shift the interference pattern of the benzene but create new interferences which can overshadow the polarization

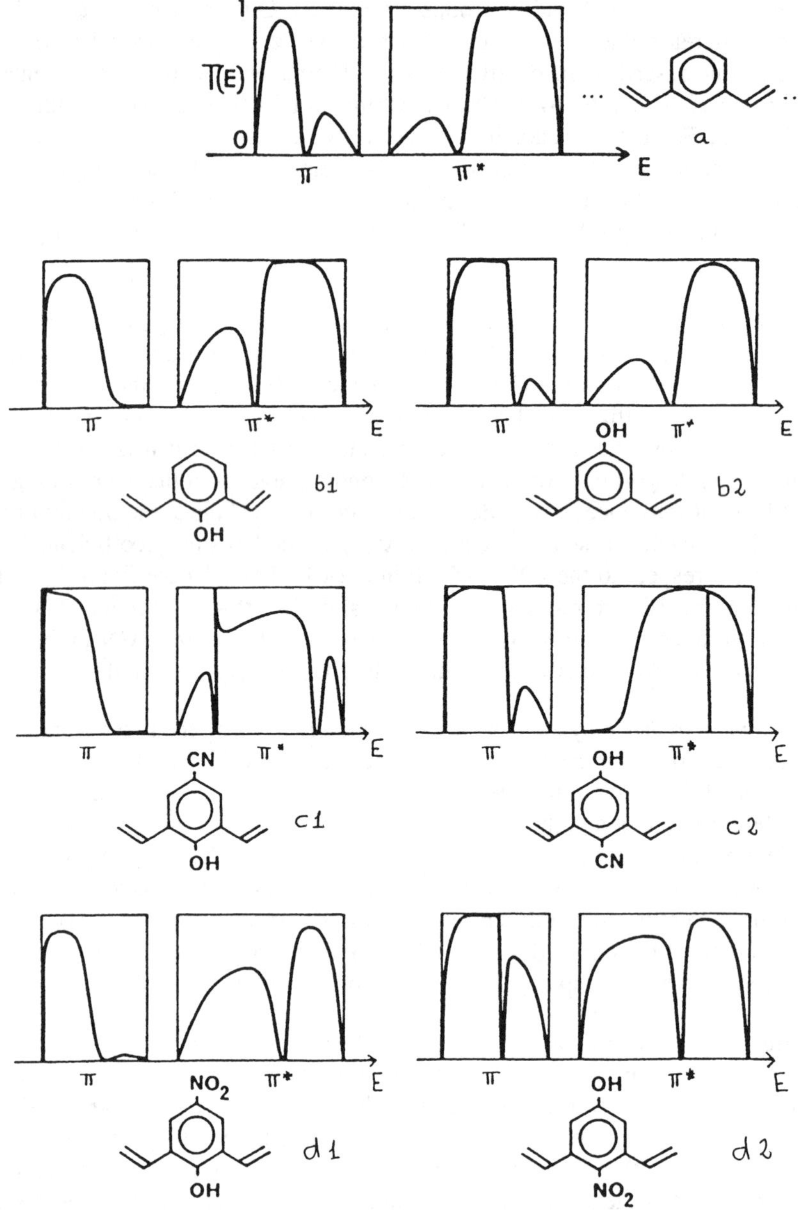

Fig. 2 - Comparison of the through benzene T(E) for different acceptor and donor combinations. The naked-benzene T(E) is recalled for reference.

effect. In such a case, a precise identification of the holes is required using molecular orbitals analysis.

4. THE CYCLOPHANE SPACER

Among the [m,m] paracyclophane series where two benzene rings are held together in a sandwich structure by two bridges of methylene $(CH_2)_m$ in the para position, the [2,2] in $\underline{1}$ is very interesting since the separation of the aromatic rings is shorter (3.1 Å) than those of benzene in solution [14]. The [2,2] paracyclophane is considerably distorted in the ground state into two boat like conformations due to a strong interaction between the π electron system of the two rings. [3,3] paracyclophanes are interesting too but the separation between the two coplanar rings is 3.30 Å. The π interaction is therefore less than for the [2,2] and no distorsion of the rings seems to be measurable.

Since the π orbitals of one ring in paracyclophanes point in direction of the π orbitals of the other ring, paracyclophanes are very attractive to study longitudinal intramolecular π electron coupling. This property was used for the synthesis of novel polymetacyclophane where electron conduction can take place via longitudinal π orbitals coupling [14]. The Taube group had also used cyclophane in its bi-nuclear ruthenium mixed valence compound to study through bond intramolecular electron transfer between the two ruthenium [16]. Here again it is the longitudinal coupling which is responsabile for the intervalence band intensity with the [2,2] paracyclophane band higher by a factor 33 than the [3,3] one.

Then, the [2,2] paracyclophane is a good candidate to be embedded in a conducting chain, a polyacetylene for example, as presented in $\underline{1}$. But there is two ways to connect [2,2] to the conducting chain. Like in the Taube experiment, one chain can be grafted on each benzene ring of the [2,2]. In that case, an electron wave coming from one conducting chain is going to tunnel first through one ring, experiment after the π orbital longitudinal coupling and tunnel through the other ring. In that case, the specific benzene ring shap attractive for interferences is not used. Therefore, to get controlable interferences, it is preferable to graft the two conducting chains on the same benzene. In this case, the connected benzene keeps its interferometer role and the other benzene is just there to control the interference via longitudinal coupling.

Due to this longitudinal intramolecular π electron coupling, the electronic spectrum of a cyclophane can be described as a dimerization of the π electron system of a distorted benzene. Each degenerate of benzene orbital gives a set of four orbitals by bonding and antibonding mixing with the neighbouring benzene unit (figure 3). Moreover, the σ electrons cannot be rigorously separated from the π ones since the planar symmetry is lost.

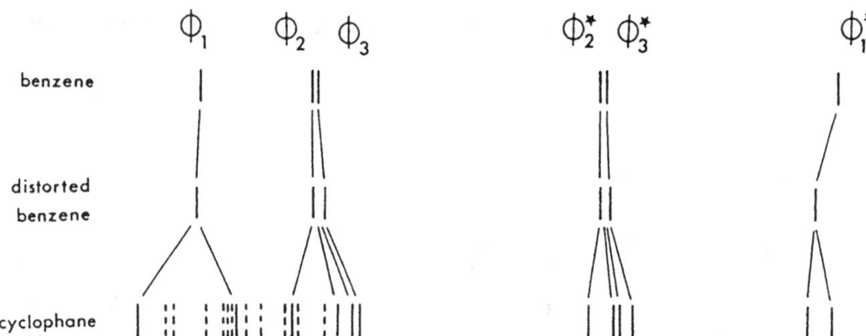

Fig. 3 - Energy and genealogy of nacked cyclophane molecular levels beginning with the benzene ones. The energy scale is the same than in figure 1. Dotted levels are the σ ones embedded in the π framework.

As a consequence, high occuped σ levels appear in the $\pi(CH)_x$ band. The corresponding molecular orbitals are nearly orthogonal to the $(CH)_x$ ones and then create many sharp interferences in the bottom of the π band (see figure 4). The four "π type" levels of cyclophane are also shown. We can focus on the π^* band for the subsequent discussion since this band is free of σ levels effect.

The dimerization of the benzene pattern introduced by the cyclophane creates two new sharp interferences, a destructive one and a constructive one that can be seen as a new narrow pic in the transmission curve. These new interferences are a direct consequence of the bonding and antibonding coupling with the second benzene. Aside from this new constructive interference in the middle of the π^* band, the overall shape of the π^* T(E) is the same than with a single benzene. This is important because we want to control the interference pattern through the ring connected to the $(CH)_x$ chains by donor and acceptor groups grafted on the other ring. Therefore, the perturbation on T(E) created by this nacked-ring must be small. This is the case in the π^* band but clearly not in the π band (figure 4).

A lot of substituant-molecular groups can be grafted on a [2,2] paracyclophane. Majority of these substituants like OH, NH_2 donor groups or CN, NO_2 acceptor groups have been used to observe effects of intramolecular charge transfer between the two rings on the spectra of the molecule [17]. In our case, they control the interference pattern of the overall cyclophane via the longitudinal coupling. But OH is a to weak donor to give interesting observable effects via these coupling. In the case of a strong donor group as NH_2 (figure 5-a), the interference pattern in the top part of the π band is shifted up toward the band edge, as was previously described in detail for the case a substituted benzene.

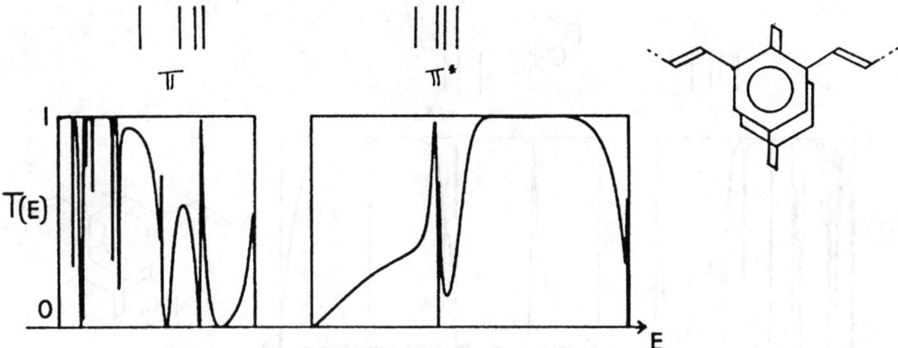

Fig. 4 - T(E) through a naked cyclophane. The ϕ_2, ϕ_3, $\phi_3{}^*$ and $\phi_3{}^*$ molecular level position are displayed to assign π and π^* interferences.

For the π^* band, only one level of the substituted ring is pushed up. Since this level is weakly coupled with the chain, only the sharp interference hole associated with it is shifted. When an acceptor like group CN is also grafted, the major effect is the appearence of a marked π^*_{CN} peak in the lower part the of the π^* band (figure 5-b). This peak could easily be displaced by an adequate chemical perturbation of the CN fragment [11].

This clearly show that controls of the through benzene interference by the substituted benzene in the cyclophane is possible. But the polarization effect of the naked benzene original interference pattern is very small, at least for the donor and acceptor group report here. Again new interferences appear which overshadow the polarization effect studied. Nevertheless, the interactions between the two rings is sufficiently low for these new interferences to be sharp and to lead to new interesting ways to control T(E).

5. CONCLUSION

The benzene offers a very versatil geometry to study interference effects and shifts in the interference patterns by a donor and acceptor polarization of the π benzene electrons. Moreover, the [2,2] cyclophane offer the opportunity to control these interferences using donor and acceptor groups not chemically connected to the benzene which plays the interferometer role. In this case, interference controls occur via longitudinal π-π coupling. This spatial separation between the propagative part and the polarization part in the cyclophane opens the way to introduce a true "third electrod" to control through benzene interferences.

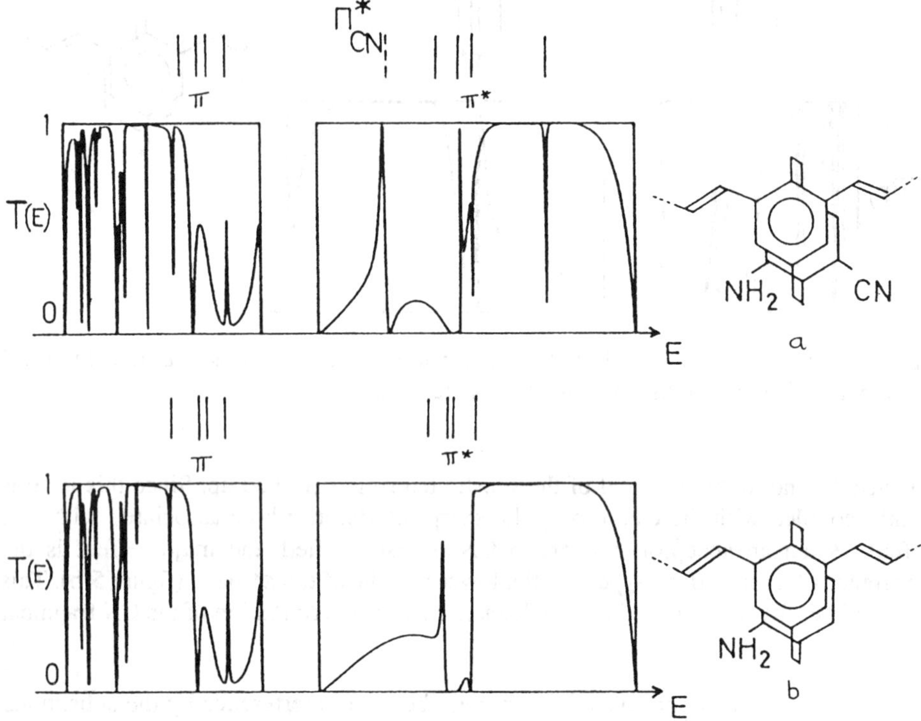

Fig. 5 - Examples of T(E) through substituted cyclophanes.
Shift in the π and π^* interference are clearly seen together with the new π^*_{CN} sharp tunnelling channel. ϕ_2, ϕ_3, ϕ_2^* and ϕ_3^* are positioned for reference.

6. REFERENCES

[1] G. Fasal, Nature, <u>338</u>, 464 (1989) and references there in.

[2] S. Washburn and R. Webb, Adv. in Phys., <u>35</u>, 412 (1986).

[3] P. Sautet, O. Eisenstein and E. Canadell, Chemistry of Materials, in press (1989).

[4] C. Joachim and P. Sautet, J. Chem. Phys., in preparation.

[5] P. Sautet and C. Joachim, Phys. Rev. B., <u>38</u>, 12238 (1988).

[6] Y. Imry, in G. Grinstein, G. Mazekko (Eds) : Direction in Condensed Matter Physics, Memorial Volume in Honor of Shang-Keng Ma, p. 101, World Scientific, Singapore, 1986.

[7]   R. Landauer, Phil. Mag., $\underline{21}$, 86 (1970) ; P.W. Anderson, D.J. Thouless, E. Abrahams and D.S. Fisher, Phys. Rev., $\underline{B22}$, 3519 (1980).

[8]   A.D. Stone and A. Szafer, IBM J. Res. Dev., $\underline{32}$, 384 (1988).

[9]   P. Sautet and C. Joachim, Chem. Phys. Lett., $\underline{153}$, 511 (1988).

[10] P. Sautet and C. Joachim, Springer Serie in Solid State Sciences, in press (1989).

[11] P. Sautet and C. Joachim, Chem. Phys., $\underline{135}$, 99 (1989).

[12] S. Washburn, H. Schmid, D.P. Kern, R.A. Webb, Phys. Rev. Lett., $\underline{59}$, 1791 (1987).

[13] See the discussion on the possible definition of a molecule conductance, in C. Joachim and J.P. Launay, J. Mol. Elec., in press (1989).

[13] S. Canuto and M.C. Zeiner, Chem. Phys. Lett., $\underline{157}$, 353 (1989).

[14] K. Tanaka, Y. Huang, S. Yamanaka and T. Yamabe, Synth. Metals, $\underline{31}$, 23 (1989).

[15] D.E. Richardson and H. Taube, J ACS, $\underline{105}$, 40 (1983).

[16] H. Allgeir , M.C. Siegel, R.C. Helgeson, E. Schmidt and D.J. Cram, JACS, $\underline{97}$, 3782 (1975).

ORGANIC CRYSTALS AND QUADRATIC NONLINEAR OPTICS :
THE TRANSPARENCY-EFFICIENCY TRADE-OFF

J. Zyss

C.N.E.T. - Laboratoire de Bagneux (UA CNRS 250)

196 avenue Henri Ravera

92220 Bagneux - FRANCE

ABSTRACT. Molecular organic crystals with enhanced quadratic nonlinear optical properties are reviewed and classified according to a transparency-efficiency scale. This scale is relevant for discussing the various fields of applications of these materials and relates to well defined molecular and crystalline structural and electronic features. Within such a framework, a predictive molecular engineering approach has proved fruitful in the three main target-families identified here namely : "coloured", "yellow" (para-nitroaniline-like) and "transparent" materials. For each of these three families, relevant chemical and physical features are discussed in relation with specific applications in the field of quadratic nonlinear optics.

1. THE EFFICIENCY-TRANSPARENCY TRADE-OFF AS A GUIDELINE IN MOLECULAR ENGINEERING FOR QUADRATIC NONLINEAR OPTICS.

The search for new organic materials with applications in nonlinear optics based on a predictive molecular engineering approach [1-6] does not aim at the sole enhancements of molecular and crystalline nonlinear susceptibilities. Additionnal transparency requirements, in accordance with specific applications, must also be taken into account and may conflict with purely polarizability related criteria. Extending the dimension of a π electron conjugated system is a natural strategy to enhance the second- or third-order molecular hyperpolarizability tensors of molecules (hereafter designated as β and γ) [7-9], while a simple free-electron model in a box [10,11] as well as experimental evidence convincingly demonstrate the ensuing lowering of the gap and increased colouration of the resulting moiety. An adequate trade-off between nonlinear efficiency versus transparency is in fact driving research in this area, in connection with various application requirements. When trying to classify the extensive catalogue of nonlinear molecules and materials which have been proposed over the past decade, and limiting the inspection to quadratic nonlinear optics, three large sets of molecular materials may be recognized and linked to three different demands in terms of transparency-efficiency trade-off. As shown in Fig.1, these three sets are identified by colour labels under the respective names of :
- *"coloured materials" refering to red, blue and dark colours (i.e. solid-state transparency cut-off typically above 550 nm),*
- *"yellow materials" of the paranitroaniline or related structure type with cut-off frequencies in the 500 nm region,*
- *"transparent materials" where the absence of coloration in the visible spectrum indicates a cut-off wavelength below the 450 nm region.*

The location of the solid-state frequency cut-off results from both intramolecular and intermolecular features underlying the linear absorption properties of the material. The emphasis is here on material transparency, which encompasses individual molecule transparency considerations (such as experimentally accessible by diluted solutions or gas phase spectroscopy), but relates also to additionnal intermolecular interactions, as applications will mostly make use of macroscopic molecular agregates such as single crystals. The nature of intermolecular interactions underlying the spectral properties will depend on the nature of the solid-state molecular assembly of interest : the delocalization over a single crystalline lattice of molecular excited states giving raise to Frenkel excitons [12] account for blue or red spectral shifts from isolated molecules to a crystalline condensed phase [13-14]. In poled functionnalized

J. L. Brédas and R. R. Chance (eds.), Conjugated Polymeric Materials:
Opportunities in Electronics, Optoelectronics, and Molecular Electronics, 545–557.

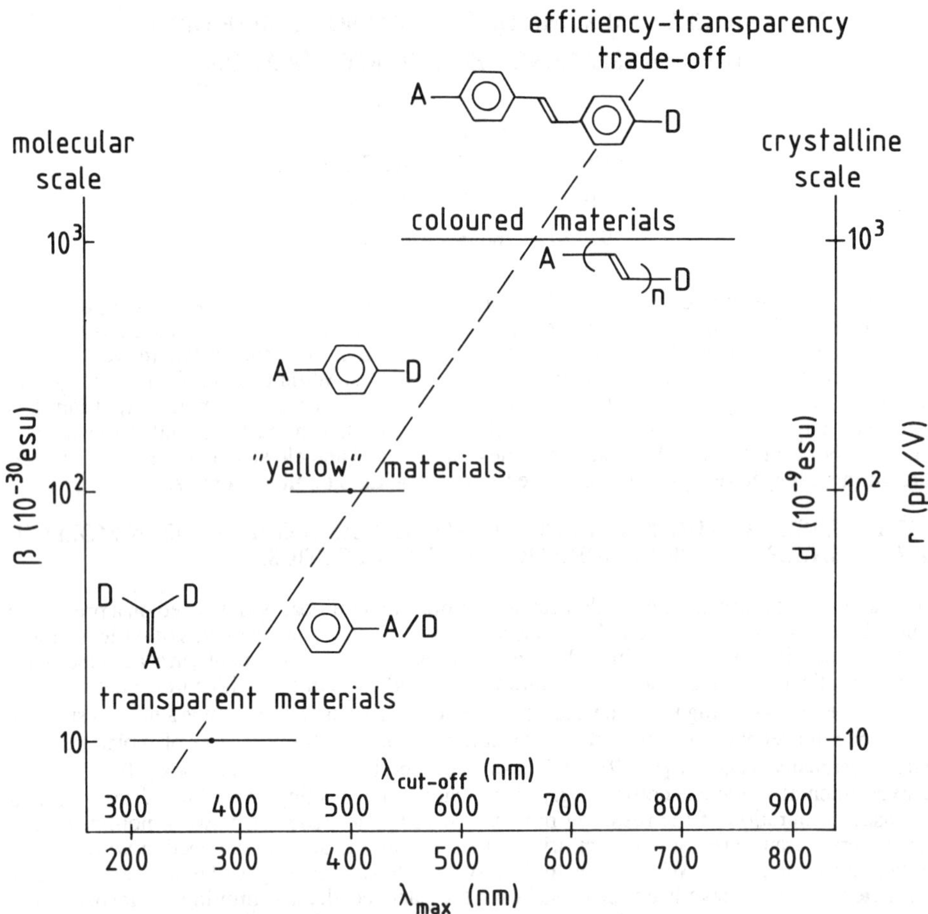

Fig.1:   In an efficiency-transparency reference frame, molecules are seen to gather in three main groups respectively labeled according to transparency features : *transparent, yellow* and *coloured.* Two vertical scales are used refering respectively to molecular quadratic hyperpolarizabilities (β) and to crystalline SHG (d) or electrooptic (r) tensor coefficients. These typical values correspond to parametric off-resonnant processes and are not then significantly enhanced by resonnant interactions. The horizontal scales refer respectively to the wavelength of maximal absorbtion in solution (λ_{max}) and to the most frequently red-shifted (by an average value of 100 nm) transparency edge in crystal ($\lambda_{cut-off}$). When the material consists of diluted or dispersed molecules (solution, or poled polymer), λ_{max} may depend on the environment (solvatochromism) : the potential of such a material can then to be discussed in terms of (β, λ_{max}). When discussing the potential of a crystal, the relevant scale is (r or d, $\lambda_{cut-off}$).

In rough terms, it may be seen from this simplified picture that one order of magnitude increase of the NLO efficiency (in β, r or d, assuming an "optimal" crystalline orientation in the two latter crystalline cases) may be expected on the average from a 200 nm red-shift of the transparency edge for charge-transfer π-conjugated systems. Molecular engineering in NLO may thus be defined as the research of the "best" trade-off in terms of efficiency-transparency. The straight dotted line approximately connecting the centers of the three groups illustrates this trade-off. Displacing upwards this line or portions of it is a constantly challenging and major problem of nonlinear optics and organics.

polymers [15], alternatively used to organize nonlinear molecules by way of covalent attachment, the dilution of the active nonlinear moiety in the polymer host will considerably screen inter-molecular interactions and spectral shift can be accounted for by guest (nonlinear side-group) - host (polymeric backbone) types of considerations such as applied in the context of nonlinear liquid solutions [16,17].

Going from coloured to yellow and on to transparent molecules and materials will correspond to applications where the emphasis is increasingly put on transparency considerations with corresponding relaxation of the nonlinear efficiency requirements.

2. COLOURED CRYSTALS

In the first set of "coloured" materials, an all-out search for molecules with record high nonlinear coefficients is undertaken, regardless of transparency considerations. Conjugated systems are favourite candidates for such competition owing to their highly absorbing optically excited eigenstates and resulting polarizabilities. Their gap will however diminish as the delocalized π electron system size increases but will generally tend to a finite limit, thus ensuring sufficient near I.R. transparency for electrooptic applications at strategic semiconductor laser wavelength (800, 1300 and 1500 nm ranges). Electrooptic (or Pockels) effect related applications will qualify here, rather than "direct" purely optical nonlinear effects (such as second-harmonic generation or I.R. to visible summ-frequency generation) as the former only, contrary to the latter, will not generate additionnal optical frequencies that will eventually get absorbed via a linear mechanism. However these materials may still be expected to satisfy the requirements for such "direct" effects when the first visible absorbtion peak, linked to the strong coloration of the substance, is separated from the next one, located at higher energy towards the blue, by a low absorbtion intermediary gap allowing for the propagation, with affordable if not totally negligible absorbtion, of a phase-matched harmonic wave. One of the attracting feature of such a confi-guration is the possibility to ensure phase-matching by anomalous index dispersion [18] rather than by birefringence compensation, a scheme of particular relevance for poled polymers and even D-C field oriented solutions [19]. A major domain of application for these molecules, in the field of purely optical nonlinear effects, is that of "inverse", phenomena where coherent photon fission phenomena (i.e. parametric amplification or emission) rather than fusion (linked to "direct" processes) take place. From the spectral point of view, 800 nm power semiconductor laser diodes are ideally located (eventually further in the I.R. if diode-pumped YAG lasers are used) with respect to the absorbtion spectra of "coloured" materials and can be used as pumping beams with emitted or amplified signal and idler waves possibly tunable in the 1 to 2 μm. I.R. spectral range centered around degenerated wavelength in the 1.6 μm domain. Such an appli-cation has not yet been demonstrated mainly because of problems linked with macroscopic organization (poled polymers, guest-host systems or Langmuir-Blodgett multilayers are probably more suited here than single crystals), problematic definition of related phase-matching schemes especially for poled polymers, and residual vibrationnal overtone absorbtions in the 1 to 2 μm range investigations. Applications of coloured materials are summed-up in <u>Table</u> 1. Most noteworthy among such molecular systems are :
stilbene derivatives [6,20,21]
diazo-molecules [22,23]
phenylhydrazones [24]
"push-pull" polyenes [25,26]
azulene derivatives [27]
dicyano-divinylidene stilbenes [23]

Some molecular or crystalline figures of merit for a sampling of well characterized representative moieties are displayed in <u>Fig</u>.2 .

	ELECTROOPTIC MODULATION	SHG	PARAMETRIC AMPL.
Coloured Materials $\lambda > 550$ nm	1.5, 1.3, 0.85 μm	$1.5 \rightarrow 0.75$ μm $1.3 \rightarrow 0.65$ μm[*] $1.06 \rightarrow 0.53$ μm[*] $0.85 \rightarrow 0.42$ μm[*]	$0.85 -> 1.7$ μm
Yellow Materials $450 < \lambda < 550$ nm	1.06, 0.85 μm	$1.5 \rightarrow 0.75$ μm $1.3 \rightarrow 0.65$ μm	$0.85 \rightarrow 1.7$ μm vis.dye $\rightarrow \sim 1.2$ μm $0.65 \rightarrow 1.3$ μm
Transparent Materials $\lambda < 450$ nm	visible laser sources	$0.85 \rightarrow 0.425$ μm (blue) $1.3 \rightarrow 0.65$ μm (red) $1.06 \rightarrow 0.53$ μm (green)[**] $1.06 \rightarrow 0.35$ μm (U.V)[**]	$0.35 \rightarrow 0.7$ μm $0.53 \rightarrow 1.06$ μm $0.425 \rightarrow 0.85$ μm

TABLE 1 : Transparency of quadratic nonlinear optical materials and some of their applications. λ corresponds to the solid-state transparency cut-off.

* : possible utilization of anomalous-phase-matching when 2ω is located in low absorbing region between $S_1 \leftarrow S_o$ and $S_2 \leftarrow S_o$ transitions.

** : third-harmonic generation via $\omega + \omega \rightarrow 2\omega$ (SHG) and $2\omega + \omega \rightarrow 3\omega$ (summ-frequency mixing).

Laser sources :
0.85, 1.3 and 1.5 μm	: semiconductor lasers
1.34 and 1.06 μm	: YAG laser
0.65 and 0.53 μm	: frequency doubled YAG
0.45 and 0.35 μm	: frequency tripled YAG
0.425 μm	: frequency doubled semiconductor laser

Styrylpyridinium cyanine dye : SPCD

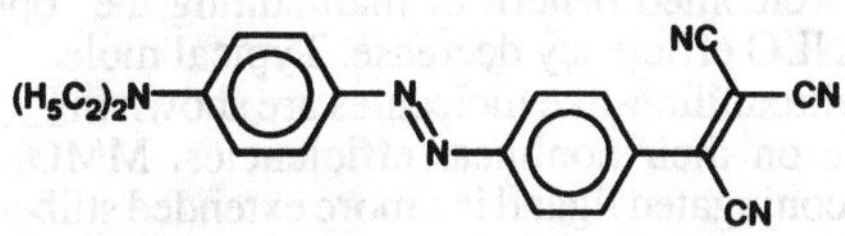

(**Meredith et al.1983,Ref.20**)
(**Yoshimura. 1987, Ref.21**)

r_{33} = 430 pm/V (633 nm) .   Lambda cut = 620 nm.

Pyrazole derivative

(**Allen et al. 1988, Ref.27**)

r_{33} = 87 pm/V (633 nm).   Lambda cut = 600 to 700 nm (dichroism).
Beta = 79 x 10⁻³⁰ esu .   Lambda max = 499 nm.

4 - 4' - Tricyanovinyldimethylaminodiazostilbene.

(**Katz et al. 1987, Ref.23**)

Beta = 154 x 10⁻³⁰ esu .   Lambda max = 580 nm.

Push - pull polyene.

(**Blanchard et al. 1988, Refs.25,26**)

Beta = 480 x 10⁻³⁰ esu .   Lambda max = 498 nm.

Phenylhydrazone derivative

(**Katz et al. 1987, Ref.23**)

Beta = 31 x 10⁻³⁰ esu.
Lambda max = 400 nm.

FIG . 2 : Structure and performance of selected coloured materials.

550

3. "YELLOW" CRYSTALS

The second set, refered to as that of "yellow" molecules has been widely studied, as para-nitroaniline-like moieties have been initially considered as prototypes of "molecular optical diodes" for quadratic nonlinear optics. The proximity of the visible absorption cut-off to the harmonic of the 1.06 μm emission of the YAG : Nd^{3+} laser has deceived early expectations as to possible applications of such yellow materials towards second-harmonic generation into the green of YAG lasers. However such materials have been recently shown to qualify for ultrafast (pico- and subpicosecond) near I.R. quadratic optical signal processing such as high-yield frequency doubling of YAG : Nd^{3+} at 1.34 μm fundamental wavelength, Optical Parametric Amplification and Sampling spectroscopy (PASS) with a 620 nm colliding pulse-mode laser as the pump [28-30], time-resolution by auto- and cross-correlation of weak near I.R. signals [31], parametric processes pumped by dye lasers emitting in the visible. An ideal match to exemplify the latter type of applications would be that of a Rhodamine 6-G dye laser tuned at 575 nm to pump a N-4-Nitrophenyl-(L)-Prolinol (NPP) crystal [32,33] in the type I -non critical phase-matching configuration with maximal degenerated emission at 1.15 μm and an effective d coefficient of the order of 200 10^{-9} e.s.u. This record high value leads to a non-saturated parametric gain surpassing that of KTP by two orders of magnitude. Although many materials in this set have been inspired and hence structurally resemble paranitroaniline as is primarily the case of extensively studied 3-Methyl-4-Nitroaniline (MNA) [34-37], interesting exceptions are worth being singled-out among which 3-Methyl-4-nitropyridine-1-oxyde (POM) [38-40] and MMONS [41,42] may stand-out. The former is more transparent than a typical parani-troaniline derivative owing to the inherently less polarizable nature of the pyridine heterocycle. A systematic strategy based on Dewar's specifications [43], has been in particular applied to derive (PNP) [44] from NPP by inserting nitrogene heteroatoms at "starred" positions in aromatic rings with the additionnal unexpected, however welcomed benefit of maintaining the "opti-mized" crystalline at a moderate price in terms of NLO efficiency decrease. Typical molecular structures belonging to the group of "yellow" paranitroaniline-like molecules are shown in Figs. 3 and 4 together with some data, when available on their nonlinear efficiencies. MMONS illustrates another interesting strategy whereby the conjugated ligand is a more extended stilbene moiety while the electron donating strength of the O-methyl substitued group is markedly weaker as compared to that of the dimethylamino group of DANS, with the same strong acceptor nitro group used in both compounds. The yellow colour of MMONS crystals clearly results from this end-group selection. Finally the NPAN family with a variable length -(CH$_2$)$_n$- spacer, with n=1, 2 or 3, linking a donating nitrogene atom to an electron accepting cyano group [45-46] illustrates a means of adjusting the transparency of a consistent family of paranitroaniline like molecules. The strong local dipole moment of the cyano group was initially intended to favour the non centrosymmetric crystalline packing consistently reported for these compounds. In general, deviations from the generic centrosymmetric paranitroaniline molecule were oriented by crystalline packing considerations meant to favour a non centrosymmetric lattice, while trans-parency considerations were less commanding. Finally, such materials should also qualify for electrooptic modulation or switching of visible or near I.R. lasers (see Table 1).

4. TRANSPARENT CRYSTALS

The third set of "transparent" molecules gathers molecular materials with transparency features allowing for coherent blue light emission by frequency doubling of 800 to 900 nm semiconductor laser diodes. Such blue emission may have been occasionnally observed from "yellow" materials like POM [39] owing to the high peak power available from pulsed picosecond fundamental beam which compensates for the harmonic absorbtion. Similarly, blue and purple emissions were observed close to absorbtion cut-off in Dicyanovinyl-anisole (DIVA) [52]. However optimization of conversion yields as well as optical damage threshold constraints have oriented research towards new families of fully transparent organic materials with no residual absorbtion in the visible. A prototype material of this category has long been urea [53-56] where clear cut conjugation features are absent with the result of an absorbtion cut-off displaced down to 200

NPAN

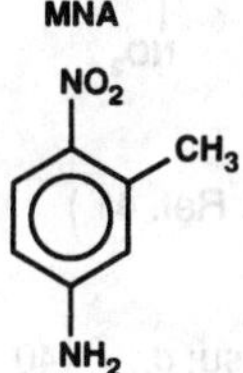

(**P.Vidakovic et al.** 1987. Ref. 45)

Fdd2
d_{eff} = 140 x 10⁻⁹ esu. Beta = 12 x 10⁻³⁰ esu.
Lambda max = 360 nm, lambda cut = 490 nm.

NPP

(**J.Zyss et al.** 1981. Ref.32)

P2₁
d_{eff} = d_{22} = 190 x 10⁻⁹ esu (1.15 x10⁻⁶ m)
Lambda max = 397 nm. Lambda cut = 500nm.
Beta₀ = 17 x 10⁻³⁰ esu.

POM

(**J.Zyss et al.** 1981. Ref.38)

P2₁2₁2₁
d_{eff} = d_{14} = 2 0 x 10⁻⁹ esu (1.32nm)
Lambda cut = 500nm. Lambda max = 325 nm.
Beta₀ = 8 x 10⁻³⁰ esu r_{52} = 5 pm/V.

MAP

(**J.L.Oudar et al.** 1977. Ref. 47)

P2₁
d_{eff} = 40 x 10⁻³⁰ esu.
Beta = 22 x 10⁻³⁰esu (1.06 x 10⁻⁶m).
Lambda cut = 500 nm.

DAN

(**Baumer et al.** 1987. Ref. 50.)

P2₁
d_{eff} = 150 x 10⁻⁹ esu (1.06 x 10⁻⁶ m).
Lambda cut = 490 nm.

MNA

(**L.Koreneva et al.** 1975. Ref. 34).

P2₁
d_{11} = 580 x 10⁻⁹ esu (1.06 x 10⁻⁶ m).
r_{11} = 70 pm/V.
Lambda cut = 500 nm, lambda max = 360 nm.

FIG. 3 : Structure and performance of selected «yellow» materials.

552

MBANP

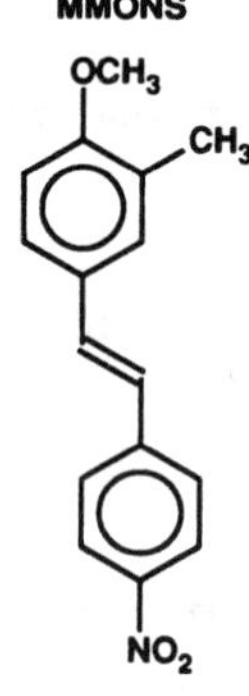

(**Twieg et al.** 1985. Ref . 51)

P2,
d = 140 x 10⁻⁹ esu.
Beta = 15 x 10⁻³⁰ esu.
Lambda max = 360 nm. Lambda cut = 500 nm.

COANP

(**Twieg et al.** 1985. Ref. 49)

Pca2,
d_{eff} = 58 x 10⁻⁹ esu.
Lambda cut = 500 nm.

MMONS

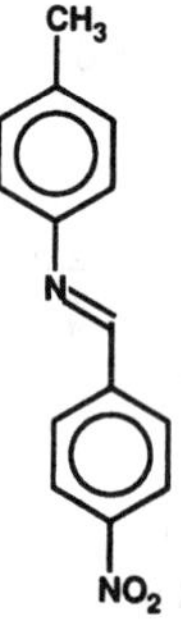

(**Jain et al.** 1989. Ref. 41)

Aba2
d_{24} = 170 x 10⁻⁹ esu; d_{33} = 440 x 10⁻⁹ esu.
Beta = 18 x 10⁻³⁰ esu. r_{33} = 40 pm/V.

NMBA

(**Filipenko et al.** 1977. Ref. 48)

Pb
d_{33} = 170 x 10⁻⁹ esu; r = 27 pm/V.

FIG. 4 : Continuation of Fig.3.

nm. Although such a considerable extension of the transparency domain may be necessary for frequency doubling or up-conversion of visible emitting laser sources (such as dye lasers) or successive up-conversions of YAG : Nd^{3+} or Neodyme doped glass lasers for laser fusion, it is not needed for "blue" emission by SHG, a present major domain of interest for applications in the field of high density optical memories.

A more adjusted target would be a material tailored for an absorbtion cut-off in the 400 nm range thus permitting to reach higher nonlinear efficiency than urea. Besides, regardless of basic molecular modifications, simple crystalline considerations have been used to show [57] that a significant improvement of the nonlinear efficiency, by up to an order of magnitude would result from an improved packing of urea molecules as compared to the far from optimal regular $\overline{4}2m$ structure of urea leading to mutual cancellation of the highest β coefficient of adjacent anti-parallel urea molecules.

Among the larger variety of molecules which have been proposed, a number of promising candidates which illustrate various molecular engineering strategies are sampled in Fig.5 such as sulfur [58,59] or silicon [60] containing aromatic molecules and 4-4' dimethylaminocya-nobiphenyl (DMACB) [61]. While the methylsulfide and trimethylsilyl endgroups are equivalent in terms of electron donor strength, when attached to a nitrobenzene moiety in para position, to that of an amino group, leading to β values similar to that of pNA, a significant blue shift of λ_{max} may be noted.

A worthwhile and realistic goal for molecular and crystalline engineering within this set of "transparent" materials should lead to a d coefficient of the order of $100 \cdot 10^{-9}$ e.s.u. surpassing that of urea by two orders of magnitude : while, as already mentionned, one order of magnitude increase can be expected from an optimal packing of urea-like molecules, another order of magnitude can further be gained from the relaxation of the near UV absorbtion cut-off from 200 nm up to 400 nm thus permitting to consider conjugated molecules with moderate intra-molecular charge transfer. For such molecules, non-resonnant β values of the order of 10^{-29} e.s.u., thus surpassing that of urea by one order of magnitude or comparable to paranitroaniline, are accessible.

Finally, it is worth noting that the analysis, detailed here in the case of quadratic properties can be applied also to cubic nonlinear properties (such as underlying third-harmonic generation or four-wave mixing experiments) where candidate molecules and polymers can be similarly classified in terms of varying transparency-efficiency trade-offs.

5. CONCLUSION

One may assign to each class of materials (coloured, "yellow" and transparent ones) a typical target for further molecular and crystalline engineering research in terms of efficiency-transparency trade-off. For transparent materials, we have seen that urea's absorption cut-off is unnecessarily blue-shifted for important applications related to 850 nm diode-laser frequency doubling, at the expense of the molecular quadratic hyperpolarizability. The unoptimized crystal packing of urea is further detrimental to its nonlinear crystalline susceptibility. A realistic goal is a material with solid-state absorption edge in the 400 nm region and a non-resonnant d coefficient of the order of 10 to 50.10^{-9} e.s.u. (i.e. 10 times the efficiency of urea in the powder SHG test). In view of this goal, urea can still serve as a standard within this class of materials.

NPP [32], with a non-critically phase-matchable d coefficient of the order of $200 \cdot 10^{-9}$ e.s.u. (corresponding to a powder SHG efficiency two decades above that of urea) and a close to optimal crystal packing [57], is representative of a similar target within the "yellow" paranitroaniline-like family. Using such a material as a standard for this class, could be more appropriate than still refering to considerably less efficient urea.

UREA

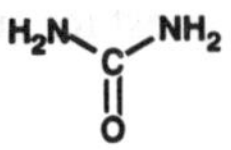

(**Kurtz et al.** 1968. Ref.53)

$P\bar{4}2_1m$
d = 3.4 x 10^{-9} esu. r_{41} = 56 x 10^{-9} esu.
Beta = 2.3 x 10^{-30} esu.
Lambda cut = 200 nm.

Organo - silicon derivative.

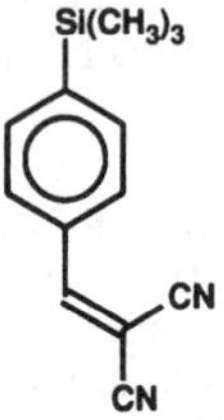

(**Mignani et al.** 1989. Ref. 60)

$Beta_0$ = 9 x 10^{-30} esu.
Lambda max = 320 nm.

4 - 4' - dimethylaminocyanobiphenyle.

DMACB

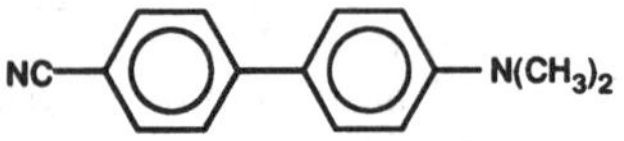

(**J.Zyss et al.** 1989. Ref. 61)

Cc
d = 500 x 10^{-9} esu (1.06 x 10^{-6} m).
$Beta_0$ = 32 x 10^{-30} esu.
I^{2w} = 20 x urea.
Lambda max = 345 nm.Lambda cut = 450 nm.

Dimethylaminophenylurea

DMNPU

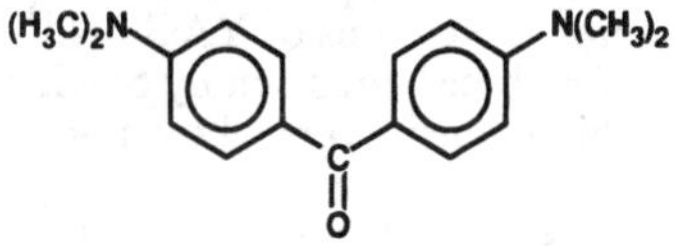

(**Jain et al.** 1981. Ref. 62)

I^{2w} = 10 x urea (powder)

Methyl-4-nitrodiphenylsulphide.

MNS

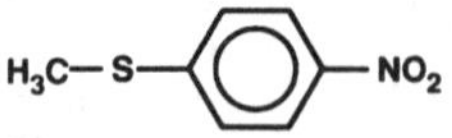

(**Barzoukas et al.** 1989. Ref. 59)

$Beta_0$ = 10 x 10^{-30} esu.
Lambda max = 335 nm.

4 - amino -4' -nitrodiphenylsulfide.

ANDS

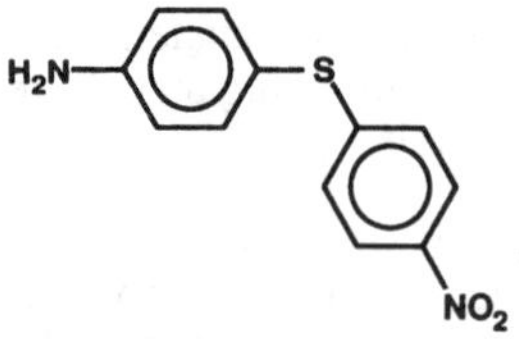

(**Cowan et al.** 1986. Ref. 58)

$Cmc2_1$
I^{2w} = 20 x urea (powder)
$Beta_0$ = 15 x 10^{-30} esu.
Lamda max = 341 nm.

L - arginine phosphate monohydrate.

LAP

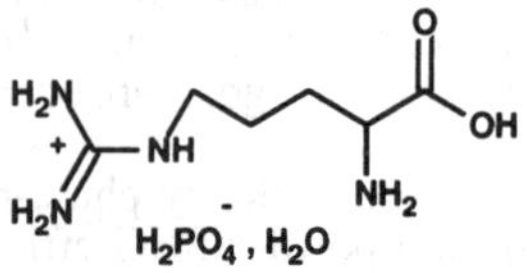

(**Xu et al.** 1983. Ref. 63)

$P2_1$
I^{2w} = 3.5 x KDP (powder).
Lambda cut < 300 nm.

FIG . 5 : Structure and performance of selected

Finally, in the case of coloured materials, a non-resonnant d coefficient of the order of 10^{-6} e.s.u., irrelevent of obvious difficulties in terms of crystal growth, can be expected from measured β values (of the order of a 10^{-28} e.s.u. that is one-order of magnitude above NPP or pNA) and an hypothetical optimal packing. There seems to be no fully convincing demonstration as yet (i.e. from bulk crystal growth down to precise d coefficients measurements) of such a crystal at present probably due to problems in crystallizing larger moeities as well as a chemical stability. Evidencing and characterizing such a crystalline structure is a challenging task towards both applications and a better understanding of nonlinear optical processes in molecular media.

6. ACKNOWLEDGMENTS

It is a pleasure to acknowledge constant fruitful interactions with R. Hierle and I. Ledoux.

7. REFERENCES

1- "Nonlinear Optical Properties of Organic Molecules and Crystals" Eds. D.S. Chemla and J. Zyss in Quantum Electronics Principles and Applications Series, 2 vols. (Academic Press, Orlando, 1987)

2- "Nonlinear Optics in Molecular Crystals" (in russian) by L. Koreneva, V Zoline and B. Davydov (Nauka, Moscow, 1985)

3- "Nonlinear Optics of Organics and Semiconductors" Ed. T. Kobayashi in Springer Proceedings in Physics (36) (Springer-Verlag, Berlin, 1988)

4- "Nonlinear Optical and Electroactive Polymers" Eds. P. Prasad and D.R. Ulrich (Plenum, 1988)

5- "Nonlinear Optical Properties of Polymers", Materials Research Society Symposium Proceedings 109 Eds. A.J. Heeger, J. Orenstein and D.R. Ulrich (MRS, Pittsburgh, 1988)

6- "Nonlinear Optical Processes in Organic Materials" Feature Issue J. Opt. Soc. Am. B 4 (6) (June 1987) Eds J. Carter and J. Zyss

7- J.L. Oudar and H. Le Person, Opt. Comm. 15, 258 (1975)

8- J.L. Oudar, J. Chem. Phys. 67, 446 (1977)

9- J. Zyss, J. Chem. Phys. 71, 909 (1979)

10- H. Kuhn, J. Chem. Phys. 16, 840 (1948)

11- K. Rustagi and J. Ducuing, Opt. Comm. 10, 258 (1972)

12- "Molecular Crystals" by J.D. Wright (Cambridge University Press, Cambridge, 1987)

13- "Theory of Molecular Excitons" by A.S. Davydov (Mc Grawhill, New York, 1962)

14- M. Kasha, H.R. Rawls and M. Ashraf El-Bayoumi, Pure Appl. Chem. 11, 371 (1965)

15- K. Singer, M. Kuzyk and J. Sohn p. 968 in Ref.6

16- K. Singer and A. Garito, J. Chem. Phys. 75, 3572 (1981)

17- I. Ledoux and J. Zyss, Chem. Phys. 73, 203 (1982)

18- J.A. Armstrong, N. Bloembergen, J. Ducuing and P.S. Pershan, Phys. Rev. 127, 1918 (1962)

19- P.A. Cahill, K.D. Singer and L.A. King, Opt. Lett. 14, 1137 (1989)

20- G. Meredith in "Nonlinear Optical Properties of Organic and Polymeric Materials" Ed. D. Williams p.27 (ACS, Washington, 1983)

21- T. Yoshimura, J. Appl. Phys. 62, 2028 (1987)

22- I. Ledoux, D. Josse, P. Vidakovic, J. Zyss, R.A. Hann, P.F. Gordon, P.D. Bothwell, S.K. Gupta, S. Allen, P. Robin, E. Chastaing, J.C. Dubois, Europhys. Lett. 3 (7), 803 (1987)

23- H. Katz, K. Singer, J. Sohn, D. Dirck, L. King and A. Gordon, J. Ann. Chem. Soc. 109, 6561 (1987)

24- D. Lupo, W. Prass, Ude Scheunemann, A. Laschewsky, H. Ringsdorf and I. Ledoux, J. Opt. Soc. Am. B 5, 300 (1988)

25- M. Barzoukas, M. Blanchard-Desce, D. Josse, J.M. Lehn and J. Zyss, Chem. Phys. 133, 323 (1989)

26- M. Barzoukas, M. Blanchard-Desce, D. Josse, J.M. Lehn and J. Zyss, Inst. Phys. Conf. Ser. N° 103 p.239 (IOP, London, 1989)

27- S. Allen, T. Mc Lean, P. Gordon, D. Bothwell, M. Hursthouse and T. Karaulov, J. Appl. Phys. 64, 2583 (1988)

28- I. Ledoux, J. Badan, J. Zyss, A. Migus, D. Hulin, J. Etchepare, G. Grillon and A. Antonetti in Ref.6

29- J. Zyss p.69 in Ref.4

30- I. Ledoux, J. Zyss, A. Migus, D. Hulin and A. Antonetti, J. Appl. Phys. 64, 3309 (1988)

31- M. Islam, G. Sucha, I. Bar-Joseph, M. Wegener, J.P. Gordon and D.S. Chemla, J. Opt. Soc. Am. B 6, 1149 (1989)

32- J. Zyss, J.F. Nicoud and M. Coquillay, J. Chem. Phys. 81, 4160 (1984)

33- I. Ledoux, J. Zyss, A. Migus, J. Etchepare, G. Grillon and A. Antonetti, Appl. Phys. Lett. 48, 1564 (1986)

34- L. Koreneva et al. First edition (1975 in russian) of Ref.2 in page 123

35- B. Levine, C. Bethea, G. Thurmond, R. Lynch and J. Bernstein, J. Appl. Phys. 50, 2523 (1979)

36- G. Lipscomb, A. Garito and R. Narang, J. Chem. Phys. 75 1509 (1981)

37- R. Morita, N. Ogasawara, S. Umegaki and R. Ito, SPIE vol.971, 260 (1988)

38- J. Zyss, D.S. Chemla and J.F. Nicoud, J. Chem. Phys. 74, 4800 (1981)

39- J. Zyss, I. Ledoux, R. Hierle, R. Raj and J.L. Oudar, IEEE J. Quant. Electron. QE-21, 1286 (1985)

40- D. Josse, R. Hierle, I. Ledoux and J. Zyss, Appl. Phys. Lett. 53, 2251 (1988)

41- W. Tam, B. Guerin, J. Calabrese and S. Stevenson, Chem. Phys. Lett. 154, 93 (1989)

42- J. Bierlein, L. Cheng, Y. Wang and W. Tam to be published in Appl. Phys. Lett.

43- M. Dewar, J. Chem. Soc. 2329 (1950) and J.F. Nicoud and R. Twieg in Ref. 1 vol.1

44- R. Twieg and C. Dirck, J. Chem. Phys. 85, 3537 (1986)

45- P. Vidakovic, M. Coquillay and F. Salin in Ref.6

46- M. Barzoukas, D. Josse, P. Fremaux, J. Zyss, J.F. Nicoud and J. Morley in Ref.6

47- J.L. Oudar and R. Hierle, J. Appl. Phys. 48, 2699 (1977)

48- O. Filipenko, V. Shigorin, V. Ponomarev, L. Atovmyan, Z. Safina and B. Tarnopolskii, Sov. Phys. Crystallogr. 22, 305 (1977)

49- J.F. Nicoud and R. Twieg page 279 vol.1 in Ref.1

50- J.C. Baumert, R. Twieg, G. Bjorklund, J. Logan and C. Dirck, Appl. Phys. Lett. 51, 1484 (1987)

51- R. Twieg, K. Jain, Y. Cheng, J. Crowley and A. Azema, Polymer Reprints 23, 147 (1982)

52- T. Wada, C. Grossman, A. Garito and H. Sasabe to be published in Mat. Res. Soc. Symp. Proc. (MRS, Boston, 1989)

53- S. Kurtz and T. Perry, J. Appl. Phys. 39, 3798 (1968)

54- C. Cassidy, J.M. Halbout, W. Donaldson and C. Tang, Optics Comm. 29, 243 (1979)

55- J. Morrell, A. Albrecht, K. Levin and C. Tang, J. Chem. Phys. 71, 5063 (1979)

56- W. Donaldson and C. Tang, Appl. Phys. Lett. 44, 25 (1984)

57- J. Zyss and J.L. Oudar, Phys. Rev. A 26, 2028 (1982)

58- H. Abdel Alim, D. Cowan, D. Robinson, F.W. Wiygel and M. Kimura, J. Phys. Chem. 90, 5654 (1986)

59- M. Barzoukas, D. Josse, J. Zyss, P. Gordon and J. Morley, Chem. Phys. 139, 359 (1989)

60- M. Barzoukas, D. Josse, J. Zyss, G. Soula, G. Mignani and R. Meyrueix submitted to J. Am. Chem. Soc.

61- J. Zyss, I. Ledoux, A. Bertaud, R. Toupet, submitted to J. Chem. Phys.

62- K. Jain, J. Crowley, G. Hewig, Y. Cheng and R. Twieg, Opt. Laser Technol. 297, (1981)

63- D. Xu, M. Jiang and Z. Tan, Acta Chim. Sin. 2, 230 (1983)

CONDUCTING POLYMER ELECTROMECHANICAL ACTUATORS

R.H. BAUGHMAN, L.W. SHACKLETTE, and R.L.
  ELSENBAUMER
Corporate Technology
Allied-Signal Inc.
Morristown, New Jersey 07962

E. PLICHTA
U.S. Army, ET & DL
Fort Monmouth, New Jersey 07703-5000

C. BECHT
Becht Engineering
22 Church Street
Liberty Corner, New Jersey 07938

ABSTRACT. Conducting polymer charge-transfer complexes are evaluated
as electromechanical materials for the direct conversion of electrical
energy to mechanical energy. Large dimensional changes upon
electrochemical doping/dedoping provide the mechanical response for
proposed extensional, pneumatic, bimorph, and micromechanical
actuators. Actuator performance is predicted using both the observed
performance of electrochromic devices, electrochemical transistors,
and batteries and observed polymer properties as a function of doping.
Conducting polymer actuators can provide more than an order of
magnitude advantage over piezoelectric polymers regarding achievable
dimensional changes, electrically generated stresses, and the work
density per cycle. Using very thin films or fibers to minimize
diffusion distances, cycle times of less than 100 ms and device
lifetimes of above 10^6 cycles should be feasible for microactuators.
Additionally, the conducting polymer actuators can operate at voltages
which are about an order of magnitude lower than for piezoelectric
actuators or electrostatic microactuators. Energy efficiencies, cycle
times, and cycle lifetimes are, however, inherently much lower than
for piezoelectric polymers.

1.   Introduction

We herein describe the application of conducting polymers for the
direct conversion of electrical energy into mechanical energy. The
concept is utilization of the large dimensional changes which occur

*J. L. Brédas and R. R. Chance (eds.), Conjugated Polymeric Materials:
Opportunities in Electronics, Optoelectronics, and Molecular Electronics, 559–582.*
© 1990 *Kluwer Academic Publishers. Printed in the Netherlands.*

upon either electrochemical donor or acceptor doping of polymers such as polyacetylene, poly(p-phenylene), polyaniline, polypyrrole, and polythiophene. These dimensional changes can provide a fractional length change ($\Delta L/L$) of over 10%, as compared with a maximum $\Delta L/L$ of about 0.1% at below depolarization voltages for the piezoelectric polymer poly(vinylidene fluoride). The much larger $\Delta L/L$ for conducting polymers, as compared with piezoelectric polymers, and the high mechanical strengths can provide a correspondingly increased work capacity per cycle. Such large dimensional changes of the conducting polymers are obtained using voltages which can be more than an order of magnitude lower than required for comparable electrostatic micromechanical actuators or for piezoelectric actuators. Conducting polymers are of special interest for micromechanical actuators because the high voltages of piezoelectric and electrostatic actuators are incompatible with many desired microcircuit applications and the thermal cycling required for shape-memory alloy microactuators limits applicability. The above advantages of the conducting polymers are countered by negative features - longer response times and lower cycle life for conducting polymers than for piezoelectric and magnetostrictive materials, since large dimensional changes of the conducting polymers require diffusion of dopant ions.

The conversions of chemical, thermal, or photon energies into mechanical energy using polymers has been of interest for a long time.[1-7] Nearly four decades ago Kuhn et al. and Katchalsky et al. reported large, reversible dimension changes of polyacrylic acid[1] and poly(vinyl phosphate)[2] with changing pH. This work has continued, with a number of twists, up to the present. Shiga et al[8] have demonstrated electromechanical effects using electric field-induced differential swelling of poly(vinyl alcohol)/poly(acrylic acid) gels. Similar effects were observed by Irie[9] for photoirradiated gels. Rossi used electrolysis to produce an acidity change that causes the dimensional changes of gel polymers for mechanical actuators, and his group has pioneered the evaluation of artificial muscle actuators.[10] Most of the gel polymers which have been evaluated for possible use in mechanical actuators have been nonconjugated polymers that are not electronic conductors. However, Yoshino et al.[11,12] demonstrated that large dimensional changes result from either changes in solvent composition or temperature for gels of poly(3-alkylthiophene)s that can be doped to provide high electrical conductivities. The chemical doping of a poly(3-octylthiophene) gel in chloroform with iodine provided a contraction to 15% of the initial volume and a conductivity increase from less than 10^{-12} S/cm to 10^{-1} S/cm. Rossi's group in Pisa is presently exploring the application in mechanical actuators of the electrochemical doping and undoping of polymer composites, such as polypyrrole in a thermally cross-linked poly(vinyl alcohol)-poly(acrylic acid) gel.[13]

The principle problem of using polymer gels for mechanical actuators is the poor mechanical properties (low moduli and low yield stresses). As a consequence of these properties, thick gels have been used in

order to support modest mechanical loads, which increases the problem
of long device response times - since diffusion distances are large.
In contrast with this approach, the materials of greatest interest in
the present work are highly conducting organic polymers that provide
dimensional changes by solid-state electrochemical transformations.
The major advantages of using solid polymers, especially chain-
oriented crystalline polymers, are unusually high obtainable
mechanical properties and electrical conductivities, which are
important property aspects for the design of useful electromechanical
devices. The major disadvantages are generally smaller fractional
deformations and lower diffusion coefficients than for gel polymers,
which requires the use of thin films or fibers in device design.

The present applications analysis is largely conceptual. Two levels
of device analyses are provided. The first level utilizes
experimentally determined materials properties and device design
concepts whose practicality has been demonstrated for related devices.
Such evaluations are made more concrete by utilizing the wealth of
information from development activities on conducting polymer
batteries, electrochemical transistors, and electrochromic displays.
The second level of evaluation is more speculative in that it requires
fabrication capabilities which are presently unavailable. This two
level strategy has been taken in order to assess what might be
achieved in the next decade, as well as to assess the ultimate
prospects for conducting polymer electromechanical actuators.
Section 2 provides general descriptions of various electromechanical
actuators based on the dimensional changes which occur during the
doping of conducting polymers. Both tensile and hydraulic devices
will be described, as well as nonelectrochemical devices which utilize
dopant-induced dimensional changes. Microelectromechanical actuators
will provide a special focus. In Section 3, observed structure-
related properties and the observed performance of electrochromic
devices and batteries are used to predict the achievable performance
of conducting polymer electromechanical actuators. Important aspects
are the dopant-concentration dependence of dimensional changes in both
covalent bonding and orthogonal directions, mechanical properties, and
electrochemical doping rates. Finally, Section 4 presents conclusions
regarding the applications potential for electrochemical mechanical
actuators which utilize conducting polymers.

2.   Description of Electromechanical Actuators

Electrochemical mechanical actuators can generally function by using
electrochemically induced (1) changes in a dimension of a conducting
polymer, (2) changes in the relative dimensions of a conducting
polymer and a counter electrode or, (3) changes in the total volume of
a conducting polymer electrode, electrolyte, and counter electrode.
In each case, a conducting polymer can serve as either only one
electrode or as both electrodes. In this section we briefly describe
general features of these different devices.

Many of the design features utilized in batteries are pertinent for
both hydraulic and extensional electromechanical actuators. For
example, optimization of electrical energy density (useable electrical
discharge energy per cycle, normalized by cell volume or cell weight)
or electrical power density (rate of energy delivery per cell volume
or cell weight) in a battery corresponds to optimization of the
related quantities of mechanical work per cycle or mechanical power
capacity (on either a volumetric or gravimetric basis) for an
electromechanical cell using the same redox transformation.
Considering different redox reactions as alternatives in device
design, there is naturally one major difference in optimization of
energy density or power density for batteries and electromechanical
actuators. The energy density and power density of a battery are
proportional to both coulomb capacity (electrical energy density) or
current capacity (electrical power density) and the appropriately
averaged cell voltage, after internal resistive losses. On the other
hand, the corresponding quantities for an electromechanical actuator
are independent of cell voltage for a specified discharge current and
depend upon the dimensional changes and electrode mechanical
properties as a function of doping, as well as being proportional to
coulomb capacity (mechanical energy density) or current capability
(mechanical power density). Consequently, electrode combinations
which would provide too low a voltage for battery applications are of
interest for electromechanical actuators.

The range of potentials for some of the electrode materials which
could be used as either electromechanical polymers or counter
electrodes is shown in Fig. 1.[14,15] These materials generally expand
upon dopant insertion and contract upon dopant expulsion, although we
will discuss (Section 3) important cases in which acceptor insertion
results in a contraction in covalent bond directions. Many of the
compositions in Fig. 1 have been extensively studied for application
in rechargeable batteries and have high demonstrated cycle lives.

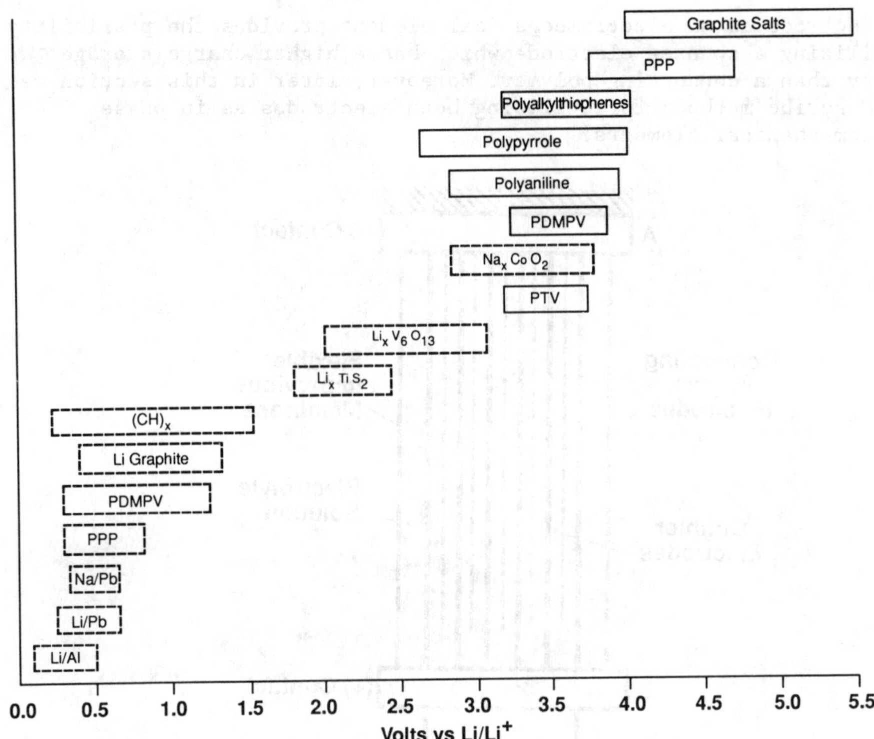

Figure 1. The voltage of various anode and cathode materials which might be used for electrochemical actuators. The voltage range indicated corresponds to the voltage change upon acceptor or donor doping. PDMPV is poly(dimethoxyphenylene vinylene), PPP is poly(p-phenylene), and PTV is poly(2,5-thienylene vinylene). The voltage ranges denoted by dashed boxes and solid boxes are for alkali metal doping and acceptor doping (BF_4^-, PF_6^-, ClO_4^-, etc.), respectively.

A mechanical actuator which operates by the tensile contraction of polymer films or fibers is illustrated schematically in Fig. 2. The device consists of conducting polymer electrode strips or fibers which are in close proximity to counter electrode strips or fibers. The conducting polymer electrodes alternate in a stack or, for the alternate design, conducting polymer electrode fibers and counter electrodes fibers are interdispersed. The conducting polymer electrode is insulated from the counter electrode for all but the movement of dopant ions, by spatial separation via either intervening electrolyte (or electrolyte and porous separator) or a layer of ionically conducting polymer surrounding the conducting polymer electrode. Only one of the electrode sets in Fig. 2 is used as the electromechanical polymer. With respect to device performance, such a design need not be disadvantageous, since the sacrifice of using only

564

one electrode as an electromechanical element provides the possibility
of utilizing a counter electrode which has a higher charge storage
density than a conducting polymer. Moreover, later in this section we
will describe methods for utilizing both electrodes as in-phase
electromechanical elements.

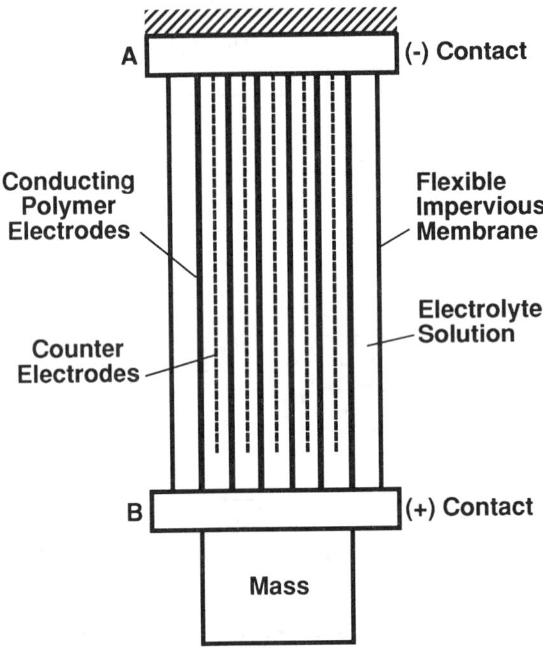

Figure 2. Schematic drawing of an extensional electromechanical
actuator which uses only one conducting polymer electrode as the
electromechanical polymer. This polymer is mechanically connected at
both A and B, and electrically connected only at B, while the counter
electrodes are mechanically and electrically connected only at A.

An electromechanical actuator which has a structure which is
superficially similar to the morphology of natural muscle is shown
schematically in Fig. 3. This design illustrates how diffusion
distances can be minimized (short interelectrode spacings and thin
electrodes) in an actuator, so as to optimize both response rate and
cycle life. The artificial muscle fibril is a hollow conducting
polymer fiber which undergoes dimensional changes in the fiber
direction as a result of electrochemical doping and dedoping. This
conducting polymer provides the sheath for a highly conducting counter
electrode, porous electrode separator, and electrolyte. These
artificial muscle fibrils are themselves enclosed in a cable-like
sheath of a metallic conductor, which is designed to minimize internal
resistive loses of the actuator by minimizing the distance current
flows in the conducting polymer. There are substantial benefits in

rate performance and cycle life from the use of parallel arrays of
numerous, small electromechanical elements. This situation is not
unsimilar to that in human striated muscle[16], where the interfacial
area between the actin and myosin filaments is about 10^6 cm^2 per cubic
cm of muscle tissue.

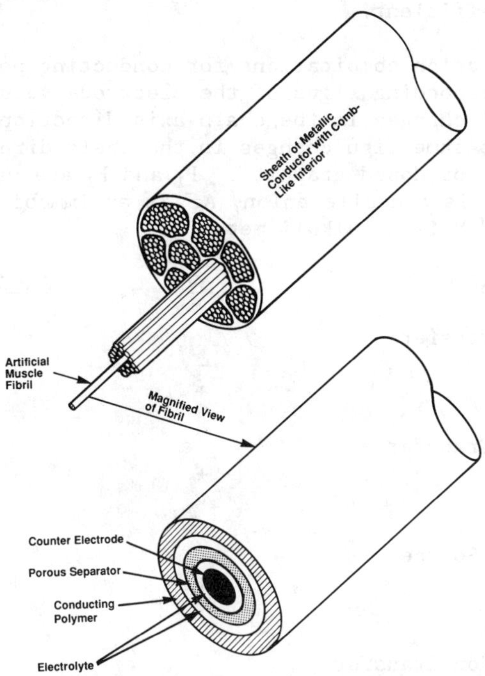

Figure 3. Schematic view of a conducting polymer electromechanical
actuator consisting of parallel arrays of "artificial muscle" fibrils
encased in a sheath of an electrochemically inert metallic conductor.
Each artificial muscle fibril consists of a hollow conducting polymer
fiber which has as its core a counter electrode wire surrounded by a
porous separator and electrolyte.

In the above discussed electrochemical cell designs, mechanical work
results from the dimensional change of only one electrode. This is
the only possibility for such cells if the oxidation of one polymer
electrode results in an oppositely directed dimensional change to that
resulting from the reduction of the second polymer electrode. This is
the case for several of the actuator types listed in Table 1, for
which electrochemical redox results in anion or cation transfer
between electrodes. These opposed dimensional changes provide the
basis for the unimorph and bimorph bending actuators. However, if the
electrochemical operation of the actuator arises from ion transfer
between both electrodes and the electrolyte, these electrodes can be
in-phase with respect to extensional behavior during the
electrochemical cycle, as indicated in Table 1. The magnitudes of

expansions and contractions would generally differ because of the generally differing expansion coefficients for anion and cation insertion. However, precise match of dimensions during the redox cycle can result from the use of thicker films or fibers (or more numerous fibers) for the electrode polymer which has the larger coulombic expansion coefficient.

Table 1. Electrode reaction combinations for conducting polymer actuators and the corresponding signs of the electrode volume changes, ΔV, and the dimensional changes in the chain-axis direction, ΔL(Chain), assuming the same sign changes in the chain direction as for doped polyacetylene or doped graphite. P_1 and P_2 are undoped conducting polymers, A^- is a mobile anion, A_p^- is an immobile anion, D^+ is a mobile cation, and M is an alkali metal.

ELECTRODE REACTIONS	ΔV	ΔL(Chain)
Interelectrode Anion Transfer		
$P_1^+A^- + e^- \,\text{---}\!\!>\, P_1 + A^-$	-	+
$P_2 + A^- - e^- \,\text{---}\!\!>\, P_2^+A^-$	+	-
Interelectrode Cation Transfer		
$P_1 + D^+ + e^- \,\text{---}\!\!>\, P_1^-D^+$	+	+
$P_2^-D^+ - e^- \,\text{---}\!\!>\, P_2 + D^+$	-	-
Metal Electrode Cation Source		
$P_1 + M^+ + e^- \,\text{---}\!\!>\, P_1^-M^+$	+	+
$M - e^- \,\text{---}\!\!>\, M^+$	-	
Electrolyte-Electrode Ion Transfer		
$P_1^+A^- + e^- \,\text{---}\!\!>\, P_1 + A^-$	-	+
$P_2^-D^+ - e^- \,\text{---}\!\!>\, P_2 + D^+$	-	-
Electrolyte-Electrode Ion Transfer		
$P_1^+A_p^- + D^+ + e^- \,\text{---}\!\!>\, P_1A_p^-D^+$	+	+
$P_2 + A^- - e^- \,\text{---}\!\!>\, P_2^+A^-$	+	-

A bimorph electromechanical cell can be designed analogously to well known bimorph structures for piezoelectric polymers. Unimorph and bimorph mechanical elements are herein defined according to the number of conducting polymer electrodes in the mechanical bender. A simple electrochemical bimorph cell consists of a polymer electrode strip and a polymer counter electrode strip cemented together by a polymeric electrolyte, which electronically separates these electrode elements. Alternately, the adhesive ion-conducting layer between electrodes can be a porous separator containing a liquid electrolyte. The major requirement for the operation of this bimorph cell is that the anode strip and cathode strip undergo differing changes in dimension upon passage of an electrochemical charge or discharge current. This is conveniently accomplished by using the same polymer as both anode and cathode strips and operating both polymer strips in the range of dopant concentrations that provides identical, but oppositely directed, transformations for anode and cathode strips during device operation. The disadvantage of using a solid-state electrolyte is the low ionic conductivities compared with those obtainable for liquid electrolytes. However, the problem of obtaining rapid device response rates for a solid-state electrolyte cell is not insurmountable, since the thickness of the electrolyte need only be sufficient to insure the absence of electronic shorts and to provide adequate mechanical bonding. Optimal performance of the bimorph electromechanical cell will generally be obtained for cell designs in which the dopant shuttles between anode and cathode during operation, in contrast with designs where the dopant ions are stored in the solid-state electrolyte. Conducting polymer unimorph or bimorph actuators can be designed for applications on microcircuits. Possible applications include microtweezers, microvalves, micropositioners for microscopic optical elements, and actuators for micromechanical materials sorting (such as the sorting of biological cells). Strategies for fabricating conducting polymer micromechanical actuators can be based on techniques presently known for both the fabrication of micron dimensioned conducting polymer electronic devices[17-26] and micromachined silicon mechanical devices[27-33].

One type of micromechanical device that could be constructed in the future is shown in Fig. 4. This device consists of a unimorph or bimorph conducting polymer strip which is in the form of a spiral. Such a device could be used to direct the flow of a gas or liquid from the inlet to either outlet I or II. Expansion of the outer layer of the spiral unimorph or bimorph element will cause clockwise rotation of the slide element, which would then block outlet II - causing a redirection of flow. This expansion could result from electrochemical redox of a bimorph strip consisting of two conducting polymer layers separated by a solid-state electrolyte. Alternately, the same device configuration could be used for the construction of an actuator which responds to a varying chemical composition of the fluid in the system. In this case, the actuator strip could be a metal layer overcoated with a conducting polymer, such as polyaniline. The device response could be, for example, the redirection of liquid flow depending upon the pH of the liquid - as a result of the pH dependent doping of the conducting polymer and resulting dimensional changes. The device scale shown in Fig. 4 has already been achieved for spiral springs on silicon substrates[30] and methods are demonstrated for the micromachining of much more complicated structures in silicon - such as two two-turn Archimedean spirals supporting a torsional resonant plate[31].

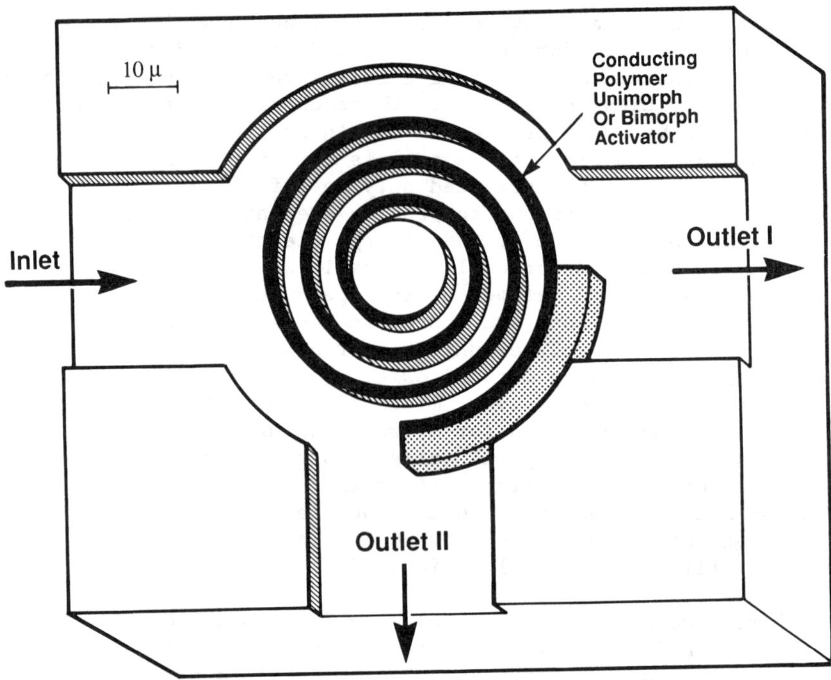

Figure 4. Micromechanical actuator device for controlling fluid flow, which utilizes unimorph or bimorph conducting polymer strips.

The application of two conducting polymers to bimorph flexors to form microtweezers is suggested by Fig. 5. The inner electrodes of both bimorph actuators are formed from one layer of conducting polymer and separated by adhesive solid-state electrolyte layers from the counter electrodes, which are the outer layers on the microtweezers. The opening and closing of the microtweezers corresponds to electrochemical dopant transfer between inner and outer electrodes. Chen et al.[33] have fabricated electrostatic microtweezers which are 200 μm long and about 2.5 μm in the orthogonal dimensions. A major advantage of the proposed conducting polymer microtweezers is the low voltage required for operation, about 1 volt or less. For comparison, a voltage of over 100 volts was required for closure of the electrostatic microtweezers, which involved only about a degree change in the angle between the two arms of the tweezers. This two orders of magnitude lower operation voltage can provide a major advantage for the electrochemical microtweezers.

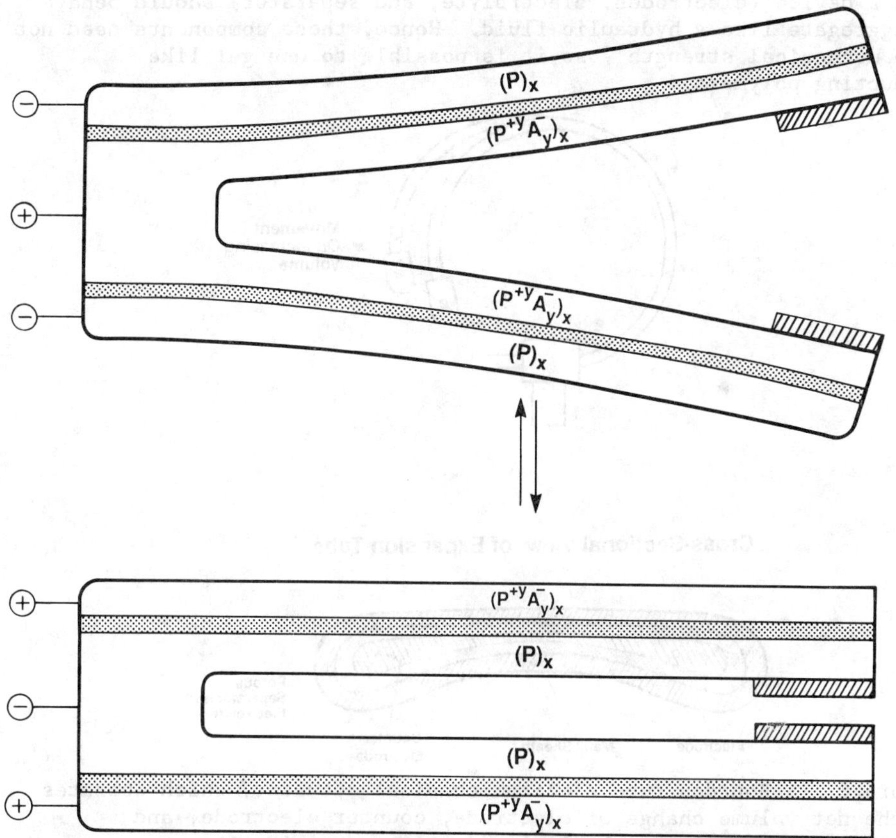

Figure 5. Paired bimorph actuators used as microelectrochemical
tweezers. Electrochemical transfer of dopant from the outer layer to
the inner layer of each bimorph causes the opening of the tweezers.

The pressure increase due to a net volume increase of anode, cathode,
and electrolyte can most simply be used to make a hydraulic
electromechanical actuator using a Bourdon tube similar to those used
in pressure control. As illustrated in Fig. 6, the Bourdon tube is a
metal tube of flattened cross-section, closed at both ends, which is
filled with the electrodes, separator, and electrolyte so that this
tube provides one electrode contact. The tube is bent during
fabrication into an arc, so that the arc radius is approximately
perpendicular to the thickness direction of the flattened tube.
Volume expansion, due to electrochemical transformations resulting
from current flow, tends to convert the flattened tube into a
cylindrical one. Consequently, the arc radius increases, causing the
tip to move upward - as illustrated in Fig. 6. The cell components in

such a device (electrodes, electrolyte, and separator) should behave
in aggregate like a hydraulic fluid. Hence, these components need not
have mechanical strength - so it is possible to use gel-like
conducting polymers.

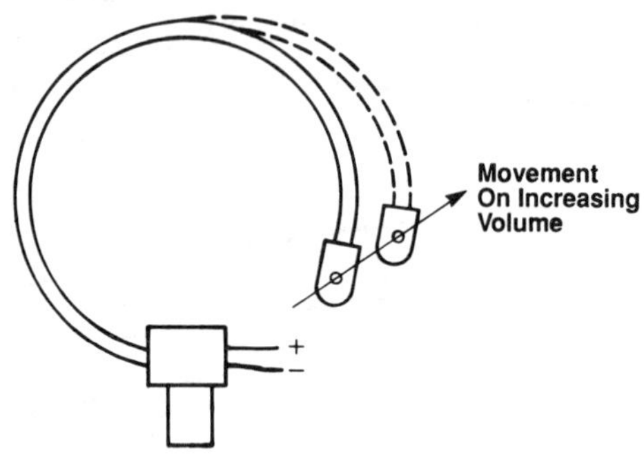

Cross-Sectional View of Expansion Tube

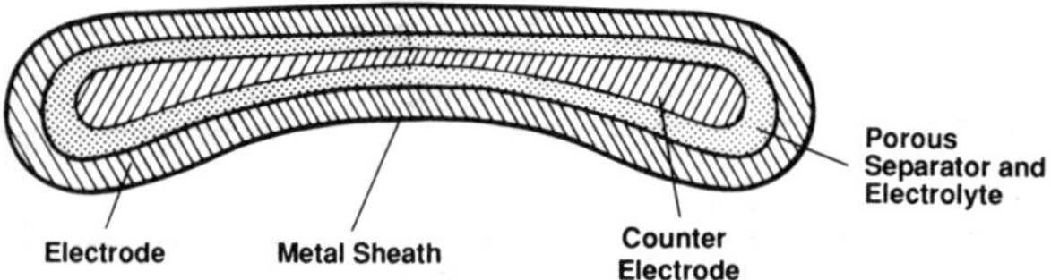

Figure 6. A Bourdon tube electromechanical actuator, which operates
by the net volume change of electrode, counter electrode, and
electrolyte upon electrochemical reduction and oxidation.

Variants of the Bourdon tube type actuator are suitable for downsizing
to a micromechanical actuator. For this purpose it is convenient to
use a flattened hollow fiber of conducting polymer as the elastic
sheath of the Bourdon tube. Such a flattened hollow fiber might be
made similarly to conventional hollow fiber spinning, but using an
elliptical orifice, or by plastic deformation of a circular hollow
fiber. In the present case, the counter electrode and electrolyte-
containing separator which is contained by this sheath becomes
equivalent to the pressurized fluid in a conventional Bourdon tube
actuator. Hence, the actuator displacement now depends upon the
difference in the electrochemical volumetric expansions of the
conducting polymer sheath electrode and the thereby contained
remaining cell components. If the doped conducting polymer is not
environmentally stable or does not have sufficiently high conductivity
to minimize cell resistance, the conducting polymer sheath can be
metallized, either by metal sputtering or electroplating.

3. Device Performance Predictions Using Observed Properties of Conducting Polymers

3.1 PROPERTIES BASIS FOR DEVICE EVALUATIONS

Section 3.1 is concerned with structure-related properties which are important for conducting polymer electrochemical actuators. These properties include the anisotropic dimensional changes resulting from doping, mechanical properties, and electrochemical doping rates. Section 3.2 uses these properties and the observed performance of conducting polymers in other electrochemical devices to predict achievable performance in actuators.

The volume change on dopant insertion in a conducting polymer can strongly depend upon the dopant concentration range. This is a consequence of the typically observed packing arrangements in conducting polymer complexes. Dopant ions often form columns or planar assemblies which are inserted between the polymer chains.[34-41] Depending upon the dopant concentration range, dopant can be accommodated by increased dopant density in preexisting dopant arrays (columns or sheets) or by the displacement of polymer chains during formation of an increased number of dopant arrays. The large dimensional changes corresponding to the compositional range of the latter process can be used in extensional actuators. The small volume changes of the conducting polymer for the former process, coupled with large associated volume changes of counter electrode or electrolyte during redox, can be used in hydraulic actuators. Polyacetylene doped with alkali metals provides important examples of both phase regimes.[34-39] For example, conversion of a structure with four polymer chains per alkali metal column ($y=0.0625$ in CHK_y) to one with two polymer chains per alkali metal column ($y=0.125$) without change in intracolumn, interion separations results in a large volume expansion (12.5 cm^3/Faraday, which is about 27% of the molar volume of potassium).[35-38] The corresponding percentage volume change for the polymer is 6.6%, or a 1.06% change in volume for a percent change in dopant concentration. This volume change per change in dopant concentration (deduced from x-ray diffraction results) is close to the volume change measured by bulk dimensional changes for sodium (1.5 by Francois et al.[42]) or for potassium (1% by Plichta[43], using a doping solution of the K^+-napthalide complex in 2-methyltetrahydrofuran). In contrast, further increases in dopant concentration (from $y=0.125$ to 0.167) results from a decreased interion separation in alkali metal columns, from 4.9 Å to 3.7 Å. Only a small volume change is associated with this increased ion density in the alkali metal ion columns. Hence, the net volume change on reduction of polyacetylene during oxidation of a potassium anode is close to the molar volume of potassium (45.9 cm^3/mole). For comparison with the above, the doping of polyacetylene with unsolvated lithium (up to $y=0.11$) produces a volume change (decrease) of the polyacetylene of only a few percent or less.[39] Consequently, the net volume change of a lithium anode and polyacetylene cathode during lithium doping is about -12 cm^3/Faraday

572

(which is close to the molar volume of lithium).

Enormous dimensional changes can result from the doping of conducting
polymers with larger dopants or with dopants which are solvated. For
example, in-situ measurements of electrode buoyancy changes by
Okabayashi et al[44] indicate that about 3 propylene carbonate molecules
are reversibly inserted with each ClO_4^- ion during the oxidation of
polyaniline in $LiClO_4$/propylene carbonate. The associate volume
change of the polyaniline is 297 cm^3/Faraday - corresponding to a
volume increase of the polyaniline by a factor of 2.2 over the total
observed doping range. The reversibility of these volume changes for
polyaniline, derived from the data of Okabayashi et al.[44], is
indicated by the results shown in Fig. 7. Despite the large volume
increase of the polyaniline (compared with the initial volume of this
polymer), the volume decrease of the electrolyte has nearly the same
value, so the change in total volume of electrodes and electrolyte is
small (about -3 cm^3/Faraday). Using the results of in-situ bulk
measurements of Slama and Tanguy,[45] the reversible volume change on
oxidation of polypyrrole in propylene carbonate/$LiClO_4$ electrolyte has
a similar value (272 cm^3/Faraday) as above discussed for polyaniline.
This large value again results from solvent cointercalation with the
anion.

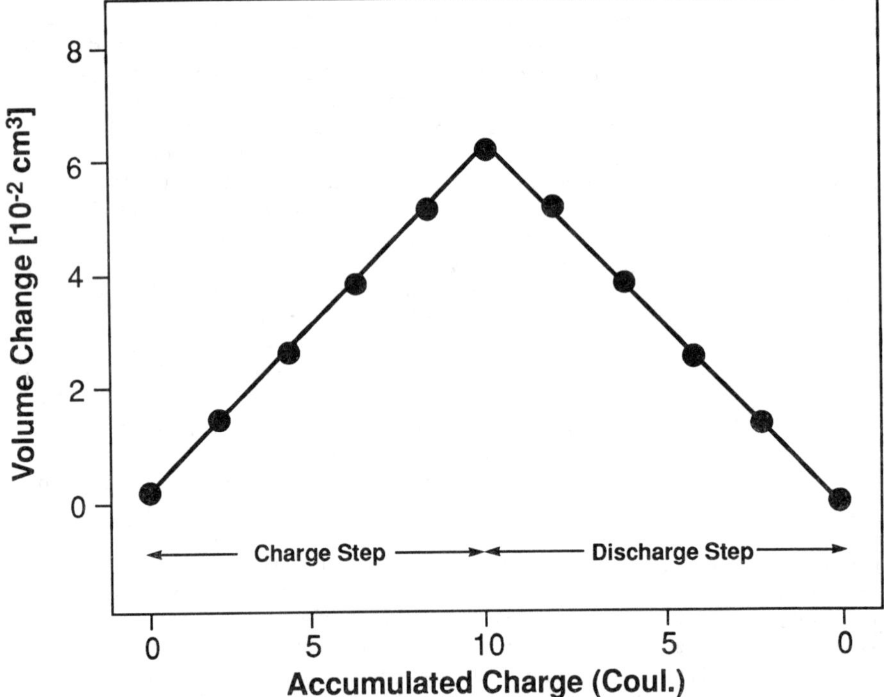

Figure 7. Reversible volume changes of polyaniline during slow
electrochemical doping and dedoping with solvated ClO_4^-.

For polymer backbones which are planar both before and after doping, the dopant-induced dimensional changes are small.[38, 46, 47] However, because of the high strength and modulus for the chain direction of highly oriented polymers, such dimensional changes can find application in electrochemical actuators. For example, the doping of polyacetylene with an electron acceptor causes chain-length contraction, while donor doping causes chain-length expansion. For lithium, sodium, and potassium the maximum change (expansion) is from 1.0 to 1.6% and for iodine the maximum change (contraction) is about -0.4%.[38,46] Fig. 8 provides the fractional expansions measured by x-ray diffraction during the electrochemical doping of polyacetylene with sodium,[46] which indicates that most of the expansion occurs at high dopant levels. Actuator response rate would, therefore, be highest at these levels (between $y = 0.10$ to 0.14), where the expansion coefficient, $(\Delta L/L)/\Delta y$, is 0.21. The corresponding coulombic coefficient is $(\Delta L/L)V/\Delta Q = 2.35$ cm^3/Faraday, where V is the molar volume of polyacetylene and ΔQ is the change in electrochemical charge which provides a fractional length change of $\Delta L/L$. While the percentage change in chain-axis length is a small fraction of the total percentage volume change for the larger dopants, this change provides a major contribution to the total volume change for polyacetylene doped with unsolvated lithium.[39] Also, because of a degree of chain misorientation and other disorder, the expansion in the orientation direction can exceed that deduced from x-ray diffraction measurements. The changes in chain-axis length of a conducting polymer can be quite large for polymers which change conformation as a consequence of doping, and can be comparable to the dimensional changes in orthogonal directions. However, while little change in per-chain modulus is expected for polymers that do not change backbone conformation during doping, major decreases in this modulus can result if a transition occurs between planar and helical backbones. This modulus decrease can decrease the achievable work per cycle in a tensile electrochemical actuator.

574

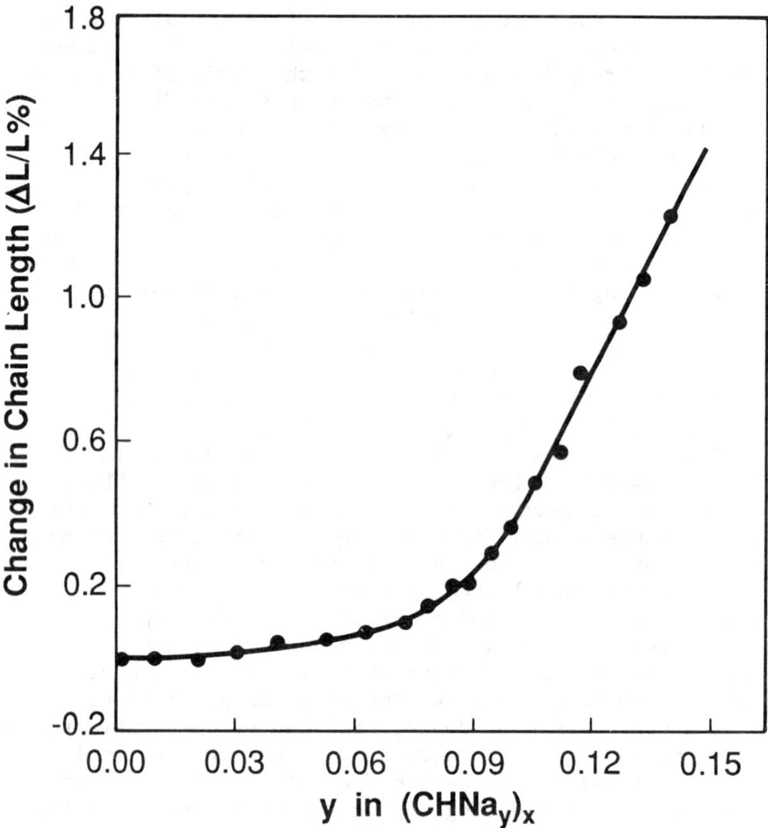

Figure 8. The x-ray diffraction measurements[46] of the chain length
expansion upon the electrochemical doping of polyacetylene with
sodium.

The effect of doping on the elastic moduli determines the dependence
of actuator displacement on mechanical stress. Also, plastic
deformation or rupture of the electromechanical polymer provides an
ultimate limit on the stress which can be employed for an
electromechanical actuator. In the calculations in this paper, the
maximum stress which can be applied to the conducting polymer in an
actuator will be approximated by 50% of the fracture stress.
Mechanical properties have not been optimized for many of the
conducting polymers, and in notable cases these properties are limited
by the low molecular weight of the polymer. Nevertheless, Akaji et
al.[48] have obtained extremely high modulus (100 GPa) and ultimate
strength (900 MPa) for highly chain-oriented trans-polyacetylene
obtained by a modification of the Naarman synthesis method. While it
is easy to choose polymer/dopant combinations which result in poor
mechanical properties for the doped polymer, examples are available
which indicate that the mechanical properties of the doped polymer can

be close to those of the undoped polymer, and in some cases exceed those of the undoped polymer. For example, Ito et al.[49] found little change in Youngs modulus (2.6 GPa doped and 3.4 GPa undoped) or tensile strength (74 MPa doped and 81 MPa undoped) upon dedoping perchlorate-doped, unoriented polythiophene film. Also, MacDiarmid et al.[50] found that the Youngs modulus decreased from 8.6 GPa to 5.0 GPa and the ultimate tensile strength decreased from 366 MPa to 176 MPa upon doping drawn fibers of polyaniline (polyemeraldine base) with HCl. For comparison with these results, tensile strengths of 50-90 MPa were reported by Abe et al.[51] for undrawn films of both undoped polyaniline and polyaniline doped with various protonic acids ($HClO_4$, HCl, H_2SO_4, and p-toluene sulfonic acid).

3.2 PREDICTED DEVICE PERFORMANCE

Based on the above discussed mechanical properties and dimensional changes, a variety of performance features of conducting polymer actuators can be calculated. The dimensional changes appropriate for device performance calculations depend upon the cycle life required, and whether or not the devices operate in hydraulic or extensional modes. In order to maximize cycle life and minimize response times, it is best to operate the actuator over a limited dopant range. One reason for such choice is that rapid doping and dedoping over large ranges of dopant concentration can result in large internal strains between regions of the electromechanical polymer which have quite different dopant concentrations. Such strains can result in degradation of the mechanical properties of the conducting polymer during cycling. Additionally, the strains developed in unimorph or bimorph actuators should not exceed the elastic limits of the component film strips or the adhesive interlayer bonding.

The maximum stress which can be developed in a reversible actuator by the electrochemical redox of a conducting polymer is the smaller of (1) the fractional change in dimension under zero load corresponding to a dopant concentration change Δy, $(\Delta L/L)_{o,\Delta y}$, multiplied by the Youngs modulus for the contracted state and (2) the limiting stress before mechanical failure, which is herein approximated as 50% of the fracture stress. Using the above quoted mechanical properties of undrawn polythiophene film, drawn polyaniline fibers, and drawn polyacetylene film, 50% of the ultimate tensile strength is reached under isometric conditions (fixed length conditions) for a $(\Delta L/L)_{0,\Delta y}$ of 1.1%, 2.1% and 0.45%, respectively. Consequently, only small changes in dopant concentrations are required under isometric conditions in order to develop high stresses (ca. 40, 180, and 450 MPa, respectively). These stresses (corresponding to 380, 1900, and 4600 kgf/cm^2, respectively) are from one to two orders of magnitude higher than the tensile stress which can be developed by application of nondestructive voltages (ie, voltages which do not cause rapid depolarization) for the piezoelectric polymer poly(vinylidene fluoride). Specifically, using reported values[52] for the in-plane modulus (Y = 3 GPa), the inverse piezoelectric constant

($d_{31} = 3 \times 10^{-11}$m/V), and the maximum electric field which can be applied without rapid depolarization (ca. $E= 3 \times 10^{7}$V/m for an alternating electric field), this stress for poly (vinylidene fluoride) is YEd_{31} or 2.7 MPa, compared with 40 to 450 MPa for the above mentioned conducting polymers. In addition to this major advantage of the conducting polymer actuator for high stress generation, the conducting polymer actuator has the significant advantage in requiring a much lower voltage for operation. Even for a film thickness as low as 1 μm, the above limiting field corresponds to a voltage of 30 volts, while the conducting polymer electrochemical actuator would require a voltage of much less than a volt to generate the higher stresses. The above derived stress generation capabilities for conducting polymer electromechanical actuators are several orders of magnitude higher than observed[10] for isometric electromechanical contractions of salt-saturated polyacrylic acid/polyvinyl alcohol gels (ca. 3kgf/cm^2 or 0.3 MPa), which are pH driven.

The mechanical work per polymer volume which can be accomplished in one electrochemical cycle provides another figure of merit which is impressive for properly designed conducting polymer electromechanical actuators. We consider here a tensile actuator operating under isotonic conditions (fixed mechanical load) and ignore changes in the elastic strain of the electromechanical polymer as a function of dopant level. The latter approximation will result in a serious overestimation of work density per cycle for a specified tensile load only when the product of polymer cross-sectional area and Youngs modulus is much lower for the contracted state than for the extended state, which is usually not the case. Using this approximation, the work density per cycle (involving a dopant concentration change Δy) is $\sigma[\Delta L/L]_{o,\Delta y,}$, where σ a stress below that required for irreversible deformation or fracture of the polymer. The value chosen for $[\Delta L/L]_{o,\Delta y}$ depends upon both the required cycle rate and cycle lifetime for the actuator, since both cycle rate and cycle lifetime generally decrease with increasing change in dopant concentration[53] and increasing fractional dimension change.

In light of these considerations, which will be further discussed later in this section, we can conservatively assume a $[\Delta L/L]_{o,\Delta y}$ of at least a few percent for unoriented conducting polymers in high cycle life electromechanical actuators. Such a dimensional change is only about 10% of that available for complete doping of conducting polymers having large coulombic expansion coefficients. Consequently, much higher work density per cycle could be achieved for actuators where high cycle life is not required. Since $\Delta L/L$ at below depolarization voltages for poly(vinylidene fluoride) is no larger than 0.1%, the conducting polymer electrochemical actuators designed for high cycle life could have more than an order of magnitude advantage compared with piezoelectric polymers in work density per cycle. Additionally, because of the high ultimate strength of chain-oriented polymers compared with that for unoriented polymers, work density per cycle at the maximum load stress can be higher for oriented conducting polymers than for unoriented conducting polymers.

Measurements on conducting polymer electrochromic devices and chemical transistors indicate that high rate capabilities are obtainable for conducting polymer actuators which utilize thin polymer films. Lacroix et al.[54] reported electrochromic switching speeds of up to 100 μs and current densities of up to 100 A/cm^2 for 1200 Å thick polyaniline films in 2M sulfuric acid. These authors suggested that the species transferred between the polyaniline and the electrolyte during redox reaction was H$^+$, so the associated volume change of the polyaniline might be quite small. However, even in the potential range where this assumption might be correct, significant dimensional changes could result for oriented polyaniline, because of conformational changes associated with protonation and deprotonation of the polymer backbone. In order to obtain such high switching rates, Lacroix et al.[54] used IR compensation methods which might be difficult to successfully apply for electrochemical actuators having high cycle life. Without using IR compensation, Lacroix and Diaz[55] obtained switching times of about 50 ms for 500 Å thick polyaniline films in aqueous solutions of various protonic acids. A much higher cycle life (above 10^6 cycles), together with an electrochromic switching time of less than 100 ms, was demonstrated by Kobagachi et al.[56] for a 500 Å thick polyaniline film switched between -0.15 and 0.4 volts (versus SCE) in 1 M HCl.

High switching speeds have also been obtained for electrochromic devices based on polypyrrole, which is especially pertinent because the electrolytes used necessitate large volume changes upon doping. Specifically, Gazard[57] reported electrochromic switching times of about 100 msec and a cycle lifetime of about 2x10^4 cycles for 1000 Å thick polypyrrole films in an electrolyte of tetraethylammonium tetrafluoroborate in acetonitrile. Also, Pickup and Osteryoung[58] obtained complete electrochemical doping in 100 ms for much thicker polypyrrole films (3μm) at 26°C in AlCl$_3$/1-methyl-(3-ethyl)-imidazolium chloride molten salt electrolyte.

The above results on switching rates can be compared with those from Wrighton's group[24-26] on the switching of conducting polymers in electrochemical transistors. Complete device turn-on, which involves a transition between insulating and conducting states, could be achieved at a frequency above 300 Hz for a several micron thick polyaniline microelectrode in 0.5M NaHSO$_4$.[24] By reducing the separation between source and drain electrodes from 1.5 μm to 50 nm, along with decreasing the overall volume of the polyaniline, an electrochemical transistor was fabricated which operated at a frequency exceeding 10 kHz. Also, Chao and Wrighton[26] demonstrated operation at a frequency of 300 Hz for an electrochemical transistor which utilizes a 5-10 μm thick polyaniline microelectrode and a solid state electrolyte of hydrated poly(vinyl alcohol)/H$_3$PO$_4$.

These results indicate that the cycle life and cycle rates obtainable

578

for electrochemical switching are sufficiently high for many
applications of conducting polymer actuators. Both cycle rate and
cycle life can be optimized by proper choice of electrolytes,
conducting polymers, polymer morphology, current collectors, and
electrode and separator thicknesses. The conducting polymer electrode
thickness and the interelectrode separation are especially important.
Although most of the above results for device response rate are for
electrochemical display applications where thin films are desirable,
the results from Chao and Wrighton[26] and Pickup and Osteryoung[58]
suggest that cycle times of less than 100 ms are obtainable for
conducting polymer electrode thicknesses of up to 10 μm. This is
about the lower limit of commercially available thicknesses for
unsupported films of conventional polymers. Cycle lifetime can suffer
from the use of electrodes which are thick[53] and the use of large
changes in the concentration of a dopant which provides a large
coulombic expansion coefficient. Also, due to the different
dependencies of doping and electrolyte degradation rates upon voltage
application time, appropriate choice of pulse shape can increase cycle
life.[53]

4. Discussion

This work shows that conducting polymer electromechanical actuators
can be designed which would have major advantages compared with prior
art technologies for the direct conversion of electrical energy to
mechanical energy. Among these advantages of the conducting polymers
compared with piezoelectric polymers are more than order of magnitude
increases for the achievable dimensional changes, the maximum
electrically generated stress, and the maximum work density per cycle.
Additionally, such performance can be achieved at voltages which can
be about an order of magnitude lower than would be required for
piezoelectric materials or, on the microscale, for electrostatic
actuators.

The major disadvantages of conducting polymers, compared with
piezoelectric polymers, are provided by limitations on cycle life and
cycle rate. Based on observed cycle lifetimes of conducting polymers
in electrochemical optical displays, cycle lifetimes in excess of 10^6
cycles should be achievable in suitably designed actuators based on
very thin films or fibers of conducting polymers. Additionally, by
limiting the amount of charge transferred during the electrochemical
cycle, such a cycle lifetime could perhaps be substantially exceeded.
However, even under the best of circumstances, the cycle lifetime of
the conducting polymer electrochemical actuator is much too low for
use in motors which operate continuously at very high frequencies.
Cycle times of about 100 ms should be feasible for conducting polymer
microactuators, corresponding to the observed electrochemical
switching times of thin conducting polymer films in electrochromic
devices. Moreover, based upon the operation frequencies observed by
Wrighton's group for microelectrochemical transistors,[25] cycle times

as short as 0.1 ms might be eventually achievable for very small microactuators.

Because cycle lifetime can be maximized and cycle time can be minimized by the use of very thin polymer films, conducting polymers will probably be of greatest interest for microactuators. Due to the likely prohibitive cost of using 10 micron or thinner films for larger actuators, and the absence of present technology for doing so, large scale actuators based on conducting polymers are likely to be useable only for applications which do not require either very high cycle life or very short cycle times. Examples of such applications are hydraulic or nonhydraulic actuators for window blinds or car door locks. Relevant for such applications, it is worthwhile noting that cycle lifetimes of about 10^3 at 30% discharge and 10^4 at a few percent discharge are claimed for nonaqueous electrolyte, polyaniline batteries manufactured by Bridgestone-Seiko.[59] However, the use of a more highly conducting electrolyte and thinner, more numerous electrodes would be required for the construction of an actuator with rate performance in the range of practical interest.

Because of the major advantages of conducting polymers regarding the stress generation capabilities, work per cycle, and low required operation voltages, these materials are of special interest for microactuators. Examples of interesting application possibilities include microtweezers, microvalves, micropositioners for microscopic optical elements, and actuators for micromechanical materials sorting. From a fabrication viewpoint, it is noteworthy that conducting polymers have already been obtained as micron or submicron patterned thin film arrays using various techniques, such as conventional photolithography.[18,20-26] Coherent films with submicron thicknesses (200 Å and less) have been obtained on substrates by various routes such as (1) solution processing of soluble undoped, doped, or precursor forms, (2) polymerization from the gas or solution phases, (3) deposition using Langmuir-Blodgett techniques, and (4) electropolymerization.[17-26] Such routes for materials manipulation and structuring can be combined with techniques similar to those already advanced for micromachining microactuators on silicon substrates.[27-33] In contrast with the case of piezoelectric polymers, no poling step is required during fabrication, which can considerably simplify device fabrication.

This analysis demonstrates that conducting polymers have considerable potential for application as electromechanical actuators. The present state of affairs is not too different from that in the early days of conducting polymer batteries.[60] The unique property combinations provided by the conducting polymers provides exciting possibilities, but the problems of achieving high performance in practical devices are clearly challenging. We hope that this analysis of both the problem and prospects for conducting polymer actuators will generate research and development activity in this new area.

References

1. Kuhn, W., Hargitay, B., Katchalsky, A., and Eisenberg, E. (1950) Nature (London) $\underline{165}$, 514.
2. Katchalsky, A. and Eisenberg, E. (1950) Nature (London) $\underline{166}$, 267.
3. Steinberg, I.Z., Oplatka, A., and Katchalsky, A. (1966) Nature (London) $\underline{210}$, 568.
4. Wasserman, A. (ed.) (1960) Size and Shape Changes of Contractile Polymers, Pergamon Press, London.
5. Osada, Y. and Sato, M. (1980) Polymers $\underline{21}$, 1057.
6. Smets, G. and DeBlauwe, F. (1974) Pure Appl. Chem. $\underline{39}$, 225.
7. Aviram, A. (1978) Macromolecules $\underline{11}$, 1275.
8. Shiga, T., Hirose, Y., Okada, A., and Kurauchi, T. (1989) Polymer Preprints $\underline{30}$, 310.
9. Irie, M. (1986) Macromolecules $\underline{19}$, 2890.
10. DeRossi, D., Domenici, C., and Chiarelli, P. (1988) in Dario, P. (ed.), Sensors and Sensory Systems for Advanced Robotics, NATO ASI Series Vol. $\underline{F43}$, Springer-Verlag, Berlin, 201.
11. Yoshino, K., Nakao, K., and Sugimoto, R. preprint.
12. Yoshino, K., Nakao, K., Onoda, M., and Sugimoto, R. (1989) Japanese J. Appl. Phys. $\underline{28}$, L682.
13. Rossi, D.D. (May, 1989) Research and Development, pp. 67-70.
14. Begenhard, J.O. and Fritz, H.P. (1974) J. Electroanal. Chem. $\underline{53}$, 329.
15. Jobert, A., Touzain, Ph., and Bonnetain, L. (1981) Carbon $\underline{19}$, 193.
16. Jacobsen, S.C., Price, R.H., Wood, J.E., Rytting, T.H., and Rafaelof, M. (1989) in Jacobsen, S.C. and Petersen, K.E. (eds.), Micro Electro Mechanical Systems, Proceedings of IEEE, 17.
17. Wuu, Y.-M., Fan, F.-R.F., and Bard, A.J. (1989) J. Electrochem. Soc. $\underline{136}$, 885.
18. Allen, P.C., Bott, D.C., Brown, C.S., Connors, L.M., Gray, S., Walker, N.S., Clemenson, P.I., and Feast, W.J. (1989) in Kuzmany, H., Mehring, M., and Roth, S. (eds.), Properties of Conjugated Polymers II, Springer, Berlin, 456.
19. Nicolau, Y.F. and Nechscheim, M. (1989) in Kuzmany, H., Mehring, M., and Roth, S. (eds.), Electronic Properties of Conjugated Polymers II, Springer, Berlin, 461.
20. Burroughes, J.H., Jones, C.A., and Friend, R.H. (1989) Synthetic Metals $\underline{28}$, C735.
21. Shimidzu, T., Iyoda, T., Ando, M., Ohtani, A., Kaneko, T., and Honda, K. (1988) Thin Solid Films $\underline{160}$, 67.
22. Kobel, W., Kiess, H., and Egli, M. (1988) Synthetic Metals $\underline{22}$, 265.
23. Meyer, W.H., Kiess, H., Binggeli, B., Meier, E., and Hanbeke, G. (1985) Synthetic Metals $\underline{10}$, 255.
24. Lofton, E.P., Thackeray, J.W., and Wrighton, M.S. (1986) J. Phys. Chem. $\underline{90}$, 6080.
25. Jones, E.T.T., Chyan, O.M., and Wrighton, M.S. (1987) J. Am. Chem. Soc. $\underline{109}$, 5526.

26. Chao, S. and Wrighton, M.S. (1987) J. Am. Chem. Soc. 109, 6627.
27. Johansson, S., Schweitz, J.-A., Tenerz, L., and Tiren, J. (1989) J. Appl. Phys. 63, 4799.
28. Muller, R.S. (1988) Acta Polytechnica Scandinavia, Electrical Engineering Series 63, 143.
29. Bart, S.F., Lober, T.A., Howe, R.T., Lang, J.H., and Schlecht, M.F. (1988) Sensors and Actuators 14, 269.
30. Jebens, R., Trimmer, W., and Walker, J. (1989) in Jacobsen, S.C. and Petersen, K.E. (eds.), Micro Electro Mechanical Systems, Proceedings of IEEE, 35.
31. Tang, W.C., Nguyen, T.H., and Howe, R.T. (1989) in Jacobsen, S.C. and Petersen, K.E. (eds.), Micro Electro Mechanical Systems, Proceedings of IEEE, 53.
32. Linden, Y., Tenerz, L., Jiren, J., and Hok, B. (1989) Sensors and Actuators 16, 67.
33. Chen, L.Y., Zhang, Z.L., Yao, J.J., Thomas, D.C., and MacDonald, N.C. (1989) in Jacobsen, S.C. and Petersen, K.E. (eds.), Micro Electro Mechanical Systems, Proceedings of IEEE, 82.
34. Baughman, R.H., Murthy, N.S., and Miller, G.G. (1983) J. Chem. Phys. 79, 515.
35. Baughman, R.H., Shacklette, L.W., Murthy, N.S., Miller, G.G., and Elsenbaumer, R.L. (1985) Mol. Cryst. Liq. Cryst. 118, 253.
36. Shacklette, L.W. and Toth, J.E. (1985) Phys. Rev. B 32, 5892.
37. Murthy, N.S., Shacklette, L.W., and Baughman, R.H., Phys. Rev. B, in press.
38. Murthy, N.S., Shacklette, L.W., and Baughman, R.H. (1987) J. Chem. Phys. 87, 2346.
39. Murthy, N.S., Shacklette, L.W., and Baughman, R.H. (1989) Phys. Rev. B 40, 12550.
40. Murthy, N.S., Miller, G.G., and Baughman, R.H. (1988) J. Chem. Phys. 89, 2523.
41. Baughman, R.H., Murthy, N.S., Miller, G.G., and Shacklette, L.W. (1983) J. Chem. Phys. 79, 1065.
42. Francois, B., Mermilliod, N., and Zuppiroli, L. (1981) Synthetic Metals 4, 131.
43. Plichta, E.J. (May, 1989) Masters Thesis, Rutgers University, New Brunswick, New Jersey.
44. Okabayashi, K., Goto, F., Abe, K., and Yoshida, T. (1987) Synthetic Metals 18, 365.
45. Slama, M. and Tanguy, J. (1989) Synthetic Metals 28, C171.
46. Winokur, M.J., Moon, Y.B., Heeger, A.J., Barker, J., and Bott, D.C., Phys. Rev. B, Rapid Communications, in press.
47. Kertesz, M., Vonderviszt, F., and Pekker, S. (1987) Chem. Phys. Lett. 90, 430.
48. Akaji, K., Soezaki, M., Shirakawa, H., Kyotani, H., Shimomura, M., and Tanabe, Y. (1989) Synthetic Metals 28, D1.
49. Ito, M., Tsurono, A., Osawa, S., and Tanaka, K. (1988) Polymer 29, 1161.
50. MacDiarmid, A.G., private communication.

51. Abe, M., Ohtani, A., Umemoto, Y., Akizuki, S., Ezoe, M., Higuchi, H., Nakamoto, K., Okuno, A., and Noda, Y. (1989) J. Chem. Soc., Chem. Commun., 1736.
52. Lee, J.K. and Marcus, M.A. (1981) Ferroelectrics 32, 93.
53. Yoshino, K., Kaneto, K., and Takeda, S. (1987) Synthetic Metals 18, 741.
54. LaCroix, J.C., Kanazawa, K.K., and Diaz, A.F. (1989) J. Electrochem. Soc. 136, 1308.
55. LaCroix, J.C. and Diaz, A.F. (1988) J. Electrochem. Soc. 135, 1457.
56. Kobayashi, T., Yonegama, H., and Tamura, H. (1984) J. Electroanal. Chem. 161, 419.
57. Gazard, M. (1986) in Skotheim, T.A. (ed.), Handbook of Conducting Polymers, Volume 1, Marcel Dekker, New York, 673.
58. Pickup, P.G. and Osteryoung, R.A. (1985) J. Electroanal. Chem. 195, 271.
59. Nakajima, T. and Kawagoe, T. (1989) Synthetic Metals 28, C629.
60. Nigrey, P.J., MacDiarmid, A.G., and Heeger, A.J. (1979) J. Chem. Soc., Chem. Commun. 594.

WORKING GROUP REPORT ON PROSPECTS IN SYNTHESIS AND PROCESSIBILITY OF CONJUGATED POLYMERS

J.P. Aimé (Exxon), R.H. Grubbs (Caltech), L. Leemans (Liège), A.G. MacDiarmid (Penn), E.W. Meijer (Philips), H. Naarman (BASF), M.F. Rubner (MIT), W.R. Salaneck (Chairman/Linköping), J.J. de Vlieger (TNO), F. Wudl (Chairman/UCSB).

I. INTRODUCTION

The field of conjugated polymers is an interdisciplinary endeavor, incorporating its own unique challenges, difficulties and opportunities. The design, synthesis, characterization and processing of materials is of key importance for this area of material science. We will address a few key issues that will be important for future development:

1. Synthesis of New Materials
2. Refinement of Present Materials
3. Detailed Characterization
4. Processibility
5. Technology in Respect with the Items 1-4.

II. SYNTHESIS OF NEW MATERIALS

New materials may not be as numerous as is desired, due to a limited number of reactions that lead to high molecular weight polymers. In spite of this, there is a strong need for new, unprecedented and challenging target molecules to guide the synthesis of new materials. New model compounds with discrete architecture are also required. A major task for everyone working in this field is the identification of target molecules with enhanced properties.

New properties can be found as well as a dramatic increase in conductivity and third-order nonlinearities can be produced through designed synthesis. Although both the latter properties are based on extended π conjugation (or even σ-conjugation, *e.g.* polysilanes), conductivity is only found in doped polymers, while large third-order susceptibilities values are to date only found in neutral polymers. Therefore, it is expected that different design and synthesis rules will exist. Special emphasis should be given to the synthesis of well defined two- and three-dimensionally conjugated polymers.

III. REFINEMENT OF THE PRESENT MATERIALS

The strong interest in studying the physics of conjugated polymers is of utmost importance to further improve the synthesis of the present polymers. Special attention should be given to attaining perfect conjugation as well as to the control of the alignment of the conjugated chains, the morphology and crystallinity.

583

J. L. Brédas and R. R. Chance (eds.), Conjugated Polymeric Materials:
Opportunities in Electronics, Optoelectronics, and Molecular Electronics, 583–585.

584

Many examples of successful approaches of this kind can be found in the history of commodity polymers. Only dedicated polymerization could lead to stereoregular polypropylene (Ziegler-Natta) and ultra-high molecular weight polyethylene (UHMWPE). With high crystallinity, the latter could be processed to a strong fiber.

Polyacetylene, as an example, has been made with an apparently almost perfect chemical structure (no sp^3 C-atoms can be detected with the characterization techniques presently available) leading to extremely high conductivities and optical nonlinearities. Polyheterocyclic polymers, to the contrary, still lack such advances in polymerization technology.

There is a great need to optimize the synthesis of the present materials. Highly advanced physical studies are only feasible and comparable with studies from others, when it is performed on well-defined (standard type) polymers.

IV. DEFINED CHARACTERIZATION

Until recently, "polyaniline", probaly the oldest known synthetic organic polymer, consisted of an ill-defined class of materials obtained by the (electro)chemical oxidative polymerization of aniline. Early studies were fraught with problems of uncertain composition and it was not until the mid-1980's with the advent of better characterized materials that significant physical studies became possible.

Therefore, it is strongly urged that the community comes to some generally accepted experimental standards with respect to synthesis and characterization. The following points are suggested:

1. The *polymerization* process should be clearly defined, and the synthesis conditions should be clearly indicated, including the purity of starting materials, the preparation of the catalysts, and the purification of the samples.

2. The product obtained should be *characterized, e.g.,* by elementary analysis, particularly of the impurities, and by spectroscopy techniques (IR, NMR, etc.).

3. *Conductivity measurements* of specimens prepared under defined conditions should be determined by four-probe method under argon atmosphere.

2. *Doping by p-oxidation or n-reduction should be carried out under standard condition, e.g.,* where appropriated, with iodine saturated in carbontetrachloride or with sodium-naphthalene in tetrahydrofuran.

5. *Stability*: The specimens should be kept in air in a standard laboratory atmosphere, and the oxygen content and the conductivity should be determined over time, *e.g.,* 1 day, 1 week, 1 month. The results should be expressed as a fraction of the original values: O_o/O_t or σ_o/σ_t (O=oxygen content, σ=conductivity S/cm).

As well as increasing the level of chemical analysis, new work is required to develop classic polymer techniques for use in characterizing the macromolecular properties of highly conjugated materials. Special problem associated with aggregation and stability must be addressed.

V. PROCESSIBILITY

Several routes are now available to increase the processibility of extended conjugated polymers. Excellent progress has been made by physical processing of precursor polymers. Other more sophisticated techniques can be applied as materials with better solubility and stability (thermal, hydrolytic and oxidative) are prepared. The techniques can also be used to form materials for various applications as discussed below. "Molecular" processing can be used to control crystallinity, morphology, and orientation. This approach can lead to materials than can be studied extensively. As a result, new fundamental insights can be obtained.

New designs with sophisticated synthesis can lead to special polymers for Langmuir Blodgett films, possible" quantum dots", superlattice structures, etc.etc..

Since morphology, solubility and processibility are related to molecular structure and hence to synthesis, future progress will result from development of new synthetic techniques for controlled synthesis of already known polymers and for the creation of new highly conjugated materials.

VI. TECHNOLOGY

The thermal, oxidative, hydrolytic, and environmental stability and processibility of conducting polymers is intimately involved with their actual and potential technological importance for some given, specific use. Depending on the use, one given property or set of properties may or may not be important. For example, environmental stability is of less concern where the polymer *is not* in contact with air, as for example, in polyaniline/Li production, but *is* of critical importance in their use for EMI shielding or as anti-stats or as blends or composites with conventional structural polymers of potential applicability as a light-weight material in aircraft as a substitute for metal.

Technical applications fall into two broad (and overlapping) categories in which they are used either as pure material or as blends or composites with conductivities in the semiconducting or metallic regimes:

(1) As new materials in their own right in which no change in their electronic, electrical and magnetic or optical properties occurs during use in batteries, electronchromic display electrodes, anti-stat coatings, replacement of metal in aircraft fuselage.

(2)In devices in which changes in some or all of the properties *are* involved during use, commonly by the application of an electric field.

In order to replace existing materials or devices it is necessary that the new product be *either* very much better and have a similar cost to the existing product *or* that it must have significantly superior properties at a similar (or greater) price. Of particular importance are ones where they do not replace existing products, *e.g.* in electrochromic "sun-screen" windows in buildings. Much progress has been made in improving processibility and associated property aspects of conducting polymers. Initially available conducting polymers generally lacked environmental stability, either solution or melt phase processibility, and useful mechanical properties. Now conducting polymers are available which have long term stability in ambient conditions and sufficient thermal stability for processing at temperatures above 200 °C. Also, compositions are known which are conveniently processed from solutions of either conducting polymer or a precursor polymer to form coatings, free-standing fibers, or high-strength fibers which are typically made conducting by post-processing doping.

WORKING GROUP REPORT ON MOLECULAR ELECTRONIC PROSPECTS

D. Bloor (Chairman/Durham), M. Hanack (Tübingen), A. Le Méhauté (Chairman/CGE),
R. Lazzaroni (Mons), J.P. Rabe (MPI Mainz), S. Roth (MPI Stuttgart), H. Sasabe
(RIKEN).

I. INTRODUCTION

Today the title "Molecular Electronics" is used for a variety of subjects. The two major
subtopics that have emerged are "Molecular Materials for Electronics" and "Electronics on
a Molecular Level". In the former case one has fairly well-defined applications in mind
and one can write down material parameters, such as conductivity or hyperpolarizability,
and compare the values of existing materials with some required thresholds. Thus it is
possible to obtain hints on where to go and how far. The second field is much more
speculative and though the long term goal is clear the steps towards that goal are less well
defined. The problem is exacerbated by the range of possible alternatives that offer some
prospects of progress.

II. MOLECULAR MATERIALS FOR ELECTRONICS

Most of these materials are dealt with in the reports of the other working groups.
Therefore they shall be only briefly touched on here.

a. Conducting Polymers

Since high conductivity polymers have already been synthesized (polyacetylene, $\sigma > 10^5$
S/cm) and moderately conducting polymers are processible and have good long term
stability, it seems to be only a matter of time till these materials will find applications in
either electronics or related areas.

b. Non Linear Optics

Organic materials with high values of $\chi^{(2)}$ are available with side chain polymers
offering good figures of merit and processibility. It seems probable that devices employing
these materials will be produced in the near future. The reports elsewhere in these
Proceedings show that for polymers with large $\chi^{(3)}$ the values are still perhaps orders of
magnitude smaller than one would like to have, in particular if one plans to work with
diode lasers.

c. Magnetooptics

To date, we dispose only of inorganic magnetooptic materials. Organic materials with
ordering temperatures even as low as 80K would be highly appreciated. In particular the

J. L. Brédas and R. R. Chance (eds.), Conjugated Polymeric Materials:
Opportunities in Electronics, Optoelectronics, and Molecular Electronics, 587–590.

588

availability of magneto-optical isolators for visible and NIR lasers, even at 80K, would be extremely valuable. Progress towards this target is discussed in the article by O. Kahn in these Proceedings.

d. Superconductors

Present T_c values of organic superconductors are around 10K. If this value rose to 80K one could think of applications. This target, however, appears to be some way off at present.

III. ELECTRONICS ON A MOLECULAR LEVEL

To structure the discussion five headings were utilised, but it soon became apparent that the fields are very strongly interconnected and that it is difficult to look at any one aspect in isolation from the others. This coincides with the experience that in projects which attempt to distinguish between "molecular wires" and "molecular switches" this distinction is soon abandoned.

a. Molecular Wires

It is still not definitely established whether a single molecular chain can "transport information". What is clear is that it will certainly behave very differently from a thin metal wire. One may ask how many parallel chains would be needed to make a "wire", i.e. one along which a current flow can be detected. Allusion was made to a paper by Hopfield et al. (Science, 241, 817 (1988)) that indicates that several hundred chains were necessary, if available electronics is to be used to detect the current. This poses a further question: If the individual molecules in the bundle were not all of exactly the same length, would there be interference effects? What about Heeger's and Epstein's statements during this meeting, that in Naarmann-polyacetylene an electron visits perhaps 100 or 1000 chains before it is inelastically scattered? Transport must involve motion between individual molecular chains as well as along them. Individual carries are not confined to individual polymer chains and the overall behaviour is that of the solid, not of independent molecular wires. Would the high conductivity persist in either an isolated bundle of a few polymer chains or a single chain? This fundamental question remains to be answered. Even if a molecular wire is shown to exist would it be of any help in connecting molecular devices to the outside world. Could they for example be addressed by a scanning tunnelling microscope? Advances in nanotechnology, occurring for other reasons, offer a prospect of tailoring interconnects to small molecular aggregates from the semiconductor technology side. Much work remains to be done to provide the compatible molecular "component". However, it is certain that the first steps towards molecular devices will be hybrids drawing heavily on current technology.

b. Molecular Switches

There are certainly many molecular switches, e.g. photochromic molecules displaying two or more stable states, are available. Some examples are in the report edited by C. Joachim and A. LeMéhauté (ARAGO 7 "L'Electronique Moléculaire: Perspectives en Matière de Traitement Moléculaire de l'Information" OFTA, Masson, Paris 1988). Up to now these molecules have not been investigated individually, but physical investigations of large ensembles (films, solutions) already give useful information (on time constants, on the selectivity of switching, etc.). Several national and international programs are running in which such measurements are complemented by quantum-chemical calculations. Furthermore, it should be pointed out that most "molecular wires" can also act as "molecular switches".

c. <u>Addressing</u>

In principle the scanning tunnelling microscope (STM) and the near-field optical microscope (NFOM) would allow the interaction of electrons and photons with individual molecules. First experiments of this kind, however, have not proved to be reproducible. To better understand the tunnelling process, charge injection studies on bulk samples, as reported by Friend at this meeting, might be helpful. STM imaging of organic regular monomolecular arrays, organic crystals, polymer chains and individual biological molecules have been reported and demonstrate the power of the technique. Combination of an STM with other techniques, e.g. emission spectroscopy, NFOM, etc., promises to provide powerful tools in investigations of individual molecules and small molecular aggregates.

Alternatively one can consider two dimensional arrays of molecules addressed on a large scale by a laser beam. This provides both a power source and the potential for self organisation. Thus one can conceive of a molecular device addressed by pulses from a laser, in which a cascade of "molecular directional couplers" could lead to transfer of information, i.e. energy, to specific molecular sites; such a device would be a molecular cellular automaton.

Such speculation reflects the ease noted earlier of defining general end goals but the difficulty in identifying specific intermediate steps towards that goal. Thus much remains to be done in studies of charge and energy transport at the microscopic scale before "real" molecular devices, or even their component part, can be "designed" with any certainty.

d. <u>Mesoscopic Structures</u>

In order to investigate how individual components could be assembled to generate a practical device, for example a cellular automaton as described above, it will be useful to look at mesoscopic structures. Such structures might be obtained by using some kinds of self-organizing processes. Studies of existing mesoscopic structures might help in finding mechanisms for self-organisation. The dimensionality of such mesoscopic structures will play an important role and one should not only look onto one, two, and three-dimensional arrays but also investigate structures with fractal or mixed dimensionalities.

e. <u>System Theory</u>

Theoretical work is not only necessary on a quantum-chemical level but also in informatics and in general physics. For example, fault tolerant algorithms will have to be developed and adapted to the molecular devices in order to take account of the probabilistic behaviour of molecular switches, and thermodynamics limits will have to be discussed (e.g. thermal losses connected to statistical processes).

Since molecules can exist in different conformations the use of molecular shape could lead to multivalued logic systems. A similar conclusion follows if ions, which may be of many different types, are used as information carriers rather than electrons. This also removes the system from the quantum limit at the expense of potential device speed. The use of a highly parallel system can, in principle, easily compensate for this.

590

IV. SUMMARY

Molecular Electronics is an intellectually very appealing field of research. In view of the large number of unsolved or not even defined problems intensive work has to be done now if any new molecular materials are to be used within the next couple of decades. For the immediate future, research in the field of "Electronics on a Molecular Level" is more likely to lead to spin-offs in "Molecular Materials for Electronics" than to commercial applications in its own field. This must always be borne in mind since even though the two areas sound disparate they are closely related. Thus, in addition to the possibility that the long term area may contribute to the near term area, as noted above, the reverse is also true. The design and synthesis of molecules for their macroscopic properties is providing vital experience in the tools needed to make progress in studies at the molecular scale. This fact should not be forgotten when looking forward to the ultimate aim of molecularly based computational systems.

WORKING GROUP REPORT ON THEORETICAL DEVELOPMENTS

J. Delhalle (Namur), J. Messier (Saclay), E. Orti (Valencia), P. Sautet (Lyon), R. Silbey (MIT), Z.G. Soos (Chairman/Princeton), and J.M. Toussaint (Mons).

I. MOLECULAR ELECTRONICS AND CONJUGATED POLYMERS

Applications of conjugated polymers to molecular electronics fundamentally involve their electronic structure. Small band gaps, high linear and nonlinear polarizabilities, and novel gap states offer many exciting possibilities, as discussed throughout the Workshop. The close interplay of theory and experiment, especially for spectroscopic and transport measurements on conjugated polymers, is typical of vigorous young fields. Phenomenological models and analogies from older fields are introduced and tested and refined. Just as the most important potential applications remain to be found, the underlying physical picture of conjugated polymers is still evolving. Conformational degrees of freedom in conventional polymers are coupled here to delocalized electronic excitations that are related both to solid-state descriptions of semiconductors and to chemical descriptions of conjugated molecules. Current theoretical activities on conjugated polymers encompass many approaches, ranging from abstract field theories primarily concerned with topology to detailed electronic structure calculations on small molecules.

In our discussions of theoretical directions, we deliberately focused on nonlinear optical (NLO) properties and on special challenges of conjugated polymers. We also anticipated major overlap with topics from the Optoelectronic Prospects and Electronic Prospects panels. There is full participation in the development and applications of theoretical models.

In broadest terms, electronic applications require explicit knowledge of several states, thus necessarily implicating excited electronic states. A few excited states are crucial to photoinduced spectra. The full spectrum of virtual states may be necessary for some NLO responses. Traditional computational methods, whether from molecular or solid-state applications, tend to be optimized for the ground state and its potential surface. Similar calibration will be needed for excited states, either individually or collectively, and a variety of theoretical methods can be extended or refined for excited states.

II. NONLINEAR OPTICAL PROPERTIES

NLO applications arose in various contexts in this Workshop. Current interests in optics and devices has led to many developments of the underlying physics. The final connection of NLO properties is inevitably to specific electronic transitions in atoms, ions, molecules, or solids. The "best" molecules or conjugated polymers must be identified before they can be designed and synthesized. An important theoretical issue is to understand the factors leading to large NLO response. The present prescriptions, of looking for small optical gaps and large transition dipoles, are quite rudimentary but clearly point to delocalized states characteristic of conjugated polymers or of large molecules. How this is achieved is left

J. L. Brédas and R. R. Chance (eds.), Conjugated Polymeric Materials:
Opportunities in Electronics, Optoelectronics, and Molecular Electronics, 591–593.

open.

Theory may nevertheless set bounds on various NLO coefficients, and reasonably accurate bounds would be extremely helpful. The Unsöld approximation for the static polarizability, to replace all higher excited states by the optical gap E_g, gives a bound based entirely on ground-state properties. Such an approximation was successfully used by London to estimate dispersion forces. Higher-order processes are naturally more difficult to reduce to ground-state properties. Bounds may also be more difficult for resonant rather than static coefficients. On the other hand, ground state wavefunctions and expectation values tend to be far more reliable.

π-Electron models often afford additional simplicity, especially at the one-electron level, and several pioneering applications have been made for noninteracting electrons in conjugated polymers and molecules. More systematic analyses for interacting electrons, probably at the Pariser-Parr-Pople level, can be anticipated. The use of sum rules, of finite bases, and of special symmetries related to the topology should give broader and more reliable results. Model Hamiltonians afford the only current method for obtaining *exact* NLO coefficients for interacting electrons.

We may safely anticipate an overlapping series of electronic structure calculations for such prototypical molecules as finite polyenes, $C_N H_{N+2}$. Rather complete all-electron treatments are possible for butadiene (N = 4); longer chains are accessible with approximate all-electron methods; exact PPP results currently go to N = 12; and approximate π-electron techniques span the entire range to polymers. A similar series of π-systems with a donor (D) and acceptor (A) at opposite ends is natural for discussing second harmonic generation. Experimental SHG studies show the relative geometries of D, A, and the π-systems to be important. The application of progressively more approximate theoretical techniques to longer segments is controlled by comparison with more reliable methods restricted to shorter segments.

In addition to such specific and systematic issues, NLO properties require a broader understanding of other topics. Local electric fields are clearly important and may not routinely be approximated for polymers by results from spherical cavities. The strong frequency dependence of the third harmonic generation in centrosymmetric polymers emphasizes the need for understanding the dynamic response. As they become available, angular anisotropies provide natural tests for theoretical ideas. Solid-state contributions will surely emerge as more systematic understanding is achieved. Interchain dispersion forces are always present, while interchain electron transfer is potentially important for close packed polymers like polyacetylene. Substituents in polydiacetylenes and polythiophenes increase the interchain separation, but may affect the exciton's energy and relaxation in ways specific to a given side group. Vibronic contributions may alter frequency dependences, among other things. Excited state lifetimes should be important close to resonance or in multiphoton absorption processes. The relative importance of such issues remains open and will initially be guided by observation. Successful theoretical work in the NLO area will certainly entail greater appreciation of experiment.

III. ELECTRONIC STRUCTURE OF CONJUGATED POLYMERS

Much has been clarified in the past several years about particular aspects of particular polymers. Such progress often affords considerable detail, as illustrated by charged photoinduced states or by conformational effects on optical absorption and emission. The models nevertheless are largely phenomenological. Different systems and different experiments evidently require different physical pictures whose overall consistency remains open. Indeed, there is considerable tension among some interpretations, even as new experiments and new ideas are modifying and refining current models.

The electronic structure of conjugated polymers requires a new, and consequently different, combination of ideas from small molecules, molecular and inorganic crystals, and from flexible chains. Important applications may be found from each area. The coupling of electronic and conformational degrees of freedom is perhaps the most distinctly associated with polymers. The identification of conformational degrees of freedom, the related potential surfaces, and the resulting statistical mechanisms offer a wide variety of challenges. Should we focus on flexible chains with some specified stiffness? Or on differentiating backbone rotations of bonds and rings and side-group motions? To what extent are the resulting states localized? How are their relaxation dynamics affected?

The traditional questions of vibronic coupling, of coherence, or of localization remain challenging in simpler systems as well. Previous studies in molecular crystals or in amorphous solids can certainly be applied to conjugated polymers. Such cross referencing should initially speed up work on polymers, but may eventually broaden understanding in both areas. Similar comments apply to electronic structure calculations, where the experience gained on small systems remains to be incorporated. Conduction mechanisms in doped polymers clearly require some tunneling steps, between chains or between crystalline and amorphous regions or between the polymer and electrodes. Both new and old problems must be addressed to determine the critical issues. As phenomenological models for transport, photoinduced states, or excited state dynamics are successfully parametrized, they will be related to and derived from more basic principles.

WORKING GROUPS REPORTS ON OPTOELECTRONICS PROSPECTS

C. Bubeck (MPI Mainz), R.R. Chance (Exxon), J.C. Dubois (Thomson), S. Etemad (Chairman/Bellcore), F. Meyers (Mons), E. Hadjoudis (Demokritos Center), A.J. Heeger (Chairman/UCSB), B.E. Kohler (UC Riverside), J. Messier (CEA), M. Nowak (Bellcore), A. Persoons (KU Leuven), G. Ruani (Bologna), J. Zyss (Chairman/CNET).

The rapid pace of progress in photonic research in general, and the observation of large optical nonlinearities in selected organic materials are the primary reasons for the recent interest in this class of materials. The general impression is that research prospects in nonlinear optical (NLO) properties of conjugated polymers is good, because many issues fundamental to both science and technology are begging answers. This report outlines the highlights of discussions among the "optoelectronic" interest group during the Workshop in Mons.

From a technological point of view, there are advantages in using conjugated polymers or organics in general. The most obvious one is the large values of $\chi^{(2)}$ and $\chi^{(3)}$. The extremely large value of $\chi^{(2)}$ in noncentrosymmetric molecular crystals has been used in a variety of applications including parametric amplification of weak infrared pulses. For example, the new subpicosecond infrared spectroscopic technique PASS (Parametric Amplification and Sampling Spectroscopy) makes unique use of the transparency, efficiency and phase-matching dispersion of tailor-made molecular crystals. Recent demonstration of a 20 fs bandwidth for auto- and cross-correlation in the 1.5 μm to 1.8 μm range in a POM crystal has opened the way for a variety of applications in this important wavelength range. It is possible and is indeed needed to grow crystals in waveguiding formats (planar waveguides, organic cored capillaries, etc.) for telecommunication applications. The possibility to grow large area thin films at relatively low cost compared to epitaxial growth techniques used in conventional semiconductor technology is remarkable. Interesting developments are taking place in the Langmuir-Blodgett films area, but further work towards reduction of scattering losses below their present level is required before we can take full advantage of the flexibility of these 2-D structures in making more exotic structures such as organic multiple quantum wells. At the present, however, the poled polymers containing $\chi^{(2)}$ molecules are the most promising examples of "organics" for optoelectronic applications. Although quasi-phase-matching has been achieved by alternation of poling pattern, phase matching in poled polymers may be a problem compared to crystalline materials.

Regarding the large $\chi^{(3)}$ in conjugated polymers, it is clear for any NLO application that is based on $\chi^{(3)}$, they are potentially the materials of choice. Recent observation of intensity dependent optical switching phenomena in a polydiacetylene based prototype device is proof of existence for the technological prospects of conjugated polymers as viable materials for NLO applications. Nonetheless, many practical issues need to be resolved before these laboratory demonstrations become useful devices.

There is also a growing support for a research direction that, instead of competition to surpass the existing silicon or III-V based technologies by higher figures of merit, focuses on the unique attributes of "organics". The possibility to co-engineer organic materials with

J. L. Brédas and R. R. Chance (eds.), Conjugated Polymeric Materials:
Opportunities in Electronics, Optoelectronics, and Molecular Electronics, 595–596.

widely different properties through molecular (organic synthesis) and/or structural (L-B technology) diversity provides umparalleled material design flexibility. For example, the possibility of designing a photorefractive macromolecule based on a hybrid of two molecules, one with large photoresponse and the other with large electrooptic coefficient, is what can be achieved through molecular design. On the negative side, however, one can begin by addressing the question to what extent the *low temperature* synthesis and processing steps associated with organics, affect long term stability and optical damage threshold?

Regarding the implementation of organics in a NLO device, we must begin by trying to form waveguides to confine intense light over sizable distances. Therefore, processability and optical clarity are as important as the large nonlinear optical response. For example, though polyacetylene has the largest $\chi^{(3)}$, it is practically useless because of high scattering losses. Hence, the scattering losses appear to be a most important issue to be addressed. Whereas crystalline materials offer a natural way of reducing scattering losses, because of their birefringence, bending of waveguide is problematic. It seems that glassy materials are a workable compromise. However, reactivity is the primary drawback of an organic material in a useful optoelectronic device. The fact that organic reactions are low temperature processes may well be the reason for the high reactivity of organics. These issues provide unique interdisciplinary research opportunities in organic chemistry.

It seems that solutions to many scientific puzzles will actually help to overcome some of the technical difficulties. For example, the multiphoton resonances play an important role in both the identification of the underlying mechanism for large $\chi^{(3)}$ and the window of operation for a device based on an organic material. As a result, measurement of the spectral response of the NLO coefficients rather than a single point determination is essential. Another measurement that is useful in theoretical interpretation is the sign of the NLO coefficient. Unfortunately, so far only a few studies close to resonance have addressed this issue.

An important parameter central to understanding of the response of a conjugated system to an external electric field is the "delocalization length", N_d. Despite the intense interest in NLO and electronic properties of conjugated systems, interpretation of N_d in the context of length dependence of $\chi^{(3)}$ in polyene molecules and $\chi^{(2)}$ in a hypothetical *"D^+- polyene - A^-"* macromolecule needs clarification.

The basic question whether conjugated polymers are large molecules or one-dimensional solids seemed to appear in different contexts. Another issue of scientific significance is the role of Coulomb correlation in conjugated polymers. This has been debated in the context of how to extrapolate the optical properties of polyenes to those of polyacetylene. In particular, a consistent evaluation of the correlation effects that determine the relative sizes of gaps for charged and neutral excitations in polyenes and in polyacetylene remains an open issue. This point is highlighted in different interpretations of the multiphoton resonances in the spectrum of $\chi^{(3)}$ in polyacetylene.

WORKING GROUP REPORT ON ELECTRONICS PROPERTIES

L. Alcacer (Lisboa), R.H. Baughman (Allied-Signal), H. Braunling (Wacker Chemie), J.L. Brédas (Mons), H. Eckhardt (Allied-Signal), A.J. Epstein (Chairman/OSU), R.H. Friend (Cambridge), G. Froyer (CNET), M. Galtier (Montpellier), E. Hannecart (Solvay), A. Monkman (Durham), H. Sixl (Chairman/Hoechst), T. Vogtmann (Bayreuth).

Though "metallic conductivity" was first reported for iodine doped polyacetylene a dozen years ago, the origin and control of mobility and conductivity in semiconducting and metallic polymers remains an issue central in the field. The initial report of doping of polyacetylene was followed by many experiments confirming its "high" conductivity and the evolution of conductivity with doping. Recent discoveries include the ultrahigh conductivity of iodine doped new-polyacetylene approaching that of copper, unusual polymeric semiconductor active devices based on polyacetylene and other polymers, and unusual microwave frequency response (including large losses) for polyaniline and other polymers. These discoveries demonstrate that conducting polymers have not yet reached the limits of their potential and that new scientific opportunities and potential technological applications will continue to occur.

Initial progress was rapid, identifying especially for polyacetylene that the polymer was a semiconductor that could be either p or n doped. Numerous charge conduction mechanisms, including exotic ones, were proposed. At very high doping levels polyacetylene developed many of the attributes of a metal (e.g., finite density of states at the Fermi energy), but it did not have a "metallic" conductivity. The more recent developments including those noted above have challenged the early concepts and go to the fundamental issue of why a conducting polymer conducts. The development of a predictive understanding of the phenomena of charge transport in electronic polymers is a worthy challenge that lies very much at the heart of the field. However, conduction phenomena are complex. An understanding requires the synthesis of detailed knowledge of many other issues including electronic structure, conformation, crystal structure, defects, and electron-phonon coupling.

Progress is based on the detailed study of the best characterized new materials with knowledge of the material crystallinity and crystal structure, degree of orientation, defects, impurity levels, molecular weight distribution, conjugation lengths, etc. In consideration of the broad existing knowledge base especially with regard to polyacetylene, the repetition of experiments on poorly characterized materials is unlikely to lead to critical new insight and the development of the sought after predictive understanding similar to the state of the art in crystalline inorganic semiconductors and metals. Rather, impact will come through (a) reinvestigation of transport properties in vastly improved versions of existing polymer systems (for example, detailed study of the new forms of polyacetylene after characterization for structure, impurity level, etc.); (b) performing new types of experiments on existing polymer systems; and (c) the thorough investigation of well characterized new polymer systems.

There are a number of issues that underlie the development of a fuller understanding of charge conduction in the metallic and semiconducting state. First it is noted that though

J. L. Brédas and R. R. Chance (eds.), Conjugated Polymeric Materials:
Opportunities in Electronics, Optoelectronics, and Molecular Electronics, 597–598.

there is general agreement as to the origin of the bandgap in the semiconducting polymers, the reason for the stability of metallic state in polymers is not well understood. The sensitivity of charge conduction of polymer metals and semiconductors to disorder, barriers (chemical and physical) and boundaries needs to be quantified together with the importance of one-, two-, or three-dimensionality of the electronic states. A related key issue for both metals and semiconductors is the control of localization (and its origin: 1D, 2D, 3D, Fermi glass, etc.) phenomena. The electronic states of the charge carriers (e.g. electrons, holes, solitons, polarons, and bipolarons), the effective mass of the charge carriers and their interaction with the off-chain dopants are central issues, together with a determination of the charge scattering mechanisms (including bond vibrations and ring flips) and the contributions of electron-electron scattering, defects and substituent effects. An open issue is the mechanism of transport when a semiconducting or "metallic" polymer is dispersed in a molecular scale composite, block copolymer or phase separated regions. Measurements on a microscopic scale (to overcome larger scale inhomogeneities) and model crystalline compounds (especially for determination of local bond and electronic structure) may lead to important insights.

Metallic polymers provide special opportunities. Recent studies of ultrahigh conductivities of doped new-polyacetylene have implied that the conductivity is still limited by barriers, though it has been suggested that the charge motion is three-dimensionally coherent. A direct measure of the Fermi surface of this metallic polyacetylene is an important challenge. Other experiments such as the determination of the structure of the density of states and other transport coefficients (including magnetotransport, thermoelectric power, etc.) are important. The estimate of a coherent mean free path of electrons of order 1000A (larger than the typical crystallite size) is a challenge to better understanding. Other relevant microscopic probes include measurement of the anisotropic plasma frequency, the infrared response, and a determination of electron dynamics using electron and nuclear spin resonance.

Intrachain and interchain charge carrier mobilities in the semiconducting polymers (and their evolution with modest doping) have been studied by a number of experiments. Photoconductivity, charge injection studies, and frequency dependent conductivity and dielectric constant are but a few of the techniques to be applied to determine the roles of mobile spins, solitons, polarons, bipolarons and electrons in charge transport. Again barriers and disorder are critical, together with an understanging of surface effects and the role of contacts.

Several theoretical concepts are critical for the development of the predictive understanding of charge transport. Among issues are the roles of inchain conjugation or coherence length and interchain bandwidth in determining one-dimensional localization or three-dimensional coherent transport. Of course, electron-phonon, electron-impurity, electron-grain boundary, and electron-electron scattering have to be accounted for together with the screening by parallel chains. The origin of infrared modes in the metallic state remains not completely resolved. More general theoretical issues include accounting for the stability of the metallic state, development of an appropriate effective medium theory to deal with the inhomogeneities of the doped polymers, exploring the potential for collective charge transport, and means of increasing charge carrier mobility in the semiconducting state.

Continued progress in understanding and control of mobilities and conductivities provide an underpinning for applications of electronic polymers in passive (static eliminators, conducting wires and cloth, electromagnetic shielding, switchable microwave absorbers,...) and active (diode devices, gated devices, solar cells, antennaes, intergrated circuitry) technologies. Considering the rapid improvements in materials, processing, measured properties, concepts, and device design, future progress should be rapid.